THE CARDIOVASCULAR SYSTEM: MORPHOLOGY, CONTROL AND FUNCTION

This is Volume 36A in the

FISH PHYSIOLOGY series
Edited by Anthony P. Farrell and Colin J. Brauner
Honorary Editors: William S. Hoar and David J. Randall

A complete list of books in this series appears at the end of the volume

THE CARDIOVASCULAR SYSTEM: MORPHOLOGY, CONTROL AND FUNCTION

Fish Physiology

A. KURT GAMPERL
Departments of Ocean Sciences and Biology
Memorial University of Newfoundland
St. John's, Newfoundland and Labrador
Canada

TODD E. GILLIS
Department of Integrative Biology
University of Guelph
Guelph, Ontario
Canada

ANTHONY P. FARRELL
Department of Zoology, and Faculty of Land and Food Systems
The University of British Columbia
Vancouver, British Columbia
Canada

COLIN J. BRAUNER
Department of Zoology
The University of British Columbia
Vancouver, British Columbia
Canada

Academic Press is an imprint of Elsevier
50 Hampshire Street, 5th Floor, Cambridge, MA 02139, United States
525 B Street, Suite 1800, San Diego, CA 92101–4495, United States
The Boulevard, Langford Lane, Kidlington, Oxford OX5 1GB, United Kingdom
125 London Wall, London, EC2Y 5AS, United Kingdom

First edition 2017

ISBN: 978-0-12-804163-5
ISSN: 1546-5098

For information on all Academic Press publications
visit our website at https://www.elsevier.com/books-and-journals

Publisher: Zoe Kruze
Acquisition Editor: Kirsten Shankland
Editorial Project Manager: Ana Claudia Garcia
Production Project Manager: Stalin Viswanathan
Cover Designer: Mark Rogers

Typeset by SPi Global, India

CONTENTS

CONTRIBUTORS

MARIA C. CERRA *(265), University of Calabria, Cosenza, Italy*

ANTHONY P. FARRELL *(155), University of British Columbia, Vancouver, BC, Canada*

HANS GESSER *(317), University of Aarhus, Aarhus C, Denmark*

ALBIN GRÄNS *(369), Swedish University of Agricultural Sciences, Skara, Sweden*

JOSÉ M. ICARDO *(1), University of Cantabria, Santander, Spain*

SANDRA IMBROGNO *(265), University of Calabria, Cosenza, Italy*

KENNETH J. RODNICK *(317), Idaho State University, Pocatello, ID, United States*

ERIK SANDBLOM *(369), University of Gothenburg, Gothenburg, Sweden*

HOLLY A. SHIELS *(55), Faculty of Biology, Medicine and Health, Division of Cardiovascular Sciences, University of Manchester, Manchester, United Kingdom*

FRANK SMITH *(155), Dalhousie University, Halifax, NS, Canada*

MATTI VORNANEN *(99), University of Eastern Finland, Joensuu, Finland*

ABBREVIATIONS

4-HT 4-hydroxytamoxifen
5-HD sodium 5-hydroxydecanoic acid
5-HT 5-hydroxytryptamine (serotonin)
2,3-DPG 2,3-diphosphoglycerate
A atrium
AA arachidonic acid
ABA afferent branchial arteries
ACE angiotensin-converting enzyme
ACH acetylcholine
ACV anterior cardinal vein
AD adrenaline
ADO adenosine
ADP adenosine diphosphate
AE anion exchange or Cl^-/HCO_3^- exchanger
AFA afferent filamental arteries
AGD amoebic gill disease
AKAP A-kinase-anchoring protein
AHR aryl hydrocarbon receptor
ALAs afferent lamellar arterioles
AMP adenosine monophosphate
AMPA alpha-amino-3-hydroxy-5-methylisoxazole-4-propionic acid
AMPK AMP-activated protein kinase
AMs adrenomedullins
ANG-1 angiopoietin 1
ANG-II angiotensin II
ANP atrial natriuretic peptide
ANRT AHR nuclear receptor translocator
A_o ventral aorta
AOP adverse outcome pathway
AP action potential

APC antigen-presenting cells
AR adrenoreceptor
ARC activity-regulated cytoskeleton-associated protein
ASCV Atlantic salmon calicivirus
ASN anterior spinal nerve
ASR aquatic surface respiration
AT_1 angiotensin 1 receptor
AT_2 angiotensin 2 receptor
ATP adenosine triphosphate
A.U. arbitrary units
AV atrioventricular
$A\text{-}VO_2$ difference between the oxygen content of arterial and venous blood. Also known as tissue oxygen extraction
AVP atrioventricular plexus
AVR atrioventricular region
AVT arginine vasotocin
AZ acetazolamide
α-AR alpha adrenoreceptor
BA bulbus arteriosus
BCT branchiocardiac nerve trunk
BKs bradykinins
$BL\ s^{-1}$ swimming speed expressed as body lengths per second
B_{max} receptor density
BMP bone morphogenic protein
BN branchial nerve
BNP brain natriuretic peptide
bpm beats per minute
BV branchial vein
BZ benzolamide
β-NHE β-adrenergic sodium-proton exchanger
β_2-AR β_2-type adrenoreceptor
β_b capacitance of blood for oxygen
C conus or compliance
CA carbonic anhydrase
Ca^{2+} calcium
$[Ca^{2+}]_i$ intracellular free Ca^{2+} concentration
CAM chorioallantoic membranes
cAMP cyclic adenosine monophosphate
C_aO_2 arterial oxygen content
CASQ2 calsequestrin 2
CAT catecholamines

CBF coronary blood flow
CBS cystathionine beta-synthase
CCO cytochrome *c* oxidase
CCK cholecystokinin
CdA caudal artery
CdV caudal vein
CD73 ecto-5′-nucleotidase
CDβ β COMMA-D cell line engineered to express β-galactosidase
CgA chromogranin A
cGMP cyclic guanosine monophosphate
CGRP calcitonin gene-related peptide
CICR Ca^{2+}-induced Ca^{2+} release
CK creatine kinase
cKit receptor tyrosine kinase
CMA coeliacomesenteric artery
CMS cardiomyopathy syndrome
cNOS constitutive nitric oxide synthase
CNP c-type natriuretic peptide
CNS central nervous system
CO carbon monoxide
CO_2 carbon dioxide
C_{O2} content of oxygen in blood
CoA coenzyme A
COX cyclooxygenase
CPCs cardiac progenitor cells
CPO cardiac power output
CPO_{max} maximum cardiac power output
CPT carnitine palmitoyltransferase
Cr creatine
CrA carotid arteries
CP creatine phosphate
COL1A1 collagen type I alpha 1 chain gene
C_s the slope of the capacitance curve
CS citrate synthase
CSE cystathionine gamma-lyase
CSPN cardiac spinal preganglionic neuron
CSQ calsequestrin
CSQ2 cardiac isoform of calsequestrin
CST catestatin
CTGF connective tissue growth factor
cTnC cardiac troponin C

cTnI cardiac troponin I
cTnT cardiac troponin T
CT_{max} critical thermal maximum
CV conal valves
C_vO_2 venous oxygen content
CVPN cardiac vagal preganglionic neuron
CVS central venous sinus
CYP1A cytochrome P450 1A
d day
DA dorsal aorta
DC Ductus of Cuvier
DDT dichlorodiphenyltrichloroethane
DIDS 4,4′-diisothiocyano-2,2′-stilbenedisulfonic acid
DLCs dioxin-like compounds
DMO 5,5-dimethyl-2,4-oxazolidinedione
DNA deoxyribonucleic acid
DPI days postinjury
DPF days postfertilization
DPT days posttreatment
dsRNA double-stranded RNA
ΔG_{ATP} Gibbs free energy per mol of hydrolyzed ATP
ΔP change in pressure (pressure difference)
ΔpH_{a-v} arterial–venous pH difference
ΔP_V pressure gradient that drives venous return
$\Delta S_{a-v}O_2$ change in arterial–venous Hb–O_2 saturation
ΔV change in volume
$E_{a-v}O_2$ extraction of O_2 at the tissues
EBA efferent branchial arteries
E-C excitation-contraction
ECG electrocardiogram
ECM extracellular matrix
ECs endocardial cushions
EDCF endothelium-derived contracting factors
EDRF endothelium-derived relaxing factors
EE endocardial endothelium,
EFAs efferent filamental arteries
EGCs eosinophilic granular cells
E_{ion} equilibrium potential
EIPA ethyl isopropyl amiloride
ELAs efferent lamellar arterioles
EM electron microscopy

E_m membrane potential
EMT epithelial-to-mesenchymal transition
EndMT endothelial-to-mesenchymal transition
eNOS endothelial nitric oxide synthase
EPA eicosapentaenoic acid
EPDCs epicardial-derived cells
EPO erythropoietin
EPOR *epo* receptor
ERG ETS-related gene
ERG Channel ether-à-go-go-related gene K^+ channel
ERK extracellular signal-regulated kinase
ET excitation-transcription
ET-1 endothelin-1
ETA eicosatetraenoic acid
F_C flow rate through chamber
FGF fibroblast growth factor
f_H frequency of the heartbeat, or heart rate
f_{Hmax} maximum cardiac frequency or maximum heart rate
FKBP12 12-kDa FK506-binding protein
f_R respiratory frequency
Fs furans
Gα13 G protein subunit Gα13
G6P glucose-6-phosphate
GA gill arch
GBF gut blood flow
GC guanylate cyclase
G_i inhibitory G protein
GLUTs glucose transporter proteins
GPCR G-protein-coupled receptor
GPI glycophosphatidylinositol
G_s stimulatory G-protein
GTP guanosine triphosphate
h hour
H^+ hydrogen ion
H_2S hydrogen sulfide
Hb hemoglobin
[Hb] hemoglobin concentration
HCN family of hyperpolarization-activated, cyclic nucleotide-gated, ion channels involved in controlling the pacemaker
Hct hematocrit
HEK cells human embryonic kidney cells

HEP high-energy phosphate
HH hedgehog
HIF-1 hypoxia-inducible factor 1
HIF 1-α hypoxia-inducible factor 1-α
HK hexokinase
HO heme oxygenase
HOAD β-hydroxyacyl CoA dehydrogenase
hpf hours postfertilization
HPV hepatic portal vein
HSMI heart and skeletal muscle inflammation
HSP heat-shock proteins
HV hepatic vein
I_{Ca} Ca^{2+} current
I_{CaL} L-type Ca^{2+} current
ICN intracardiac neuron
ICNS intracardiac nervous system
I_{CaT} current through T-type Ca^{2+} channels
I_f pacemaker current, "funny" current
IFNα interferon alpha
IFNγ interferon gamma
IGF-1 insulin-like growth factor 1
I_{K1} inward-rectifier K^+ current
I_{KACH} acetylcholine-activated inward rectifier current
I_{KATP} ATP-sensitive inward rectifier current
I_{Kr} delayed-rectifier current
I_{NCX} Na^+–Ca^{2+} exchange current
ION_e extracellular "free" ion concentration
ION_i intracellular "free" ion concentration
I_{Na} Na^+ current
iNOS inducible nitric oxide synthase
IP_2 inositol diphosphate
IP_5 inositol pentaphosphate
IRK2 inwardly rectifying potassium channel 2
ISA infectious salmon anemia
ISG15 interferon-stimulated gene 15
ISL1 transcription factor islet 1
ISO isoproterenol
JAK1 janus kinase 1
JNK C-jun NH_2-terminal kinase
JV jugular vein
K^+ potassium ion

$[K^+]_e$ extracellular potassium concentration
$[K^+]_i$ intracellular potassium concentration
K^+/Cl^- potassium chloride cotransport
K_{ATP} ATP-sensitive K^+ channels
K_{Ca} Ca^{2+}-activated potassium channels
K_{eq} equilibrium constant
K_{cat} turnover number
K_d receptor binding affinity
kDa kilodalton
k_i inhibition constant
Kir inward rectifier K^+ channels
L vessel length
La lamella
LCV lateral cutaneous vein
LDA length-dependent activation
LDH lactate dehydrogenase
L-NMMA L-N^G-monomethyl-L-arginine
LTCC L-type Ca^{2+} channels
M_2R muscarinic type-2 receptor
MAPK mitogen-activated protein kinase
max dP/dt_{dia} minimum rate of pressure change during diastole
max dP/dt_{sys} maximal rate of pressure change during systole
M_b body mass
MCFP mean circulatory filling pressure
MCT monocarboxylate transporters
MHC major histocompatibility complex
min minute
miRNAs microRNAs
MMPs matrix metalloproteinases
$\dot{M}O_2$ rate of oxygen consumption
$\dot{M}O_{2max}$ maximum rate of oxygen consumption
MPP5 membrane palmitoylated protein 5
mRNA messenger RNA
M_V mass of ventricle
MY million years
Na^+ sodium
NAD noradrenaline
NANC nonadrenergic noncholinergic
NCX Na^+/Ca^{2+} exchanger
NECs neuroepithelial cells
NeKA neurokinin A

NF-κB nuclear factor kappa light-chain-enhancer of activated B cells
η_H Hill coefficient
NHE sodium-proton exchanger
NKA Na^+–K^+-ATPase
NMDA *N*-methyl-D-aspartate
NPR natriuretic peptide receptor (A, B, C, D, V)
NPs natriuretic peptides (ANP, BNP, CNP, CNP, VNP)
NO nitric oxide
NO_2^- nitrite
NOAA National Oceanic and Atmospheric Association
NOK novel oncogene with kinase domain
NOS nitric oxide synthase
NTP nucleotide triphosphate
NPY neuropeptide Y
OEC oxygen equilibrium curve
OFT outflow tract
P atrioventricular plug
P_{50} partial pressure of oxygen at which 50% of hemoglobin is bound to oxygen
P_A arterial blood pressure
PACs polycyclic aromatic compounds
PACA plasma-accessible carbonic anhydrase
P_aCO_2 arterial partial pressure of carbon dioxide
P_a50 arterial P_{50}
PAF1 polymerase-associated factor 1
PAHs polycyclic aromatic hydrocarbons
PAK p-21-activated kinase
P_aO_2 arterial partial pressure of oxygen
pCa_{50} pCa for half-maximal activation
PCBs polychlorinated biphenyls
PCDD polychlorinated dibenzo-*p*-dioxins
PCNA proliferating cell nuclear antigen
PCO_2 partial pressure of carbon dioxide
PCR polymerase chain reaction
PCS posterior cardinal sinus
PCV posterior cardinal veins
P_{CV} central venous blood pressure
PD pancreas disease
P_{DA} dorsal aortic pressure
PDH pyruvate dehydrogenase
PE phenylephrine

PFK 6-phosphofructokinase
PGE2 prostaglandin E2
pH_a arterial pH
pH_e extracellular pH
pH_i intracellular pH
PHZ phenylhydrazine
PICA plasma inhibitors of carbonic anhydrase
P_i inorganic phosphate
P_{in} input pressure
PI-PLC phosphatidylinositol-specific phospholipase C
PK pyruvate kinase
PKC protein kinase C
PKG protein kinase G
PLB phospholamban
PMCA plasma membrane Ca^{2+}-ATPase
PMCV piscine myocarditis virus
PO_2 partial pressure of oxygen
P_{out} output pressure
proANF atrial natriuretic factor prohormone
PRV piscine reovirus
PTx pertussis toxin
PUFA polyunsaturated fatty acids
PV pulmonary vein
P_V plasma volume
PVA blood pressure in the ventral aorta
P_{ven} venous pressure
P_VO_2 venous partial pressure of oxygen
$\dot{Q}$ cardiac output
Q_{10} temperature quotient; the ratio of a rate function over a 10°C temperature difference
$\dot{Q}_{max}$ maximum cardiac output
qPCR quantitative polymerase chain reaction
r vessel radius
R vascular resistance
RA retinoic acid
RAS renin–angiotensin system
RBC red blood cell
R_{cor} vascular resistance of the coronary circulation
RD2 retinaldehyde dehydrogenase 2
ReA renal artery

R_{gill} vascular resistance of the gill circulation
RMP resting membrane potential
RNA-Seq high-throughput RNA sequencing
ROS reactive oxygen species
RQ respiratory quotient
R_{sys} vascular resistance of the systemic circulation
rTNFα recombinant tumor necrosis factor-α
R_{tot} total peripheral vascular resistance
RyR ryanodine receptor
Rv resistance to venous return
RVM relative ventricular mass (mass of the ventricle relative to body mass, expressed as a percentage)
RVR resistance to venous return
S1P lysosphingolipid sphingosine-1-phosphate
S1Pr2 lysosphingolipid sphingosine-1-phosphate receptor 2
SA sinoatrial
SA node sinoatrial node
sAC soluble adenylyl cyclase
SAP sinoatrial plexus
SAR sinoatrial region
SAV salmonid alphavirus
$S_{a\text{-}v}O_2$ difference in arterial–venous Hb–O_2 saturation
SBV stressed blood volume
SCA subclavian artery
SCP salmon cardiac peptide
SCV subclavian vein
SD sleeping disease
SDS sodium dodecyl sulfate
SEM scanning electron microscope
SERCA sarco–endoplasmic reticulum Ca^{2+}-ATPase
SeV subepithelial vein
SG sympathetic ganglia
SGC soluble guanylate cyclase
SIV supraintestinal vein
SL sarcolemmal
SL_n sarcomere length
SNP sodium nitroprusside
SNS sympathetic nervous system
SPN spinal preganglionic neurons
SO_2 Hb–O_2 saturation (arterial S_aO_2 and venous S_vO_2)
SP substance P

SR sarcoplasmic reticulum
SUR sulfonylurea receptor
SV sinus venosus
TBX T-box transcription factor
TCA cycle tricarboxylic acid cycle
TCDD 2,3,7,8-tetrachlorodibenzo-*p*-dioxin
TCR T cell receptor
TDEE temperature-dependent deterioration of electrical excitation
TdT dUTP nick-end labeling
TEFs toxicity equivalence factors
TGF-β_1 transforming growth factor-beta 1
TIMP tissue inhibitor of metalloproteinase
TM tropomyosin
TMAC transmembrane adenylyl cyclase
TnC troponin C
TNFα tumor necrosis factor-alpha
TO_{2max} maximum arterial oxygen transport
TUNEL terminal deoxynucleotidyl transferase
U_{crit} critical swimming speed of a fish
UDP uridine diphosphate
UI urotensin I
UII urotensin II
Us urotensins
USBV unstressed blood volume
UTP uridine triphosphate
UTR 3′ untranslated region
V ventricle
V_b blood volume
$\dot{V}_b$ blood flow
VC vena cava
VEGF vascular endothelial growth factor
VIP vasoactive intestinal peptide
V_m resting membrane potential
VNP ventricular natriuretic peptide
$\dot{V}O_2$ oxygen uptake per unit time
V_R total ventilation volume
V_s stroke volume, volume of blood pumped with each heartbeat
$V_{s\,max}$ maximum stroke volume
VS vasostatin
VS-1 vasostatin 1
VS-2 vasostatin 2

WAF water-accommodated fraction
W_S stroke work
WT1 Wilms tumor protein 1
X cardiac vagus rami
X1 lateral vagal motor neuron group
Xm medial vagal motor neuron group
Xmc caudal components of the medial vagal motor neuron group
Xmr rostral components of the medial vagal motor neuron group
Φ Bohr coefficient
η fluid viscosity

PREFACE

The cardiovascular system plays a critical role in fishes, as it does in all vertebrates. It transports oxygen, carbon dioxide, metabolic substrates and wastes, hormones, and various immune factors throughout the body. As such, it is one of the most studied organ systems in fishes, and significant insights into the evolution of this taxonomic group have emerged through this research. The cardiovascular system also determines, in part, the ecological niche that can be occupied by many fish species and how they will potentially be impacted by future abiotic and biotic challenges. For example, over the past decade considerable evidence has accumulated that metabolic scope (the difference between resting and maximum metabolic rate) is a key determinant of fish thermal tolerance, and that this parameter is largely determined by the capacity of the fish to deliver enough oxygen to the tissues to meet the rising metabolic demand as temperature increases. Likewise, the heart needs its own oxygen supply to function properly, and so, the increasing examples of aquatic hypoxia are of great concern for fish distributions.

Twenty-five years ago, two volumes (Volumes 12A and 12B) titled *The Cardiovascular System* were published in the *Fish Physiology* series. These volumes provided a synthesis of the state of the field and played a significant role in driving this research area forward; indeed they are required reading for any scientist interested in comparative cardiovascular physiology. However, molecular and cellular approaches have tremendously advanced the field, and consequently, we now have a much more complete understanding of the mechanistic underpinnings of cardiovascular function. It is, therefore, critical that we take stock of the current state of knowledge in this very active and growing field of research. The two current volumes (Volumes 36A and 36B) contain 14 contributions from 25 world experts in various aspects of fish cardiovascular physiology. As such, they provide an up-to-date and comprehensive coverage and synthesis of the field. The 14 chapters, collectively, highlight the tremendous diversity in cardiovascular morphology and function among the various fish taxa and the anatomical and physiological plasticity shown by this system when faced with different abiotic and biotic challenges. These chapters also integrate molecular and cellular data with the growing body of knowledge on heart (i.e., whole organ) and *in vivo* cardiovascular function and, as a result, provide insights into some of the most interesting, and important, questions that remain to be answered in this field.

The two volumes are organized to complement each other. The first volume (36A) titled *The Cardiovascular System: Morphology, Control and Function* summarizes our current understanding of the fish heart and vasculature, and how they work. The first chapter of Volume 36A provides an extensive review of the morphology and structure of the fish heart and highlights the tremendous diversity in form and function displayed by various taxa. The second chapter focuses on the morphology and function of cardiomyocytes, and relates form and function to the mechanistic basis of cardiac contraction. Chapter 3 describes, in significant detail, the electrical excitability of the fish heart. This chapter's major focus is on recent advances in our understanding of cardiomyocyte ion currents, their molecular basis and their regulation by the autonomic nervous system, and on how this enables the heart to respond to changes in physiological requirements. The fourth chapter provides a comprehensive account of the form, function, and physiology of the fish heart, including information on cardiac innervation and the intracardiac nervous system, and provides a general overview of how this organ responds to various physiological and environmental challenges. Chapter 5 reviews what is known of how hormones and locally released (autacoid) substances regulate cardiac function under both normal and stress-induced conditions, and the biochemical pathways that underpin this control. The sixth chapter examines our current understanding of the heart's energy requirements, and the biochemical pathways utilized by various fish species to meet cardiac energy demand and to maintain cardiac energy state. Finally, the seventh chapter of this volume provides a review of what is currently known about the vasculature in different fish species and the mechanisms that regulate its function.

The second volume (36B) titled *The Cardiovascular System: Development, Plasticity and Physiological Responses* is focused on how cardiac phenotype and the cardiovascular system change during development and in response to variations in physiological and environmental conditions. It covers the impact of a variety of challenges from development, to environmental stressors to disease on cardiovascular function, and the mechanisms that provide this system with the plasticity and capacity to respond to such challenges. The first chapter is focused on the role of the erythrocytes in maintaining O_2 delivery to the tissues under a range of physiological conditions. Further, it describes the heterogeneous distribution of plasma-accessible carbonic anhydrase, and the role that this enzyme may play in enhancing tissue oxygen extraction in various tissues and species. The second chapter summarizes our current understanding of cardiovascular development and function in embryonic and larval fishes, as well as the plasticity shown during this process, and how it is influenced by environmental change. In Chapter 3, the authors discuss the hypoxic preconditioning response of the fish heart, as well as the remodeling and/or regenerative capacity of the hearts of some fish species following exposure to various physiological stressors. Chapter 4 summarizes the influence of acute and chronic changes in temperature on cardiovascular function and the capacity of some fish species to compensate for changes in their thermal environment. Chapter 5 details cardiovascular responses to oxygen limitation and their regulation in water-breathing and air-breathing fishes that are intolerant of hypoxia, as well as the strategies that hypoxia/anoxia-tolerant species use to survive prolonged periods of oxygen deprivation. The sixth chapter summarizes the consequences of toxicant exposure for the development and function of the heart, with an emphasis on the effects of dioxins, polychlorinated biphenyls, polycyclic aromatic hydrocarbons, and crude oil exposure. Finally, Chapter 7 describes the effects of morphological abnormalities, pathological conditions, and various parasites and pathogens (i.e., diseases) on the function of the fish heart.

These volumes represent the collective effort of a large number of individuals. We wish to thank all the authors for the significant work that was needed to write their respective chapters, and all the anonymous reviewers who ensured that each chapter was of the highest quality. We also acknowledge the staff at Elsevier for their assistance in constructing these volumes and efforts to keep us on schedule. Finally, we recognize the tremendous contributions that numerous individuals have made to this very interesting and important discipline of fish physiology, and encourage them to keep asking key questions and challenging dogma in the field, and to keep inspiring those who will follow in their footsteps.

Todd E. Gillis and A. Kurt Gamperl

1

HEART MORPHOLOGY AND ANATOMY

JOSÉ M. ICARDO[1]

University of Cantabria, Santander, Spain
[1]Corresponding author: icardojm@unican.es

This chapter covers morphological aspects of the fish heart. It starts with the recognition of the existence of six heart components that are present during embryonic development, and in the adult, and that are anatomically arranged in sequence: sinus venosus, atrium, atrioventricular canal, ventricle, conus arteriosus, and bulbus arteriosus. The anatomy and structure of all the cardiac components are then described and compared between the different fish groups. Data from the basal Cyclostomata, chondrichthyans, basal Osteichthyes, and more advanced teleosts are included and analyzed. Then, data on blood supply to the heart, nerves, and localization of the heart

The Cardiovascular System: Morphology, Control and Function, Volume 36A
FISH PHYSIOLOGY

DOI: http://dx.doi.org/10.1016/bs.fp.2017.05.002

pacemaker are discussed. A separate section at the end of the chapter is committed to the lungfish heart. Lungfishes share many anatomical and functional characteristics with both freshwater fish and amphibians, and the lungfish heart shows some unique and very peculiar characteristics. The enormous diversification of morphological characteristics of the fish heart is emphasized throughout this chapter.

1. INTRODUCTION

As the main organ of the cardiovascular system, the vertebrate heart has been the focus of a myriad of morphological, functional, and molecular studies. The heart of fishes has not escaped from this stream of interest, and a wealth of information has accumulated on this subject over the past 50 years. Many excellent reviews have summarized this breadth of knowledge (Burggren et al., 1997; Farrell, 2007, 2011a; Farrell and Jones, 1992; Icardo, 2012; Santer, 1985; Satchell, 1991), and recent studies have added new information and points of discussion, further stimulating interest in this field of research. This chapter reviews the main morphological features of the fish heart and attempts to cover all of the major fish groups, from the basal hagfishes to the more advanced teleosts. Emphasis is placed on general features of the heart, but minor details are also highlighted where appropriate.

2. FISH HEART CHAMBERS: A REASSESSMENT

Classical descriptions have considered the piscine heart to be formed by four chambers arranged in series: the sinus venosus, atrium, ventricle, and conus arteriosus (or bulbus arteriosus) (see Chapter 4, Volume 36A: Farrell and Smith, 2017). The sinus venosus is the most caudal portion of the heart and collects the venous blood; the atrium is a large sac with a relatively thin muscular wall; the ventricle is a muscular chamber responsible for developing the pressure needed to propel the blood into the circulation and shows considerable variability in shape and inner architecture; the conus arteriosus also has a muscular wall and constitutes a major part of the heart's outflow tract (OFT) in basal species; finally, the bulbus arteriosus has an arterial-like structure since it contains collagen, elastin, and smooth muscle cells and constitutes the predominant portion of the OFT in most teleosts. However, there is a growing body of evidence, indicating that this description is incomplete, and that other muscular and nonmuscular portions have to be included to adequately define the morphology of the fish heart.

For example, recent studies indicate that the atrioventricular (AV) area is a distinct, ring-like segment formed by myocardium and connective tissue that establishes structural and functional connections between the atrium and the ventricle (Icardo and Colvee, 2011). The presence of the AV segment has often been overlooked, but its existence and shape become evident when the atrium is simply pulled away from the ventricle. Detailed structural characteristics of the AV segment have been described in teleosts (Icardo and Colvee, 2011), but many features can be extended to other fish groups (see Section 3.3).

Recent studies have also modified our view of the most cranial portion of the fish heart. A muscular portion of variable length, the conus arteriosus, was considered to be the single chamber situated between the ventricle and the ventral aorta in chondrichthyans, primitive bony fishes, and basal teleosts, whereas it was replaced by the bulbus arteriosus in more recently evolved teleosts. However, this morphological classification appears to be too simplistic. Examination of a wide range of species indicates that more advanced teleosts display a more or less discrete myocardial segment, the conus arteriosus, interposed between the ventricle and the distal, arterial-like, bulbus arteriosus (Grimes et al., 2006; Icardo, 2006; Schib et al., 2002). Recent studies also indicate a complex situation in basal species. Chondrichthyans (Durán et al., 2008), chondrosteans (Durán et al., 2014; Icardo et al., 2002a), and holosteans (Grimes et al., 2010) display a discrete arterial-like segment, the bulbus arteriosus, interposed between the conus and the ventral aorta. It should be underscored that only a small number of species have been examined, but in all the cases, a conus and a bulbus can be recognized within the pericardial cavity.

As a result of these observations, it can be stated that the fish heart consists of six components arranged in series within the pericardium: sinus venosus, atrium, AV segment, ventricle, conus arteriosus, and bulbus arteriosus. The latter two portions are collectively included under the term "OFT" (Icardo, 2006). As an exception, OFT components could not be discerned in the heart of Cyclostomata (hagfishes and lampreys) (Farrell, 2007; Icardo et al., 2016a; Kardong, 2006). In addition, the AV region in lungfishes lacks any distinct segmental characteristics (see Section 7).

With the exceptions identified earlier, developmental analyses indicate that all six heart components are present in the developing fish (Fig. 1), and this has been confirmed by scanning electron microscope (SEM) studies in Acipenseriformes (Icardo et al., 2004) and observations in juvenile teleost species (Icardo, 2006). This situation is similar to what occurs in tetrapods. However, unlike in tetrapods where the bulbus and conus fail to persist, all the components can also be structurally identified in adult fishes (Fig. 2). Uncertainty as to where the six heart components occur in some adult fish species may be due to a specific portion of the heart having reached a morphological predominance or having undergone more substantial changes than others, and

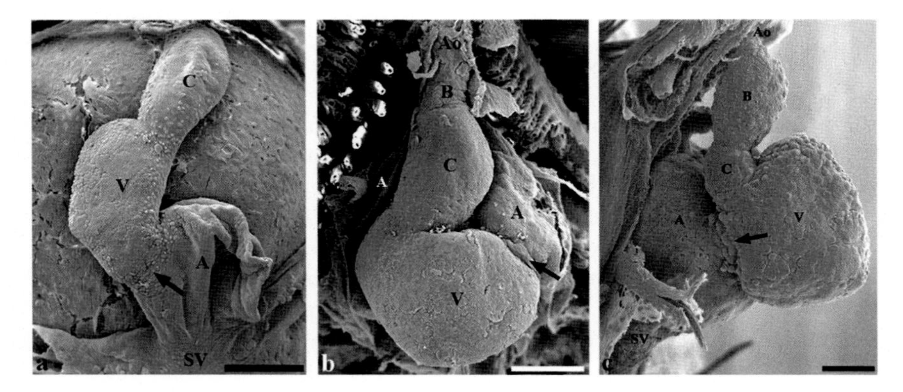

Fig. 1. Early heart development. *A*, atrium; *Ao*, ventral aorta; *B*, bulbus; *C*, conus; *SV*, sinus venosus; *V*, ventricle. *Arrow* in (A)—(C) indicates atrioventricular segment. (A) *Acipenser naccarii* (Adriatic sturgeon), SEM. *Frontal view*. Three days post-fertilization (dpf). Most heart segments are already present. The developing outflow tract (OFT) is solely formed by the conus. The atrium is caudal and dorsal to the ventricle. (B) *Acipenser naccarii* (Adriatic sturgeon), SEM. *Frontal view*, 15 dpf. The OFT is now formed by a long conus and a short bulbus. Ventricle and atrium have established adult relationships. (C) *Sparus auratus* (Gilthead seabream), 16 dpf. *Right lateral view*. The OFT is composed of a proximal conus and a distal, more prominent, bulbus. Scale bars: A, 200 μm; B, 250 μm; C, 25 μm. Panel (C): From Icardo, J.M., 2006. Conus arteriosus of the teleost heart: dismissed, but not missed. Anat. Rec. A 288, 900–908.

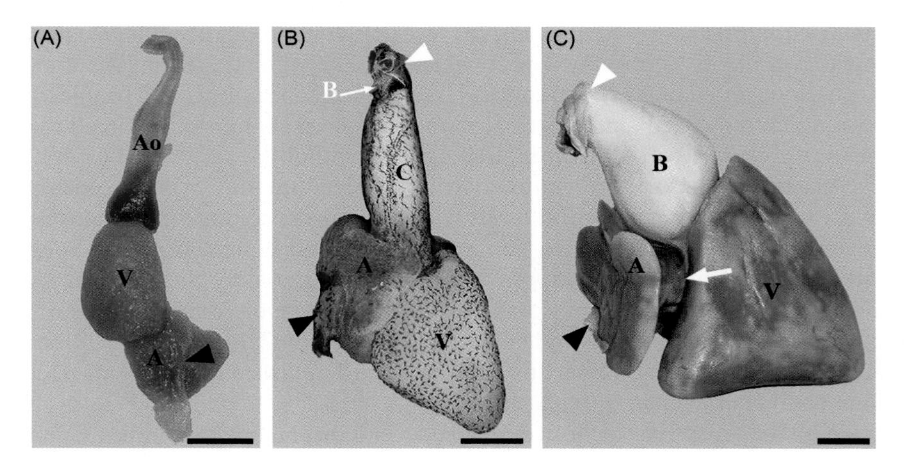

Fig. 2. External heart morphology. *A*, atrium; *Ao*, ventral aorta; *B*, bulbus; *C*, conus; *V*, ventricle. *Black arrowhead* in (A)–(C) indicates sinus venosus. *White arrowhead* in (B)–(C) indicates bulbus–aorta transition. (A) *Eptatretus stoutii* (Pacific hagfish). *Left view*. The atrium is dorsal and caudal to the ventricle. The sinus venosus ends at the *left side* of the atrium. The aorta is very long and shows a swollen base. Outflow tract components are lacking. (B) *Polypterus senegalus* (Gray bichir). *Right view*. All the six chambers except the atrioventricular (AV) canal are apparent. Note the length of the conus and its relationship to the short bulbus. A coronary trunk runs along the *right side* of the conus. (C) *Thunnus albacares* (Yellowfin tuna). *Right view*. The robust bulbus dominates the outflow tract. The conus is hidden under the bulbus. A coronary trunk is visible along the posterior midline of the bulbus surface. *White arrow* indicates the AV segment. Scale bars: A, 0.5 cm; B, 0.1 cm; C, 1 cm. Panel (A): From Icardo, J.M., Colvee, E., Schorno, S., Lauriano, E.R., Fudge, D.S., Glover, C.N., Zaccone, G., 2016. Morphological analysis of the hagfish heart. I. The ventricle, the arterial connection and the ventral aorta. J. Morphol. 277, 326–340.

this is blurring the awareness of the persistence of these structures. Additional discrepancies may arise from the fact that the heart chambers arise sequentially, and that all the chambers may not be present at a given developmental stage (see Guerrero et al., 2004; Icardo et al., 2004). Fig. 1A and B illustrate this point.

While it must be emphasized that the fish heart is divided into six morphologically distinct components and that these six components can be recognized in both developing and adult hearts, it must also be appreciated that the AV canal, the bulbus in basal species, and the conus in more advanced teleosts are all relatively short in length and are functionally better defined as connecting segments. They are ring-like portions that establish structural connections between the other much larger chambers (or with the ventral aorta).

A note of caution should be introduced at this point. The terminology used here assumes homology between the different components of the fish heart. While this may be a safe assumption for chambers like the atrium and the ventricle, the assumption of homology at the heart's venous and arterial poles may be more complicated. For instance, Section 3 discusses the possibility that the sinus venosus of Cyclostomata may not be homologous to that of the rest of the fishes. This is despite the fact that the sinus venosus always collects the venous blood. On the other hand, a recent study casts doubts on segment homology at the OFT level. As stated earlier, the heart of the primitive Actinopterygians displays an arterial-like segment interposed between the myocardial conus and the ventral aorta. Morphological identification of this segment was taken to indicate the existence of a bulbus arteriosus in the hearts of basal fishes, homologous to the bulbus arteriosus in more advanced teleosts. However, recent molecular studies appear to indicate that the bulbus arteriosus is a synapomorphy, i.e., a novel structure whose appearance depends on the phenomenon of whole-genome duplication and the generation of two elastin genes (Moriyama et al., 2016). It has been suggested that neofunctionalization and subspecialization of one of these genes, the teleost-specific *elnb*, determine smooth muscle identity in the bulbus, block cardiac muscle differentiation, and contribute significantly to the morphological and functional characteristics of the teleost bulbus (Moriyama et al., 2016). If the teleost bulbus is indeed an evolutionary novelty, the use of this term to define the most distal portion of the OFT in basal fishes would have to be reconsidered. However, in the absence of a more appropriate definition, and to avoid further confusion, the use of this term has been maintained.

3. SEQUENTIAL ANALYSIS OF THE HEART: A COMPARATIVE APPROACH

This section analyzes the structure of the different heart components and tries to establish commonalities, while simultaneously highlighting the presence of specific features. However, a distinction should be made between the Cyclostomata and the rest of the fishes since they appear to represent different stages of vertebrate evolution. The Cyclostomata (hagfishes and lampreys) are jawless fishes that stand at the base of the chordate tree. The branchial heart of hagfishes shows distinct atrial and ventricular chambers joined by a long AV segment (Fig. 2A). However, there are a number of differences that distinguish these hearts from the general arrangement observed in other fishes (Farrell, 2007, 2011a; Icardo et al., 2016a,b). First, the atrium and ventricle keep a dorsocaudal spatial relationship (Fig. 2A). This contrasts with the dorsal position of the atrium seen in most fishes (Fig. 2B and C), and with the cranial position of the atrium observed, for instance, in avian and mammalian hearts. In addition, the hagfish branchial heart appears to lack any OFT components. Cranial to the ventricle, only a long ventral aorta supplying the numerous gill pouches can be observed (Fig. 2A). The absence of an OFT and the spatial relationship between the atrium and ventricle also appear to be inter-related.

The development of the vertebrate heart involves the sequential addition (and the accompanying molecular differentiation) of the different chambers. The progressive incorporation of cardiac precursors results in bending and torsion of the primitive heart tube (a process known as looping) (Fig. 1). Formation of the cardiac loop brings about modifications of the spatial relationships between the different portions of the heart and, thus, a progression to the adult configuration (e.g., compare the atrium–ventricle relationships in Fig. 1A and B). Icardo et al. (2016a) suggested that the absence of OFT components would result in an incomplete looping, with the adult heart maintaining the embryonic spatial relationships (Fig. 1A exemplifies early relationships between atrium and ventricle, i.e., compare Figs. 1A and 2A).

At the venous pole, the hagfish sinus venosus terminates into the left side of the atrium (Fig. 2A). This contrasts with the serial arrangement of the sinus venosus in all jawed fishes, where a dorsal position with respect to the atrium is common (Fig. 2B and C). The situation in hagfishes is similar to that seen in amphioxus (see Simoes-Costa et al., 2005; Xavier-Neto et al., 2010) and suggests that the hagfish sinus venosus is analogous (i.e., has the same function), but not homologous (not of similar origin), to the vertebrate sinus venosus. Although some uncertainty remains, the hagfish branchial heart appears to be the only craniate heart with two large chambers, atrium and ventricle. The two chambers are connected by an elongated AV segment. It also seems to represent the transition between the heart of the basal amphioxus and that of the vertebrates (Icardo et al., 2016b).

Morphological details of the heart in the other Cyclostomata group (the lampreys) are not well described. In lampreys, the atrium and ventricle appear to keep a lateral relationship, which is likely the result of a discrete looping movement during development. Also, the relationship of the sinus venosus to the atrium is similar to that seen in hagfishes (Richardson et al., 2010). The major difference reported, to date, is the presence in lampreys of a conus arteriosus, which has a pair of mounds or cushions that project into the lumen (Wright, 1984). However, detailed examination of Wright's report indicates that the lamprey conus does not contain myocardium and that its structure is very similar to that of the aorta upstream of the "conus". This indicates that the "conus" is part of the ventral aorta. The presence of the two cushions above the arterial valves remains as a distinct feature of the lamprey circulatory system.

Within Gnathostomata (jawed fishes), the sarcopterygian (lungfishes and coelacanths) heart also differs in specific ways from the general pattern of the fish heart. However, the gross morphological features of the lungfish heart will be treated separately at the end of this chapter (Section 7). This brings us to the actinopterygians whose hearts can be grouped into two external, broad morphological categories: those having a long conus arteriosus (Fig. 2B) and those having a prominent bulbus arteriosus (Fig. 2C). In chondrichthyans and basal actinopterygians, the heart has an elongated to saccular ventricle from which emerges an OFT formed by a long, cylindrical conus and a short, ring-like bulbus (Figs. 2B and 3A). The atrium enters into the dorsal side of the ventricle, and the sinus venosus is a short segment that runs dorsal to the atrium and ends at the level of the septum transversum (Fig. 2B). By contrast, the heart of more advanced teleosts shows an OFT dominated by the presence of a robust bulbus arteriosus that usually shows an expanded base (Figs. 2C, 3B, and 4), and a conus that, while hidden by the bulbus expansion, can be visualized in many species by pulling the bulbus cranially. Furthermore, it is clearly observed in histological sections (Figs. 3B and 4A). The shape of the bulbus varies from elongate to thick, whereas the ventricular shape ranges from tubular to pyramidal (see below). The atrium is located dorsal to the ventricle, while the sinus venosus–atrium relationship is similar (Fig. 4B) to that observed in basal fish. The gross morphology and the structure of the individual chambers are discussed in the following sections.

3.1. The Sinus Venosus

The sinus venosus is the most caudal portion of the heart and collects all of the venous blood *via* the ducts of Cuvier and hepatic veins. With the exception of cyclostomes (see above), the sinus venosus is in series with the rest of the heart chambers (Figs. 2 and 4B), generally displays a more or less tubular (inverted truncated cone; Santer, 1985) shape, and extends between the atrium and the transverse septum. The wall of the sinus venosus is generally thin and appears to be mostly formed by scattered muscle bundles embedded in a connective tissue. However, the amount of muscle appears to vary widely between

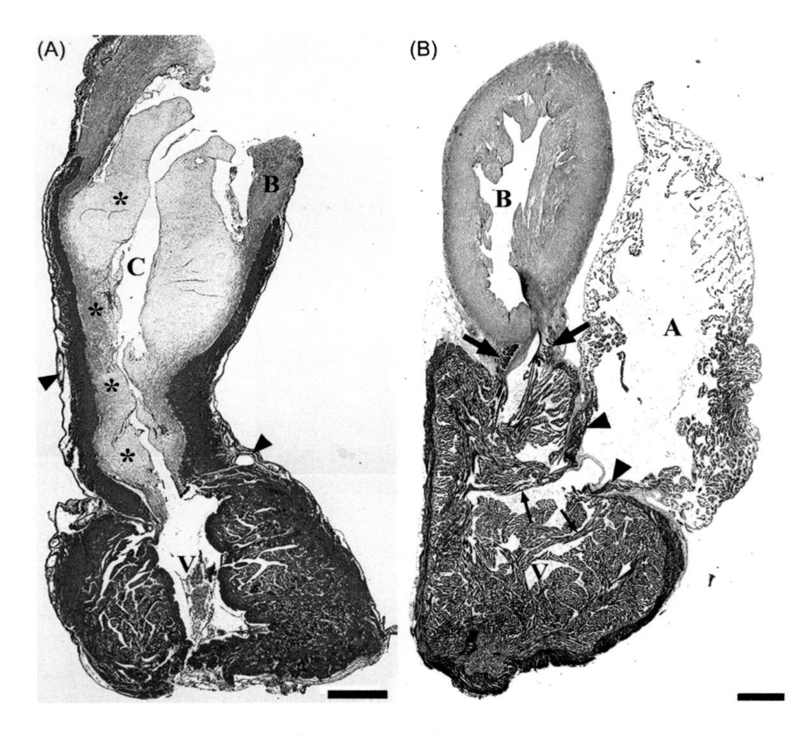

Fig. 3. Inner heart structure. *B*, bulbus; *C*, conus; *V*, ventricle. (A) *Raja clavata* (Thornback ray). Paraffin sectioning, Martin's trichrome staining. Parasagittal section. The long muscular conus contains four valve rows (*asterisks*). The most distal valves are larger than the others. The short bulbus continues the conus. The ventricle displays compact and spongy layers. *Arrowheads* indicate subepicardial coronary trunks. (B) *Anguilla anguilla* (European eel). Paraffin sectioning, Martin's trichrome staining. Parasagittal section. Note the size relationship between the bulbus and the conus (*thick arrows*). In the atrium, the trabeculae are denser near the atrioventricular (AV) region. *Arrowheads* indicate the AV myocardium. *Thin arrows* indicate trabecular connections between the AV region and the ventricle. Scale bars: A, 1 mm; B, 100 μm. Panel (B): From Icardo, J.M., 2013. Collagen and elastin histochemistry of the teleost bulbus arteriosus: false positives. Acta Histochem. 115, 185–189.

species. In elasmobranchs, the amount of myocardium appears to be fairly constant (Saetersdal et al., 1975; Yamauchi, 1980). In Polypteriformes, the wall of the sinus venosus contains myocardial bundles oriented in several directions (unpublished observation). In teleosts, the amount of myocardium in the sinus venosus varies from being almost absent (as in the zebrafish, *Danio rerio*), to scarce (as in the European plaice, *Pleuronectes platessa*) to robust (as in the European eel, *Anguilla anguilla*; Fig. 5A) (see Farrell and Jones, 1992; Santer, 1985; Yamauchi, 1980). However, it should be underscored that the

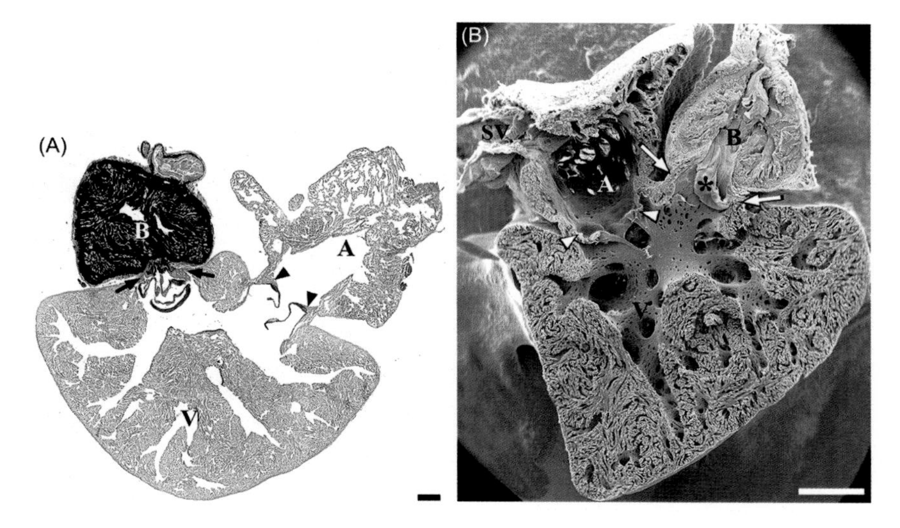

Fig. 4. The teleost heart. *A*, atrium: *B*, bulbus: *C*, conus; *SV*, sinus venosus; *V*, ventricle. (A) *Balistes carolinensis* (Grey triggerfish). Paraffin sectioning, Orcein staining. Parasagittal section. The bulbus stains intensely for orcein. The conus (*arrows*) and the atrioventricular (AV) region (*arrowheads*) are discrete segments formed by compact myocardium. The ventricle is sac-like and entirely trabeculated. Trabeculae in the atrium are denser near the AV region. (B) *Sparus auratus* (Gilthead seabream), SEM. The ventricle is entirely trabeculated. Note the central lumen and radiating, smaller, luminae. The conus is indicated by *arrows*. *Asterisk*, left conal valve leaflet. *Arrowheads*, AV region and valves. The sinus venosus enters the dorsal side of atrium. Scale bars: A, 100 μm; B, 1 mm. Panel (A): From Icardo, J.M., 2012. The teleost heart: a morphological approach. In: Sedmera, D., Wang, T. (Eds.), Ontogeny and Phylogeny of the Vertebrate Heart. Springer, New York, pp. 35–53.

amount of myocardium is generally greater more proximal to the atrium, and that the exact location of most previous observations is unknown. Muscular cells in the sinus venosus are typical myocardial cells, although they appear to be lacking, at least in elasmobranchs (Saetersdal et al., 1975) and Polypteriformes (unpublished observation), myocardial granules that are the storage site for atrial natriuretic peptides. Further, striated myocardial cells can coexist with smooth muscle cells in teleost species such as *A. anguilla* (Yamauchi, 1980) and the Atlantic cod *Gadus morhua* (unpublished observation), whereas smooth muscle cells appear to have replaced the myocardium in species such as the goldfish (*Carassius auratus*) and the common carp (*Cyprinus carpio*) (see Yamauchi, 1980).

The walls of the sinus venosus also contain numerous nerve bundles and large neuronal bodies (see Chapter 4, Volume 36A: Farrell and Smith, 2017). Nerves and neurons form plexus-like structures. These plexuses

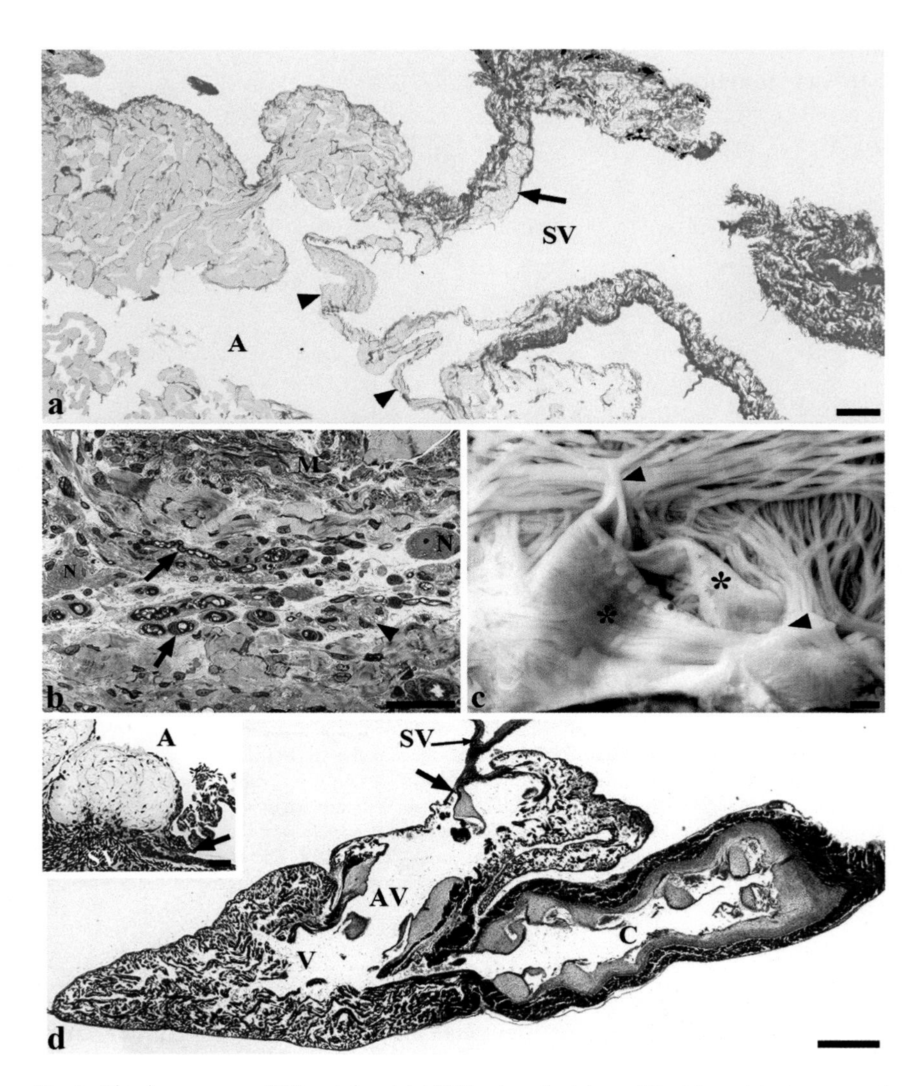

Fig. 5. The sinus venosus (SV) and sinoatrial (SA) valve. *A*, atrium; *AV*, atrioventricular segment; *C*, conus; *V*, ventricle. (A) *Anguilla anguilla* (European eel). Paraffin sectioning, Sirius red staining. Longitudinal section. The SV wall shows a robust muscular component (*arrow*). The SA valve leaflets (*arrowheads*) project into the atrium and have a thick muscular core. (B) *Acipenser naccarii* (Adriatic sturgeon). Semithin section, toluidine blue staining. The SV wall shows myocardial (M) cells, neurons (N), and myelinated (*arrows*) and unmyelinated (*arrowhead*) nerve fibers. (C) *Raja clavata* (Thornback ray). This dissection shows the atrial side of the SA leaflets (*asterisks*) and the valve commissures (*arrowheads*). Note slit-like opening of the SA aperture. Groups of thick atrial trabeculae run in several directions. (D) *Lepisosteus oculatus* (Spotted gar). Paraffin sectioning, Martin's trichrome staining. Sagittal section. The SA (*arrow*) and the AV leaflets are large and bulky. The ventricle is pyramidal and lacks a compacta. Valves in the conus are arranged into *transverse rows* and show bulky leaflets. *Inset*: *Erpetoichthys calabaricus* (Reedfish). Semithin section, toluidine blue staining. The two bulky leaflets of the SA valve are present. The wall of the atrium is continuous (*arrow*) with the SV wall. Muscle is present in the SV wall and at the base of the leaflets. Scale bars: A, 100 μm; B, 50 μm; C, 200 μm; D, 300 μm; *inset* of (D), 50 μm.

differentiate from the cardiac ganglia of birds and mammals in that they are loosely organized and are not enclosed by a connective capsule (Santer, 1985). Complex nerve fiber-neuron formations may also be recognized along the wall of the sinus venosus in elasmobranchs (Saetersdal et al., 1975; Yamauchi, 1980), sturgeons (Fig. 5B), and teleosts, but are most likely found in the subepicardium of the sinoatrial (SA) junction (see Sections 5 and 6).

From a functional point of view, the sinus venosus conveys the blood into the atrium (Yamauchi, 1980). The contraction of the sinus venosus has been said to contribute to *vis-a-tergo* atrial filling. However, this is difficult to recognize as a general feature due to the sparse muscular content of the sinus wall in many species, and due to the paucity of venous valves to prevent backflow (Farrell and Jones, 1992; Santer, 1985).

The sinus venosus is separated from the atrium by the SA valve. In general, the fish SA valve is composed of two, left and right, flap-like leaflets that delimit a slit-like opening with dorsal and ventral commissures (Fig. 5C). The leaflets attach to the SA orifice and project into the atrium. The leaflet edge may be completely free (Fig. 5C), or it may be attached to atrial trabeculae such as in hagfishes (Icardo et al., 2016b) and in several teleosts (Santer, 1985). The two leaflets of the SA valve are usually similar in size, but they may also be of unequal length. The SA valve may even be absent (Ishimatsu, 2012; Munshi et al., 2001). For unknown reasons, air-breathing teleosts appear to display the widest variety of SA valve morphologies (see Santer, 1985).

The leaflets of the SA valve show a variable structure. In hagfishes, they are formed by a collagenous-rich connective tissue with few interstitial cells (Icardo et al., 2016b). A collagenous structure is also observed in several species of elasmobranchs (Hamlett et al., 1996), whereas a thick muscular core has been reported in the SA leaflets of other chondrichthyans such as the dogfish (*Scyliorhinus canicula*; Gallego et al., 1997). In teleosts, the SA leaflets of many species contain a muscular core (Santer, 1985) that appears to be especially prominent in *A. anguilla* (Fig. 5A). The presence of muscle has been related to pacemaker activity (Haverinen and Vornanen, 2007; Newton et al., 2014; Stoyek et al., 2015). This will be commented on later (see Section 6).

The presence of relatively thin SA leaflets, with or without a muscular core, is not a general feature among all fish groups. Unpublished observations indicate that the SA valve of Polypteriformes and Lepisosteiformes is formed by two bulky leaflets of unequal size (Fig. 5D) that contain stellated fibroblasts embedded in a loose matrix. In addition, the leaflets in the polypteriform *Erpetoichthys calabaricus* contain numerous granulocytes (Fig. 5D, inset). In this species, muscular components are only observed at the area of leaflet insertion (Fig. 5D). It is presently unknown whether other basal actinopterygians display a similar SA valve structure.

3.2. The Atrium

The atrial chamber follows the sinus venosus and is generally located in a dorsal position relative to the ventricle. The shape of the atrium is variable, although a clear account of atrium morphology is difficult to provide since it usually collapses after fixation. The atrium shows a variable relationship with the rest of the chambers. For instance, the distance between the inflow and outflow orifices of the ventricle loosely defines the atrium's relationship with the OFT. When the distance is longer, like in elasmobranchs (Santer, 1985), the atrium remains far from the OFT. When the distance is shorter, the atrium contacts the OFT and frequently extends ventral prolongations that embrace the lateral aspects of the OFT. This is more evident in air-breathing species (Munshi and Mishra, 1974; Santer et al., 1983). In addition, the atrial wall may be attached to the OFT wall by a connective tissue. This can be observed, for instance, in the chondrostean *Polypterus senegalus*, in the holostean *Lepisosteus oculatus*, in the Asian swamp eel *Monopterus albus* (unpublished observation), and in lungfishes (Section 7). All of these species are air breathers, although they display different types of breathing organs. Perhaps this situation is related to specific cardiovascular adaptations to air breathing.

The atrial wall is formed by myocardial trabeculae, the pectinate muscles that generally appear to radiate from (or converging to) the AV orifice. This arrangement has been observed in cyclostomes (Icardo et al., 2016b) and in several teleost species (Icardo, 2012; Santer, 1985). By contrast, groups of robust trabeculae course in different directions in the atrium of the thornback ray *Raja clavata* (Fig. 5C). Atrial trabeculae are thin and slender in cyclostomes (Icardo et al., 2016a), appear thicker in elasmobranchs (Figs. 5C and 6A) and sturgeons, and are thin in Polypteriformes (Fig. 5D), holosteans (Fig. 6B), and most teleosts (Figs. 3B and 4). The possible relationship between the trabecular appearance of the atrium and hemodynamic factors, or with other factors such as body mass, is presently unknown.

It should be highlighted that the mode of preparation of the specimens, and the contractile state at fixation, may modify the appearance of the trabecular system (both in the atrium and in the ventricle) significantly. For instance, the atrium of the specimen shown in Fig. 6A was fixed immediately after capture, and the atrium appears contracted and collapsed. By contrast, the specimen shown in Fig. 6B was fixed in the laboratory, the atrium being gently distended by perfusion of fixative through the lumen. In this case, the trabecular system has a spongy appearance. Only when the collapse of the atrium is prevented may the real arrangement of the trabeculae be discerned. Note that, in Fig. 6B, the wall thickness is limited by the presence of trabeculae running in parallel with the outer atrial surface. The same can be observed in histological sections

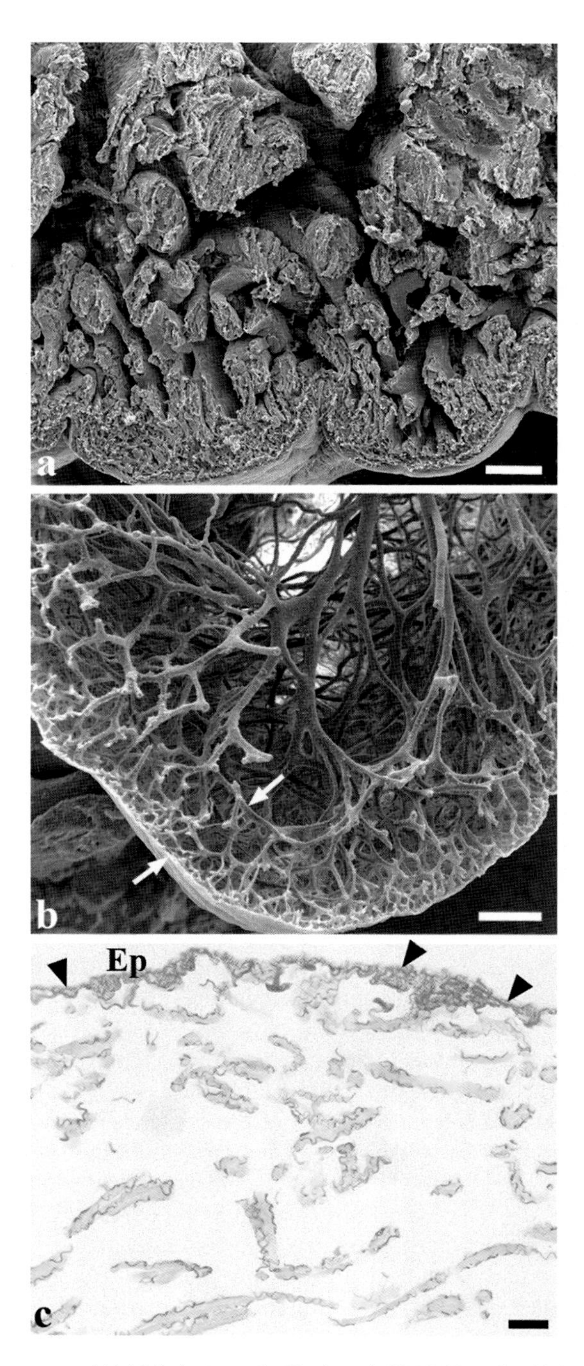

Fig. 6. Atrium structure. (A) *Myliobatis aquila* (Eagle ray), SEM. Note the thick atrial trabeculae. The wall has collapsed during fixation. (B) *Polypterus senegalus* (Gray bichir), SEM. Atrium distended in fixation. Note the delicate trabecular pattern. *Arrows* indicate the limits of the atrial wall. (C) *Amia calva* (Bowfin). Paraffin sectioning, Sirius red staining. Collagen accumulates under the trabecular subendocardium. Note the discontinuities (*arrowheads*) in the external myocardial layer. The epicardium (Ep) is indicated. In (B) and (C), most peripheral trabeculae are thinner than those located centrally. Scale bars: A, 100 μm; B, 200 μm; C, 50 μm.

(Figs. 3B and 4). In addition, atrial trabeculae appear to be thicker and to form a denser network near the AV region in most species (Figs. 3B, 4A, and 5D). In general, trabeculae appear to be thicker near the lumen than closer to the wall's outer surface (Fig. 6B).

Atrial trabeculae are composed of a myocardial core surrounded by endocardium. Collagen is an important component of the trabecular subendocardium, but little collagen is seen between the myocardial cells (Fig. 6C). Also, collagen forms a continuous layer under the atrial epicardium. Noticeably, the outer myocardial layer of the atrium is a discontinuous mantle in many species (Fig. 6C; see Icardo et al., 2016b). This is probably reminiscent of the situation found in early development when atrial cardiomyocytes lose intercellular connections and acquire stellated shapes that transform the continuous atrial muscle wall into a webbed myocardium (Foglia et al., 2016). This may facilitate atrial expansion, but results in myocardial discontinuities that appear to persist in adult specimens in many cases. In the areas where the most peripheral layer of the atrial muscle is lacking, subepicardial collagen contributes to the structural closure of the atrial wall (Icardo et al., 2016b).

3.3. The AV Segment

Classical morphological accounts describe the AV region as being formed by a ring of cardiac tissue supporting the AV valves (Farrell and Jones, 1992; Santer and Cobb, 1972). The presence of connective tissue surrounding the AV muscle results in electrical insulation and explains the presence of a delay in the electrical conduction of the heart (Satchell, 1991; Sedmera et al., 2003). However, the AV region appears to be more complex than anticipated. Recent studies in teleosts indicate that the AV region constitutes a distinct heart segment containing muscle, connective tissue, and vessels (Icardo and Colvee, 2011). In addition, it carries a complex network of nerves and neurons (see Section 5).

In teleosts, the AV muscle forms a ring of compact myocardium formed, in most cases, by the accretion of muscle bundle layers (Figs. 3B and 4; Icardo and Colvee, 2011). This is independent of the existence of a ventricular compacta (see Section 3.4 for a definition of compacta). Thus, the compact AV myocardium may be inserted between two completely trabeculated chambers. In most cases, the AV myocardium is supplied by coronary vessels. In a few cases, however, the coronaries are replaced by endocardial sinuses that are open to the circulating blood and supply the AV myocardium directly (Icardo and Colvee, 2011). The AV muscle is also surrounded by a ring of connective tissue, that is rich in collagen and extends into the AV muscle. Depending on the species, this collagen may assemble into thick bundles, may form loosely arranged strands, or may even be almost absent. In general, both the AV muscle and the AV's ring of connective tissue are more robust and better vascularized in more active species like tuna, and less prominent in less active fishes or those with highly specialized hearts like the Antarctic teleosts (see Icardo and Colvee, 2011; more information is provided in Section 4). The presence of connective tissue within the AV ring partially isolates the AV muscle from the atrium and ventricle. However, isolation is not complete since the AV myocardium remains connected both to the atrial and to the ventricular musculature (Figs. 3B and 4).

Classifying the AV segment of the teleost heart as a distinct component raises the question of whether it also is a distinct component in the heart of other groups. In hagfishes, the AV region is a long, funnel-like, segment and its wall contains loosely arranged myocardial bundles embedded in a collagenous matrix (Icardo et al., 2016b). Also, it is avascular. This appears to be the single case in which the AV segment maintains separation between the atrium and ventricle instead of being a discrete, junctional segment. The extended length and the poor muscular component of the AV segment in hagfishes explain the long delay in the ECG (Farrell, 2007). Very little information is available for other fish groups. Unpublished observations indicate that in elasmobranchs (Fig. 7A and B) and in sturgeons, the AV region is formed by a thick myocardial ring. Compact, vascularized myocardium is also observed in the AV region of other basal fishes such as the Polypteriformes and the holosteans (Fig. 5D), whose ventricles lack a compact layer. In these groups, the AV myocardium is partially isolated from the atrial and ventricular musculature by a connective tissue rich in collagen (Fig. 7A and B), but the thickness of the connective tissue and the amount of collagen vary widely. Elasmobranchs (Fig. 7A and B) and sturgeons display a thick compacta, whereas bichirs and gars show completely trabeculated ventricles (Fig. 5D).

Another basic structural feature of the AV segment is that it supports the AV valves. However, very little attention has been devoted to this topic. In hagfishes, the AV valve is formed by two pocket-like leaflets that have organized collagen bundles and a few interstitial cells (Icardo et al., 2016b). In sharks, the AV valve has been described as being composed of two leaflets, chordae tendineae, and papillary muscles (Hamlett et al., 1996). In the heart of rays, the AV leaflets may be bulky, contain numerous interstitial cells (Fig. 7A), or may be much thinner and have a predominantly collagenous structure (Fig. 7B). Chordae tendineae, mostly formed by wavy collagen bundles, are present in all cases (Fig. 7A and B). The chordae extend from the ventricular wall to the under surface and to the free margin of the leaflets. However, papillary muscles are not observed. Chordae tendineae are also present in sturgeons (Fig. 7C). The collagenous component of the chordae spreads under the ventricular endocardium and attaches directly to the ventricular musculature (Fig. 7A–C). Thus, no papillary muscles exist.

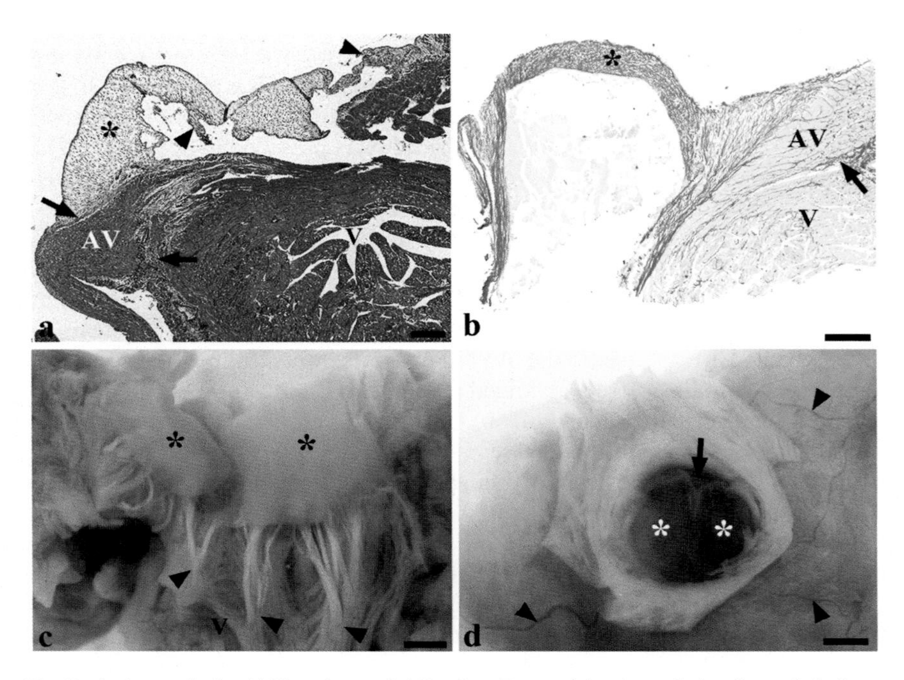

Fig. 7. Atrioventricular (AV) region and AV valve. *V*, ventricle. *Asterisks* in all panels indicate valve leaflets. (A) *Raja clavata* (Thornback ray). Paraffin sectioning, Martin's trichrome staining. Parasagittal section. The muscular AV region is flanked by collagen (*arrow*). Note the bulky structure of leaflet. *Arrowheads* indicate chordae tendineae. (B) *Myliobatis aquila* (Eagle ray). Paraffin sectioning, Sirius red staining. Parasagittal section. The leaflet is thin and shows a collagenous structure. Collagen (*arrow*) separates the AV region from the ventricle. (C) *Acipenser naccarii* (Adriatic sturgeon). Two large leaflets are anchored to the ventricular muscle (*arrowheads*) by chordae tendineae. The crimped appearance of the chordae is due to the wavy collagen bundles that form the chordal core. Papillary muscles are absent. (D) *Trigla lucerna* (Yellow gounard). Detail of the AV orifice after dissecting the atrium. Cranial and caudal leaflets, and a small lateral leaflet (*arrow*), can be seen. Despite the spongy ventricle, coronary vessels (*arrowheads*) run in the subepicardium around the AV orifice. Scale bars: A, 200 μm; B, 50 μm; C, 1 mm; D, 500 μm.

Much less is known of the characteristics of the AV valves in other basal fishes. Polypteriformes and holosteans (Fig. 5D) show two bulky, cell-rich leaflets and a system of chordae tendineae that is less developed than in elasmobranchs and sturgeons. Again, papillary muscles are absent. The overall absence of papillary muscles in basal species raises the question of whether the identification of papillary muscles in sharks (Hamlett et al., 1996) is an overestimation due to the presence of sample deformations during processing.

In more advanced teleosts, the AV valve is formed by cranial and caudal leaflets attached to the AV muscle ring (Fig. 7D). AV valves lack papillary muscles and chordae tendineae (Figs. 3B and 4). The AV leaflets contain a dense core formed by numerous cells and some collagen and elastin fibrils. The leaflets also exhibit an atrial fibrosa rich in collagen and elastin and may display a poorly developed ventricular fibrosa. However, this is a general description and many variations can be observed. One or two small additional leaflets may be present in a lateral position. This has been observed, for instance, in *Sparus auratus* and in *Trigla lucerna* (Fig. 7D). Within the leaflet, interstitial cells may be densely packed or form a loose network, the amount of collagen and elastin is variable, and the fibrosa may be thin or may appear to be formed by several thick collagenous layers. All of these variations, with an indication of the species, have been described earlier (Icardo and Colvee, 2011). In general, more active fishes have thicker leaflets with higher cell numbers and a thicker fibrosa. Also, species without distinct dense cores appear to have compensated for the lower cell number with an increase in leaflet collagen. This has been related to the maintenance of the valve's resilience (Icardo and Colvee, 2011). The leaflet fibrosa is continuous with the extracellular matrix at the areas of leaflet attachment, forming a kind of fibrous hinge that facilitates valve movement.

3.4. The Ventricle

The ventricle is responsible for generating the heart's output and central arterial blood pressure. The ventricle of the fish heart has been studied from many morphological and functional points of view, and several classifications have been developed. However, there is considerable species variability making categorization difficult in many cases.

From the external point of view, the ventricle can be elongated, sac-like, or pyramidal. The ventricle appears elongated in cyclostomes (Fig. 2A) and sac-like in elasmobranchs (Santer, 1985). In chondrosteans (Fig. 2B) and holosteans (Fig. 5D) the ventricle is pyramidal showing rounded borders and a base smaller than the lateral faces. Compared to other fish, more advanced teleosts show the greatest shape variability, with the ventricle ranging from elongated (Fig. 3B), to sac-like (Fig. 4A), to pyramidal (Figs. 2C and 4B; Farrell and Jones, 1992; Icardo, 2012; Santer, 1985). The pyramidal ventricles of teleosts are usually regular triangular pyramids (Figs. 2C and 4B) with sharper edges than those observed in chondrosteans and holosteans. However, the meaning of all these different morphologies is unclear. Elongated ventricles have been associated with an elongated fish shape, but this is not always the case (Santer, 1985). Sac-like ventricles (Fig. 4A) have been associated with fish with restricted locomotory activity, but this is difficult to reconcile with the lifestyle of many elasmobranchs. Pyramidal ventricles (Fig. 2C) have been associated with a robust and muscularized ventricular wall

(Fig. 8), but many teleost ventricles with pyramidal shape (Fig. 4B) do not fit into this classification (see Farrell and Jones, 1992; Icardo, 2012; Santer, 1985). Noticeably, the ventricular external shape is not totally fixed since it may be modified in response to rearing conditions, as occurs in salmonids (Gamperl and Farrell, 2004).

The external shape of the ventricle has also been related to the inner ventricular architecture, although the relation is imprecise in many cases. Internally, the ventricle of the fish heart can be divided into two broad categories: those having completely trabeculated (spongy) ventricles, and those having an inner spongiosa and an outer compacta of variable thickness. The presence and characteristics of the compacta have been used to classify the fish ventricle into four different types (Types I–IV) (Davie and Farrell, 1991; Farrell et al., 2012; Tota, 1989; Tota et al., 1983). This classification relates the inner ventricular architecture and the degree of vascularization, and will be discussed in Section 4. In all the cases, the ventricle is surrounded by the epicardium and by a subepicardial layer containing cellular and extracellular elements, vessels, and nerve fibers in variable amounts.

Spongy ventricles (Type I) are observed in hagfishes (Icardo et al., 2016a), in several Holocephali (Durán et al., 2015), in bichirs and gars (Fig. 5D), and in somewhere between 50% and 80% of all the teleost species (see Farrell et al.,

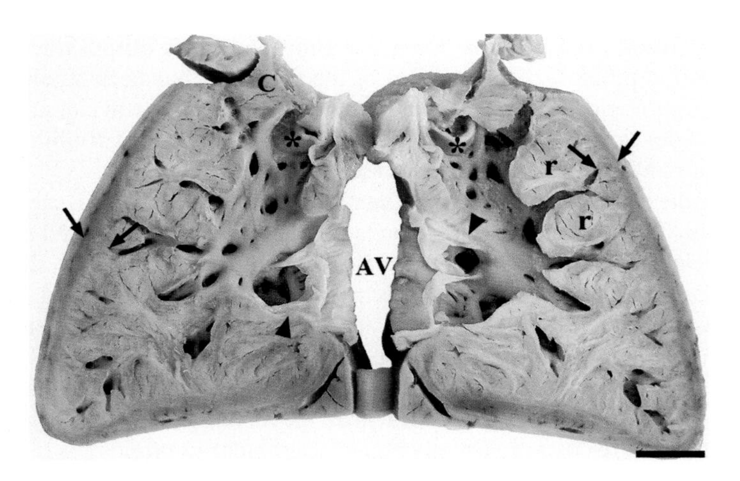

Fig. 8. Ventricular structure. *Thunnus albacares* (Yellowfin tuna). Sagittal section of the ventricle. The two halves are present. In each ventricular half, the smooth surface corresponds with the main ventricular lumen. The orifices in the surface communicate with radiating lumina. The secondary luminae are separated by compact ridges (r). Paired arrows indicate the compacta thickness. In the conus (C), *left* and *right leaflets* (*asterisks*) are clear. A third, posterior leaflet appears cut into half. In the atrioventricular (AV) region, the AV leaflets attach directly (*arrowheads*) to the ventricular wall. Scale bar: 1 cm.

2012; Santer and Greer Walker, 1980). In these ventricles (Figs. 4 and 5D), the trabeculae organize into trabecular sheets that delimit small luminae that radiate out from the main ventricular lumen (Fig. 4B; Icardo, 2012; Icardo et al., 2005a; Munshi et al., 2001). The trabecular sheets show numerous orifices that allow the blood to be squeezed in and out, and to contact individual trabeculae. At the ventricular periphery, however, the trabeculae do not organize into sheets, but form arch-like systems, the vane of the arches being occupied by other trabeculae running in different directions (see Icardo et al., 2005a, 2016a). Thus, the ventricular periphery only contains the trabecular arches and the corresponding lacunary spaces (Fig. 4). Unfortunately, detailed descriptions of the ventricular architecture in spongy ventricles are restricted to a few species, mostly within the teleost group (Icardo, 2012; Icardo et al., 2005a; Munshi et al., 2001). Thus, it is difficult to say whether this arrangement may be shared by many other species. Of note, trabecular sheets joined by collagenous chords and peripheral, arch-like, trabecular systems have recently been described in the spongy ventricle of hagfishes (Icardo et al., 2016a). In addition, the spongy component of the salmonid ventricle that also has a compacta appears to be organized similarly (Pieperhoff et al., 2009). This information raises questions about whether this is a general pattern in the spatial distribution of the ventricular trabeculae, and what the functional significance of such an architecture might be.

From the functional point of view, the existence of a central lumen and a discrete number of small luminae radiating from the center convert the ventricle into a multichambered structure. The main lumen presumably would support the highest amount of stress, with the stress being progressively attenuated toward the periphery. This would protect the outermost myocardium, avoiding disruption and maintaining the integrity of the ventricular wall (Icardo et al., 2005a).

Spongy ventricles are typical of fish with low/middle levels of locomotory activity and have generally been considered unable to maintain high levels of cardiac performance. This morphofunctional generalization can be applied, for instance, to the hagfish heart (see Farrell, 2007) and to the extremely specialized Antarctic teleosts (Tota et al., 1997). However, studies in the gilthead seabream *S. auratus* have challenged this general view. The ventricle of *S. auratus* is entirely trabeculated (Fig. 4B). Despite this, the heart is able to maintain high heart rates and high levels of cardiac output (Icardo et al., 2005a), similar to those observed in salmonids (Gattuso et al., 2002; Graham and Farrell, 1989). In comparison, salmonids are more active, athletic fishes that show robust hearts and a ventricular compacta. Thus, the maintenance of high levels of cardiac performance in *S. auratus* is entirely due to the trabecular organization. It is likely that other spongy hearts are also able to sustain similar levels of performance.

In this context, further studies of the architecture of the fish ventricle may be helpful. SEM studies of the ventricular trabeculae have mostly been based on parasagittal or frontal sections (see Fig. 4B). However, transverse sections may reveal unsuspected patterns of organization. In the elongated ventricle of juvenile specimens of *A. anguilla*, many large trabecular sheets appear that are oriented circumferentially (Fig. 9). Furthermore, trabecular sheets become confluent and delimit the central lumen. It is also clear that secondary luminae, marked by exposure to the free surface of trabecular sheets, radiate from a central lumen, like in *S. auratus* (Fig. 4B). However, it is not clear whether trabecular sheets in pyramidal ventricles may orientate transversely. It is also unclear whether all the elongated ventricles display the same trabecular architecture. Nonetheless, Fig. 9 makes it easy to envision that coordinated contraction from the heart apex would completely squeeze the ventricle, and result in a low end-systolic volume and a high ejection fraction.

The ventricle of many fishes shows both spongiosa and compacta (ventricle Types II–IV). A compact layer is observed in all elasmobranchs

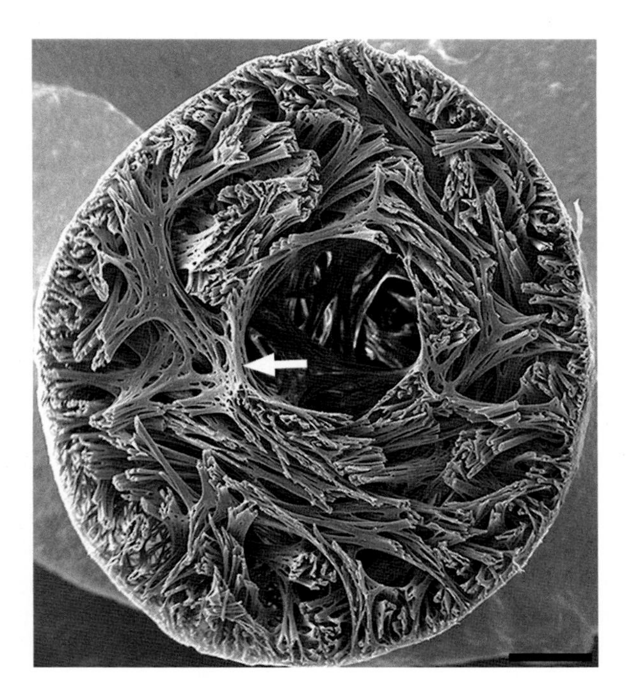

Fig. 9. Ventricular structure. *Anguilla anguilla* (European eel), SEM. Transverse section at the midventricular level. Several large trabecular sheets are circumferentially oriented and delimit the central lumen. *Arrow* indicates communication between the central lumen and the radiating, secondary luminae. The compacta is underdeveloped in juvenile specimens. Scale bar: 0.5 mm.

(Figs. 3A and 7A), in sturgeons (Icardo et al., 2002a), in the holostean *Amia calva* (unpublished observation), and in many teleost species (Figs. 3B and 8), especially the Clupeidae, Salmonidae, Anguillidae, Carangidae, and Scombridae families (see Santer and Greer Walker, 1980). A general characteristic of these teleost species is that they are active, migrating fish that are able to sustain high levels of cardiac performance (see Farrell and Jones, 1992). Thus, it is not surprising that the possession of a compacta has been directly linked to athletic, very active fish. However, the proportion of the compacta varies widely between species, and a direct correlation between compacta thickness and fish activity cannot always be established. Nor is there any direct relationship between the presence of a compacta and ventricular shape. Most saccular ventricles in elasmobranchs display a compacta, and many pyramidal ventricles in higher teleosts are entirely trabeculated (see Icardo, 2012). In any case, the compacta constitutes only a small part of the ventricular mass in most fishes. The estimated proportion of compacta ranges between 5% in the rabbitfish *Chimaera monstrosa* and 74% in the bigeye tuna *Thunnus obesus* (Farrell and Jones, 1992; Santer and Greer Walker, 1980). However, so far, only in tunas, in the Pacific tarpon (Farrell et al., 2007), and in anchovies (Santer and Greer Walker, 1980) has the proportion of the compacta been reported to surpass that of the spongiosa. Of note, the proportion of compacta is not a fixed value since it has been shown to vary, for instance, with smoltification, changing seasons (Farrell and Jones, 1992), and growth (Cerra et al., 2004; Farrell et al., 1988, 2007; Gamperl and Farrell, 2004).

One challenge to determining the proportion that each of the two muscular compartments comprises of the ventricle is the method by which the specimens are fixed and prepared for study. Contracted spongy ventricles may erroneously give the impression of possessing a compacta (see Section 3.2). Also, the thickness of the compacta is not regular and undergoes variations from apex to base (Icardo, 2012; Santer, 1985). More importantly, compact and spongy are purely intuitive terms. We lack a good definition for either of these two tissue compartments. What defines ventricular compacta? In a previous paper, a two- to three-cell-thick layer, as observed in the lesser weever *Echiichthys vipera*, was considered to be compacta (Icardo, 2012). However, a compacta may be better defined as an outer, vascularized, myocardial layer (Farrell et al., 2012). This involves a more complete definition of the characteristics of a compacta, although it may often require careful histological analyses. Also, it may be anticipated that full categorization will be difficult in several cases, such as in the Australian lungfish *Neoceratodus forsteri* (see Section 4). Nonetheless, an agreement is needed on this point, since the attribution of a compacta to any fish species has consequences for basic definitions and classifications.

A definitional problem may also arise with the spongiosa. For instance, the spongiosa of elasmobranchs (Cox et al., 2016; Tota, 1989), sturgeons

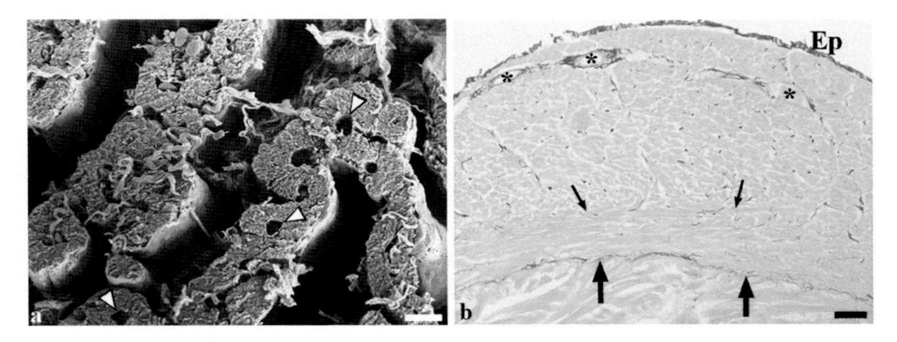

Fig. 10. Ventricular structure. (A) *Acipenser naccarii* (Adriatic sturgeon), SEM. Spongy compartment. Detail of trabeculae. *Arrowheads* indicate trabecular capillaries. (B) *Scomber japonicus* (Chub mackerel). Paraffin sectioning, Sirius red staining. Under the epicardium (Ep), the thick compacta displays two muscle layers (*small arrows* mark the boundary between layers). *Asterisks* indicate a large vessel. The compacta–spongiosa boundary is enriched in collagen (*large arrows*). The spongiosa lacks vessels. Scale bars: A, 50 μm; B, 50 μm.

(Fig. 10A), and tuna (Icardo, 2012; Tota, 1989) has a coronary arterial supply similar to that observed in the outer compacta. The presence of vessels in the spongy myocardium is a substantial difference that has to be defined adequately. In fact, detailed accounts of the architecture of the spongy layer in basal species are still lacking. Within teleosts, the spongiosa appears to be formed by trabecular sheets and luminae that radiate out from the central lumen. This can clearly be observed in the sparid *S. auratus* (Fig. 4B) and also in the ventricle of the yellowfin tuna (*Thunnus albacares*) (Fig. 8). However, careful comparison of the spongiosa shows profound differences between the two species. In *S. auratus*, the separation between luminae is made by trabecular sheets that allow the blood to squeeze in and out and to bathe individual trabeculae. In *T. albacares*, the separation between luminae is made by compact muscular ridges (Fig. 8). These ridges are formed by the accretion of trabeculae and appear heavily vascularized (Fig. 5 of Icardo, 2012, exemplifies the histological arrangement in the ventricle of the albacore tuna (*Thunnus alalunga*)). Could this organization be defined as spongy? On the other hand, the ventricular compacta is a relatively thick layer that could be considered to be limited internally by the numerous luminae that connect with the main central lumen (Fig. 8). Notably, the luminae also delimit, in the spongy ventricle, an outer zone exclusively formed by single trabeculae and lacunary spaces (see above). This corresponds to the area occupied by the vascularized compacta. It is proposed that the level of penetration of the secondary luminae should be taken as the inner boundary for the ventricular compacta. If so, Fig. 8 shows a compacta/spongiosa ratio much lower than that generally reported for tuna (Santer and Greer Walker, 1980). Under this paradigm, the proportional values of compacta and spongiosa would have to be reassessed in many species.

The development of a compacta is accompanied by the progressive organization of the myocardium into discrete bundle layers. Myocardial cell layers oriented in several directions can be discerned in most hearts possessing a compacta (Fig. 10B). This organization becomes more evident in large, athletic, fish such as swordfishes and tunas (Sánchez-Quintana and Hurle, 1987). In these species, two or three layers of well-defined myocardial fiber bundles loop around the ventricle or form circular bands around this chamber's vertices. This arrangement may provide mechanical advantages during systolic contraction (Farrell and Jones, 1992).

In hearts with ventricular compacta, the interface between the compacta and the spongiosa is of interest because the function of the two compartments has to be coordinated. For instance, the thickness of the compact myocardium and the collagen content of the ventricle need to be modified in a coordinated manner to maintain cardiac output during thermal acclimation (Johnson et al., 2014; Keen et al., 2017; Klaiman et al., 2011). In teleost hearts, a layer of collagen of variable thickness is always present at the compacta–spongiosa boundary (Fig. 10B), and this was identified as the structural link between compartments (Poupa et al., 1974; Tota, 1978). More recently, it has been suggested that the concentration of myocardial junctions at this boundary creates the attachment force necessary to hold the two compartments together (Pieperhoff et al., 2009). Since both cellular and extracellular elements are present (Midttun, 1983), they likely act in concert to maintain the integrity of the ventricular architecture.

3.5. The Outflow Tract

3.5.1. Basal Gnathostomata

With the exception of the Cyclostomata, the OFT of basal fishes is formed by a proximal conus arteriosus that is wrapped in cardiac muscle which is electrically coupled to the ventricular myocardium, and by a distal bulbus arteriosus (Figs. 2B, 3A, and 5D). The conus is a long cylinder that arises from the ventricular chamber. The dimensions of the cylinder are not regular, being thicker in the middle. This occurs, at least, in rays (Fig. 3A), sturgeons (Icardo et al., 2002a) and bichirs (Fig. 2B). The wall of the conus is organized into layers. There is an external layer formed by the epicardium and the subepicardial tissue. In general, the subepicardium contains fibroblastic cells embedded in a collagenous matrix, vessels (see Section 4), and nerves (see Section 5). However, this tissue may be much more complex. In sturgeons, the subepicardium contains nodular structures formed by a lymphohemopoietic, thymus-like tissue. These nodes are likely involved in the establishment and maintenance of the immune response (Icardo et al., 2002b). The subepicardium may also contain melanocytes. These cells appear in low numbers in the subepicardial tissue of several species (Guerrero et al., 2004; Santer, 1985), but may be very abundant in others. For instance, melanocytes are uniformly distributed throughout the subepicardium in *P. senegalus* (Fig. 2B; Reyes-Moya et al., 2015). The role of these cells remains unclear.

Under the subepicardium, the middle layer of the conus is formed by compact, vascularized myocardium (Fig. 3A; see Grimes and Kirby, 2009; Santer, 1985; Satchell, 1991; Yamauchi, 1980). Histological studies in sturgeons have shown that conus myocardial cells are arranged into bundles that vary in number (one to three) and have different orientations along the conus length (Icardo et al., 2002a). Myocardial organization may be different in other species and groups. For instance, the outer layer of the conus myocardium is formed by transversely oriented fibers in the shortfin mako shark, *Isurus oxyrinchus* (Sánchez-Quintana and Hurle, 1987). However, it is unknown whether myocardial bundles orientate differently under that layer. In all the cases, myocardial bundles are separated by loose connective tissue septa that carry numerous vessels that supply the conal myocardium (Icardo et al., 2002a; Zummo and Farina, 1989). These vessels arise from subepicardial coronary arteries.

Under the middle layer of the conus, there is an inner layer formed by endocardium and subendocardial tissue. The subendocardium is thick and contains interstitial cells embedded in a loose matrix rich in collagen and elastin (Fig. 3A; see Grimes and Kirby, 2009; Icardo et al., 2002a). In general, this tissue is denser under the myocardium than under the endocardium.

A fundamental characteristic of the conus arteriosus is that it supports the conal valves. Conal valves are semilunar in shape and consist of a leaflet and the supporting sinus. Valves are arranged into transverse rows or tiers that contain from two to six valves, the number of tiers varying from one to eight (see below). Nonetheless, the terms conus or conal cannot be applied to the early Cyclostomata due to the absence of all OFT components. In these fishes, there are two (right and left) arterial valves that arrange into a single valve row. The valve leaflets show a collagenous structure and are anchored to the ventricular myocardium (Icardo et al., 2016a). By contrast, the conal valves in basal Gnathostomata are arranged into a variable number of transverse and longitudinal rows and display variable histological characteristics. In elasmobranchs, the number of transverse tiers usually ranges between three (as in the Port Jackson shark, *Heterodontus portusjacksoni*—Satchell and Jones, 1967) and six (as in the angelshark, *Squatina squatina*—Santer, 1985). Four tiers may be more common, as in *R. clavata* (Fig. 3A), but several species of the Scyliorhinidae family, such as the blackmouth catshark (*Galeus melastomus*), show only two (Bertin, 1958; Parsons, 1930; White, 1936). Two transverse rows have also been described in chimaeras (White, 1936), with three to five valves composing each transverse tier. However, usually, only three of them are well developed. In addition, intraspecific variations are frequent. Longitudinal valve tiers have also been described, but this arrangement is less consistent in many species owing to the irregular development of the individual valves.

Within the chondrosteans, the sturgeon conus has three valve rows. The most distinct characteristic of the sturgeon conus is the existence of a wide gap, devoid of valves, between the two proximal rows and the third distal row (Icardo et al., 2002a; Parsons, 1930). Each valve row is composed of four to six valves that display very variable morphologies (Fig. 11A). The conus of the Polypteriformes, such as that of *P. senegalus*, shows eight valve tiers, each one consisting of six valves of different sizes (Grimes et al., 2010; Parsons, 1930). Within the holostean group, the bowfin *A. calva* has three transverse tiers with four valves each, whereas the garfishes (Fig. 5D) show eight transverse rows with eight valves each, half of the valves being very reduced in size. The number of conal valves is greatly reduced in basal teleosts. Only two transverse tiers, containing a total of four to six valves, have been described in members of the Albulidae, Elopidae, and Megalopidae families that have been analyzed (Bertin, 1958; Parsons, 1930; Santer, 1985). In contrast, the osteoglossiform silver arowana (*Osteoglossum bicirrhosum*) has a single transverse valve row containing right and left semilunar valves (Lorenzale et al., 2017) in a pattern that resembles that of more advanced teleosts.

Contrasting with the regular semilunar appearance of the outflow valves in Cyclostomata, conal valve development in basal Gnathostomata is quite irregular. Many leaflets are of small size, are not fully excavated, or may appear as irregular mounds or cords bulging from the conus wall (Figs. 5D and 11A). These valves display collagenous chords that resemble chordae tendineae. The chordae arise from the free border and from the parietal side of the leaflets, and extend distally to attach to the conus wall and to the adjacent cranial valves. This occurs in elasmobranchs (Parsons, 1930) and sturgeons (Fig. 11A; Icardo et al., 2002a). In Polypteriformes and holosteans, collagenous chords jump from one leaflet to another forming longitudinal systems that extend along the entire length of the conus (Bertin, 1958; Parsons, 1930; unpublished observation). In general, valves in the distal row are larger and better developed and show a more regular and uniform shape than the rest of the valves. This occurs in elasmobranchs (Fig. 3A) and in basal Osteichthyes (Bertin, 1958; Icardo et al., 2002a; Parsons, 1930).

From a structural point of view, the semilunar leaflets in Cyclostomata show a thin, uniform, collagenous core covered by endothelium. By contrast, conal leaflets in basal Gnathostomata are bulky structures formed by a loose

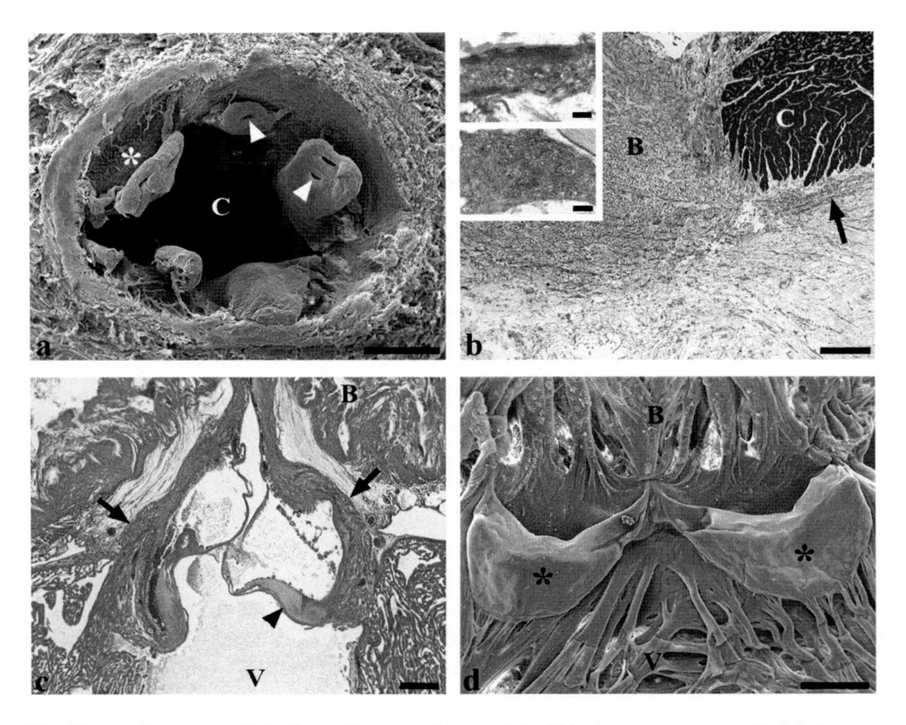

Fig. 11. Outflow tract. *B*, bulbus; *C*, conus; *V*, ventricle. (A) *Acipenser naccarii* (Adriatic sturgeon). SEM. Transverse section of conus at the level of the second valve row. Note the various valve morphologies. Well-excavated valves show a large sinus (*asterisk*). *Arrowheads* indicate holes in the leaflet structure. (B) *Amia calva* (Bowfin). Semithin section, toluidine blue staining. Note the abrupt histological transition between bulbus and conus. A system of elastin and collagen fibers (*arrow*) runs under the conus myocardium. *Upper inset*: Bulbus. Smooth muscle cell with filament bundles and associated elastin. *Lower inset*: Myofibroblast. This cell lacks filament bundles and associates mostly with collagen. (C) *Spondyliosoma cantharus* (Black seabream). Paraffin sectioning, Martin's trichrome staining. The compact conus myocardium (*arrows*) stands between the bulbus and ventricle. *Arrowhead* indicates the leaflet's stout body and ventricular fibrosa. (D) *Pagellus acarne* (Spanish seabream), SEM. The *left* and *right conus leaflets* are indicated by *asterisks*. The bulbus shows longitudinal ridges covered by endocardium. Scale bars: A, 300 μm; B, 100 μm; *upper inset*, 500 μm; *lower inset*, 500 μm; C, 200 μm; D, 300 μm.

connective tissue (Figs. 3A and 5D) that contain numerous stellated fibroblasts and collagen and elastin bundles embedded in a ground matrix (Icardo et al., 2002b; Sans-Coma et al., 1995). The leaflets show a wide base of insertion that blends with the conus subendocardium (Fig. 3A). The leaflets also show a well-developed fibrosa on the luminal side and a less-developed fibrosa on the parietal side (Hamlett et al., 1996; Icardo et al., 2002b; Lorenzale et al., 2017; Sans-Coma et al., 1995).

The exact function of both the conus and the conal valves is still under discussion. Contraction of the conus myocardium appears to add to the ejection fraction. However, the extent of this contribution is surely small due to the reduced conus lumen (Johansen et al., 1966). The valve system can hardly be considered competent due to the array of morphologies. Only the distal valve row that contains larger, more uniform valves may be functionally competent (Icardo et al., 2002a), passively closing due to backflow from the ventral aorta (see Jones and Braun, 2011). It has also been suggested that coordinated contraction of the conus would draw all the leaflets together, closing the conus lumen to prevent ventricular backflow. In fact, when the conus muscle is paralyzed, significant backflow occurs and is primarily due to ineffectiveness of the lower valve row (Satchell and Jones, 1967).

While the structural features of the muscular conus are well studied, several aspects of the distal portion of the OFT are still controversial. Early studies in basal Gnathostomata showed that the muscular covering of the conus stopped short of the pericardial boundary. The wall of the distal portion was formed by an elastic tissue similar to that of the aorta. Also, it was found to be of variable length, usually being shorter in elasmobranchs and longer in bowfins. The presence of muscular and nonmuscular segments was considered proof of the existence of conus and bulbus in the heart of the species studied (Bertin, 1958; Parsons, 1930). These early observations were overlooked for years, but have recently been revisited. In sturgeons, a transitional segment that bears no myocardium is present between the uppermost rim of the conal myocardium and the pericardial insertion (Icardo et al., 2002b). This segment contains smooth muscle cells and elastin, but its organization is different from that observed in the ventral aorta. More recently, a similar segment has been recognized in chondrichthyans, chondrosteans, and holosteans (Durán et al., 2008, 2014; Grimes et al., 2010). Thus, the current view is that the OFT of basal Gnathostomata shows both a conus arteriosus and a bulbus arteriosus. The existence of the two segments has been analyzed in the context of comparative developmental anatomy (Durán et al., 2015; Grimes et al., 2010; Guerrero et al., 2004). The appropriateness of the use of the term bulbus arteriosus to define the distal OFT segment in basal Gnathostomata has been discussed earlier (see Section 2).

Histological studies have shown the existence of structural elements common to the ventral aorta, bulbus, and conus. For instance, a system of elastic fibers originating in the aorta runs along the bulbus and enters the conus where it courses under the myocardium extending down to, approximately, the level of the most cranial valves (Fig. 3A; Grimes et al., 2010; Icardo et al., 2002b). In addition, nitric oxide (NO) synthase activity was localized by the DAF-2DA antibody at the same OFT levels (Durán et al., 2008; Grimes and Kirby, 2009; Grimes et al., 2010). This antibody also marked the teleost bulbus (rich in smooth muscle cells—see below) and regions in the developing vertebrate heart that ultimately express a smooth muscle phenotype (Grimes et al., 2006). The idea that the DAF-2DA antibody specifically marks smooth muscle cells made these authors conclude that smooth muscle and myocardium colocalize at the same OFT level. Furthermore, since the upper part of the conus is richer in elastin, and the lower part of the conus is richer in collagen (see below), it was concluded that three histomorphological regions along the OFT exist (Durán et al., 2015; Grimes et al., 2010). However, the degree of specificity of the DAF-2DA antibody has never been examined. In sturgeons, the ventral aorta and the bulbus contain smooth muscle cells. However, the conus area marked by DAF-2DA (Grimes et al., 2010) lacks smooth muscle cells. Instead, it contains myofibroblasts and fibroblasts (Icardo et al., 2002b). Myofibroblasts share smooth muscle cell properties and may be responsible for the antibody positivity. The same structural shift occurs in the bowfin (*A. calva*; Fig. 11B). When the cellular phenotypes are investigated, smooth muscle cells are present in the bulbus (Fig. 11B, upper inset), but only myofibroblasts can be identified in the upper part of the conus (Fig. 11B, lower inset). Something similar occurs in the polypteriform *E. calabaricus* (unpublished observation) and in the teleost *S. auratus* (Icardo et al., 2003). In *E. calabaricus*, however, the transition at the ultrastructural level is less marked. Thus, the view that smooth muscle cells and myocardium colocalize at the level of the conus is not supported by the available structural evidence and should be examined in more detail. Additionally, colocalization of myocardium and smooth muscle could not be observed in the silver arowana (Lorenzale et al., 2017), an osteoglossiform, although ultrastructural studies to identify cell phenotypes are still lacking. Of note, signals generated in the extracellular matrix preclude the coexpression of smooth muscle and myocardial phenotypes in the teleost bulbus. This is achieved by promoting the differentiation of precursor cells into smooth muscle cells (Moriyama et al., 2016). Conversely, intrinsic myocardial signals may regulate the characteristics of the underlying tissue, excluding smooth muscle cell differentiation (Icardo et al., 2002b).

Along the same lines, the division of the OFT into three distinct histomorphological regions appears to be artificial. It is true that in sturgeons (Icardo et al., 2002b) and in other basal species (Durán et al., 2015; Grimes et al., 2010), that the upper and lower parts of the conus have a slightly different cellular and extracellular composition. However, these parts show no clear boundaries. The elastin richness and the presence of myofibroblasts in the distal part of the conus may simply represent the structural transition between bulbus and conus, two heart components with different composition and functional capabilities.

3.5.2. More Advanced Teleosts

The OFT of more advanced teleosts is also formed by a conus and a bulbus. The difference is that the bulbus is more developed (Figs. 2C, 3B, and 4), whereas the conus appears as a very short segment interposed between the bulbus and the ventricle (Figs. 3B and 4; Genten et al., 2009; Grimes et al., 2006; Icardo, 2006; Icardo et al., 2003; Schib et al., 2002).

3.5.2.1. The Conus Arteriosus and the Conus Valves. In more advanced teleosts, the conus is a short muscular ring of compact and vascularized myocardium (Fig. 11C) formed by the accretion of muscle bundles. The compactness of the conus myocardium is independent of the ventricular structure. As it occurs in the AV canal, the conal myocardium is supplied by coronary vessels in most cases (Fig. 11C), even when the adjacent ventricular myocardium is not vascularized (Icardo, 2006, 2012; Schib et al., 2002). In hearts with spongy ventricles, the compact structure of the conus stands out against the ventricular trabeculae (Fig. 11C). The conus may be more difficult to recognize in hearts with ventricular compacta. However, the conus myocardium has thicker basement membranes; higher collagen, elastin, and laminin content (Garofalo et al., 2012; Icardo, 2006; Schib et al., 2002); and different lectin-binding properties (Icardo et al., 2003) than the ventricular myocardium. Visibility of the conus also depends on conal shape. While the long conus of *A. anguilla* is very clear (Fig. 3B), the height and thickness of the conus can vary widely between species (Icardo, 2006). Similar to what occurs in the AV canal, the conus of a few species, including several Antarctic teleosts, lacks vessels. Instead, large intercellular spaces communicating with the ventricular lumen give access to blood for oxygen supply (Icardo, 2006).

As in any other species, teleosts have semilunar conal valves. The main differences with basal species are the uniformity in number and location, and the structural arrangement. In higher teleosts, a single transverse row containing two (left and right) valves is common (Fig. 11D). Nonetheless, one additional valve of small size, usually located in a dorsal position, may be observed (Icardo et al., 2003; Parsons, 1930; Santer, 1985). In extreme cases, as in tuna, three valves of roughly equal size are present (Fig. 8; Icardo, 2012). The valve leaflets are inserted into the lower end of the conus proximally, extending distally to the bulbus arteriosus (Fig. 11C). However, the structural relationships of the conus and conal valves with the ventricle are variable (Smith, 1918). The conus and the caudal origin of the valves may be recessed into the ventricle (as in the giant mudskipper, *Periophthalmodon schlosseri*), may be situated at the level of the ventricular base (as in the black bream, *Spondyliosoma cantharus*) (Fig. 11C), or may be located above the ventricular orifice (as in *A. anguilla*) (Fig. 3B). The location of the conus with respect to the ventricular base dictates the morphological transition between the ventricle and the bulbus. In the first case, the connection is made solely by a ring of connective tissue (and the outer epicardium) (see Greer Walker et al., 1985); in the other two cases the connection includes, partially (Fig. 11C) or totally (Fig. 3B), the conal myocardium (see Garofalo et al., 2012; Icardo, 2006).

From a structural viewpoint, each conal valve leaflet is formed, in most species, by a stout proximal body and a flap-like distal region (Fig. 11C). The stout body consists of densely packed cells embedded in an extracellular matrix rich in collagen and elastin, whereas the flapping portion is mostly formed by collagen. There is also a strong luminal fibrosa that may be formed by several collagenous layers, and a more loosely organized parietal fibrosa. As in the AV valves, collagen continuity at the leaflet attachment creates a fibrous joint that should facilitate valve movement. In other species, however, the stout proximal body is lacking and the leaflets are reduced to thin structures formed by the apposition of several collagenous layers. This occurs, for instance, in the yellow gounard (*T. lucerna*), in the Asian swamp eel (*M. albus*), and in several Antarctic species (Icardo, 2006).

3.5.2.2. The Bulbus Arteriosus. The bulbus arteriosus is the morphologically predominant portion of the OFT in more advanced teleosts. It acts as an elastic reservoir, expanding during ventricular contraction to accommodate a large part of the stroke volume, a "windkessel" function (Farrell, 1979; Johansen, 1962). Subsequent elastic recoil gradually releases this volume to prevent gill damage due to high arterial systolic blood pressures, and to provide a more even perfusion of gill lamellae for gas exchange during each cardiac cycle (Burggren et al., 1997; Braun et al., 2003a,b; Farrell and Jones, 1992; Priede, 1976; Satchell, 1991). This function depends on the cellular and extracellular components of the bulbus wall as noted earlier. The chemical nature of the bulbar elastin, the arrangement of the collagen, and the activity of the smooth muscle cells all combine to control bulbar dilation, allowing it to accommodate different stroke volumes with little variation in inner blood pressure (Braun et al., 2003a,b).

The external shape of the bulbus varies from pear-shaped, to elongated, to thick and robust (Fig. 2C). Internally, the lumen of the bulbus is characterized by the presence of longitudinal ridges (Fig. 11D) that occupy the entire bulbar length. These ridges follow the bulbar shape, being thicker at the base and tapering at the cranial end (Farrell and Jones, 1992; Priede, 1976), and vary greatly in thickness between species. For instance, ridges are very prominent in tuna and very attenuated in icefishes (Icardo et al., 2000a), and this difference is associated with a large disparity in mean arterial blood pressure (see Chapter 4, Volume 36A: Farrell and Smith, 2017).

Structurally, the bulbus is organized into outer, middle, and inner layers. The outer layer is formed by the epicardium and the subepicardial tissue. The subepicardium is formed by loose connective tissue, rich in collagen that likely controls dilation. The subepicardium also contains fibroblasts, vessels, and nerves (Icardo et al., 1999a,b, 2000a,b). However, the subepicardial tissue may be more complex in several species, where it appears to be the place of origin of humoral immune responses: dendrite-like cells, lymphocytes, macrophages, and plasma cells have been reported to be organized into structures reminiscent of primitive germinal centers (Icardo et al., 1999b).

The middle layer of the bulbus is formed by smooth muscle cells, elastin and collagen. These elements are common to arterial walls, and despite the fact that the structural organization is quite different, the term "arterial-like" has gained wide usage. Smooth muscle cells are arranged in layers with different orientations. These cells have a corkscrew nucleus, obliquely oriented microfilament bundles and pinocytotic vesicles (Icardo et al., 1999a,b, 2000a,b), and are enmeshed in a filamentous meshwork of mostly elastin fibers. In fact, the teleost bulbus stains intensely for orcein (Fig. 4A), a dye specific for elastic fibers. However, the presence of elastin is not a universal feature. Structural studies in several Antarctic teleosts indicate that elastin fibers are replaced by a filamentous material (Icardo et al., 1999a,b). The absence of elastin was considered to be a specificity related to heart adaptation to subzero temperatures, and perhaps their lower arterial blood pressures (see Chapter 4, Volume 36A: Farrell and Smith, 2017). Curiously, however, the bulbus of Antarctic fishes also stains intensely with orcein. This raises the question of whether the filaments are formed by elastin precursors that have not aggregated into mature fibers (Icardo et al., 1999a), or alternatively, that they are the result of glycosaminoglycan aggregates (see Isokawa et al., 1988). Orcein positivity could also constitute a histochemical artifact in these species (Icardo, 2013). Different trichrome stains for collagen have demonstrated positive reactivity with the entire bulbar wall (Fig. 11C; Genten et al., 2009; Hu et al., 2001). However, a high collagen content would make the bulbar wall very stiff, decreasing its elastic capabilities and severely compromising its "windkessel" function. In this regard, a recent study has shown that trichrome staining for bulbar collagen is an artifact, probably due to cross-reaction between the dyes and some matrix components (Icardo, 2013). Ultrastructural analyses, and staining with Sirius red, appear to be more reliable methods for localizing collagen in the bulbus. In most species, collagen accumulates in the subepicardium and at the subepicardium-middle layer boundary, whereas discrete amounts of collagen may appear in the subendocardium (Icardo et al., 1999a,b, 2000a,b). Thus, most of the bulbar middle wall is collagen free (Fig. 3B; Garofalo et al., 2012; Icardo et al., 2000b; Yamauchi, 1980).

Nonetheless, the situation is different in several other species. For instance, the bulbar middle layer in *A. anguilla* shows elastin material alternating with discrete collagen layers (Icardo et al., 2000a; Yamauchi, 1980). Also, connective bundles containing collagen, vessels, and nerves originate in the subepicardium and penetrate the bulbar middle layer of tuna (Icardo, 2013). These bundles are likely present in other active species.

The inner layer of the bulbus is formed by longitudinal ridges covered by endocardium (Fig. 11D). These ridges contain stellated cells that are usually clustered into small groups surrounded by dense matrix. Ridge cells do not show clear phenotypes, but display discrete microfilament bundles and pinocytotic-like vesicles (Icardo et al., 1999a,b, 2000a). The morphology of the endocardium is very variable. Bulbar endocardial cells may be squamous, cuboidal, or columnar and may contain moderately dense bodies that have been implicated in secretion (Benjamin et al., 1983, 1984; Icardo et al., 2000a; Leknes, 2009). Other activities of the teleost endocardium have recently been reviewed (Icardo, 2007; see Chapter 5, Volume 36A: Imbrogno and Cerra, 2017).

The general features of the bulbus arteriosus described earlier should not negate the fact that the bulbus is widely variable in shape, structure, and ultrastructure. In fact, the morphology of the endocardial cells, the presence and content of endocardial secretory granules, the arrangement of cells in the ridges and in the middle layer, the variable structure of the elastin material, and the distribution of collagen appear to almost be species specific. While we partially understand the collagen and elastic distributions (Jones and Braun, 2011), the exact reasons for this range variation are unclear, but may be related to several factors such as phyletic position, lifestyle, ecophysiology, and cardiovascular dynamics (Icardo, 2012; Icardo et al., 2000a).

4. BLOOD SUPPLY TO THE HEART CHAMBERS

Venous blood passing through the cardiac chambers constitutes the first, and most direct, mode of oxygen and nutrient supply to the heart. Indeed, this is the single mode of supply for the avascular hearts of Cyclostomata. However, evolutionary modifications and the increasing complexity of the heart's design called for additional and complimentary modes of blood supply. Although the precise factors (increase in the oxygen diffusion distance, myocardial hypoxia, increased workload) involved in the appearance of the coronary arteries are still under discussion (Farrell et al., 2012; Chapter 4, Volume 36A: Farrell and Smith, 2017), the development of a coronary circulation ensured adequate oxygen supply to the myocardium. The heart of gnathostomes uses the two modes of supply (the luminal venous supply and a dedicated coronary supply) in varying degrees that have been mostly associated with the morphology of the ventricle (Davie and Farrell, 1991; Farrell et al., 2012; Grant and Regnier, 1926; Santer and Greer Walker, 1980; Tota, 1983; Tota et al., 1983).

Coronary arteries in fishes do not originate from the base of the aorta or form a "crown" around the ventricular base, as in birds and mammals. In the absence of a better term, the use of "coronary" simplifies descriptions and facilitates the establishment of homologies; i.e., it is a direct supply of oxygenated blood from the efferent side of the gills. The coronary arteries in fishes have a double origin. In the simplest form, they can originate cranially from the efferent branchial arch arteries, where one to three bilateral branches meet medially to form a pair of hypobranchial arteries. These arteries give raise to right and left coronary trunks that often merge and reach the cranial pole of the heart. These cranial coronaries may be complemented by a second set of coronary vessels, the pectoral or caudal coronaries. They originate from the coracoid artery, which branches from the dorsal aorta as its first branch, runs ventrally, and enters the pericardium caudally. The branches of the coracoid artery reach the dorsal heart surface *via* pericardial ligaments.

The two hypobranchial arteries may occupy a lateral or a dorsal–ventral position on the ventral aorta or merge into a single artery on the dorsal or ventral trunk of the OFT. The mode and extent of heart vascularization also varies considerably. In chondrichthyans, the coronary trunks descend along the OFT but may occupy the lateral positions, the dorsal and ventral sides, or run in parallel along the OFT's ventral side, while having lateral branches to supply the conal myocardium (de Andrés et al., 1990, 1992; Durán et al., 2015; Tota, 1989). The number of coronary trunks may also vary from one to four, depending on the species (Grant and Regnier, 1926). After reaching the base of the ventricle, the coronary trunks divide into several branches that spread over the ventricular and atrial surfaces (de Andrés et al., 1990, 1992).

In basal Osteichthyes, the coronary pattern is poorly known. In general, coronary trunks originate cranially and descend along the OFT to supply the conal myocardium, before dividing into ventricular and atrial branches after reaching the ventricular base (Santer, 1985). Fig. 2B shows the presence of a coronary trunk descending along the right side of the OFT in *P. senegalus*.

In more advanced teleosts, a single coronary trunk courses along the ventral or dorsal (Fig. 2C) surface of the OFT (Grant and Regnier, 1926). Again, variations are frequent since two descending lateral trunks have also been described (Simoes et al., 2002). In any case, the descending coronary trunks give rise to small vessels that run in the bulbus subepicardium but do not penetrate the middle, muscular layer of the bulbus. As an exception, coronary vessels accompanied by nerve fibers have been detected in the bulbus middle layer of tuna (Icardo, 2013). It is possible that they are also present in the bulbus of other large, very active teleosts. When the coronary trunks reach the ventricular base, they divide into several arteries that distribute over the ventricular and sometimes atrial surfaces. In the ventricle, the main branches run along the dorsal–ventral or the lateral surfaces (Grant and Regnier, 1926; Santer, 1985; Simoes et al., 2002; Watson and Cobb, 1979).

In many studied species, the main coronary trunks may anastomose at the ventricular base and form vascular rings from which superficial coronaries arise. Vascular rings are also formed at the AV junction in some species. From the AV ring, smaller branches reach the atrium and ventricle. The origin and distribution of fish coronaries appear to follow a general pattern. However, the distribution of individual coronaries has been found to exhibit remarkable variation, despite the fact that only a limited number of species (out of the ~30,000 fish species) have been analyzed (see Farrell et al., 2012).

Caudal (pectoral) coronaries reach the heart *via* the ligaments that connect the pericardium with the heart's dorsal surface. These arteries supply most of the ventricle, including the lateral surfaces and the apex. However, they are only present when pericardial ligaments are found. Caudal coronaries have been described in rays and in several teleosts such as the eel and the swordfish (Davie and Farrell, 1991; Grant and Regnier, 1926). Although caudal coronaries do not appear to constitute a common supply route, the presence of pericardial ligaments may have been overlooked.

With respect to the final coronary distribution, small coronaries running in the subepicardium give out branches that may penetrate the underlying myocardium. From a phyletic point of view, large coronary branches and a coronary microvasculature are very prominent in the ventricle of chondrichthyans (Cox et al., 2016) and other basal fishes (Grant and Regnier, 1926; Halpern and May, 1958), whereas there is a tendency toward simplification of the coronary tree in more advanced teleosts (Farrell et al., 2012). In other words, vascular patterns are more similar between basal fishes and higher vertebrates than with the teleost group. The evolutionary reduction of the coronary circulation is in line with the specialization of the teleost heart and the architecture of the ventricular chamber. Indeed, different types of ventricular architecture correspond to different vascularization patterns. As stated in Section 3.4, the fish heart ventricle has been classified into Types I–IV according both to the myocardial architecture and to the distribution of the coronaries (Davie and Farrell, 1991; Farrell, 2011b; Farrell et al., 2012; Tota, 1989; Tota et al., 1983).

Type I ventricles (entirely trabeculated) lack a coronary supply. However, this does not mean the absence of coronaries. Ventricular subepicardial vessels are almost always present, even if they are of small size and appear in small numbers. Coronary vessels must reach the conus and the AV canal since these segments contain compact, vascularized myocardium in most cases. Thus, a coronary circulation, although reduced (Fig. 7D), should be present. Type II ventricles (having compacta and spongiosa) have coronary vessels restricted to the compacta. Ventricular Types III and IV are essentially similar to Type II, but the compacta is generally thicker and the spongiosa is vascularized. The proportion of the compact myocardium related to the total heart mass has arbitrarily been taken as a hallmark to establish the separation between Types III and IV. Type III ventricles have less than 30% of compact myocardium, whereas Type IV ventricles have more than 30% (Farrell et al., 2012; Tota, 1983). Adjustments to this classification have recently been proposed and are discussed in Chapter 4, Volume 36A: Farrell and Smith (2017).

The distribution of the different ventricular types along the evolutionary scale is nonlinear. The problem is that we do not fully understand the mechanisms responsible for this variation. For instance, the basal Cyclostomata have a Type I heart. Curiously, a few capillaries have been observed in the subepicardium of the ventricular base in the New Zealand hagfish, *Eptatretus cirrhatus* (Icardo et al., 2016b). These vessels reach the heart surface accompanying the loose connective tissue that surrounds the ventral aorta. Clearly, they do not constitute a functional coronary system. Rather, their presence has been interpreted as an evolutionary attempt to supply the heart from a cephalad origin (Icardo et al., 2016b). The ventricular type in chondrichthyans is variable. Most elasmobranchs have a Type III ventricle; i.e., they have a vascularized spongy layer. Curiously, the degree of vascularization of the spongiosa appears to be greater than that of the compacta, as shown in the spiny dogfish, *Squalus suckleyi* (Cox et al., 2016). While other elasmobranch ventricles were initially considered to be Type IV (see Farrell et al., 2012), the need for a distinction between Types III and IV has recently been questioned (Farrell and Smith, 2017; Chapter 4, Volume 36A). Studies in Holocephali place the ventricle of these species within Type I or III (Durán et al., 2015). Additionally, the ventricle of the Atlantic chimaera (*Hydrolagus affinis*) has vascularized trabeculae in the absence of a compact layer (Durán et al., 2015). This situation was not contemplated in the original classification and has been discussed in relation to the evolution of the compacta and its vascular supply (Durán et al., 2015). However, previous studies have reported the existence of a thin compacta in *H. affinis* (Santer and Greer Walker, 1980). This raises again the need for careful specimen fixation (see Section 3.2). Discrepancies could also be better addressed if a clear definition for the compact and spongy layers was definitively established (see Section 3.4).

In the chondrostean group, sturgeons have vascularized trabeculae (Fig. 10A) and, thus, they can be classified within Type III. The presence of trabecular capillaries, however, does not depend on the thickness of the individual trabeculae (see Santer, 1985). Observations in polypteriforms indicate that they have Type I ventricles. Within holosteans, gars have a Type I ventricle (Fig. 5D), while *A. calva* displays a loosely organized, vascularized, compacta and can be ascribed to Type II. These observations in holosteans are at odds with a previous report (Santer and Greer Walker, 1980), but they have carefully been checked on semithin sections (unpublished observation). The situation in basal teleosts is unclear. Morphological studies on these species are scarce, and none of them have reported on the presence of trabecular vessels. In more advanced teleosts, most ventricles are Types I and II. It is considered that between 50% and 80% of all teleosts have a Type I ventricle (Farrell et al., 2012; Santer, 1985). In the teleosts having a Type II ventricle, the thickness of the compacta is generally relatively low, occupying about 20% of the ventricular cross-section (Santer and Greer Walker, 1980). Active teleosts like tuna show a thicker compacta (up to 74% of the ventricular cross-section: Santer and Greer Walker, 1980) and a vascularized spongiosa (Icardo, 2012; Tota, 1983) and have classically been characterized as having Type IV ventricles (see Chapter 4, Volume 36A: Farrell and Smith, 2017). It should be underscored that the presence of a thick compacta does not appear to be exclusive to large fish. The small sprats (*Sprattus sprattus*) and anchovies (*Engraulis encrasicolus*) have been reported to have a proportion of compacta similar to that found in several tuna (Santer and Greer Walker, 1980). It is unknown whether the spongiosa of these fish contains vessels. As a comparison, the ventricle of the chub mackerel *Scomber japonicus*, which shows a compacta/spongiosa ratio close to 50% (Santer and Greer Walker, 1980), lacks vessels in the spongiosa (Fig. 10B).

The atrium can also receive coronary vessels. It is assumed that, owing to the delicate structure of the atrial wall, these coronary vessels are scarce and remain on the atrial surface. However, small arteries can be observed in the atrial wall, running between the pectinate muscles in several species such as rays (Fig. 12A), sturgeons (Fig. 12B), and tuna (unpublished observation). As in the ventricular chamber, small capillaries run inside the atrial trabeculae

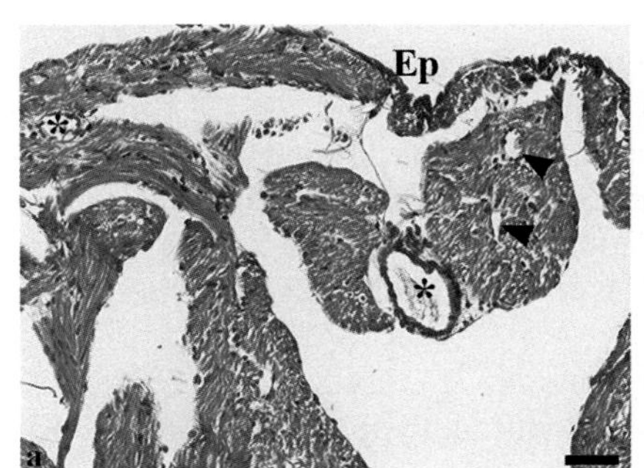

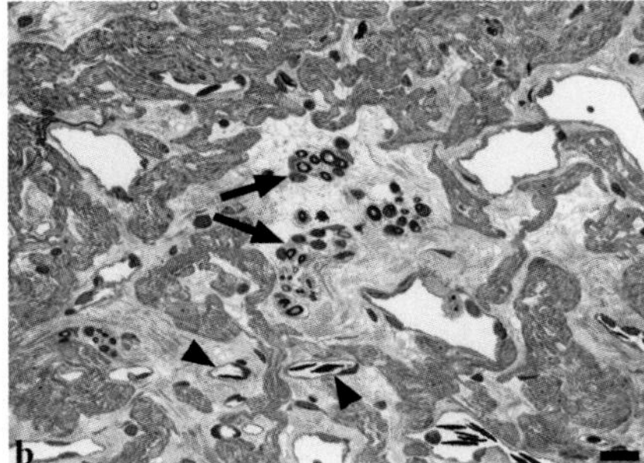

Fig. 12. Atrial vessels. (A) *Raja clavata* (Thornback ray). Paraffin sectioning, Martin's trichrome staining. In the atrium, small arterioles (*asterisks*) and capillaries (*arrowheads*) can be observed. (B) *Acipenser naccarii* (Adriatic sturgeon). Semithin section, toluidine blue staining. Myocardial cells in the atrium are supplied by lacunary spaces and capillaries (*arrowheads*). *Arrows* indicate myelinated nerve fibers. Scale bars: A, 50 μm; B, 20 μm.

of these species. This is more patent in the portion of the atrium closer to the AV canal (Fig. 12A and B), where the trabecular density is higher.

A venous coronary system, collecting venous blood from the ventricular muscle, has been described in the subepicardium of several species (Tota et al., 1983). A similar venous system may be assumed for the compact myocardium of the conus in basal species. Finally, numerous communications between the smaller coronary vessels and the heart lumen have been reported (Grant and Regnier, 1926; Tota et al., 1983). These communications include arterioluminal vessels, venous channels, and myocardial sinusoids, but their functional significance is unknown.

5. CARDIAC NERVES

The nerves reaching the heart belong to the autonomic nerve system and have a dual origin: the spinal (sympathetic) and the cranial (parasympathetic) autonomic limbs (see Nilsson, 2011). Nerves from the cranial limb reach the heart *via* the paired cardiac branches of the vagus nerve, whereas nerves from the spinal sympathetic limb reach the heart by joining the vagus nerves (hence the term vagosympathetic trunks) or through independent spinal nerves (Laurent et al., 1983; Nilsson and Holmgren, 1992). In addition, nerves of the spinal limb may also course with the coronary arteries (Zaccone et al., 2009a). In cyclostomes, the heart of hagfishes is considered to be aneural (see Icardo et al., 2016a), while the lamprey heart is solely innervated by the vagus. In elasmobranchs, the vagus nerves reach the sinus venosus and form a plexus together with resident ganglion cells (see Santer, 1985). The vagal branches only contain myelinated axons, suggesting that the spinal limb of the autonomic nervous system is absent. By contrast, the chondrostean (Zaccone et al., 2009a,b), holostean (Zaccone et al., 2012), and teleostean (Nilsson and Holmgren, 1992) hearts receive input from the two limbs of the autonomic system. However, adrenergic innervation may not be universal, since several studies have failed to find cardiac adrenergic nerves in the sturgeon *Huso huso* and in several teleost species (reviewed in Laurent et al., 1983; Nilsson and Holmgren, 1992; Santer, 1985).

Nerves entering the heart course along the wall of the sinus venosus toward the SA junction. In elasmobranchs and bony fishes, the nerves interact with resident, intracardiac neurons and form a dense SA plexus that circumscribes the SA orifice (Haverinen and Vornanen, 2007; Newton et al., 2014; Yamauchi, 1980; Zaccone et al., 2009a,b, 2010). In some bony fishes, the plexus is more developed at the commissures of the SA valve (Stoyek et al., 2015). From the SA plexus, nerve fibers enter the atrium bilaterally reaching the AV canal and the ventricle (Laurent et al., 1983; Newton et al., 2014;

Santer, 1985). Upstream from the SA plexus, innervation is rich in the AV canal, where a plexus has also been described, but it appears to be sparse in the wall of both the atrium and ventricle (Laurent et al., 1983; Newton et al., 2014; Zaccone et al., 2009a, 2012). The teleost ventricle may also be innervated by nerve branches entering the heart *via* the bulbus arteriosus (Stoyek et al., 2015). Nerves of cranial origin also appear to be present in chondrosteans (Zaccone et al., 2009a). In general, nerve distribution within the heart appears to depend on both the area of the heart and the species studied (Abraham, 1969; Laurent, 1962; Taylor, 1992). Whether several of the differences reported may depend, at least partially, on the different methodologies used is unclear.

Cardiac nerves contain both myelinated and unmyelinated axons. Classically, large myelinated nerve fibers have been considered to be sensory (cholinergic afferent nerves), whereas small myelinated fibers were thought to be preganglionic cholinergic efferent (vagal) nerves, with inhibitory effects on both heart rate and inotropism. As an exception, vagal innervation in lampreys increases the heart rate and the force of contraction, probably due to a local release of catecholamines (Nilsson and Holmgren, 1992; Taylor, 1992). In general, many unmyelinated nerve fibers have been considered to be spinal adrenergic efferent (post-ganglionic) nerves with stimulatory effects on heart function. A second population of unmyelinated fibers appears to be formed by post-ganglionic parasympathetic (vagal) fibers traveling from the heart plexuses to their final destination (Laurent et al., 1983; Satchell, 1991). The fact that the axons establish synaptic contacts at the pacemaker sites, and with muscle fibers in the atrium (Fig. 12B) and ventricle, indicates that heart activity is controlled both at the cardiac pacemaker and by direct innervation of the myocardium. Non-neural control of the myocardium is addressed in Chapters 4 and 5, Volume 36A: Farrell and Smith (2017) and Imbrogno and Cerra (2017), respectively.

Nonetheless, the long-established concept that there is strictly dual control of cardiac performance by adrenergic and cholinergic fibers has recently been challenged. Nerve fibers may release, alone or along with the classical transmitters, a number of neurotransmitters collectively named regulatory peptides. These are nonadrenergic noncholinergic peptides that include purine derivatives, such as serotonin (5-hydroxytryptamine), the vasoactive intestinal polypeptide, bombesin, neuropeptide Y, galanin, substance P, and others (Nilsson and Holmgren, 1992; Satchell, 1991; Taylor, 1992; Zaccone et al., 2010). Recently, other substances such as NO have been added to the list of neurotransmitters that affect cardiac function (Garofalo et al., 2015; Chapter 5, Volume 36A: Imbrogno and Cerra, 2017; Imbrogno et al., 2011; Zaccone et al., 2009a,b, 2012). Similarly, neuronal bodies in the heart plexuses respond positively to regulatory peptides other than the classical neurotransmitters (Nilsson and Holmgren, 1992). Although the coexistence of several markers in both nerves and neurons indicates that neural regulation of the heart's activities is very complex, the functional implications of this complexity are still to be fully established.

The origins of the axonal components of the nerves, the myocardial effectors, and the intracardiac nervous system, together with their functional implications and the mode of regulation, are also reviewed in Chapter 4, Volume 36A: Farrell and Smith (2017).

6. THE HEART'S PACEMAKER AND CONDUCTION SYSTEM

Numerous physiological studies indicate that the SA region is the site of the cardiac pacemaker in fish (Jensen et al., 2014; Vornanen et al., 2010). In teleosts, where it has mostly been studied, spontaneous action potentials typical of vertebrate pacemaker cells are recorded at the SA junction (Haverinen and Vornanen, 2007). The SA junction shows a rich plexus of nerve fibers and neuronal somata (see Section 5; Farrell and Smith, 2017). In addition, the SA junction contains a ring of densely innervated myocardium that follows the base of the SA leaflets (Haverinen and Vornanen, 2007; Newton et al., 2014; Santer, 1985). Myocardial cells in this ring are continuous with the "true" atrial myocardium, but have some specialized features such as fewer striations and the virtual absence of specific granules (Satchell, 1991). In some species, such as in the rainbow trout (*Oncorhynchus mykiss*), the SA ring may be incompletely limited by collagen (Haverinen and Vornanen, 2007). On the other hand, the SA myocardial ring in *D. rerio* expresses the transcription factors Islet 1 and HCN4 (Newton et al., 2014; Stoyek et al., 2015; Tessadori et al., 2012). Islet-1 is expressed in cardiac progenitor cells during early mammalian development, the expression becoming restricted to the sinus node in adult hearts. HCN4s are hyperpolarization-activated cyclic nucleotide-gated ionic channels that are expressed by pacemaker cells in the vertebrate heart. Of note, the body of the leaflets of the SA valve contains variable amounts of myocardial cells. This myocardium was thought to be part of the pacemaker tissue, but it appears to only elicit secondary (not primary) pacemaker potentials (Haverinen and Vornanen, 2007). Although the heart's pacemaker in most teleosts appears to be located at the SA junction (Santer, 1985), the situation may not be universal. For instance, pacemaker potentials could not be found at the SA junction in the European plaice *P. platessa*, and instead pacemaker cells were located scattered throughout the atrial wall (Santer and Cobb, 1972). In the common eel, the pacemaker has been localized at the boundary between the sinus venosus and the ducts of Cuvier. This may be related to the presence of a thick layer of myocardium in the eel sinus venosus (Fig. 5A).

The location of the pacemaker in other fish groups is less clear. In hagfishes, the presence of specialized pacemaker tissue in the sinus wall or at the SA junction could not be demonstrated (Icardo et al., 2016b). Also, electrical recordings have shown depolarization of the sinus venosus in some species (Davie et al., 1987), but not in others (Arlock, 1975; Satchell, 1986). Nonetheless, hagfish myocytes throughout the atrium and ventricle exhibit a high degree of automatism and electrical characteristics of pacemaker cells (Arlock, 1975; Jensen, 1965). In addition, they are rich in HCN channels (Wilson and Farrell, 2013). These channels appear to control the different atrial and ventricular intrinsic beating rates. Furthermore, a higher concentration of HCN channels in the atrium ensures that atrial contraction precedes ventricular contraction, thus allowing for chamber synchrony (Wilson and Farrell, 2013; Wilson et al., 2013). It has been proposed that HCN channel density could be a precursor of the vertebrate cardiac conduction system (Wilson et al., 2013). If this is confirmed, the presence of a morphologically defined pacemaker would not be necessary in hagfishes (Icardo et al., 2016b). In elasmobranchs, the pacemaker appears to be imprecisely assigned to near the SA region (Yamauchi, 1980).

Electrical waves of depolarization and repolarization, related to heart contractile events, can be recorded as positive and negative deflections in the electrocardiogram. In the fish heart, delays at the AV canal (and at the ventricle–conus junction in elasmobranchs and basal bony fishes) provide for coordinated chamber contraction (Farrell and Jones, 1992; Satchell, 1991). While chamber synchrony is achieved by specialized conduction traits in the mammalian heart, studies have failed to demonstrate the presence of fast-conducting fibers in the fish heart (see Sedmera et al., 2003). However, preferential conduction pathways must exist. In teleosts, propagation of the electrical impulses starts at the dorsal side of the atrium, follows from the apex to the base of the ventricle, and ends at the base of the bulbus arteriosus (Satchell, 1991; Sedmera et al., 2003). In many teleosts, trabecular sheets are often seen extending between the atrioventricular ring and the ventricular apex (see above; Fig. 3B). It has been hypothesized that these sheets constitute a preferential way for the conduction of the electrical impulses from the AV canal to the heart's apex (see Icardo and Colvee, 2011; Sedmera et al., 2003). If so, myocardial cell arrangement, junctional complexes, and ionic channels should act in concert to play equivalent functional roles to those of the Purkinje system.

7. LUNGFISH HEART: A SPECIAL CASE

Extant lungfishes constitute a special biological case. They are considered to be the closest living relatives to tetrapods (Amemiya et al., 2013; Johanson and Ahlberg, 2011), and while being aquatic, most species are able to live out of the water for extended periods of time (aestivation), some species for years. Thus, they constitute a model to study the adaptive changes involved in the evolutionary transition from aquatic to terrestrial life. Regarding the heart, the possibility of shifting from gill/lung respiration to exclusive air breathing implies the need for oxygenated and deoxygenated blood stream separation within the heart and onto the general circulation. Specifically, the development of heart septation and the presence of other morphological features would make the lungfish heart anatomically closer to the tetrapod heart than to the heart of other fish. The presence of unique features indicates the specificity of the evolutionary pathways followed by lungfishes. General features of the lungfish heart and circulation have been known for years (Bugge, 1961; Burggren, 1988; Burggren and Johansen, 1986; Burggren et al., 1997; Graham, 1997; Johansen and Burggren, 1980), and more recent studies have added to this information (reviewed in Icardo et al., 2016c).

The morphological complexity of the lungfish heart is evident in Fig. 13. Externally, the connective sinus venosus reaches the dorsal side of the atrium. The atrium is a large sac that shows separate left and right caudal appendages. The left and right sides of the atrium course along the conoventricular sulcus, embrace the proximal portion of the OFT, and appear attached to the dorsal side of the heart by connective tissue. The ventricle is sac-like and shows a rounded apex. The OFT arises from the cranial side of the ventricle following a complex 270-degree path (Fig. 13A, inset). It can be divided into conus and bulbus portions, giving origin distally to three pairs of branchial arches.

7.1. The Sinus Venosus

The sinus venosus in lungfish receives, as in any other fish, the systemic, venous blood. In addition, the sinus venosus receives the oxygenated blood carried from the lungs by a single pulmonary vein. This vein enters the sinus venosus, attaches to the inner, dorsal wall of the sinus, and, in this position, courses toward the AV region. At the AV region, the wall of the pulmonary vein fuses with a membranous fold, the pulmonalis fold, and disappears as an anatomic entity (Bugge, 1961; Icardo et al., 2005a). This process of fusion creates a pulmonary channel (Fig. 13B) that courses toward the left and empties the oxygenated blood directly into the left side of the atrium, bypassing the sinus venosus (Szidon et al., 1969). The formation of the pulmonary channel has a secondary effect. It allows the sinus venosus to direct the systemic venous return toward the right side of the atrium (Bugge, 1961; Icardo et al., 2005a), as shown by angiographic studies where contrast material was injected (Szidon et al., 1969).

The wall of the sinus venosus contains myocardium that contracts actively to aid in atrial filling (Szidon et al., 1969). The sinus venosus receives branches

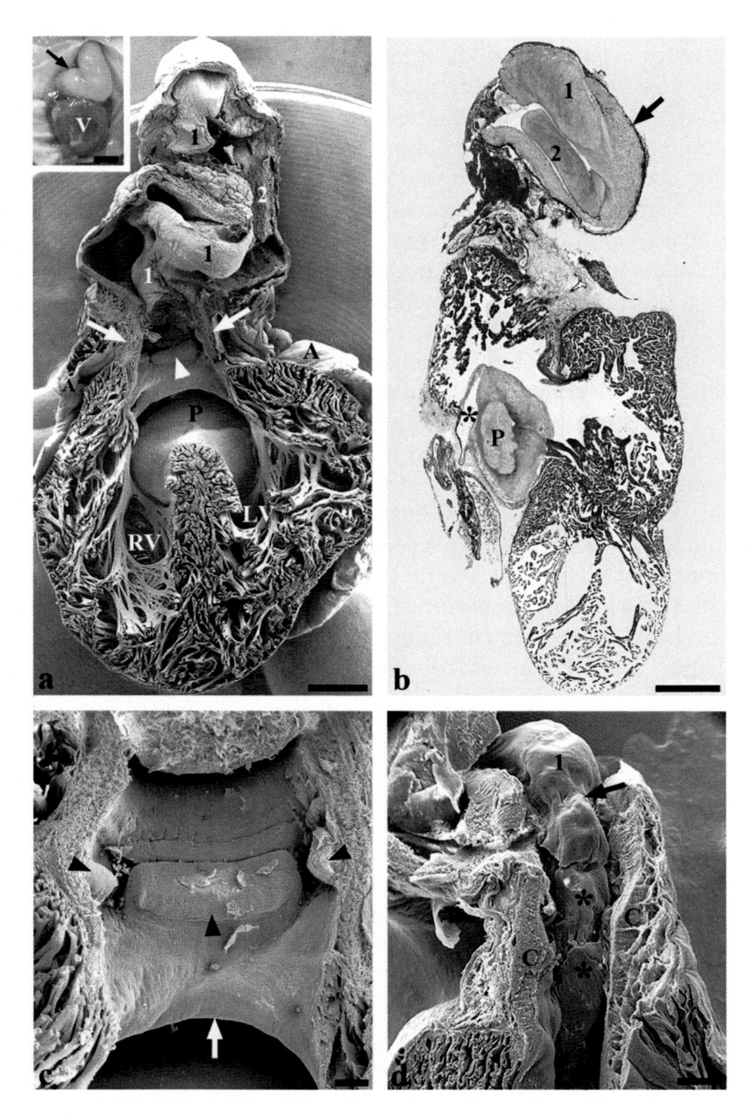

Fig. 13. The lungfish heart. *A*, atrium; *C*, conus; *P*, atrioventricular (AV) plug; V, ventricle; 1, spiral fold; 2; bulbar fold. (A) *Protopterus dolloi* (Slender lungfish), SEM. Frontal dissection (dorsal segment seen from the front). The entirely trabeculated ventricle is partially divided into *left* (LV) and *right* (RV) *chambers*. The AV plug sits atop the ventricular septum. The outflow tract (OFT) lumen is occupied by the spiral and bulbar folds. The conus (*arrows*) is muscular and supports the conal valves (*arrowhead*). The atrium is attached to the conal wall. *Inset*: *Protopterus annectens* (West African lungfish), frontal view. Note the saccular ventricle and the complex

OFT course (*arrow*). (B) *Lepidosiren paradoxa* (South American lungfish), Paraffin sectioning, Martin's trichrome staining. Parasagittal section. The OFT displays spiral and bulbar folds. A layer of myocardium (*arrow*) overlies the bulbus' elastic tissue. The pulmonary channel (*asterisk*) is dorsal to the AV plug. (C) *Protopterus annectens* (West African lungfish), SEM. The *arrow* indicates the proximal border of the conus. *Arrowheads* indicate valves of the proximal valve row. The distal row is a mere ridge. (D) *Neoceratodus forsteri* (Australian lungfish), SEM. Proximal and distal conal valves are indicated by *asterisks*. The proximal end (*arrow*) of the spiral fold looks like an additional valve. Scale bars: A, 500 μm; *inset* of A, 900 μm; B, 2 mm; C, 200 μm; D, 500 μm. Panel (A): From Icardo, J.M., Brunelli, E., Perrotta, I, Colvée, E., Wong, W.P., Ip, Y.K., 2005. Ventricle and outflow tract of the African lungfish Protopterus dolloi. J. Morphol. 265, 43–51.

---

of the vagus nerves. However, cardiac ganglion cells could not be demonstrated in the South American lungfish (Gibbins, 1994). An adrenergic system appears to be absent in all lungfishes (see Donald, 1998), but aggregations of chromaffin tissue have been detected in the walls of the posterior cardinal veins and atrium in *Protopterus* (Gibbins, 1994). This situation resembles the distribution of chromaffin tissue in hagfishes and elasmobranchs (see Gibbins, 1994). Also, in the African lungfishes, the heart pacemaker has been located in an upstream position, at the junction between the sinus venosus and the venous trunks that reach the heart (Satchell, 1991).

7.2. The Atrium, the AV Region and the Ventricle

The lungfish atrium is a large sac with a thin wall formed by a delicate trabecular system. The atrial wall is attached to the OFT wall by a connective tissue (Icardo et al., 2005b). This feature is shared with other air breathers (see Section 3.2). Externally, the atrium is divided caudally into two lobes that are located dorsal to the ventricle. The atrium shows a common part above the sinus venosus. Internally, the atrium is partially divided by a semilunar septum that originates from the dorsal surface and is located over the AV plug. The atrial septum appears to be formed by the accretion of atrial trabeculae, has a crescent shape, and cleaves the dorsal part of the atrium into left and right sides.

The AV region is quite special. The lungfish heart appears to be unique in that it lacks a well-defined AV segment and regular AV valves (Fig. 13A and B). Instead, the AV region is a wide communication occupied by a large, crescent-shaped hyaline cartilage, the AV plug. This cartilage sits atop the dorsal part of the incomplete ventricular septum largely occluding the communication between the atrial and ventricular cavities. The AV plug likely acts as a valve, regulating the size of the AV aperture and avoiding backflow during ventricular contraction (Bugge, 1961; Burggren et al., 1997; Icardo et al., 2005b).

The lungfish ventricle is partially divided into left and right chambers by a vertical septum that originates at the heart's apex and extends upward over half of the ventricular length. The septum shows a smooth free border located under the opening to the OFT. Both the septum and the free ventricular walls are entirely trabeculated (Fig. 13A and B). Thus, most lungfishes have a Type I ventricle. Despite the delicate ventricular trabeculae, a few subepicardial coronary vessels can always be observed in the ventricle. As an exception, the Australian lungfish *Neoceratodus forsteri* ventricle displays a thin, loosely organized outer layer of myocardium that could be defined as a compacta (difficulties inherent to the definition of a compacta are discussed in Section 4). In this species, vascular profiles are very numerous in the subepicardium, but only a few capillaries have been observed within the compacta (unpublished observation). Gas exchange with the ventricular trabeculae likely benefits from the oxygenated blood that is carried in the pulmonary vein.

7.3. The Outflow Tract

The OFT follows a very unusual course. After originating from the ventricular base, the OFT is first directed cranially; then, it bends forward and courses to the left; finally, it bends backward to course cranially to its end (Bugge, 1961; Icardo et al., 2005b,c). The two bending zones create three distinct and different portions for the lungfish's OFT (Fig. 13A and B). The proximal portion corresponds to the conus, whereas the middle and distal portions correspond to the bulbus (Icardo et al., 2005c, 2016c). As occurs in all jawed fishes, the conal wall is encased by compact, vascularized myocardium (Icardo et al., 2005c). Coronary vessels enter the heart cranially and descend along the OFT (Foxon, 1950; Szidon et al., 1969). It is presently unclear whether caudal coronaries may also reach the heart *via* the *gubernaculum cordis*, the ligament joining the ventral ventricular surface with the anterior pericardium (see Icardo et al., 2005c).

The muscular conus supports the conal valves. However, there are important specific differences. The OFT valves in the African and South American lungfishes are represented by transverse ridges that barely protrude into the conal lumen (Fig. 13C). Thus, they look like vestigial or underdeveloped valves. Typically, they are arranged into two transverse rows. Valves in the proximal row are more prominent, appear elongated in the transverse plane, and show little excavation. There are usually 3–4 valves per row. However, the valves may be fragmented by vertical separations increasing the number to 7–8 units per row. Valves in the distal row present a similar division, although they are most often reduced to mere wall ridges. By contrast, the conus of the Australian lungfish (*N. forsteri*) displays two rows of bulky semilunar valves, each row containing three to four discrete valves (Fig. 13D). Short chords may connect the inner side of the valve leaflets to the conal wall.

Downstream of the conus, the OFT is formed by a long bulbus that corresponds to the middle and distal portions of the OFT. The bulbus is about twice the length of the conus. Structurally, the bulbar wall contains an arterial-like tissue formed by ordered collagenous and elastic fibers, and by smooth muscle cells. In addition, the bulbus contains a thin layer of myocardium under the subepicardial tissue, unlike the much thicker layer of conal myocardium that it is continuous with (Grimes et al., 2010; Icardo et al., 2005c). This myocardial layer contains small vascular profiles and covers the elastic component of the bulbus up to its cranial end. The overlapping of myocardium and smooth muscle cells in the OFT's bulbus appears unique to lungfishes (see Section 3.5). Obviously, the presence of myocardium, albeit very thin, brings into question the definitions of conus and bulbus when applied to the lungfish OFT. However, given the overwhelming presence of the arterial-like tissue, and in the absence of a better definition, the use of the term bulbus has been maintained.

Another morphological feature exclusive to the lungfish heart is the presence of two mounds or cushions located along the inner surface of the OFT (Fig. 13A and B). From a comparative viewpoint, a spiral valve, separating oxygenated and deoxygenated blood streams, is present in the arterial conus of the amphibian heart (see Kardong, 2006). Two spiraling folds of mesenchymal tissue are also present in the OFT of birds and mammals during development. These folds are transient structures that are involved in OFT septation and disappear with further development (for review, see Icardo, 1984). Whether these folds are homologous to the lungfish cushions (see Icardo et al., 2005c) is unclear. In the lungfish, the cushions have classically been termed bulbar and spiral folds (Bugge, 1961; Burggren et al., 1997; Graham, 1997). The bulbar fold occupies the middle and distal OFT portions, while the spiral folds extend along the entire OFT length. The two folds are formed by a loose connective tissue that arises independently from the OFT wall and fuses at their distal ends. The surface of the folds may appear irregular showing protrusions and depressions. For instance, Fig. 13D shows two valve rows and the spiral fold in *N. forsteri*. In this heart, the most caudal part of the spiral fold looks like a third valve. It is possible that similar morphologies may also occur distally. While these irregularities could be fixation artifacts, they have been previously used as evidence for the presence of additional valves. While the number and distribution of the OFT valves in the *Protopterus* and *Lepidosiren* genera appear to be well established, the situation in *Neoceratodus* is more complex and has yet to be fully assessed. On the other hand, observations like those presented in Fig. 13 question whether folds and valves develop from longitudinal cushions that may be partitioned later in development to form separate adult structures, or from cushions that arise independently along the OFT wall (see Robertson, 1914).

The presence of the folds partially divides the OFT lumen establishing, together with the ventricular and atrial septa, the morphological basis to partially separate the blood streams from the pulmonary vein and systemic venous return as they move through the heart. The incompleteness of the septa should allow for some mixing between the oxygenated and the deoxygenated streams, but when the heart is injected with opaque media, the two streams advance with little mixing (Szidon et al., 1969). Furthermore, the OFT folds appear better aligned during aestivation than in aquatic conditions (Icardo et al., 2008, 2016c), probably making for a better separation of the blood streams. Noticeably, cardiac septation is less developed in the Australian lungfish (*N. forsteri*) (Foxon, 1950), the single lungfish species that does not aestivate. The OFT folds could also play an important role in heart dynamics. In the absence of competent OFT valves, contraction of the OFT myocardium would result in apposition of the folds preventing backflow during ventricular diastole (see Section 3.5.1).

8. SUMMARY AND FUTURE DIRECTIONS

The fish heart, with exclusion of the cyclostomes, can be divided into six different components that are arranged in sequence and can be recognized both during embryonic development and in adult specimens: sinus venosus, atrium, AV segment, ventricle, conus arteriosus, and bulbus arteriosus. All of these regions are entirely or partially muscular. However, the most cranial portion, the bulbus, contains only smooth muscle and no striated muscle (except in lungfishes). This chapter has reviewed the morphological and structural features of all these cardiac components. The vascular supply, the nerves, and the characteristics of the pacemaker have also been detailed. Wherever possible, literature data have been complemented with unpublished observations. Besides data compilation and the addition of specific details, one of the most significant features that results from this review is the recognition of the extraordinary morphological diversity at both the gross anatomical and structural levels. Furthermore, many particular features, such as the architecture of the ventricular myocardium, the extent of vascularization, or the number of conus valves, appear to have shifted back and forth across the evolutionary scale. This comes as no surprise given the tremendous diversification of fishes that has resulted in the 30,000 plus species that exist today. Morphological diversity is quite obvious, but it often gets overlooked when specific studies try to extrapolate findings and extend them to families and orders. It is not possible to study every species, but diversity should have to be present when formulating general categorizations.

On the other hand, precise knowledge of the fish heart at the morphological, physiological, and molecular levels is of increasing interest due to the use of several fish species, specially zebrafish and medaka, as models to study many biological processes. The successful identification of gene regulatory networks has generally been taken as evidence of homology across vertebrates and invertebrates as well. However, the inherent diversity among fishes makes the use of fish models particularly dangerous in establishing generalities. Having recognized this, the analysis of embryonic development, the progression of diseases, the regeneration of tissues, and many other processes will surely continue to benefit from studies of the fish heart.

ACKNOWLEDGMENTS

The author is indebted to all the colleagues who, along the years, contributed with the precious specimens needed to accomplish this work.

REFERENCES

Abraham, A., 1969. Innervation of the Heart and Blood Vessels in Vertebrates Including Man. Pergamon Press, Oxford.

Amemiya, C.T., Alfoldi, J., Lee, A.P., Fan, S., Philippe, H., Maccallum, I., Braasch, I., Manousaki, T., Schneider, I., Rohner, N., Organ, C., Chalopin, D., Smith, J.J., Robinson, M., Dorrington, R.A., Gerdol, M., Aken, B., Biscotti, M.A., Barucca, M., Baurain, D., Berlin, A.M., Blatch, G.L., Buonocore, F., Burmester, T., Campbell, M.S., Canapa, A., Cannon, J.P., Christoffels, A., De Moro, G., Edkins, A.L., Fan, L., Fausto, A.M., Feiner, N., Forconi, M., Gamieldien, J., Gnerre, S., Gnirke, A., Goldstone, J.V., Haerty, W., Hahn, M.E., Hesse, U., Hoffmann, S., Johnson, J., Karchner, S.I., Kuraku, S., Lara, M., Levin, J.Z., Litman, G.W., Mauceli, E., Miyake, T., Mueller, M.G., Nelson, D.R., Nitsche, A., Olmo, E., Ota, T., Pallavicini, A., Panji, S., Picone, B., Ponting, C.P., Prohaska, S.J., Przybylski, D., Saha, N.R., Ravi, V., Ribeiro, F.J., Sauka-Spengler, T., Scapigliati, G., Searle, S.M., Sharpe, T., Simakov, O., Stadler, P.F., Stegemann, J.J., Sumiyama, K., Tabbaa, D., Tafer, H., Turner-Maier, J., van Heusden, P., White, S., Williams, L., Yandell, M., Brinkmann, H., Volff, J.N., Tabin, C.J., Shubin, N., Schartl, M., Jaffe, D.B., Postlethwait, J.H., Venkatesh, B., Di palma, F., Lander, E.S., Meyer, A., Lindblad-Toh, K., 2013. The African coelacanth genome provides insights into tetrapod evolution. Nature 496, 311–316.

Arlock, P., 1975. Electrical activity and mechanical response in the systemic heart and portal vein heart of *Myxine glutinosa*. Comp. Biochem. Physiol. 51A, 521–522.

Benjamin, M., Norman, D., Santer, R.M., Scarborough, D., 1983. Histological, histochemical and ultrastructural studies on the bulbus arteriosus of the sticklebacks, *Gasterosteus aculeatus* and *Pungitius pungitius* (Pisces: Teleostei). J. Zool. 200, 325–346.

Benjamin, M., Norman, D., Scarborough, D., Santer, R.M., 1984. Carbohydrate-containing endothelial cells lining the bulbus arteriosus of teleosts and the conus arteriosus of elasmobranchs (Pisces). J. Zool. (Lond.) 202, 383–392.

Bertin, L., 1958. Appareil circulatoire. In: Grassé, P.-P. (Ed.), Traité de Zoologie. Masson, Paris, pp. 1399–1458.

Braun, M.H., Brill, R.W., Gosline, J.M., Jones, D.R., 2003a. Form and function of the bulbus arteriosus in yellowfin tuna (*Thunnus albacares*), bigeye tuna (*Thunnus obesus*) and blue marlin (*Makaira nigricans*): static properties. J. Exp. Biol. 206, 3311–3326.

Braun, M.H., Brill, R.W., Gosline, J.M., Jones, D.R., 2003b. Form and function of the bulbus arteriosus in yellowfin tuna (*Thunnus albacares*): dynamic properties. J. Exp. Biol. 206, 3327–3335.

Bugge, J., 1961. The heart of the African lungfish, Protopterus. Vidensk. Meddr. Dansk. Natuhr. Foren. 123, 193–210.

Burggren, W.W., 1988. Cardiac design in lower vertebrates: what can phylogeny reveal about ontogeny. Experientia 144, 919–930.

Burggren, W.W., Johansen, K., 1986. Circulation and respiration in lungfishes (Dipnoi). J. Morphol. (Suppl. 1), 217–236.

Burggren, W.W., Farrell, A., Lillywhite, H., 1997. Vertebrate cardiovascular systems. In: Dantzler, W.H. (Ed.), Handbook of Physiology, Sect. 13, Comparative Physiology, vol. 1. Oxford University Press, New York, pp. 215–308.

Cerra, M.C., Imbrogno, S., Amelio, D., Garofalo, F., Colvee, E., Tota, B., Icardo, J.M., 2004. Cardiac morphodynamic remodelling in the growing eel (*Anguilla anguilla* L.). J. Exp. Biol. 207, 2867–2875.

Cox, G.K., Kennedy, G.E., Farrell, A.P., 2016. Morphological arrangement of the coronary vasculature in a shark (*Squalus sucklei*) and a teleost (*Oncorhynchus mykiss*). J. Morphol. 277, 896–905.

Davie, P.S., Farrell, A.P., 1991. The coronary and luminal circulations of the myocardium of fishes. Can. J. Zool. 69, 1993–2001.

Davie, P.S., Forster, M.E., Davison, B., Satchell, G.H., 1987. Cardiac function in the New Zealand hagfish, *Eptatretus cirrhatus*. Physiol. Zool. 60, 233–240.

de Andrés, A.V., Muñoz-Chapuli, R., García, L., Sans-Coma, V., 1990. Anatomical studies of the coronary system in elasmobranchs. I. Coronary arteries in lamnoid sharks. Am. J. Anat. 187, 303–310.

de Andrés, A.V., Muñoz-Chapuli, R., Sans-Coma, V., García-Garrido, L., 1992. Anatomical studies of the coronary system in elasmobranchs: II. Coronary arteries in hexanchoid, squaloid, and carcharhinoid sharks. Anat. Rec. 233, 429–439.

Donald, J.A., 1998. The autonomic nervous system. In: Evans, D.H. (Ed.), The Physiology of Fishes, second ed. CRC Press, Boca Raton, pp. 407–439.

Durán, A.C., Fernández, B., Grimes, A.C., Rodríguez, C., Arqué, J.M., Sans-Coma, V., 2008. Chondrichthyans have a bulbus arteriosus at the arterial pole of the heart: morphological and evolutionary implications. J. Anat. 21, 597–606.

Durán, A.C., Reyes-Moya, I., Fernández, B., Rodríguez, C., Sans-Coma, V., Grimes, A.C., 2014. The anatomical components of the cardiac outflow tract of the gray bichir, *Polypterus senegalus*: their evolutionary significance. Fortschr. Zool. 117, 370–376.

Durán, A.C., Lopez-Unzu, M.A., Rodríguez, C., Fernández, B., Lorenzale, M., Linares, A., Salmerón, F., Sans-Coma, V., 2015. Structure and vascularization of the ventricular myocardium in Holocephali: their evolutionary significance. J. Anat. 226, 501–510.

Farrell, A.P., 1979. The Wind-Kessel effect of the bulbus arteriosus in trout. J. Exp. Zool. 208, 169–174.

Farrell, A.P., 2007. Cardiovascular system in primitive fishes. In: McKenzie, D.J., Farrell, A.P., Brauner, C.J. (Eds.), Primitive Fishes. Elsevier, Amsterdam, pp. 53–120.

Farrell, A.P., 2011a. Accessory hearts in fishes. In: Farrell, A.P. (Ed.), Encyclopedia of Fish Physiology. From Genome to Environment. Circulation, Design and Physiology of the Heart, vol. 2. Academic Press, New York, pp. 1073–1076.

Farrell, A.P., 2011b. The coronary circulation. In: Farrell, A.P. (Ed.), Encyclopedia of Fish Physiology. From Genome to Environment. Gas Exchange, Internal homeostasis and Food Uptake, vol. 2. Academic Press, New York, pp. 1077–1084.
Farrell, A.P., Jones, D.R., 1992. The heart. In: Hoar, W.S., Randall, D.J., Farrell, A.P. (Eds.), Fish Physiology. Part A, The Cardiovascular System, vol. XII. Academic Press, New York, pp. 1–88.
Farrell, A.P., Smith, F., 2017. Cardiac form, function and physiology. In: Gamperl, A.K., Gillis, T.E., Farrell, A.P., Brauner, C.J. (Eds.), Fish Physiology. In: The Cardiovascular System: Morphology, Control and Function, vol. 36A. Academic Press, San Diego, pp. 155–264.
Farrell, A.P., Hammons, A.M., Graham, M.S., Tibbits, G.F., 1988. Cardiac growth in rainbow trout, *Salmo gairdneri*. Can. J. Zool. 66, 2368–2373.
Farrell, A.P., Simonot, D.L., Seymour, R.S., Clark, T.D., 2007. A novel technique for estimating the compact myocardium in fishes reveals surprising results for an athletic air-breathing fish, the Pacific tarpon. J. Fish Biol. 71, 389–398.
Farrell, A.P., Farrell, N.D., Jourdan, H., Cox, G.K., 2012. A perspective on the evolution of the coronary circulation in fishes and the transition to terrestrial life. In: Sedmera, D., Wang, T. (Eds.), Ontogeny and Phylogeny of the Vertebrate Heart. Springer, New York, pp. 75–102.
Foglia, M.J., Cao, J., Tornini, V.A., Poss, K.D., 2016. Multicolor mapping of the cardiomyocyte proliferation dynamics that construct the atrium. Development 143, 1688–1696.
Foxon, G.E.H., 1950. A description of the coronary arteries in Dipnoan fishes and some remarks on their importance from the evolutionary standpoint. J. Anat. 84, 121–131.
Gallego, A., Durán, A.C., de Andrés, M.V., Navarro, P., Muñoz-Chapuli, R., 1997. Anatomy and development of the sinoatrial valves in the dogfish (*Scyliorhinus canicula*). Anat. Rec. 248, 224–232.
Gamperl, A.K., Farrell, A.P., 2004. Cardiac plasticity in fishes: environmental influences and intraspecific differences. J. Exp. Biol. 207, 2539–3550.
Garofalo, F., Imbrogno, S., Tota, B., Amelio, D., 2012. Morpho-functional characterization of the goldfish (*Carassius auratus* L.) heart. Comp. Biochem. Physiol. 163A, 215–222.
Garofalo, F., Amelio, D., Icardo, J.M., Chew, S.F., Tota, B., Cerra, M.C., Ip, Y.K., 2015. Signal molecule changes in the gills and lungs of the African lungfish *Protopterus annectens*, during the maintenance and arousal phases of aestivation. Nitric Oxide 44, 71–80.
Gattuso, A., Mazza, R., Imbrogno, S., Sverdrup, A., Tota, B., Nylund, A., 2002. Cardiac performance in *Salmo salar* with infectious salmon anaemia (ISA): putative role of nitric oxide. Dis. Aquat. Organ. 52, 11–20.
Genten, F., Terwinghe, E., Danguy, A., 2009. Atlas of Fish Histology. Science Publishers, Enfield.
Gibbins, I., 1994. Comparative anatomy and evolution of the autonomic nervous system. In: Nilsson, S., Holmgren, S. (Eds.), Comparative Physiology and Evolution of the Autonomic Nervous System. Hardwood Academic Publishers, Chur (Switzerland), pp. 1–67.
Graham, J.B., 1997. Air-Breathing Fishes. Evolution, Diversity and Adaptation. Academic Press, San Diego.
Graham, J.B., Farrell, A.P., 1989. The effect of temperature acclimation and adrenalin on the performance of a perfused trout heart. Physiol. Zool. 62, 38–61.
Grant, R.T., Regnier, M., 1926. The comparative anatomy of the cardiac coronary vessels. Heart 13, 285–317.
Greer Walker, M., Santer, M., Benjamin, M., Norman, D., 1985. Heart structure of some deepsea fish (Teleostei: Macrouridae). J. Zool. (Lond.) 205, 75–89.
Grimes, A.C., Kirby, M.L., 2009. The outflow tract of the heart in fishes: anatomy, genes and evolution. J. Fish Biol. 74, 963–1036.
Grimes, A.C., Stadt, H.A., Sheperd, I.T., Kirby, M.L., 2006. Solving an enigma: arterial pole development in the zebrafish heart. Dev. Biol. 290, 265–276.

Grimes, A.C., Durán, A.C., Sans-Coma, V., Hami, D., Santoro, M.M., Torres, M., 2010. Phylogeny informs ontogeny: a proposed common theme in the arterial pole of the vertebrate heart. Evol. Dev. 12, 552–567.

Guerrero, A., Icardo, J.M., Durán, A.C., Gallego, A., Domezain, A., Colvee, E., Sans-Coma, V., 2004. Differentiation of the cardiac outflow tract components in alevins of the sturgeon *Acipenser naccarii* (Osteichthyes, Acipenseriformes). Implications for heart evolution. J. Morphol. 260, 172–183.

Halpern, M.H., May, M.M., 1958. Phylogenetic study of the extracardiac arteries to the heart. Am. J. Anat. 102, 469–480.

Hamlett, W.C., Schwartz, F.J., Schmeinda, R., Cuevas, E., 1996. Anatomy, histology, and development of the cardiac valvular system in elasmobranchs. J. Exp. Zool. 275, 83–94.

Haverinen, J., Vornanen, M., 2007. Temperature acclimation modifies sinoatrial pacemaker mechanism of the rainbow trout heart. Am. J. Physiol. 292, R1023–R1032.

Hu, N., Yost, H.J., Clark, E.B., 2001. Cardiac morphology and blood pressure in the adult zebrafish. Anat. Rec. 264, 1–12.

Icardo, J.M., 1984. The growing heart: an anatomical perspective. In: Zak, R. (Ed.), Growth of the Heart in Health and Disease. Raven Press, New York, pp. 41–79.

Icardo, J.M., 2006. Conus arteriosus of the teleost heart: dismissed, but not missed. Anat. Rec. A Discov. Mol. Cell Evol. Biol. 288, 900–908.

Icardo, J.M., 2007. The fish endocardium. A review on the teleost heart. In: Aird, W.C. (Ed.), Endothelial Biomedicine. Cambridge University Press, Cambridge, pp. 79–84.

Icardo, J.M., 2012. The teleost heart: a morphological approach. In: Sedmera, D., Wang, T. (Eds.), Ontogeny and Phylogeny of the Vertebrate Heart. Springer, New York, pp. 35–53.

Icardo, J.M., 2013. Collagen and elastin histochemistry of the teleost bulbus arteriosus: false positives. Acta Histochem. 115, 185–189.

Icardo, J.M., Colvee, E., 2011. The atrioventricular region of the teleost heart. A distinct heart segment. Anat. Rec. 294, 236–242.

Icardo, J.M., Colvee, E., Cerra, M.C., Tota, B., 1999a. Bulbus arteriosus of Antarctic teleosts. I. The white-blooded *Chionodraco hamatus*. Anat. Rec. 254, 396–407.

Icardo, J.M., Colvee, E., Cerra, M.C., Tota, B., 1999b. Bulbus arteriosus of Antarctic teleosts. II. The red-blooded *Trematomus bernacchii*. Anat. Rec. 256, 116–126.

Icardo, J.M., Colvee, E., Cerra, M.C., Tota, B., 2000a. The bulbus arteriosus of stenothermal and temperate teleosts: a morphological approach. J. Fish Biol. 57 (Suppl. A), 121–135.

Icardo, J.M., Colvee, E., Cerra, M.C., Tota, B., 2000b. Light and electron microscopy of the bulbus arteriosus of the European eel (*Anguilla anguilla*). Cells Tissues Organs 167, 184–198.

Icardo, J.M., Colvee, E., Cerra, M.C., Tota, B., 2002a. The structure of the conus arteriosus of the sturgeon (*Acipenser naccarii*) heart. II. The myocardium, the subepicardium and the conus-aorta transition. Anat. Rec. 268, 388–398.

Icardo, J.M., Colvee, E., Cerra, M.C., Tota, B., 2002b. Structure of the conus arteriosus of the sturgeon (*Acipenser naccarii*) heart. I. The conus valves and the subendocardium. Anat. Rec. 267, 17–27.

Icardo, J.M., Schib, J.L., Ojeda, J.L., Durán, A.C., Guerrero, A., Colvee, E., Amelio, D., Sans-Coma, V., 2003. The conus valves of the adult gilthead seabream (*Sparus auratus*). J. Anat. 202, 537–550.

Icardo, J.M., Guerrero, A., Durán, A.C., Domezain, A., Colvee, E., Sans-Coma, V., 2004. The development of the sturgeon heart. Anat. Embryol. 208, 439–449.

Icardo, J.M., Imbrogno, S., Gattuso, A., Colvee, E., Tota, B., 2005a. The heart of *Sparus auratus*: a reappraisal of cardiac functional morphology in teleosts. J. Exp. Zool. 303A, 665–675.

Icardo, J.M., Ojeda, J.L., Colvee, E., Tota, B., Wong, W.P., Ip, Y.K., 2005b. The heart inflow tract of the African lungfish *Protopterus dolloi*. J. Morphol. 263, 30–38.

Icardo, J.M., Brunelli, E., Perrotta, I., Colvée, E., Wong, W.P., Ip, Y.K., 2005c. Ventricle and outflow tract of the African lungfish *Protopterus dolloi*. J. Morphol. 265, 43–51.

Icardo, J.M., Amelio, D., Garofalo, F., Colvee, E., Cerra, M.C., Wong, W.P., Tota, B., Ip, Y.K., 2008. The structural characteristics of the heart ventricle of the African lungfish *Protopterus dolloi*: freshwater and aestivation. J. Anat. 213, 106–119.

Icardo, J.M., Colvee, E., Schorno, S., Lauriano, E.R., Fudge, D.S., Glover, C.N., Zaccone, G., 2016a. Morphological analysis of the hagfish heart. I. The ventricle, the arterial connection and the ventral aorta. J. Morphol. 277, 326–340.

Icardo, J.M., Colvee, E., Schorno, S., Lauriano, E.R., Fudge, D.S., Glover, C.N., Zaccone, G., 2016b. Morphological analysis of the hagfish heart. II. The venous pole and the pericardium. J. Morphol. 277, 853–865.

Icardo, J.M., Tota, B., Ip, Y.K., 2016c. Anatomy of the heart and circulation in lungfishes. In: Zaccone, G., Dabrowski, K., Hedrick, M.S., Fernandes, J.M.O., Icardo, J.M. (Eds.), Phylogeny, Anatomy and Physiology of Ancient Fishes. CRC Press, Boca Raton, pp. 133–150.

Imbrogno, S., Cerra, M.C., 2017. Hormonal and autacoid control of cardiac function. In: Gamperl, A.K., Gillis, T.E., Farrell, A.P., Brauner, C.J. (Eds.), Fish Physiology. The Cardiovascular System: Morphology, Control and Function, vol. 36A. Academic Press, San Diego, pp. 265–315.

Imbrogno, S., Tota, B., Gattuso, A., 2011. The evolutionary functions of cardiac NOS/NO in vertebrates tracked by fish and amphibian paradigms. Nitric Oxide 25, 1–10.

Ishimatsu, A., 2012. Evolution of the cardiorespiratory system in air-breathing fishes. Aqua-BioSci. Monogr. 5, 1–28.

Isokawa, K., Takagi, M., Toda, Y., 1988. Ultrastructural cytochemistry of trout arterial fibrils as elastic components. Anat. Rec. 220, 369–375.

Jensen, D., 1965. The aneural heart of the hagfish. Ann. N. Y. Acad. Sci. 127, 443–458.

Jensen, B., Boukens, B.J.D., Wang, T., Moorman, A.F.M., Christoffels, V.M., 2014. Evolution of the sinus venosus from fish to human. J. Cardiovasc. Dev. Dis. 1, 14–28.

Johansen, K., 1962. Cardiac output and pulsatile aortic flow in the teleost *Gadus morhua*. Comp. Biochem. Physiol. 7, 169–174.

Johansen, K., Burggren, W.W., 1980. Cardiovascular function in lower vertebrates. In: Bourne, G. (Ed.), Hearts and Heart-Like Organs. Academic Press, New York, pp. 61–117.

Johansen, L., Franklin, D.L., Van Citters, R.L., 1966. Aortic blood flow in free-swimming elasmobranches. Comp. Biochem. Physiol. 19, 151–160.

Johanson, Z., Ahlberg, P.E., 2011. Phylogeny of lungfishes. In: Jorgensen, J.M., Joss, J. (Eds.), The Biology of Lungfishes. Science Publishers, Enfield, pp. 43–60.

Johnson, A.C., Turko, A.J., Klaiman, J.M., Johnston, E.F., Gillis, T.E., 2014. Cold acclimation alters the connective tissue content of the zebrafish (*Danio rerio*) heart. J. Exp. Biol. 217, 1868–1875.

Jones, D.R., Braun, M.H., 2011. The outflow tract from the heart. In: Farrell, A.P. (Ed.), Encyclopedia of Fish Physiology. From Genome to Environment. Circulation, Design and Physiology of the Heart, vol. 2. Academic Press, New York, pp. 1015–1029.

Kardong, K.V., 2006. Vertebrates: Comparative Anatomy, Function, Evolution, fourth ed. McGraw-Hill, New York.

Keen, A.N., Klaiman, J.M., Shiels, H.A., Gillis, T.E., 2017. Temperature-induced cardiac remodelling in fish. J. Exp. Biol. 220, 147–160.

Klaiman, J.M., Fenna, A.J., Shiels, H.A., Macri, J., Gillis, T.E., 2011. Cardiac remodeling in fish: strategies to maintain heart function during temperature change. PLoS One 6, e24464.

Laurent, P., 1962. Contribution à l'étude morphologique et physiologique de l'innervation du coeur des télèostéens. Arch. Anat. Microsc. Morphol. Exp. 51 (Suppl. 3), 337–458.

Laurent, P., Holmgren, S., Nilsson, S., 1983. Nervous and humoral control of the fish heart: structure and function. Comp. Biochem. Physiol. 76A, 525–542.

Leknes, I.L., 2009. Structural and histochemical studies on the teleostean bulbus arteriosus. Anat. Histol. Embryol. 38, 424–428.

Lorenzale, M., López-Unzu, M.A., Fernández, M.C., Durán, A.C., Fernández, B., Soto-Navarrete, M.T., Sans-Coma, V., 2017. Anatomical, histochemical and immunohistochemical characterisation of the cardiac outflow tract of the silver arowana, *Osteoglossum bicirrhosum* (Teleostei: Osteoglossiformes). Fortschr. Zool. 120, 15–23.

Midttun, B., 1983. Ultrastructure of the junctional region of the fish ventricle. Comp. Biochem. Physiol. 76A, 471–474.

Moriyama, Y., Ito, F., Takeda, H., Yano, T., Okabe, M., Kuraku, S., Keeley, F.W., Koshiba-Takeuchi, K., 2016. Evolution of the fish heart by sub/neofunctionalization of an elastin gene. Nat. Commun. 7, 10397.

Munshi, J.S.D., Mishra, N., 1974. Structure of the heart of *Amphipnous cuchia* (Ham.) (Amphipnoidae, Pisces). Zool. Anz. 193, 228–239.

Munshi, J.S.D., Olson, K.R., Roy, P.K., Ghosh, U., 2001. Scanning electron microscopy of the heart of the climbing perch. J. Fish Biol. 59, 1170–1180.

Newton, C.M., Stoyek, M.R., Croll, R.P., Smith, F.M., 2014. Regional innervation of the heart in the goldfish, *Carassius auratus*: a confocal microscopy study. J. Comp. Neurol. 522, 456–478.

Nilsson, S., 2011. Comparative anatomy of the autonomic nervous system. Auton. Neurosci. 165, 3–9.

Nilsson, S., Holmgren, S., 1992. Cardiovascular control by purines, 5-hydroxytriptamine, and neuropeptides. In: Hoar, W.S., Randall, D.J., Farrell, A.P. (Eds.), Fish Physiology. Part B. The Cardiovascular System, vol. XII. Academic Press, New York, pp. 301–341.

Parsons, C.W., 1930. The conus arteriosus in fishes. Quart. J. Microsc. Sci. 73, 145–176.

Pieperhoff, S., Bennett, W., Farrell, A.P., 2009. The intercellular organization of the two muscular systems in the adult salmonid heart, the compact and the spongy myocardium. J. Anat. 215, 536–547.

Poupa, O., Gesser, H., Johnson, S., Sullivan, L., 1974. Coronary-supplied compact shell of ventricular myocardium in salmonids: growth and enzyme pattern. Comp. Biochem. Physiol. A Comp. Physiol. 48, 85–95.

Priede, I.G., 1976. Functional morphology of the bulbus arteriosus of rainbow trout (*Salmo gairdneri* Richardson). J. Fish Biol. 9, 209–216.

Reyes-Moya, I., Torres-Prioris, A., Sans-Coma, V., Fernández, B., Durán, A.C., 2015. Heart pigmentation in the gray bichir, *Polypterus senegalus* (Actinopterygii: Polypteriformes). Anat. Histol. Embryol. 44, 475–480.

Richardson, M.K., Admiraal, J., Wright, G.M., 2010. Developmental anatomy of lampreys. Biol. Rev. 85, 1–33.

Robertson, J.I., 1914. The development of the heart and vascular system of *Lepidosiren paradoxa*. Quart. J. Microsc. Sci. 59, 53–132.

Saetersdal, T.S., Sorensen, E., Mykeblust, R., Helle, K.B., 1975. Granule-containing cells and fibres in the sinus venosus of elasmobranches. Cell Tissue Res. 163, 471–490.

Sánchez-Quintana, D., Hurle, J.M., 1987. Ventricular myocardial architecture in marine fishes. Anat. Rec. 217, 263–273.

Sans-Coma, V., Gallego, A., Muñoz-Chápuli, R., De Andrés, A.V., Durán, A.C., Fernández, B., 1995. Anatomy and histology of the cardiac conal valves of the adult dogfish (*Scyliorhinus canicula*). Anat. Rec. 241, 496–504.

Santer, R.M., 1985. Morphology and innervation of the fish heart. Adv. Anat. Embryol. Cell Biol. 89, 1–102.

Santer, R.M., Cobb, J.L.S., 1972. The fine structure of the heart of the teleost, *Pleuronectes platessa* L. Z. Zellforsch. 131, 1–14.
Santer, R.M., Greer Walker, M., 1980. Morphological studies on the ventricle of teleost and elasmobranch hearts. J. Zool. Lond. 190, 259–2372.
Santer, R.M., Greer Walker, M., Emerson, L., Witthames, P.R., 1983. On the morphology of the heart ventricle in marine teleost fish (Teleostei). Comp. Biochem. Physiol. 76A, 453–457.
Satchell, G.H., 1986. Cardiac function in the hagfish, *Myxine* (Myxinoidea: Cyclostomata). Acta Zool. (Stockh.) 67, 115–122.
Satchell, G.H., 1991. Physiology and Form of Fish Circulation. Cambridge University Press, Cambridge.
Satchell, G.H., Jones, M.P., 1967. The function of the conus arteriosus in the Port Jackson shark, *Heterodontus portusjacksoni*. J. Exp. Biol. 46, 376–382.
Schib, J.L., Icardo, J.M., Durán, A.C., Guerrero, A., López, D., Colvee, E., de Andrés, A.V., Sans-Coma, V., 2002. The conus arteriosus of the adult gilthead seabream (*Sparus auratus*). J. Anat. 201, 395–404.
Sedmera, D., Reckova, M., De Almeida, A., Sedmerova, M., Biermann, M., Volejnik, J., Sarre, A., Raddatz, E., McCarthy, R.A., Gourdie, R.G., Thompson, R.P., 2003. Functional and morphological evidence for a ventricular conduction system in zebrafish and *Xenopus* hearts. Am. J. Physiol. 284, H1152–H1160.
Simoes, K., Vicentini, C.A., Orsi, A.M., Cruz, C., 2002. Myoarchitecture and vasculature of the heart ventricle in some freshwater teleosts. J. Anat. 200, 467–475.
Simoes-Costa, M.S., Vasconcelos, M., Sampaio, A.C., Cravo, R.M., Linhares, V.L., Hochgreb, T., Yan, C.Y.I., Davidson, B., Xavier-Neto, J., 2005. The evolutionary origin of cardiac chambers. Dev. Biol. 277, 1–15.
Smith, W.C., 1918. On the process of disappearance of the conus arteriosus in teleosts. Anat. Rec. 15, 65–71.
Stoyek, M.R., Croll, R.P., Smith, F.M., 2015. Intrinsic and extrinsic innervation of the heart in zebrafish (*Danio rerio*). J. Comp. Neurol. 523, 1683–1700.
Szidon, J.P., Lahiri, S., Lev, M., Fishman, A.P., 1969. Heart and circulation of the African lungfish. Circ. Res. 25, 23–38.
Taylor, E.W., 1992. Nervous control of the heart and cardiorespiratory interactions. In: Hoar, W.S., Randall, D.J., Farrell, A.P. (Eds.), Fish Physiology. Part B. The Cardiovascular System, vol. XII. Academic Press, New York, pp. 343–387.
Tessadori, F., van Weerd, J.H., Burkhard, S.B., Verkerk, A.O., de Pater, E., Boukens, B.J., Vink, A., Christoffels, V.M., Bakkers, J., 2012. Identification and functional characterization of cardiac pacemaker cells in zebrafish. PLoS One 7, e47644.
Tota, B., 1978. Functional cardiac morphology and biochemistry in Atlantic bluefin tuna. In: Sharp, G.D., Dizon, A.E. (Eds.), The Physiological Ecology of Tunas. Academic Press, New York, pp. 89–112.
Tota, B., 1983. Vascular and metabolic zonation in the ventricular myocardium of mammals and fishes. Comp. Biochem. Physiol. 76A, 423–437.
Tota, B., 1989. Myoarchitecture and vascularization of the elasmobranch heart ventricle. J. Exp. Zool. 122–135.
Tota, B., Cimini, V., Salvatore, G., Zummo, G., 1983. Comparative study of the arterial and lacunary systems of the ventricular myocardium of elasmobranch and teleost fishes. Am. J. Anat. 167, 15–32.
Tota, B., Cerra, M.C., Mazza, R., Pellegrino, D., Icardo, J.M., 1997. The heart of the Antarctic icefish as a paradigm of cold adaptation. J. Therm. Biol. 22, 409–417.

Vornanen, M., Halinen, M., Haverinen, J., 2010. Sinoatrial tissue of crucian carp heart has only negative contractile responses to autonomic agonists. BMC Physiol. 10, 10.

Watson, A.D., Cobb, J.L.S., 1979. A comparative study on the innervation and the vascularization of the bulbus arteriosus in teleost fish. Cell Tissue Res. 196, 337–346.

White, E.G., 1936. The heart valves of the elasmobranch fishes. Am. Mus. Nov. 838, 1–21.

Wilson, C.M., Farrell, A.P., 2013. Pharmacological characterization of the heartbeat in an extant vertebrate ancestor, the Pacific hagfish, *Eptatretus stoutii*. Comp. Biochem. Physiol. 164A, 258–263.

Wilson, C.M., Stecyk, J.A.W., Couturier, C.S., Nilsson, G.E., Farrell, A.P., 2013. Phylogeny and effects of anoxia on hyperpolarization-activated cyclic nucleotide-gated channel gene expression in the heart of a primitive chordate, the Pacific hagfish (*Eptatretus stoutii*). J. Exp. Biol. 216, 4462–4472.

Wright, G.M., 1984. Structure of the conus arteriosus and ventral aorta in the sea lamprey, *Petromyzon marinus*, and the Atlantic hagfish, *Myxine glutinosa*: microfibrils, a major component. Can. J. Zool. 62, 2445–2456.

Xavier-Neto, J., Davidson, B., Simoes-Costa, M.S., Castro, R.A., Castillo, H.A., Sampaio, A.C., Azambuja, A.P., 2010. Evolutionary origins of hearts. In: Rosenthal, N., Harvey, R.P. (Eds.), Heart Development and Regeneration. Academic Press, New York, pp. 3–45.

Yamauchi, A., 1980. Fine structure of the fish heart. In: Bourne, G. (Ed.), Heart and Heart-like Organs. vol. 1. Academic Press, New York, pp. 119–148.

Zaccone, G., Mauceri, A., Maisano, M., Gianneto, A., Parrino, V., Fasulo, S., 2009a. Distribution and neurotransmitter localization in the heart of the ray-finned fish, bichir (*Polypterus bichir bichir* Geoffroy St. Hilaire, 1802). Acta Histochem. 111, 93–103.

Zaccone, G., Mauceri, A., Maisano, M., Fasulo, S., 2009b. Innervation of lung and heart in the ray-finned fish, bichirs. Acta Histochem. 111, 217–229.

Zaccone, G., Mauceri, A., Maisano, M., Gianneto, A., Parrino, V., Fasulo, S., 2010. Postganglionic nerve cell bodies and neurotransmitter localization in the teleost heart. Acta Histochem. 112, 328–336.

Zaccone, D., Grimes, A.C., Farrell, A.P., Dabrowski, K., Marino, F., 2012. Morphology, innervation and its phylogenetic step in the heart of the longnose gar *Lepisosteus osseus*. Acta Zool. (Stock.) 93, 381–389.

Zummo, G., Farina, F., 1989. Ultrastructure of the conus arteriosus of *Scyliorhinus stellaris*. J. Exp. Zool. 252 (S2), 158–164.

2

CARDIOMYOCYTE MORPHOLOGY AND PHYSIOLOGY

HOLLY A. SHIELS[1]

Faculty of Biology, Medicine and Health, Division of Cardiovascular Sciences, University of Manchester, Manchester, United Kingdom
[1]Corresponding author: holly.shiels@manchester.ac.uk

This chapter describes the morphology and physiology of the cells that dominate the composition of the working chambers of the fish heart—the atrial and ventricular myocytes. Contraction and relaxation of the fish atrium and ventricle are achieved through communication between, and the coordination of, these cardiomyocytes. Gap junctions provide electrical continuity between myocytes and propagate the electrical impulse that coordinates the firing of myocyte action potentials. The action potential initiates a sequence of events within the myocyte, which couple the electrical changes in the cell membrane with the mechanical contraction of the myofilaments (actin and myosin) through the transient rise

The Cardiovascular System: Morphology, Control and Function, Volume 36A
FISH PHYSIOLOGY

DOI: http://dx.doi.org/10.1016/bs.fp.2017.04.001

and fall of intracellular Ca^{2+}. This phenomenon is called excitation–contraction coupling (EC coupling). Mechanical connections between cardiomyocytes, like adherens junctions, help transform cellular contractions and relaxations into the coordinated contraction and relaxation of the heart's chambers. The strength and rate of myocyte contraction is determined by the magnitude and rate of cellular Ca^{2+} cycling and by the Ca^{2+} sensitivity of the myofilaments. These in turn can be altered by post-translational modifications like phophorylation, by the cellular environment (incl. temperature, pH and energetics), and in the case of myofilament Ca^{2+} sensitivity, by sarcomere length. This interplay between the Ca^{2+} cycling and length-dependent mechanisms that contribute to cardiac force may be important for protecting fish heart function in an environment where conditions are variable.

1. INTRODUCTION

The vertebrate heart is a syncytium of cells that work in unison to pump blood around the circulation. The pumping of the fish heart, like that of other vertebrates, is accomplished through the contraction and relaxation of individual muscle cells (cardiomyocytes) that make up the contractile chambers of the heart. The fish heart consists of two or three contractile chambers, one atrium and one ventricle, and is some species, a conus arteriosus (see Chapter 1, Volume 36A: Icardo, 2017). Blood from the body enters the atrium from the sinus venosus, and then enters the ventricle where it is pressurized and propelled into the bulbus or conus arteriosus before reaching the gills where it is oxygenated. Oxygenated blood then leaves the gills and flows to the rest of the body, before returning to the heart in a single circuit.

Contraction and relaxation of the atria and ventricle are initiated and synchronized by myogenic electrical impulses that arise spontaneously from a region of the heart called the sinoatrial node (SA node) that sits at the junction between the sinus venous and the atrium (see Chapter 3, Volume 36A: Vornanen, 2017). These impulses are myogenic as they originate from the muscle cells that make up the SA node rather than being triggered by the nervous system. However, their activity is still strongly influenced by both sympathetic and parasympathetic output from the autonomic nervous system (see Chapter 3, Volume 36A: Vornanen, 2017; Chapter 4, Volume 36A: and Farrell and Smith, 2017). These electrical impulses are transmitted between cardiomyocytes *via* gap junctions, which are pores between cells that permit the transmission of this electrical excitability during systole (described more fully later).

The fish heart also contains cardiac fibroblasts, endothelial cells, connective tissue and vascular tissue, as well as the specialized myocytes that make up the nodal (pacemaker) tissue of the heart. Fibroblasts are small in size, but abundant in number. Current dogma from the mammalian literature indicates that fibroblasts make up the largest cell population of the heart. However, this appears to vary between species and has not been determined for fish. Fibroblasts are known to play a critical role in maintaining normal cardiac function as well as in cardiac remodeling during pathological conditions (Polyakova et al., 2004). They have numerous other functions, including synthesis and deposition of extracellular matrix, and we are just starting to understand their role in cell–cell communication and intracellular signaling cascades (Gourdie et al., 2016). Fibroblasts, nodal cells, vascular cells, and epithelial cells will not be discussed in detail in this chapter. See Chapter 3, Volume 36A: Vornanen (2017) for information on nodal tissue and pace-making in the fish heart. The focus of this chapter is the morphology and physiology of myocytes of the working myocardium—the atrial and ventricular cardiomyocytes. Cardiomyocyte morphology is strongly linked with the physiology and function of the fish heart. Expression of key proteins in the fish cardiomyocyte, such as contractile proteins, ion pumps, and ion channels, is malleable and can change with pathology and be remodeled to meet the cardiac demands of a changing environment (Keen et al., 2015; Vornanen et al., 2002).

2. GROSS MYOCYTE MORPHOLOGY

Despite large differences in the gross morphology of the heart across the >30,000 extant fish species, the morphology of the cardiomyocytes that make up these hearts is remarkably uniform. The dimensions of cardiomyocytes from different fish species are rather similar, despite huge variations in body size and heart mass. Indeed, ventricular myocytes from a 0.55 g zebrafish (heart mass = 0.001 g, Fig. 1B) are only marginally smaller than a 1400 g bluefin tuna (heart mass = 5.2 g, Fig. 1F). In all fish studied to date, the majority of the working cardiomyocytes from both the atria and the ventricle are spindle shaped with an extended length: width ratio (see Fig. 1A and Table 1). Both atrial and ventricular myocytes are long (~100–180 μm) and thin (~3–15 μm), and ellipsoid in cross section (see Fig. 6A and Table 1). Interestingly, there are a small number (<10%) of working ventricular myocytes that are not long, thin, elliptical "tubes" (as in Fig. 1B), but rather have a more "sheet-like" morphology. This is shown in Fig. 1C, where the cell's dimensions are shorter and wider, but the depth is still narrow (<5 μm). The universal narrowness of fish cardiomyocytes translates into a low cytosolic volume and a greater surface area to volume ratio (SA:V) compared with mammalian myocytes (Table 1). On average, the SA:V of the fish myocyte is more than twice that of the wider and thicker mammalian myocyte. This has important implications for the functional physiology of these cells.

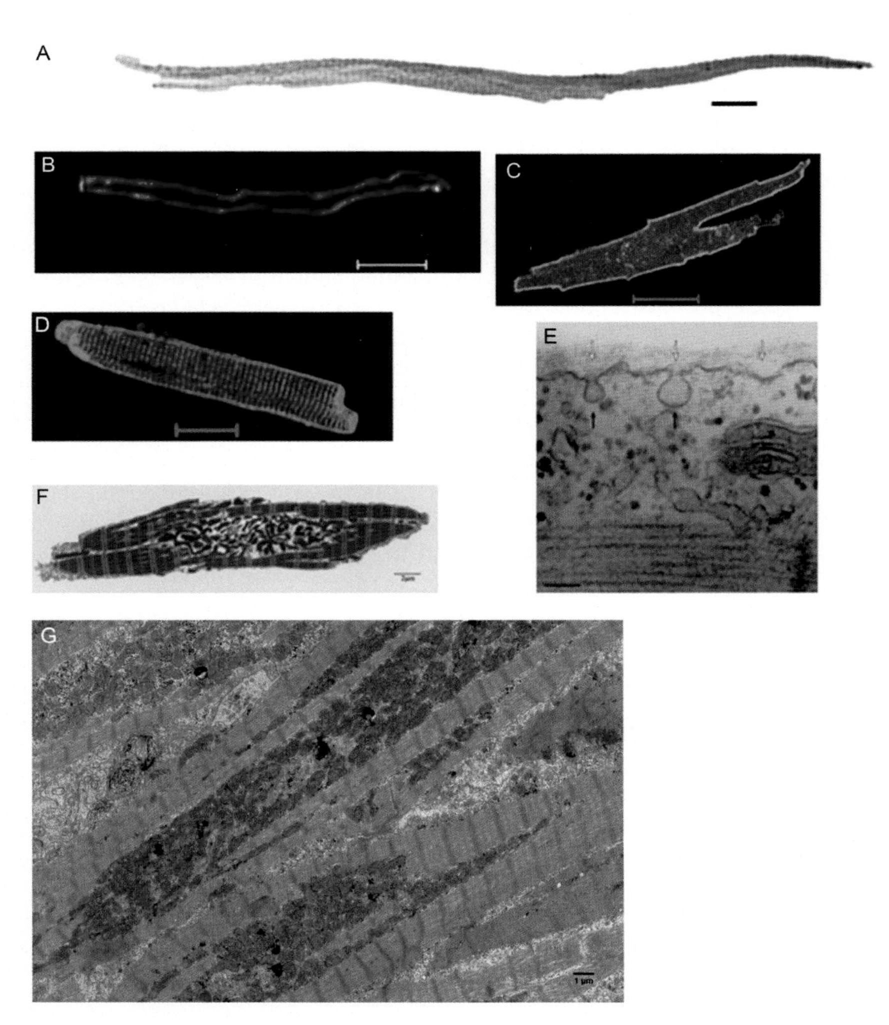

Fig. 1. Morphology of fish myocytes. (A) Light micrograph of an atrial myocyte from a Pacific mackerel (*Scomber japonicus*) with visible sarcomeric banding (scale bar 15 µm). (B) Confocal image of an adult zebrafish ventricular myocyte, the sarcolemma (SL) is visualized with the membrane dye di-8-ANEPPS (bar 20 µm). (C) Sheet-like ventricular myocyte from the rainbow trout (*Oncorhynchus mykiss*) colabeled with the Ca^{2+} dye fluo 4AM (cytosol) and di-8-ANEPPS (SL). Notice the absence of T-tubular invaginations in B and C. (D) Confocal image of an adult rat ventricular myocyte with the SL visualized with di-8-ANEPPS (bar 20 µm). Notice the T-tubular invaginations of the SL membrane which appear as a banding pattern. (E) Caveolae (*black arrows*) in burbot (*Lota lota*) ventricle. Caveolae are small folds that project inward from the SL to increase the membrane surface area. The glycocalyx (*white arrows* at *top*) covers the outer side of the SL. Original magnification, 50,000 ×, scale bar = 0.16 µm. (F) Single isolated ventricular myocyte from a Pacific bluefin tuna (*Thunnus orientalis*); note the sarcomeric pattern of the myofibrils

and the internal organization of the organelles along the longitudinal axis of the cell. (G) Transmission EM of the ventricle from a 14°C acclimated rainbow trout showing the ultrastructural organization of the mitochondria and sarcomeres. Panels (A) and (B): Courtesy of H.A. Shiels (unpublished). Panels (C) and (D): Adapted with permission from Shiels, H.A., White, E., 2005. Temporal and spatial properties of cellular Ca^{2+} flux in trout ventricular myocytes. Am. J. Physiol. Regul. Integr. Comp. Physiol. 288, R1756–R1766. Panel (E): Adapted from Tiitu, V., Vornanen, M., 2002. Morphology and fine structure of the heart of the burbot (Lota lota), a cold stenothermal fish. J. Fish Biol. 61, 106–121, with permission. Panel (F): Adapted from Di Maio, A., Block, B., 2008. Ultrastructure of the sarcoplasmic reticulum in cardiac myocytes from Pacific bluefin tuna. Cell Tissue Res. 334, 121–134 with permission. Panel (G): Courtesy of Alexander Holsgrove, unpublished, with permission.

Fig. 2 is a schematic diagram of a fish cardiomyocyte, showing the key components required to understand its contraction and relaxation. Key features will be mentioned briefly here and then discussed in detail later on. The sarcolemma (SL) is the outer cell membrane and the interface between the extracellular and intracellular space (cytosol). Embedded in this cell membrane are a number of ion channels and transporters that permit an ordered and regulated passage of ions between the extracellular space and the inside of the cell in both directions.

Ca^{2+} is the key ion in myofilament contraction, whereas Na^+ and K^+, in addition to Ca^{2+}, are important in the electrical excitability of the myocyte and its action potential (AP). The primary SL membrane proteins involved in moving Ca^{2+} into the fish myocyte are the L-type Ca^{2+} channels (LTCCs) and the Na^+/Ca^{2+} exchangers (NCXs). Another key intracellular structure involved in the cycling of Ca^{2+} is the sarcoplasmic reticulum (SR). This organelle acts as a site for Ca^{2+} storage, Ca^{2+} uptake, and Ca^{2+} release. Ca^{2+} release from the SR occurs primarily through proteins called ryanodine receptors (RyRs). Ca^{2+} uptake into the SR occurs *via* a Ca^{2+} pump called SERCA (sarco-endoplasmic reticular Ca^{2+} ATPase). SERCA consumes ATP to actively pump Ca^{2+} into the SR lumen. Ca^{2+} is buffered inside the SR by a protein called calsequestrin (CSQ). There are a number of other intracellular organelles like the mitochondria and the nucleus (not shown) that can accumulate Ca^{2+}, but this Ca^{2+} is not directly involved in cellular contraction and relaxation. However, this Ca^{2+} may regulate gene expression (Bers, 2014) and cellular energetics (Wei et al., 2015). The actin and myosin myofilaments are also illustrated in Fig. 2. Ca^{2+} must bind to the myofilaments for contraction to occur and be removed from the myofilaments for relaxation to proceed (shown in detail in Fig. 6). Lastly, there are a number of buffers in the cytosol that bind Ca^{2+} with different affinities and rates. These buffers will impact the rate of Ca^{2+} diffusion inside the cell and, thus, the amplitude and kinetics of the Ca^{2+} transient.

Table 1
Comparative morphometric data for vertebrate cardiomyocytes

	Lamprey[a]	Trout	Zebrafish[b]	Frog	Turtle[c]	Lizard[d]	Turkey[e]	Neonatal rabbit[f]	Rat[g]
Cell length (μm)	323	159.8[c]	100	300[h]	189.1	151.2	136	70	141.9
Cell width (μm)	11.9	9.9[c]	4.6	5[h]	7.2	5.9	8.7	8.2	32.0
Cell depth (μm)	—	5.7[c]	6.0	—	5.4	5.6	—	—	13.3
Capacitance (pF)	220	46[i]	26.6	75[j]	42.4	41.2	25.9	19	289.2
Cell volume (pL)	22.6	2.5[i]	2.2	2.9[k]	2.3	2.3	—	—	34.4
SA/V ratio (pF/pL)	10	18.2[i]	12	25.8[k]	18.3	18.2	—	—	8.44
T-tubular system	No	No[l]	No	No[j]	No[c]	No[d]	No[m]	No	Yes[l]

Data are means, but SEM (when known) has been left out for clarity. All data are from adult ventricular myocytes except the neonate, which are from 10-day-old atrial myocytes.

[a]Vornanen and Haverinen (2012, 2013).
[b]Brette et al. (2008).
[c]Galli et al. (2006).
[d]Galli et al. (2009b) and Galli (2006).
[e]Kim et al. (2000).
[f]Ye Sheng et al. (2011).
[g]Satoh et al. (1996).
[h]Goaillard et al. (2001).
[i]Vornanen (1998).
[j]Bean et al. (1984).
[k]Derived from (h) assuming an elliptical cross-sectional area.
[l]Shiels and White (2005).
[m]Jewett et al. (1971) for finch and humming bird (not turkey).

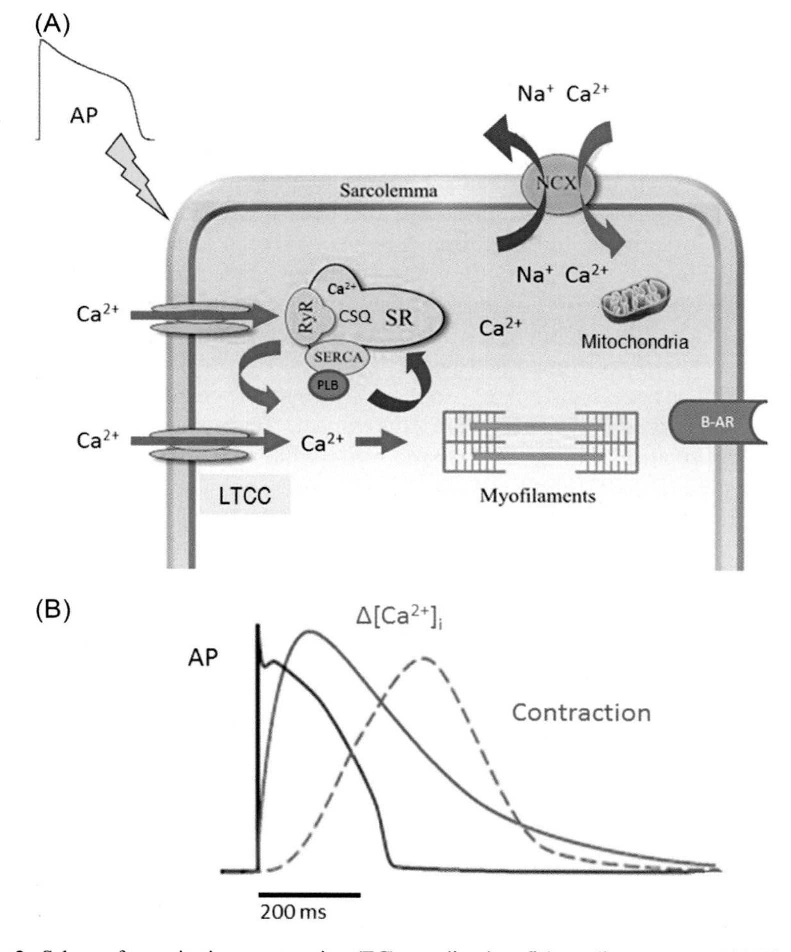

Fig. 2. Schema for excitation–contraction (EC) coupling in a fish cardiac myocyte. (A) The sarcolemma (SL) being excited by an action potential (AP), which opens L-type Ca^{2+} channels (LTCCs) allowing Ca^{2+} influx (*red arrows*) down its concentration gradient. Ca^{2+} can also enter the cell *via* reverse-mode Na^{+}–Ca^{2+} exchange (NCX). Ca^{2+} influx can trigger Ca^{2+} release from the sarcoplasmic reticulum (SR) through ryanodine receptors (RyR) in a process called Ca^{2+}-induced Ca^{2+} release (CICR). Together, these Ca^{2+} influxes cause a transient rise in Ca^{2+} ($\Delta[Ca^{2+}]_i$) that initiates the contraction of the myofilaments. Relaxation occurs when Ca^{2+} is removed from the cytosol (*purple arrows*) either back across the SL *via* forward-mode NCX or back into the SR *via* the SR Ca^{2+} pumps (SERCA). The activity of SERCA is influenced by phospholamban (PLB) (see text for details). All of these proteins can be regulated by intracellular signaling cascades following activation of β-ARs. Also shown are the myofilaments, the end effector in EC coupling and the mitochondria. (B) The time course of the key events in EC coupling during a single contraction cycle—the AP depolarizes the SL membrane (*black line*) leading to the transient rise in Ca^{2+} (*blue line*), followed by the contraction and relaxation of the myofilaments (*red line*). Panel (B): Adapted from Bers, D.M., 2002. Cardiac excitation-contraction coupling. Nature 415, 198–205.

2.1. Sarcolemmal and Cell–Cell Interactions

The sarcolemma (SL) is the lipid membrane that encompasses the cardiomyocyte. In mammalian ventricular myocytes, the SL forms deep invaginations from the surface into the cell interior called the transverse tubular system (T-tubules) (see Fig. 1D). These T-tubules are vital for activating Ca^{2+} release deep inside the thicker mammalian ventricular myocytes (see Table 1 for dimensions). T-tubules have also been observed in atrial myocytes from larger mammals, but are absent, or very rare, in atrial myocytes from rodents (Richards et al., 2011). In contrast, the SL does not form deep membrane invaginations into the center of fish atrial or ventricular myocytes (see zebrafish ventricular myocyte, Fig. 1B). As fish myocytes are thinner and the myofilaments are positioned just beneath the SL (Fig. 6A), the T-tubule system may not be necessary for Ca^{2+} to reach the myocyte center for effective EC coupling (described in detail later). In this respect, fish myocytes are more similar to those of amphibians, reptiles, and birds, which are also narrow in diameter and devoid of T-tubules (Table 1). However, the fish SL does possess smaller invaginations called caveolae (Fig. 1E) (Di Maio and Block, 2008; Tiitu and Vornanen, 2002). Caveolae, which increase the cell SA:V ratio, are important for intracellular compartmentalization and cell signaling in mammals (Harvey and Calaghan, 2012) and have been implicated in nitric oxide signaling in the fish heart (Garofalo et al., 2009).

Embedded in the SL are a multitude of proteins that are vital for cell–cell interactions. They fall into two main categories: structures that mechanically couple adjacent cells and structures that electrically couple adjacent cells.

2.1.1. Mechanical Connections Between Cells

There are two main structures responsible for cell–cell adhesion in the fish heart-fascia adherens and desmosomes (Cobb, 1974). Fascia adherens and desmosomes form junctions (sometimes referred to as adhesion junctions) between the SL of neighboring myocytes at the spindly poles and also along the lateral sides (Fig. 3A). By contrast, adult mammalian cardiomyocytes are connected primarily at two poles *via* structures also composed of fascia adherens and desmosomes called intercalated disks (Bers, 2001). Internally, fish fascia adherens are associated with the actin filaments of the Z-disks and desmosomes are associated with intermediate filaments (Fig. 3B; Cobb, 1974; Icardo, 2012; Icardo et al., 2016). These structures provide strong mechanical linkages that are important in transmitting contractile force across the heart and may play a role in mechanical signaling (Borrmann et al., 2006; Pieperhoff et al., 2009). Because many fish hearts contain two distinct myocardial layers, the outer compact layer and the inner spongy layer, there is interest in how the myocytes at the interface of the two layers interact to support the

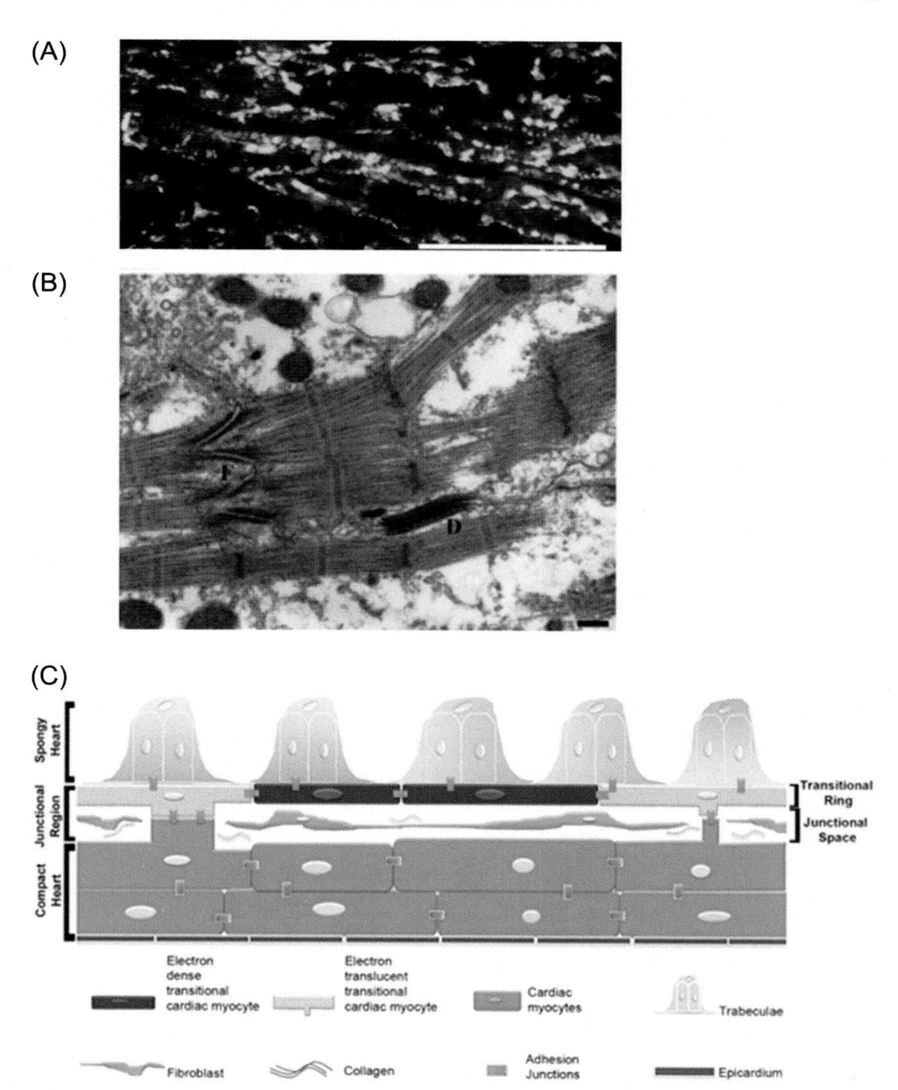

Fig. 3. Cell–cell connections in the fish heart. (A) Double-labeled immunofluorescence image of a cryostat section through the heart of the eel (*Anguilla anguilla*), showing desmosomes [*via* staining for desmoplakin (*green*)] and fascia adherens [*via* staining for N-cadherin (*red*)]. Note the regions of colocalization indicated by the *yellow–orange color* in numerous, but not all, junctions. Scale bar is 0.5 μm. (B) Transmission electron micrograph showing junctional complexes containing fascia adherens (F) and desmosomes (D) in the hagfish heart (*Eptatretus cirrhatus*). Note the typical bands of myofibrils. Scale bar is 300 nm. (C) Schematic representation of mechanical cell–cell connections in the compact and spongy myocardium of the adult zebrafish heart. Contact areas

between cells display complex adhesions junctions, which contain desmosomes and fascia adherens. Also illustrated is the interface region between the compact and spongy myocardium. The interface is complex with an interstitial space populated by a fibroblast network and collagen fibrils. A ring of flattened cardiac myocytes (both electron dense and electron translucent) forms the base of the trabeculae projecting into the ventricular lumen. At discrete intervals, these cells also make contact with the compact myocardium through bridges across the junctional space. Panel (A): Adapted from Borrmann, C.M., Grund, C., Kuhn, C., Hofmann, I., Pieperhoff, S., Franke, W.W., 2006. The area composita of adhering junctions connecting heart muscle cells of vertebrates. II. Colocalizations of desmosomal and fascia adhaerens molecules in the intercalated disk. Eur. J Cell Biol. 85, 469–485, with permission. Panel (B): Adapted with permission from Icardo, J.M., Colvee, E., Schorno, S., Lauriano, E.R., Fudge, D.S., Glover, C.N., Zaccone, G., 2016. Morphological analysis of the hagfish heart. I. The ventricle, the arterial connection and the ventral aorta. J. Morphol. 277, 326–340. Panel (C): Adapted from Lafontant, P.J., Behzad, A.R., Brown, E., Landry, P., Hu, N., Burns, A.R., 2013. Cardiac myocyte diversity and a fibroblast network in the junctional region of the zebrafish heart revealed by transmission and serial block-face scanning electron microscopy. PLoS One 8, e72388, with permission.

---

structural and physiological demands of the fish heart. Early studies with tuna suggested that these layers were attached to each other *via* connective tissue (Breisch et al., 1983; Sanchez-Quintana et al., 1996), a finding that was substantiated *via* TEM using species such as the Atlantic mackerel (*Scomber scombrus*), Atlantic herring (*Clupea harengus*), sprat (*Sprattus sprattus*), pike (*Esox lucius*) (Midttun, 1983), and most recently the zebrafish (*Danio rerio*) (Lafontant et al., 2013; Fig. 3C). There are conflicting reports in salmonids, however, with some authors reporting a lack of a connective tissue boundary layer (sockeye salmon, *Oncorhynchus nerka* and rainbow trout, *Oncorhynchus mykiss*) (Pieperhoff et al., 2009), while others report connective tissue-rich staining at the junction of the two layers (Klaiman et al., 2011) that may vary with thermal acclimation (Keen et al., 2015).

2.1.2. Electrical Connections Between Cells

The SL also contains gap junctions that connect cardiomyocytes electrically. In fish, as in other vertebrate hearts, gap junctions are formed by two hemichannels of connexon from opposing cells that come together to form a large conductance pore (a connexin). These form a low-resistance electrical pathway that allows for the passage of ions, signaling molecules, and metabolites between cells (Beyer et al., 1988; Severs et al., 2008). In mammals, approximately 1000 connexins are required to form the gap junction (Bruzzone et al., 1996), which is concentrated in the intercalated discs, and thus, colocalized with both fascia adherens and desmosomes. The mammalian heart predominantly expresses Cx43, Cx40, and Cx45, with Cx43 being by far the most prominent (Boyett et al., 2006). The numeral in the name of each connexin refers to their molecular weight. The working myocytes of the ventricle and atrium are strongly interconnected by Cx43 gap junctions, whereas Cx40 and Cx45 are more strongly expressed in the conduction system, and Cx30 is important during development (Valiunas et al., 2000). Several functional differences have been observed between various connexin subtypes, including pore conductance, size selectivity, charge selectivity, voltage gating, and chemical gating (Campbell et al., 2014). It is the combination of these factors, and their specific distribution within cardiac tissue, that allow connexins to control conduction velocity and action potential propagation through the heart (Johnstone et al., 2009; Valiunas, 2002; Valiunas and Weingart, 2000; Valiunas et al., 2000).

Connexins are highly conserved between species, and although studies are still limited, similarities between fish and mammalian connexins have been documented in terms of structure, function, and regulation; however, the names can vary due to minor differences in molecular weight. As in mammals, fish connexins can form heterologous as well as homologous gap junctions (Bolamba et al., 2003). The expression pattern of Cx43 in the developing zebrafish heart is very similar to that in mammals (Chatterjee et al., 2005). In the adult zebrafish, Cx45.6 (analogous to mammalian Cx40) is present in the atrium and ventricle (Christie et al., 2004). Interestingly, the zebrafish mutant *ftk* exhibits a severe cardiac developmental phenotype related to a loss of function of Cx36.7 (Sultana et al., 2008), and disrupting Cx48.5 expression results in severe cardiovascular deficiencies in embryonic and adult zebrafish (Cheng et al., 2004). Moreover, zebrafish studies have shown that a Cx46 mutation in the heart not only leads to disturbed electrical conductance in the ventricle, but this conduction impairment can also induce cardiac remodeling (Chi et al., 2010). Thus, information on the role of connexins in developmental patterning, and to a lesser extent in activation sequences, is limited in fish and exclusive to zebrafish hearts (Jensen et al., 2013). Currently, nothing is known about how variable expression and the regulation of connexins affect the conduction velocity between myocytes in other fishes.

The SL membrane also contains the ion channels and transporters responsible for ion flow during EC coupling, as illustrated in Fig. 2 and discussed in detail later.

2.2. Mitochondria

Mitochondria contain the enzymes and proteins necessary to provide energy in the form of ATP *via* cellular respiration. Cardiomyocytes allocate most of their energy to fuel the contraction/relaxation cycle, which is energetically expensive. Thus, it is not surprising that fish cardiomyocytes contain a large number of mitochondria (Di Maio and Block, 2008; Driedzic and Gesser, 1994; Shiels et al., 2011). Mitochondrial volume, as a proportion of myocyte volume, in fish compares well with mammals and ranges from 15% to 45% (Driedzic and Gesser, 1994), with the amount correlating well with metabolic activity/aerobic capacity. Mitochondria are located just under the myofibrils and toward the center of the myocytes (Figs. 1F and G, and 6A). As with the other key aspects of myocyte morphology, mitochondrial volume and number are not fixed in time. They can be modulated by changes in metabolic demand triggered by either the environment or pathological conditions (Guderley, 2011; Hilton et al., 2010; Iftikar and Hickey, 2013; Seebacher et al., 2010). Mitochondria in rainbow trout myocytes are found linked together in chains that run between the myofibrils (Fig. 1G). However, they have a less orderly arrangement than in rat ventricular myocytes where intermyofibrillar mitochondria are precisely arranged in parallel strands (Birkedal et al., 2006). The role of mitochondria and energy utilization in fish cardiac function are covered in detail in Chapter 6, Volume 36A: Rodnick and Gesser (2017).

3. EXCITATION–CONTRACTION COUPLING

Contraction of the fish heart is initiated and controlled by the flow of ions through ion channels in the SL membrane. The process of EC coupling links SL excitation *via* the AP with the mechanical contraction of the myofilaments through the Ca^{2+} transient. The schematic diagram in Fig. 2A shows the main features of fish cardiomyocyte EC coupling, while Fig. 2B shows the time course of changes in the AP, the intracellular Ca^{2+} transient, and force development during one contractile cycle. The concerted opening and closing of ion channels during the AP controls the amount and timing of the flow of Na^+, K^+, and Ca^{2+} ions (among others) across the membrane which are key for electrical excitability and electrical stability. These features are discussed in detail in the chapter "Cardiac Excitability and Its Autonomic Regulation" by Vornanen (Chapter 3, Volume 36A; Vornanen, 2017). Most aspects of EC coupling in the fish heart are plastic. Variable gene expression and isoform switching allow electrical and contractile properties to adjust to changing environmental conditions within and across fish species (Alderman et al., 2012; Genge et al., 2013; Vornanen et al., 2002).

3.1. Extracellular Ca^{2+} Influx

The large SA:V of fish myocytes means that Ca^{2+} influx across the SL results in larger changes in cytosolic Ca^{2+} as compared with mammals. In other words, the higher SA:V results in an increase in the efficacy of extracellular Ca^{2+} flux pathways for EC coupling. This feature may contribute to the reduced role of the SR in fish EC coupling compared with mammals (discussed later). Most of the information on cellular Ca^{2+} flux in fish hearts comes from studies on isolated myocytes where ion movement is recorded either *via* a florescent imaging system or electrically using voltage/current clamp technology. Interestingly, a recent study used a cell population-based approach to study Ca^{2+} dynamics in populations of freshly isolated cardiac myocytes from European seabass (*Dicentrarchus labrax*) (Ollivier et al., 2015). This technique may prove useful in identifying the cellular determinants of inter-individual variability in a fish's capacity for environmental adaptation. However, the majority of the studies discussed in this chapter were conducted on single isolated cardiomyocytes. For a detailed description of this technique, see Standen et al. (1987).

3.1.1. The L-Type Ca^{2+} Channel

The main Ca^{2+} influx pathway across the SL is the LTCC. The LTCC is a voltage-gated ion channel that is closed when the membrane is at rest, but opens during depolarization (e.g., during an action potential) allowing Ca^{2+} to flow into the cardiomyocyte. This influx of Ca^{2+} (the Ca^{2+} current (I_{Ca})) is registered as a downward deflection of the current amplitude measured during whole-cell voltage clamp, as seen in Fig. 4A. There is a large driving force for Ca^{2+} influx when these channels open because, at rest, the concentration of extracellular Ca^{2+} (mM range) is more than 1000-fold greater than the concentration of intracellular Ca^{2+} (<1 μM). This large electrochemical driving force, the high number of LTCC in the SL, and the large cell SA:V are responsible for the rapid rising phase of the Ca^{2+} transient. This rapid rising phase is seen in the cytosolic Ca^{2+} transients from a bluefin tuna (*Thunnus orientalis*) cardiac myocyte (Fig. 4B and C). Electrophysiological studies show that between 20 and 80 μmol L^{-1} (non-mitochondrial cell volume) of Ca^{2+} enters the fish myocyte (range depending on fish species and method of investigation) through the LTCCs during each beat (Vornanen et al., 2002). I_{Ca} generates a Ca^{2+} transient that is alone sufficient to initiate the contraction of the myocyte in many fish species including carp (Vornanen, 1997) and some salmonids (Hove-Madsen and Tort, 2001). I_{Ca} has an additional role in the hearts of some fish. It can act as a trigger for the release of Ca^{2+} stored in the SR (Hove-Madsen and Tort, 1998; Hove-Madsen et al., 1999). This process, termed *Ca^{2+}-induced Ca^{2+} release (CICR)*, will be discussed in more detail later.

LTCCs stop conducting Ca^{2+} soon after opening [e.g., ~0.3 s at 22°C (Vornanen, 1998)], but open time can vary depending on a number of intracellular and extracellular conditions and the membrane voltage. Inactivation of I_{Ca} depends on two features: the repolarization of the membrane during the AP (voltage-dependent inactivation), and the fact that Ca^{2+} itself can stop the channel from conducting more Ca^{2+} (Ca^{2+}-dependent inactivation) (Cros et al., 2014; Vornanen, 1998). The cessation of Ca^{2+} conductance is important

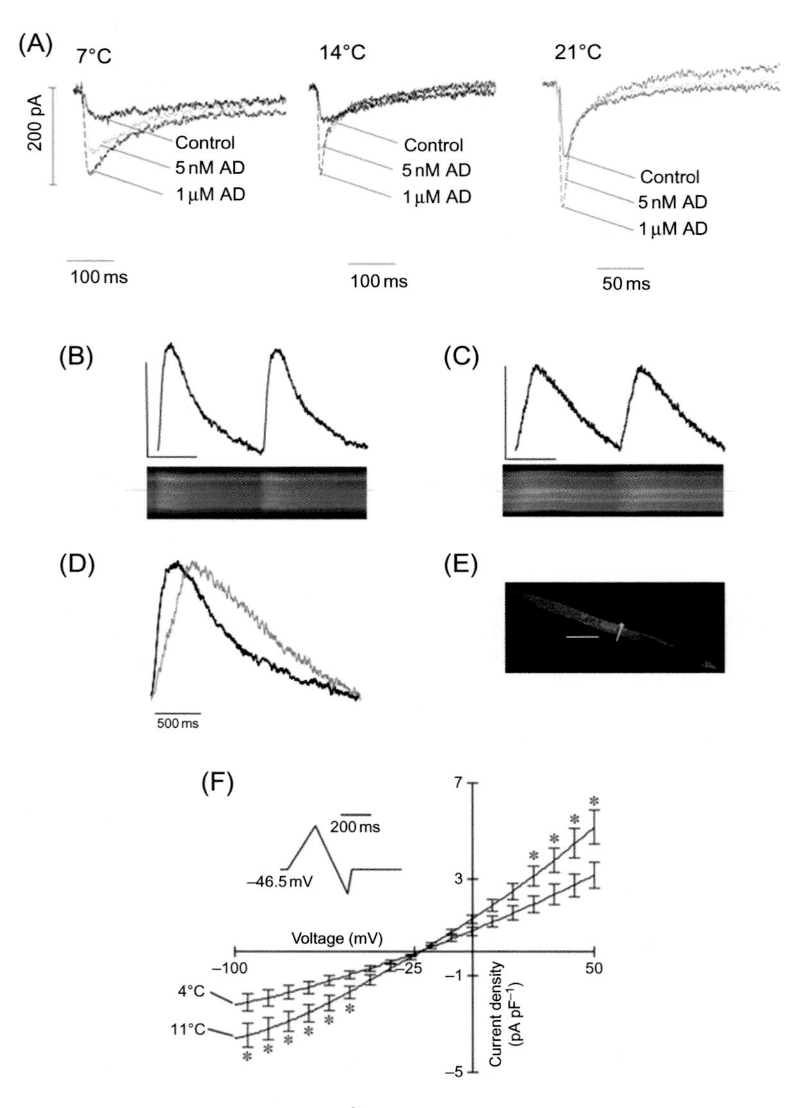

Fig. 4. Extracellular and intracellular Ca^{2+} cycling in fish myocytes. (A) Effect of adrenergic (AD) stimulation and temperature (7°C, 14°C, and 21°C) on peak L-type Ca^{2+} channel (LTCC) current (I_{Ca}) from a trout atrial myocyte. Currents were elicited by 500 ms depolarizations from −70 to 0 mV using a whole-cell voltage clamp. Note: The increase in current with increasing dose of AD and the temperature dependence of the adrenergic response. Also note the expanded timescale at 21°C. (B) The Ca^{2+} transient ($\Delta[Ca^{2+}]_i$) measured using confocal line scan imaging across the width of a 14°C acclimated ventricular myocyte from Pacific bluefin tuna under control conditions, and (C) after sarcoplasmic reticular (SR) Ca^{2+} cycling was inhibited with ryanodine

(5 μM) and thapsigargin (2 μM). Representative time courses (*top*) and corresponding raw line scan images (*below*) show temporal and spatial characteristics of Ca^{2+} flux. Scale is 100 nM $[Ca^{2+}]$ by 1 s. Line scans are 2500 lines, 512 pixels. (D) Amplitude in (B) and (C) normalized to emphasize the effect of SR inhibition on the kinetics of $\Delta[Ca^{2+}]_i$. (E) Tuna ventricular myocyte loaded with the Ca^{2+} indicator dye fluo-3 showing the position of the transverse line scan. (F) Na^+–Ca^{2+} exchange current (I_{NCX}) measured during the hyperpolarizing phase of the ramp pulse (*inset*) in ventricular myocytes from the burbot (*Lota lota*). Values are means ± SEM from eight cells at 4°C and 11°C. *The density of I_{NCX} is increased significantly by acute warming to 11°C (Student's *t*-test, $P < 0.005$). Panel (A): Adapted with permission from Shiels, H.A., Vornanen, M., Farrell, A.P., 2003. Acute temperature change modulates the response of I_{Ca} to adrenergic stimulation in fish cardiomyocytes. Physiol. Biochem. Zool. 76, 816–824. Panels (B)–(E): Adapted with permission from Shiels, H.A., Di Maio, A., Thompson, S., Block, B.A., 2011. Warm fish with cold hearts: thermal plasticity of excitation-contraction coupling in bluefin tuna. Proc. R. Soc. B Biol. Sci. 278, 18–27. Panel (F): Adapted with permission from Shiels, H.A., Paajanen, V., Vornanen, M., 2006. Sarcolemmal ion currents and sarcoplasmic reticulum Ca^{2+} content in ventricular myocytes from the cold stenothermic fish, the burbot (Lota lota). J. Exp. Biol. 209, 3091–3100.

---

because cytosolic Ca^{2+} needs to fall to a certain level for the heart to be able to relax and fill with blood between beats. The flow of Ca^{2+} through the LTCC is temperature sensitive, with the current amplitude and kinetics greater at warm temperatures and reduced at cooler temperatures (Fig. 4A) (Galli et al., 2011; Kubly and Stecyk, 2015; Shiels et al., 2000, 2015). As I_{Ca} is the main Ca^{2+} influx route, and also the trigger for release of SR Ca^{2+} stores, changes in this ion current could have large impacts on EC coupling during temperature change. I_{Ca} is also increased by adrenergic stimulation in a variety of species [e.g., rainbow trout, Fig. 4A (Shiels et al., 2003; Vornanen, 1998), crucian carp (Vornanen, 1998), Pacific bluefin tuna (Shiels et al., 2015)], which may be protective during temperature changes (Farrell et al., 1996; Graham and Farrell, 1989; Shiels et al., 2003, 2015) or during other periods of stress (Cros et al., 2014; Farrell and Milligan, 1986; Farrell et al., 1983). However, some species like the cod, flounder, sea bass (Farrell et al., 2007), and tilapia (Lague et al., 2012), and some elasmobranchs, are less responsive to adrenergic stimulation (Axelsson, 1988; Lurman et al., 2012; Mendonca et al., 2007). Some of these differences may be temperature dependent (see Section 4).

There is another inward Ca^{2+} current present in the fish heart called the T-type Ca^{2+} current (I_{CaT}). This current is activated at lower voltages than I_{CaL} and is functionally important in mammalian pace-making and during neonatal development (Nilius and Carbone, 2014). It has been recorded in working cardiomyocytes in some fish, for example, zebrafish (Nemtsas et al., 2010) and the Siberian sturgeon (*Acipenser baerii*) (Haworth et al., 2014).

However, the significance of I_{CaT} in the fish heart is not clear. Further discussion on the modulation of I_{CaL} and I_{CaT} is provided in Chapter 3, Volume 36A: Vornanen (2017).

3.1.2. The Na^+–Ca^{2+} Exchanger

The NCX also brings Ca^{2+} across the SL membrane and, thus, can contribute to the Ca^{2+} transient. The NCX transfers three Na^+ for each Ca^{2+} ion in opposite directions across the SL membrane, and because of the unequal charge transfer, there is a net movement of positive charge in the direction of Na^+ transport. The activity of NCX is complex because the electrochemical gradient that determines the direction of ion transport depends on both membrane potential and ion concentrations, and the net gradient changes during the cycle. During the relaxation phase of the Ca^{2+} transient, the exchanger moves Ca^{2+} out of the cell and brings Na^+ in. This is called forward-mode exchange. During the upsweep of the action potential, when the membrane is depolarized, the exchanger reverses direction and brings Ca^{2+} into the cell and moves Na^+ out (reverse-mode exchange) (see Fig. 4B). There is only a small range of membrane voltages (~0–10 mV) during a normal action potential which favor Ca^{2+} influx through the NCX (reverse mode), but due to the high surface area to volume ratio and high intracellular Na^+ concentration (~13 mM) in fish myocytes (Birkedal and Shiels, 2007), reverse-mode NCX can play a significant role in the systolic Ca^{2+} transient in fish. If Ca^{2+} influx through the NCX is inhibited, it reduces the contractility of trout myocytes by 30%–50% (Hove-Madsen et al., 2000). In carp (*Carassius carassius*), ~70 μmol L^{-1} (non-mitochondrial cell volume) Ca^{2+} can be brought into the cell during the upsweep of an AP *via* reverse-mode NCX (Vornanen, 1999). Thus, almost 50% of the total SL Ca^{2+} influx occurs in this species *via* reverse-mode NCX. Together, the NCX and the LTCC can bring in ~150 μmol L^{-1} Ca^{2+} (nonmitochondrial cell volume) from the extracellular space, and this generates a Ca^{2+} transient of sufficient magnitude to activate a full-scale contraction (Vornanen et al., 2002). In most mammalian myocytes, this Ca^{2+} influx pathway is not very important; the NCX primarily works to remove Ca^{2+} from the cell during the relaxation phase of the Ca^{2+} transient (Vornanen et al., 2002).

Similar to the LTCC, there are processes that stop Ca^{2+} influx through the NCX, and thereby limit the size and duration of the rising phase of the Ca^{2+} transient. The first is repolarization of the membrane potential, which removes the permissive membrane voltage window for Ca^{2+} influx. Second, as the Ca^{2+} transient rises, the concentration of Ca^{2+} in the cytosol rises, and this reduces the favorable concentration gradient for Ca^{2+} influx through the NCX. Together, the repolarization of the membrane and the rising phase of the Ca^{2+} transient are what cause the NCX to change direction and start moving

Ca^{2+} out of the cell (see Fig. 4B). Thus, the inactivation of the LTCCs and the change in the direction of the NCX limit the contribution of extracellular Ca^{2+} influx to the rising phase of the Ca^{2+} transient.

The prominent role of the NCX in fish cardiomyocyte relaxation is evidenced by the fact that NCX activity alone (i.e., with SR function inhibited) can clear the cytosol of Ca^{2+} and allow full relaxation in trout myocytes (Hove-Madsen and Tort, 2001; Hove-Madsen et al., 2000). Inhibiting the trout NCX also severely retards relaxation, and inhibiting both the NCX and the SR completely abolishes relaxation (Hove-Madsen et al., 2003). This is because Ca^{2+} levels remain elevated and the myofilaments stay in a contracted state.

I_{NCX} is temperature sensitive in the fish heart (Fig. 4F), but the degree of sensitivity varies between species. For example, although an oocyte expression study of purified channels cloned from trout or tilapia found similar I_{NCX} current properties at 30°C, they displayed very different responses to decreasing temperature. At 7°C, trout NCX maintained ~60% of the activity measured at 30°C, whereas tilapia NCX sustained only ~10% of its activity (Marshall et al., 2005). Fig. 4F shows the effect of warming on burbot I_{NCX} in native myocytes (i.e., not in channels cloned and expressed in an expression system). The Q_{10} (2–2.5) for the trout and burbot I_{NCX} is similar to that for I_{CaL} (Llach et al., 2001; Shiels et al., 2006b).

3.2. Intracellular Ca^{2+} Cycling and the SR

The SR is a specialized endoplasmic reticulum that acts as an intracellular Ca^{2+} storage and Ca^{2+} release site in muscle cells. Quantitative measurements of the SR in cardiomyocytes are only available for two species of fish and range from 2.6% to 6.1% of total cell volume (Bowler and Tirri, 1990; Di Maio and Block, 2008). This is generally less than in mammals where it ranges from 2% and 10% (Bers, 2001). The role of the SR in Ca^{2+} cycling during EC coupling is rather limited in most fish (Shiels and Galli, 2014). In the majority of fish species, atrial and ventricular myocyte contraction can be supported exclusively by transsarcolemmal Ca^{2+} flux. Regardless of its functional role, all fish studied to date possess SR, and on average, there is a greater complement and functional activity of the SR in atrial compared with ventricular myocardium (Shiels and Galli, 2014), a feature conserved across all vertebrates (Bers, 2002; Genge et al., 2012). Moreover, there is greater reliance on the SR as an important Ca^{2+} source in fish with higher heart rates and high levels of activity (like salmonids and scombrids) compared with more sedentary species (e.g., carp, flounder) (Shiels and Galli, 2014). Tuna, which are considered the elite athletes among the teleosts, have a well-developed SR network (Di Maio and Block, 2008), and SR inhibition significantly decreases the contractile force [e.g., in bigeye tuna ventricle by 30% (Galli et al., 2009a), in yellowfin tuna atrium by 50% (Shiels et al., 1999), and in skipjack tuna atrium by 30% (Keen et al., 1992)]. This can be seen by comparing the Ca^{2+} transients measured in single isolated myocytes using confocal microscopy before and after SR Ca^{2+} cycling is inhibited pharmacologically: compare panels B and C in Fig. 4 (and their overlay in Fig. 4D) where the depressive effect of SR inhibition on the amplitude and the rate of the cytosolic Ca^{2+} transient ($\Delta[Ca^{2+}]_i$) can be seen. For further details on this technique, please see Shiels and White (2005).

A link between phylogeny and SR dependence has also been suggested for neotropical freshwater fish species from the order Characiformes, superorder Ostariophysi, which utilize SR Ca^{2+} during EC coupling independent of activity level or temperature (Costa et al., 2004; Rocha et al., 2007). The reasons underlying this phylogenetic SR dependence are still unclear.

Environmental temperature also impacts the relative role of the SR in cellular Ca^{2+} cycling in fishes. Cold acclimation increases SR density (e.g., perch and tuna ventricle: Bowler and Tirri, 1990; Di Maio and Block, 2008), whereas acute warming increases SR Ca^{2+} cycling during the contraction/relaxation cycle [e.g., trout (Hove-Madsen, 1992; Shiels and Farrell, 1997)]. Why the majority of fish species studied to date do not rely strongly on SR Ca^{2+} is unclear. It is not due to low concentrations of Ca^{2+} in the fish SR as studies with caffeine (which releases SR Ca^{2+} stores by affecting the open probably of the RyR) show that the fish SR holds very large amounts of Ca^{2+} (Fig. 5A). This dichotomy between the actual amount of Ca^{2+} stored in the SR and what proportion is released, and the apparent malleability of SR involvement with changing environments, has made the fish SR an area of intense research over the years (for review, see Galli and Shiels, 2012; Shiels and Galli, 2014; Shiels and Sitsapesan, 2015). One idea under investigation is that the SR acts as a "safety mechanism" in the fish heart, storing Ca^{2+} but only releasing it to augment SL Ca^{2+} influx during times of stress (Farrell et al., 1983; Shiels and Galli, 2014). Recently, there has been some support for this hypothesis (Cros et al., 2014). Thus, the next section will discuss in some detail the routes of Ca^{2+} flux through the fish SR.

3.2.1. Ultrastructure Studies of the SR

Each region of the SR is specialized to perform specific functions during EC coupling. The majority of SR is "longitudinal" or "free" SR, which forms a loose interconnected network surrounding the myofibrils and mitochondria, and is not closely associated with the SL membrane system (Santer, 1974). This SR is enriched in the Ca^{2+} ATPase pumps (SERCA) that are responsible for the rapid uptake of Ca^{2+} into the SR during myocyte relaxation. There are also regions of "junctional" SR that form close associations with the SL membrane called peripheral couplings (Di Maio and Block, 2008;

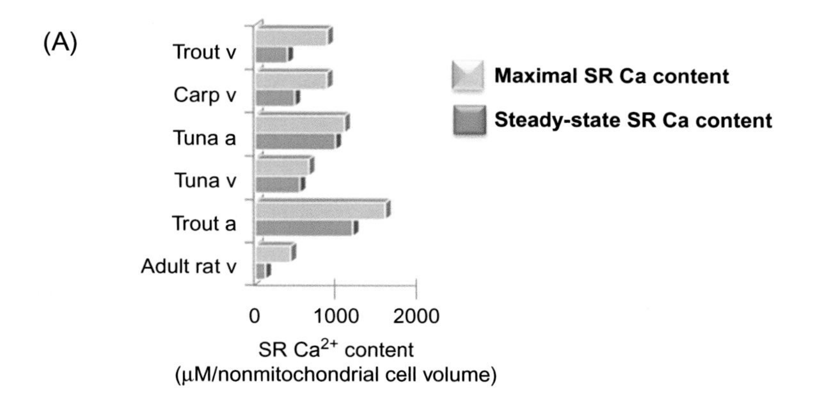

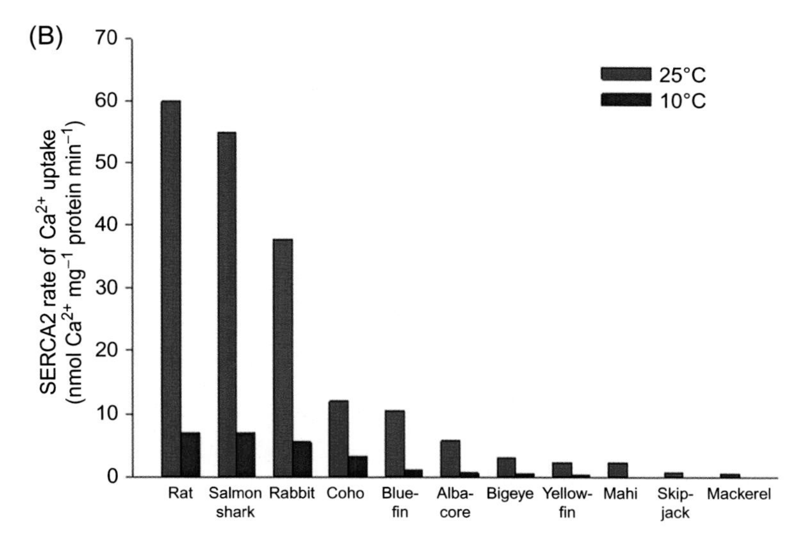

Fig. 5. Sarcoplasmic reticulum (SR) Ca^{2+} content and SR Ca^{2+} cycling. (A) Maximal and steady-state SR Ca^{2+} content expressed as μM Ca^{2+}/non-mitochondrial cell volume. The figure is compiled from a number of studies, using various methods to assess the source of Ca^{2+} for contraction, and the SR Ca^{2+} content. Thus, the figure is illustrative rather than quantitative. (B) The rate of SR Ca^{2+} uptake by SERCA in SR vesicles from a variety of fishes and a mammal at 10°C and 25°C. As above, data are illustrative rather than absolute as data from different papers are pooled. Panel (A): Data adapted from adult rat ventricle (Negretti, N., O'neill, S.C., Eisner, D.A., 1993. The relative contributions of different intracellular and sarcolemmal systems to relaxation in rat ventricular myocytes. Cardiovasc. Res. 27(10), 1826–1830); tuna atrium/ventricle (Galli, G.L., Warren, D.E., Shiels, H.A., 2009. Ca^{2+} cycling in cardiomyocytes from a high-performance reptile, the varanid lizard (Varanus exanthematicus). Am. J. Physiol. Regul. Integr. Comp. Physiol. 297, R1636–R1644; Galli, G.L.J., Shiels, H.A., Brill, R.W., 2009. Temperature sensitivity of cardiac function in pelagic fishes with different vertical mobilities: yellowfin tuna (Thunnus albacares), bigeye tuna (Thunnus obesus), mahimahi (Coryphaena hippurus), and swordfish (Xiphias gladius). Physiol. Biochem. Zool. 82, 280–290); trout atrium (Shiels, H.A., Vornanen, M., Farrell, A.P., 2002. Temperature dependence of cardiac sarcoplasmic reticulum function in rainbow trout myocytes. J. Exp. Biol. 205, 3631–3639); and trout/carp ventricle (Haverinen, J., Vornanen, M., 2009. Comparison of sarcoplasmic reticulum calcium content in

atrial and ventricular myocytes of three fish species. Am. J. Physiol. Regul. Integr. Comp. Physiol. 297, R1180–R1187). Panel (B): Data are from Castilho, P.C., Landeira-Fernandez, A.M., Morrissette, J., Block, B.A., 2007. Elevated Ca^{2+} ATPase (SERCA2) activity in tuna hearts: comparative aspects of temperature dependence. Comp. Biochem. Physiol. A Mol. Integr. Physiol. 148, 124–132; Da Silva, D., Costa, D.C., Alves, C.M., Block, B.A., Landeira-Fernandez, A.M., 2011. Temperature dependence of cardiac sarcoplasmic reticulum Ca(2)(+)-ATPase from rainbow trout Oncorhynchus mykiss. J. Fish Biol. 79, 789–800; Landeira-Fernandez, A., Morrisette, J.M., Blank, J.M., Block, B.A., 2004. Temperature dependence of Ca^{2+}-ATPase (SERCA2) in the ventricles of tuna and mackerel. Am. J. Physiol. 286, R398–R404. This figure was complied with the help of GLJ Galli.

Shiels et al., 2011; Tiitu and Vornanen, 2002). Peripheral couplings are the site of CICR (discussed later). Peripheral couplings appear to be more widely separated in fish cardiac myocytes than birds or mammals (Di Maio and Block, 2008), which would reduce the likelihood of propagation of the Ca^{2+} signal between clusters of SR Ca^{2+} release channels (RyRs). The percentage of total SR membrane that forms peripheral couplings is low in fish [0.98% for atria and 0.40% for ventricle, as quantified for bluefin tuna (Di Maio and Block, 2008)]. In mammals, peripheral couplings form at the SL surface membrane, but also throughout the T-tubular system in structures called dyads, and this greatly increases the opportunity and incidence of CICR.

The lumen of the SR contains Ca^{2+} binding proteins that facilitate Ca^{2+} storage. Calsequestrin (CSQ2 is the cardiac isoform) is a low-affinity, high-capacity Ca^{2+} buffer (Korajoki and Vornanen, 2009), and CSQ2 has been observed with electron microscopy (EM) in burbot SR (Tiitu and Vornanen, 2002). Recent estimates suggest that 75% of Ca^{2+} taken up by the SR is bound to CSQ2, and thus, CSQ2 content is a major determinant of SR Ca^{2+} content. The steady-state (~500 μM) and maximal (>1000 μM) Ca^{2+} content of the fish SR (Galli et al., 2011; Haverinen and Vornanen, 2009; Shiels et al., 2002) greatly exceeds that of mammals (50–200 μM) when both are assessed by the application of 10 mM caffeine (Negretti et al., 1995; Venetucci et al., 2006) and expressed relative to non-mitochondrial cell volume (Fig. 5A). This suggests that there are fundamental differences between fish and mammals in their lumenal Ca^{2+} storage mechanisms. Protein levels of CSQ2 have been quantified in the trout heart, but they did not vary between cardiac tissues or with thermal acclimation (Korajoki and Vornanen, 2009), thus calling into question whether another buffer may work together with CSQ2 to facilitate Ca^{2+} storage in the fish heart (Shiels and Sitsapesan, 2015). Based on CSQ2 levels, Korajoki and Vornanen (2009) calculated the total Ca^{2+} buffering capacity within the SR of trout to be 15.8–21.1 mmol L^{-1} (396–528 μmol L^{-1} myocyte volume), which is slightly greater than in the mammalian heart (Trafford et al., 1999).

3.2.2. Ryanodine Receptors

RyRs are large (~2.2 MDa) ion channels that gate the release of SR Ca^{2+}. They are homotetrameric assemblies with a large cytoplasmic head that protrudes from the SR membrane into the cytosol of the myocyte. They also possess a transmembrane region that forms the channel pore (Sutko and Airey, 1996). The C-terminus of the RyR protein resides in the lumen of the SR where it can interact with lumenal proteins. The RyR isoform in the fish heart is not resolved, but it is often referred to as RyR2 (i.e., the cardiac isoform). [^{3}H]-ryanodine binding data have been used to estimate the number of RyR2 in fish hearts. RyR2 binding sites in carp heart (B_{max} 0.18 pmol mg protein^{-1}; Chugun et al., 2003) and burbot heart (B_{max} 0.26 pmol mg protein^{-1}; Vornanen, 2006) are low compared with skeletal muscle from fish [B_{max} 1.57 pmol mg protein^{-1} for blue marlin (*Makaira nigricans*) heart *vs* 3.7 pmol mg protein^{-1} for the blue marlin white swimming muscle and 2.18 pmol mg protein^{-1} for extraocular muscle (O'Brien et al., 1995]. These data suggest that there are low levels of RyR channels in fish cardiac muscle. This conclusion is supported by studies with antibodies, which indicate that RyR protein density is 80% lower in the fish (trout, carp, and zebrafish) ventricle compared with mammals (rabbit, rat) (Birkedal et al., 2009; Bovo et al., 2013; Chugun et al., 2003).

Because RyRs are responsible for the release of Ca^{2+} from the SR, RyR opening is a tightly controlled process with regulators acting on both the cytosolic and lumenal face of the channel. Ca^{2+} is the primary ligand responsible for opening the RyRs (Meissner, 2010). Using [^{3}H]-ryanodine binding to indirectly predict the open probability of RyR channels, Vornanen (2006) showed that the Ca^{2+} sensitivity of the rat RyR ($K_d \sim 0.16$ μM) was similar to the cold stenothermic burbot (~0.19 μM), but considerably greater than that of the lamprey (~0.35 μM), rainbow trout (~0.83 μM), and crucian carp (~1.10 μM). The low RyR Ca^{2+} sensitivity may also explain the rarity of spontaneous SR Ca^{2+} release events (i.e., Ca^{2+} sparks) in ectotherm myocytes. Ca^{2+} sparks are tiny Ca^{2+} signals that arise from the activation of a cluster of RyRs (Cheng et al., 1993), which summate in space and time to form the systolic Ca^{2+} transient. Ca^{2+} sparks are generally absent in quiescent myocytes (zebrafish ventricle; Bovo et al., 2013) or in low abundance (trout atrium; Shiels and White, 2005; but see Llach et al., 2011). A rarity of Ca^{2+} sparks in fish hearts is surprising in relation to the large SR Ca^{2+} content. However, the reduced Ca^{2+} sensitivity of the fish RyRs means that higher levels of systolic Ca^{2+} are required to open the channel compared to those in the mammalian heart. Clearly, the low Ca^{2+} sensitivity of the RyR, coupled with a lower RyR density, is a major factor underlying the reduced role of the SR in Ca^{2+} cycling during the contraction–relaxation cycle in the fish heart.

Importantly, however, the efficacy of SR Ca^{2+} release through the RyRs can be modulated. Adrenergic stimulation has been shown to increase triggered Ca^{2+} release from the SR in rainbow trout (Cros et al., 2014), which supports the idea that the SR acts as a safety mechanism that contributes to Ca^{2+} cycling during periods of stress (Farrell et al., 1983). SR release may also be affected *via* interaction of the RyR with proteins including FK-506 binding proteins (FKBPs) and calmodulin on the cytosolic face and junctophilin, triadin, and CSQ2 on the lumenal face (see Shiels and Sitsapesan, 2015 for a review). The role of FKBPs and their interaction with RyRs is conserved across vertebrate species (Jeyakumar et al., 2001), and FKB12 expression in the trout heart was found to be higher in atrial than ventricular tissue, and after cold acclimation (Korajoki and Vornanen, 2014). FKBP12 and FKBP12.6 regulate different isoforms of RyR (Galfre et al., 2012; Venturi et al., 2014), and at present, it is not known whether FKBP12 is an activator or an inhibitor of the fish RyR2. Modification of RyR opening in mammals is also induced through phosphorylation and oxidation of specific residues, and the presence of agents such as Mg^{2+} and ATP (Meissner, 2010). To date, the influence of these pathways has not been investigated in the fish heart.

3.2.3. SR Ca^{2+} Content, SERCA, and Phospholamban

SR Ca^{2+} content is determined by the activity of SERCA, which pumps Ca^{2+} from the cytosol into the SR lumen. SR Ca^{2+} content is also determined by SR luminal Ca^{2+} buffering (Korajoki and Vornanen, 2009) and SR density (Di Maio and Block, 2008). In mammals, SERCA activity largely determines the rate of myocyte relaxation. When a fish myocyte is contracting at a steady state, the Ca^{2+} entering the cell across the SL with each beat moves back out across the SL *via* the NCX. Because transsarcolemmal Ca^{2+} cycling is more prominent than SR Ca^{2+} cycling in the fish heart, SERCA contributes a lesser amount to relaxation in the fish heart than it does in the mammalian heart. However, if SERCA is inhibited, it will slow the decay of the Ca^{2+} transient and the relaxation of the fish myocyte (see Fig. 4B *vs* C, and the normalized amplitude overlay in panel D, which highlights the kinetic effects of SR inhibition). SERCA activity is temperature dependent (see Fig. 5B). For example, in rat and trout, SERCA activity is almost equivalent at room temperature, but at their respective physiological temperatures, the rat heart (37°C) can pump three times more Ca^{2+} than the trout heart (Aho and Vornanen, 1998). All studies to date show that SERCA activity is robust in active fish like trout (Aho and Vornanen, 1998) and the tunas (Landeira-Fernandez et al., 2004) (Fig. 5B). However, the maximal or steady-state SR Ca^{2+} content does not necessarily correlate with SERCA activity (Fig. 5A).

Fish myocytes can store very large amounts of Ca^{2+} in their SR, and thus, the steady-state SR Ca^{2+} load in fishes is higher than in mammalian cardiomyocytes (see Fig. 5A). SR Ca^{2+} content in the ventricle of the trout, burbot, crucian carp, Pacific mackerel, Pacific bluefin tuna, yellowfin tuna, and bonito is also considerably higher than in mammalian cardiomyocytes (Galli et al., 2011; Haverinen and Vornanen, 2009). Indeed, there is a striking dichotomy between the large SR Ca^{2+} stores and the diminished role of the SR in Ca^{2+} cycling in some species. These species differences may be underpinned by variable lumenal Ca^{2+} buffering within the SR, but this has not been investigated. Interestingly, large SR Ca^{2+} loads and insensitivity to ryanodine are also characteristics of mammalian neonatal cardiomyocytes that share morphological similarities with fish myocytes (see Table 1; Bers, 2001; Louch et al., 2015), suggesting a possible link between SR Ca^{2+} handling and cardiomyocyte morphology across vertebrates. Moreover, from the studies described earlier, it appears that the difference between fish or neonates, and the mammalian SR, resides in the ability to release Ca^{2+}, rather than sequester it (Galli and Shiels, 2012).

Phospholamban (PLB) is a regulatory protein associated with SERCA in the longitudinal/free SR. The presence of PLB has been confirmed in a range of vertebrates and its molecular structure is well conserved (Cerra and Imbrogno, 2012). In its dephosphorylated state, PLB acts as an inhibitor (brake) of SERCA, slowing the rate of Ca^{2+} pumping into the SR. Phosphorylation of PLB removes this brake and increases the rate of SR Ca^{2+} uptake. Therefore, SR Ca^{2+} content can also be regulated by altering either PLB expression or phosphorylation status. Phosphorylation occurs in response to β-adrenergic stimulation and CAMKII and accelerates SR Ca^{2+} uptake. Recent work with zebrafish ventricular myocytes shows that stimulation with PKA increases SR Ca^{2+} content and leads to greater CICR (Bovo et al., 2013), indicating that the SR is regulated *via* similar pathways in fish and mammals. Furthermore, Korajoki and Vornanen (2012) found higher SERCA expression and a lower PLB/SERCA2 ratio in the atrium *vs* the ventricle of the trout heart, which correlates with the faster Ca^{2+} uptake in trout atrial muscle. Thus, the large SR Ca^{2+} stores of fish may also be due, in part, to the phosphorylation status of PLB.

3.2.4. Ca^{2+}-Induced Ca^{2+} Release

In mammals, Ca^{2+} is released from the SR through a process called CICR, whereby a small Ca^{2+} influx across the SL triggers the release of a large amount of Ca^{2+} from the SR through RyRs. In mammals, the SR trigger is almost exclusively I_{Ca} *via* the LTCC. In contrast, studies show that NCX can induce CICR in the fish cardiomyocyte (Hove-Madsen et al., 2003). The role of CICR is hotly debated for fish cardiac myocytes (Haverinen and Vornanen, 2009). Ca^{2+} is not consistently released during routine contractions despite the high SR Ca^{2+} content in fish (Fig. 5A; Shiels and Sitsapesan, 2015), and CICR is almost completely absent in some fish species, usually (but not always) slow and sluggish fish like carp. In contrast, athletic fish with fast heart rates and high blood pressures (like salmon and tuna) often show more CICR. This is illustrated for bluefin tuna myocytes in Fig. 4D and E. Clearly, inhibiting the SR (and thus CICR) reduced the overall Ca^{2+} transient amplitude and slowed the rising phase of the Ca^{2+} transient in this species. Interestingly, the relatively minor steady-state CICR (<10% of SR stores) can increase *via* adrenergic stimulation in trout (Cros et al., 2014) and zebrafish (Bovo et al., 2013). Reasons why CICR is much less important in fish *vs* mammalian cardiomyocytes may include a reduced Ca^{2+} sensitivity of the fish RyR to trigger Ca^{2+} (Haverinen and Vornanen, 2009; Korajoki and Vornanen, 2009; Vornanen et al., 2005a) and/or inadequate spatial organization of RyRs (Birkedal et al., 2009; Chugun et al., 2003). Regardless of the mechanism, the large SA:V ratio of the fish myocyte, the peripheral location of myofilaments (Fig. 6A), and the high efficacy of SL Ca^{2+} flux pathways may collectively negate the need for a large SR Ca^{2+} release during EC coupling under normal conditions. This may be helped by the low heart rate and reduced contractile strength of the fish heart (when compared with mammals). That some fish species are able to overcome functional and structural limitations, and recruit SR Ca^{2+} cycling to augment transsarcolemmal flux and elevate cardiac performance when required, is fascinating and an area under active study (see Shiels and Sitsapesan, 2015).

3.3. The Ca^{2+} Transient ($\Delta[Ca^{2+}]_i$) and Myocardial Relaxation

The Ca^{2+} transient ($\Delta[Ca^{2+}]_i$) is defined as the transient rise and fall in cytosolic Ca^{2+} concentration during a contraction. Intracellular Ca^{2+} rises quickly (within ~100 ms) and then decays more slowly (within ~400 ms). The magnitude of the Ca^{2+} transient is important in setting the strength of contraction and is used as an index of myocardial force (Yue, 1987). As discussed earlier, the source of Ca^{2+} can be of extracellular origin (coming in across the SL) or of intracellular origin (released from the SR). In fish, generally, the extracellular Ca^{2+} route is by far the most important (Tibbits et al., 1992).

It is important to note that the Ca^{2+} transient actually refers to the change in *free* Ca^{2+} inside the cytosol. A large amount of Ca^{2+} is not free inside the cell but rather is bound to Ca^{2+} buffers, and therefore, much larger amounts of *total* Ca^{2+} are required to achieve a *free* Ca^{2+} transient of a given magnitude. The amount of Ca^{2+} that a buffer can bind and the strength of the binding vary greatly (see Bers, 2001). However, very little is known about the cytosolic buffering power of the fish cardiac myocyte.

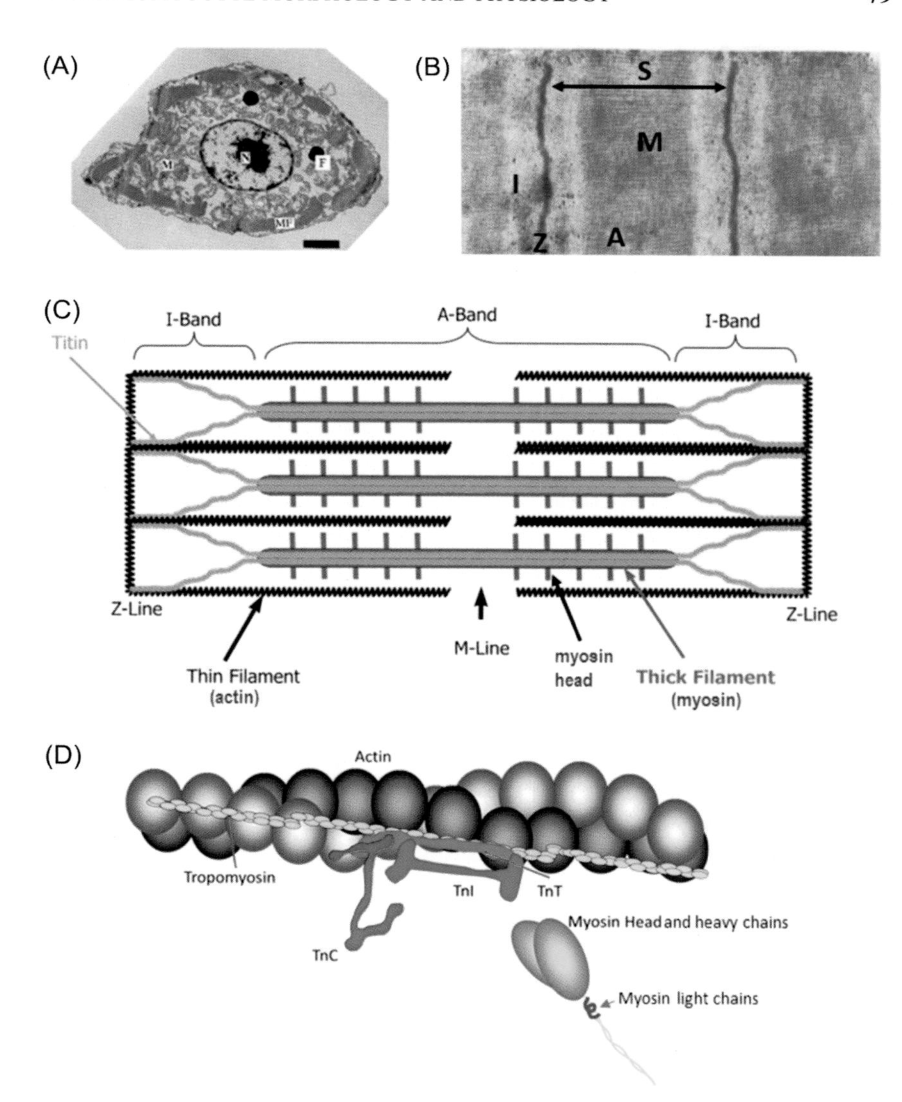

Fig. 6. Myofilament organization. (A) Electron micrographs of a ventricular myocyte from rainbow trout in cross section (*F*, dark droplets are fat; *M*, mitochondria; *MF*, myofibrils; *N*, nucleus). Scale bar is 2 μm. (B) Longitudinal section of sturgeon (*Acipenser stellatus*) atrium, showing the sarcomeric banding. Z-, I-, and A-bands are clearly seen and labeled. The M-line (M) is more diffuse. Magnification = 35,700×. The sarcomere (S) is the distance between two Z-lines. (C) Schematic diagram of a cardiac sarcomere for comparison with micrograph above. The sarcomere consists of a central A-band (*thick filaments*) and two halves of the I-band (*thin filaments*). The I-band from two adjacent sarcomeres meets at the Z-line. The central portion of the A-band is

the M-line, which does not contain actin. The figure shows the positioning of the major filament systems that compose the sarcomere: titin, actin (*thin*), and myosin (*thick*) filaments, and shows where they interact for cross-bridge generation. (D) Detail of myofilaments showing the proteins involved in cross-bridge formation between the actin and myosin filaments. Associated with the actin thin filament is a group of proteins called the regulatory unit that is responsible for initiating and controlling the activation of force generation. Shown is tropomyosin lying in the actin groove, cardiac troponin I (TnI), T (TnT), and C (TnC). Ca^{2+} binds to TnC resulting in a conformational change through the component proteins of the thin filament. The end result is the movement of tropomyosin across actin and the exposure of a myosin binding site. Also shown are the two myosins, the heavy chains in *orange/yellow* and the light chains in *gray*. There are four light chains in total—two on each heavy chain but only one set is shown here for clarity. Panel (A): Adapted with permission from Vornanen, M., 1998. L-type Ca^{2+} current in fish cardiac myocytes: effects of thermal acclimation and beta-adrenergic stimulation. J. Exp. Biol. 201, 533–547. Panel (B): Adapted with permission from Myklebust, R., Kryvi, H., 1979. Ultrastructure of the heart of the sturgeon Acipenser stellatus (Chondrostei). Cell Tissue Res. 202, 431–438. Panel (C): Adapted with permission from Shiels, H.A., White, E., 2008. The effect of mechanical stimulation on vertebrate hearts: a question of class. In: Kamkin, A., Kiseleva, I. (Eds.), Mechanosensitivity in Cells and Tissues. Mechanosensitive Ion Channels. Springer Verlag, pp. 323–342; Shiels, H.A., White, E., 2008. The Frank-Starling mechanism in vertebrate cardiac myocytes. J. Exp. Biol. 211, 2005–2013. (D) Figure adapted from by TE Gillis, with permission.

Relaxation of the cardiac myocyte is dependent on Ca^{2+} removal from the cytosol. This Ca^{2+} efflux leads to the falling (decay) phase of the Ca^{2+} transient that can be seen schematically (Fig. 2B) and in the tuna myocyte (Fig. 4B and C). This Ca^{2+} removal can occur *via* either of the two main efflux pathways illustrated in Fig. 2A. Ca^{2+} can either be pumped back across the SL membrane *via* the NCX or into the SR *via* SERCA. Because the majority of Ca^{2+} enters the fish cardiac myocyte across the SL membrane, the majority of Ca^{2+} must also leave the cell *via* this route to keep the cell in a steady state. Similar to Ca^{2+} influx, the large surface area to volume ratio of the fish myocyte aids in the efficacy of SL Ca^{2+} efflux. The NCX operating in the forward mode transports Ca^{2+} out of the cell. This is the primary Ca^{2+} removal pathway and is the main cause of the decay in the Ca^{2+} transient in fish cardiac myocytes. Forward-mode NCX is favorable during the onset of relaxation because of the initially high intracellular Ca^{2+} concentration and the repolarization of the membrane potential. However, the SR may also play a role in the decay of the Ca^{2+} transient in certain species. This is illustrated by the significant slowing of the Ca^{2+} transient when SR uptake is inhibited in the bluefin tuna ventricular myocyte (Fig. 4B *vs* C). An alternative route for Ca^{2+} efflux across the SL is the plasma membrane Ca^{2+} ATPase (PMCA). However, this pathway is expected to make a very minor contribution (if any) to the decay of the Ca^{2+} transient, although its activity has not been directly measured in fish myocytes (Bers, 2001).

4. β-ADRENERGIC RECEPTORS

Also embedded in the SL membrane are adrenergic receptors (ARs), which belong to the seven-transmembrane domain G-protein-coupled receptor (GPCR) family, and are responsible for transducing the effects of adrenaline and noradrenaline in the stress response. Currently three β-AR subtypes have been identified in the mammalian heart: β_1, β_2, and β_3, each of which has distinct downstream effects. The β_1-AR is the major receptor subtype in the mammalian heart and is responsible for the majority of the positive inotropic effects that result from β-AR stimulation. The β_1-AR is coupled to the stimulatory G-protein (G_s) which becomes activated upon agonist stimulation of the receptor. The α subunit of the G_s protein then activates adenylyl cyclase resulting in increased levels of cyclic adenosine monophosphate (cAMP) and subsequent activation of protein kinase A (PKA) (Lefkowitz et al., 1983). PKA in turn phosphorylates many intracellular targets including: PLB, PLB phosphorylation reduces the inhibition on SERCA, increasing Ca^{2+} uptake into the SR and increasing the rate of myocyte relaxation; and the LTCC, which increases calcium flux into the cell upon the arrival of an action potential, and thus, increases the amplitude of the systolic calcium transient and the force of myocyte contraction.

Early work proposed that cardiac myocytes from the rainbow trout express only a β_2-AR orthologue (Ask, 1983; Gamperl et al., 1994), in contrast to the three β-AR subtypes present in mammalian myocardium. The β_2-AR is coupled to both G_s and inhibitory G-proteins (G_i) in mammals (Brodde et al., 2001). G_i proteins antagonize the effects of G_s by negatively regulating adenylyl cyclase to decrease cAMP levels and reduce PKA activation. The rise in cAMP mediated by the β_2-AR subtype is restricted to the SL in mammals, which is in contrast to the global increase observed in the β_1-AR response (Nikolaev et al., 2006). This restriction prevents PKA phosphorylation of distant targets such as SERCA and troponin I and is thought to be mediated in part by the location of the receptor in caveolae (Nikolaev et al., 2006). We do not know if the signaling pathways transducing the β_2-AR response in fish are the same as mammals, but the restricted spread of a β_2-AR signal may not be limiting in fish due to their narrow morphology and the large number of caveolae present in the SL. Certainly, the end effect, increased cardiac output, following β-AR stimulation is similar in both groups of vertebrates, despite the fact that the β_2-AR is found at a much lower density in mammalian cardiac tissue than the β_1-AR.

More recent studies have detected the presence of a β_3-AR orthologue, β_{3a}, in the trout myocardium (Nickerson et al., 2003). Both trout β_2- and β_{3a}-ARs have been classified into mammalian subgroups using a combination of phylogenetic analysis and radioligand binding assays. However, as the β_2- and

β_{3a}-ARs share only 63% (Nickerson et al., 2001) and 52.8% (Nickerson et al., 2003) amino acid identity with mammalian β_2- and β_3-ARs, respectively, it is possible that the regulation and downstream signaling pathways may not be identical to those observed in mammals. Previous studies have demonstrated that nonspecific β-AR agonists such as adrenaline and isoprenaline do induce positive inotropic, chronotropic, and lusitropic effects when applied to teleost myocytes (Milligan, 1991), which suggests that the end targets of AR stimulation are similar. However, the signaling pathways transducing the stimulation may differ. Stimulation of the β_3-AR has been shown to generate a negative inotropic effect in human ventricular myocardium *via* activation of the endothelial nitric oxide synthase pathway (Gauthier et al., 1998) and similar findings are reported in the heart of the eel (Garofalo et al., 2009; Imbrogno et al., 2006). For additional information on β_3-Rs and their physiological effects, see Chapter 5, Volume 36A: Imbrogno and Cerra (2017).

Importantly for this chapter, activation of the β-AR signaling cascade is capable of modulating all of the key proteins in EC coupling, and hence, provides a mechanism that increases heart rate (chronotropy), contractile force (inotropy), and relaxation rate (lusitropy) after sympathetic stimulation in most fish (discussed earlier) and a wide range of vertebrates. Additionally, similar to other players in fish EC coupling, ARs in fish are modulated by temperature. Chronic cooling has been shown to increase the number of β_2AR on the myocyte surface in rainbow trout (Keen et al., 1993) and chinook salmon (*Oncorhynchus tshawytscha*) (Gamperl et al., 1998), whereas warming was shown to increase cell surface expression in another study (Eliason et al., 2011), though this may have been a stress response to short-term warming. In rainbow trout, acute cooling increases the efficacy of a given degree of adrenergic stimulation on I_{Ca} (Shiels et al., 2003), suggesting rapid cycling of β-AR or acute temperature-dependent changes in affinity. Interestingly, the opposite was observed in the bluefin tuna (Shiels et al., 2015) where adrenaline provoked greater effects when exposed acutely to acutely warm *vs* cold temperatures. Thus, variation in β_2-AR expression in the SL may explain the heightened cardiac sensitivity to adrenaline in the cold in some, but not all, species. The effect of chronic temperature change on β_3-AR expression has not been studied, but pharmacological activation of the β_3-AR pathway suggests shifts in receptor affinity or density in the common carp and the channel catfish (*Ictalurus punctatus*) following thermal acclimation (Petersen et al., 2015).

5. THE MYOFILAMENTS

The myofilaments are the contractile machinery of the cardiomyocyte. In fish cardiomyocytes, they are located peripherally, forming a single cylindrical shaped layer directly below the SL (Figs. 1F and 6A). This arrangement contrasts with mammals, which have numerous layers of myofibrils arranged longitudinally throughout the cell. Fish cardiac tissue, like that of all vertebrates, is striated, formed from the arrangement of actin and myosin filaments, which overlap to form the contractile functional unit of the cell called the sarcomere (Fig. 6B). In fish, the overall organization and periodicity of the sarcomeres are more heterogenetic than it is in mammals, often appearing less organized, and they sometimes do not extend from one end of the cell to the other (Shiels et al., 2006a). A schematic of a sarcomere and its composite proteins is provided in Fig. 6C. The morphology of the fish sarcomere is similar to that of the mammalian sarcomere with an actin filament length of approximately 0.95 μm in ventricular myocytes (Shiels et al., 2006a). Sarcomere length, which is approximately 1.6–2.0 μM, is defined as the distance between the Z-lines and changes as the myocyte stretches and contracts during the cardiac cycle. These changes in sarcomere length during systole and diastole underlie stretch (or length) regulation of contractile force and the Frank–Starling law of the heart (Shiels and White, 2008b).

5.1. Myosin ATPase Activity

The capacity of heart to express different isoforms of the thick and thin filaments provides plasticity in the cardiac response to changes in cardiovascular demands (Vornanen et al., 2002, 2005b). Cardiac myosin consists of four myosin light chains and two myosin heavy chains (Fig. 6C). Variation in amount, regulation, and in the amino acid sequence of these chains contributes strongly to the variation in muscle properties across species (McGuigan et al., 2004). These can be appreciated functionally as differences in myosin ATPase activity. Such differences have been observed between myocardial layers in the carp (*Cyprinus carpio*), where the compact muscle had greater maximal Ca^{2+}-activated myosin ATPase activity than the spongiosa (Bass et al., 1973), which may underlie its greater force generation. Similarly, in rainbow trout, atrial ATPase activity is greater than that of the ventricle at both warm (17°C) and cold (4°C) acclimation temperatures, which may underlie the faster kinetics and greater force of contraction in atrial compared with ventricular muscle (Aho and Vornanen, 1999).

A now classic study by Vornanen (1994) demonstrated how temperature can trigger a change in the isoform expression of myosin ATPase. In winter, the ventricle only expressed one isoform of myosin heavy chain. In the summer, however, a second isoform was also expressed comprising about 50% of the total isoform pool. The "summer" isoform had higher ATPase activity, which functionally enabled higher heart rates with faster and shorter contraction times in summer fish. This supports the idea of inverse thermal compensation in crucian carp; in the winter, they are slow and dormant, and the heart rate is slow and the kinetics are satisfied by the slower "winter" isoform; but in the summer, myosin remodels to the faster "summer" isoform to permit higher cardiac frequencies (Vornanen, 1994). The relationship between temperature and myosin isoform expression is not as straightforward in other fish such as common carp (Nihei et al., 2006) or has not been quantified directly (trout) (Vornanen et al., 2005a).

Although studies have cloned and sequenced the myosin light chains (Krasnov et al., 2003) in trout, their link to functional differences in muscle function remains unknown.

5.2. Ca^{2+} and Myofilament Contraction

During myocyte contraction, the energy provided by myosin ATPase draws the thick and thin filaments past each other by the cyclic attachment, ratchetting, and detachment of the myosin heads (Fig. 6C). In the fish myocyte, as in all vertebrates, physical interaction of cross-bridges from the myosin filaments and the actin filaments is controlled by the reversible binding of Ca^{2+} to the troponin complex. Contraction is initiated by the rise in the Ca^{2+} transient that increases cytosolic Ca^{2+} (from external and/or internal sources). This Ca^{2+} binds to cardiac troponin C (cTnC), resulting in a conformational change that pulls cardiac troponin I (cTnI) away from the actin and uncovers the myosin binding sites—allowing cross-bridge formation between actin and myosin (Fig. 6D). An increase in the amplitude of the Ca^{2+} transient causes an increase in the contractile force because more Ca^{2+} binds to TnC, which, in turn, opens up more sites for actin–myosin interaction and cross-bridge cycling (see Gillis and Tibbits, 2002 for review of these processes in fish).

The contractile apparatus in fish has a number of interesting adaptations that permit function across a range of environments. Fish myofilaments are inherently more Ca^{2+} sensitive than mammalian myofilaments (Churcott et al., 1994). This is due in part to modifications in the cTnC protein sequence (Gillis et al., 2003). Gillis and colleagues have shown that these modifications combat the reduction in myofilament Ca^{2+} sensitivity that is known to occur in the cold (Gillis and Tibbits, 2002). Fish are also able to modulate the interaction between Ca^{2+} and the contractile element though adrenergic stimulation, which increases the open probability of the LTCCs and enhances the Ca^{2+} transient amplitude. Nonetheless, the lusitropic effects of adrenergic stimulation may be more limited in fish compared with mammals because of a truncation in cTnI. This means that the protein lacks PKA/PKC phosphorylation sites (Kirkpatrick et al., 2011; Shaffer and Gillis, 2010). For a thorough discussion of the evolution of regulatory control of contraction in the fish heart, see Gillis (2012).

5.3. The Frank Starling and Cellular Length–Tension Relationships

The force of contraction can be modulated by altering the delivery of Ca^{2+} to the myofilaments, but it can also be modulated by changing the way that those myofilaments respond to Ca^{2+}. Myofilament length has a profound impact on myofilament Ca^{2+} sensitivity as revealed through the cellular length–tension relationship. Myofilament length changes as the heart fills (stretches) and empties (shortens) during the cardiac cycle, and this change in the sarcomere length is the cellular equivalent of the Frank–Starling response in the whole heart (Farrell et al., 1992; Shiels and White, 2008a). The Frank–Starling response is an intrinsic property of vertebrate hearts that ensures that an increase in venous return, which stretches the myocardium, results in a more forceful contraction. Some fish are able to double or even triple stroke volume to modulate cardiac output under certain circumstances (e.g., during exercise; see Chapter 4, Volume 36A: Farrell and Smith, 2017).

The exquisite stretch sensitivity of the fish heart has been demonstrated through the use of an *in situ* preparation (Farrell, 1991; Farrell et al., 1986; Graham and Farrell, 1989), where the sinus venosus and ventral aorta are cannulated, without disrupting the rigid pericardial cavity, and the response of the heart to different filling pressures (preload) and output pressures (afterload) is controlled. These studies show that over the *in vivo* range of filling pressures, only very small increments in preload (0.2–0.3 kPa) result in very large (threefold) increases in stroke volume (Farrell and Olson, 2000). Much larger increases in filling pressure are required for small increases in stroke volume in mammals. Thus, the fish heart has a large capacity for volume regulation (Farrell and Jones, 1992). The cellular demonstration of the Frank–Starling response can be seen for a single trout ventricular myocyte in Fig. 7A and B. Fig. 7A shows the large increase in active tension, in the form of a twitch, with stretch in an individual trout myocyte. Fig. 7B shows the corresponding ascending limb, the peak, and the beginning of a descending limb of the cellular length–tension relationship.

Fig. 7C shows a schematic of the cellular length–tension relationship for mammalian cardiac muscle: as the sarcomere is stretched, force increases up to a point, after which, further stretch causes a decline in force. A large proportion of the increase in force with decreased muscle length on the descending limb of the sarcomere length–tension relationship can be explained by myofilament overlap, as the degree of overlap between thick and thin filaments determines the potential availability of cross-bridges to generate tension. The insets show how changes in sarcomere length affect myofilament overlap, and thus, tension generation. Indeed, in skeletal muscle, the majority of the sarcomere length–tension relationship is explained in this way. However, in cardiac

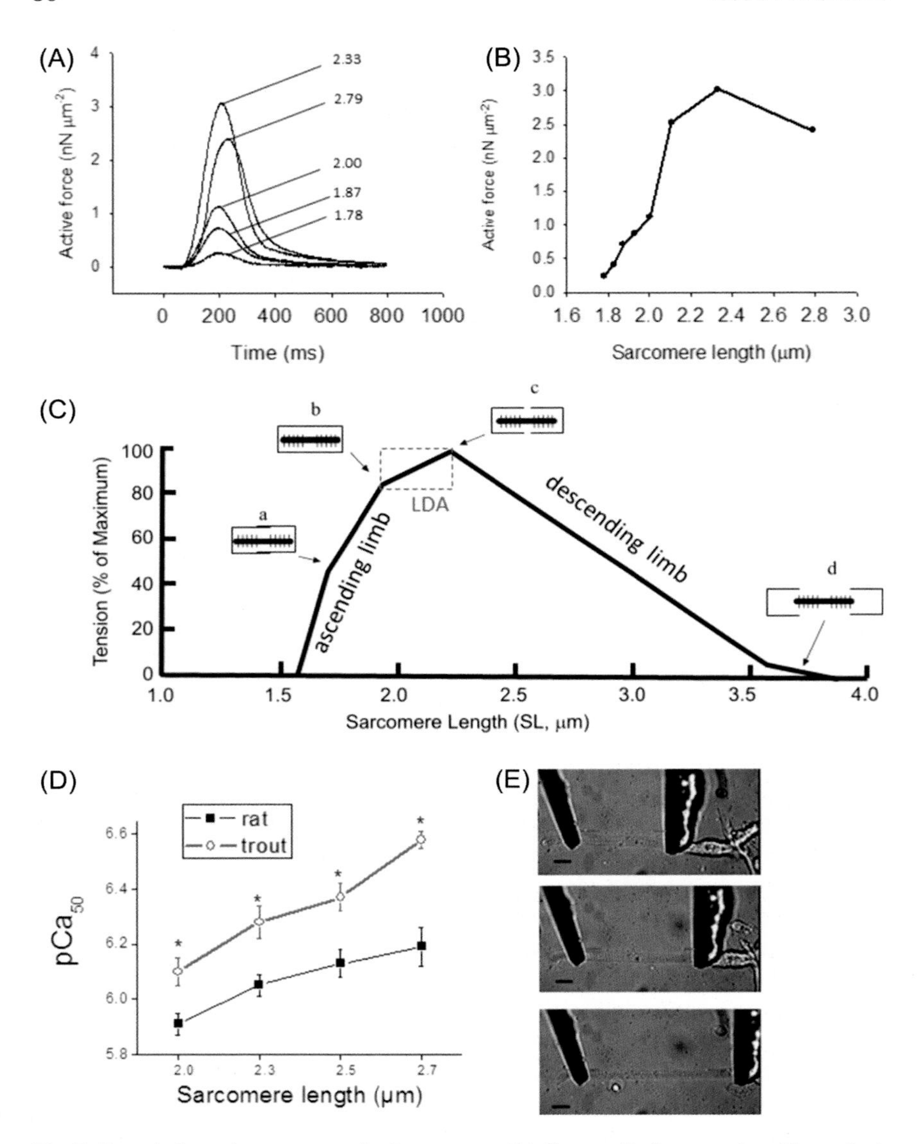

Fig. 7. Length-dependent processes in the myocyte. (A) Contractile force produced by a single trout myocyte as it is stretched. The figure shows contractions at each sarcomere length (indicated by *numbers* in μm) and how increasing sarcomere length increases force generation (up to a point). (B) Sarcomere length–tension relationship from the same fish myocyte. Notice the similarity between the sarcomere length–tension relationship in the single trout myocyte and the schematic diagram in (C). (C) Schematic diagram of the sarcomere length–tension relationship for mammalian heart. The *insets* (a–d) show the role of myofilament overlap in the length-dependent increase in force. (a) This inset shows the position of actin and myosin at short sarcomere lengths when myosin comes in contact with the Z-line and there is a rapid decline in tension as sarcomere

lengths decrease (to the *left of the arrow*). Sarcomere lengths to the *left of inset* (b) show how tension decreases when the thin filaments from the opposite ends of the sarcomere overlap at the M-line. The region between (b) and (c) is the range of sarcomere lengths where the potential availability of cross-bridges remains constant during sarcomere stretch because the central cross-bridge head-free zone of the myosin filament (M-line) is progressively uncovered. *Inset* (d) shows how tension declines toward zero when the sarcomere is stretched such that there is no overlap between thick and thin filaments. (D) The sarcomere length dependency of myofilament Ca^{2+} sensitivity in rat and trout permeabilized ventricular myocytes. The pCa for half-maximal activation (pCa_{50}) is an index of myofilament Ca^{2+} sensitivity of the contractile machinery. *Data that are significantly different. (E) Stretch of a single trout ventricular myocyte held between two carbon fibers. *Images* show the myocyte at slack length (*top*; SL 1.80 μm), and following stretches to SL 2.16 μm (*middle*) and 2.50 μm (*bottom*). Notice the compression, or narrowing, of the myocyte with stretch. Scale bar is 10 μm in each image. Panels (B) and (E): Adapted with permission from Shiels, H.A., Calaghan, S.C., White, E., 2006. The cellular basis for enhanced volume-modulated cardiac output in fish hearts. J. Gen. Physiol. 128, 37–44. Panel (C): Adapted from Bers, D.M., 2001. Excitation-contraction coupling and contractile force. Kluwer Academic Publishers, Dordrecht, The Netherlands. Panel (D): Adapted with permission from Patrick, S.M., Hoskins, A.C., Kentish, J.C., White E., Shiels, H.A., Cazorla, O., 2010. Enhanced length-dependent Ca^{2+} activation in fish cardiomyocytes permits a large operating range of sarcomere lengths. J. Mol. Cell. Cardiol. 48(5), 917–924.

---

muscle, other mechanisms contribute to the length-dependent increase in force (Moss and Fitzsimons, 2002). Here, a significant increase in active tension occurs over the range of sarcomere lengths where the potential availability of cross-bridges remains constant. This is shown in Fig. 7C by the increase in tension between inset (b) and (c), demarked by the square. Over this sarcomere length range, an increase in length does not equate to an increase in potential cross-bridge sites as the central zone of the myosin filament (near the M-line, see Fig. 6B) does not have any myosin heads attached. Hence, myofilament overlap cannot fully account for the sarcomere length–tension relationship in cardiac muscle, and factors such as increased myofilament Ca^{2+} sensitivity and a process known as length-dependent activation (LDA) may play a role.

5.4. Myofilament Ca^{2+} Sensitivity and Length-Dependent Activation

Stretch of the myocyte increases myofilament Ca^{2+} sensitivity and this increases contractile force. This phenomenon is illustrated for a trout cardiac myocyte in Fig. 7D. Here stretch-induced increases in myofilament Ca^{2+} sensitivity are quantified by the $[Ca^{2+}]_i$ required for half-maximal activation of tension (pCa_{50}) at different sarcomere lengths (Patrick et al., 2010). As sarcomere length increases, so does the Ca^{2+} sensitivity of the myofilaments as illustrated by the increase in pCa_{50}. In this figure, it is immediately apparent that the trout myofilaments are more sensitive to Ca^{2+} than those of the rat at any given sarcomere length. This confirms earlier work (Churcott et al., 1994; Gillis et al., 2003) and is partially due to differences in TnC between fish and mammals (Gillis and Tibbits, 2002). These differences in Ca^{2+} sensitivity may be required to offset the negative effects of cold temperature on myofilament Ca^{2+} sensitivity in trout (Gillis et al., 2000; Gillis and Tibbits, 2002). However, and importantly, the slope of relationship between sarcomere length and pCa_{50} is greater in the trout heart than in the rat heart, and this indicates greater LDA (Patrick et al., 2010).

Enhanced LDA is due to length-dependent changes in myofilament Ca^{2+} sensitivity or in length-dependent changes in the force generated by the cross-bridges. However, the mechanisms underlying this important feature of cardiac muscle are still unresolved. One possibility involves lattice compression. When muscle is stretched, in addition to an increase in length, there is a decrease in cross-sectional area (notice the decreasing width of the trout myocyte during stretch in Fig. 7E). This is thought to encourage the recruitment of strong force producing cross-bridges between the thick and thin filaments (Gillis, 2012). However, the precise role of filament spacing remains controversial in mammalian studies and has not been investigated in fish. Thin filament activation, or cooperativity between functional units along the thin filament, has also been put forward as a contributor to length-dependent changes in myofilament Ca^{2+} sensitivity in mammals (Gillis et al., 2005; Moss and Fitzsimons, 2002), but has not been explored in fish hearts.

The consequence of the greater length-dependent Ca^{2+} sensitivity in fish (Patrick et al., 2010) is that it permits a twofold extension of the functional ascending limb of the trout ventricular sarcomere length–tension relationship (Shiels et al., 2006b) when compared with mammals [i.e., the difference in sarcomere length where peak tension is achieved; compare Fig. 7B ($\sim$2.4 μm, fish) and C ($\sim$ 2.3 μm, mammals)]. This extension of the ascending limb of the fish length–tension relationship may be explained by greater LDA (Patrick et al., 2010). Indeed, fish myofilaments are not only strongly activated by Ca^{2+}, but that activation increases with length by a greater proportion in fish than in mammalian myofilaments. These features make contractility of the fish heart very responsive to stretch and may underlie their robust Frank–Starling response. Fish myocytes also develop more titin-based passive tension than rat myocytes, which could also account for the greater LDA (Patrick et al., 2010). However, further studies are required to test this hypothesis. For a more thorough discussion of these relationships and their impact on the passive properties of the fish heart, see Shiels and White (2008b).

6. CONCLUSIONS

The fish cardiomyocyte is structurally similar to that of reptiles, amphibians, and birds. Its high SA:V ratio, due to its narrow diameter, enhances the efficacy of SL Ca^{2+} cycling and may preclude the requirement for routine use of intracellular Ca^{2+} stores of the SR during EC coupling. Transsarcolemmal Ca^{2+} flux is modulated *in vivo* by adrenergic stimulation, temperature, and other aspects of the myocyte's environment, and this allows for adjustments in the amplitude of the Ca^{2+} transient to the contractile demands of the heart (Shiels et al., 2015). In some species, this may include the recruitment of SR Ca^{2+} *via* the process of CICR (Bovo et al., 2013; Cros et al., 2014). Fish are also able to modulate myocyte contractile force *via* changes in sarcomere length due to enhanced LDA. This provides a separate mechanism for altering contractile force, independent of altering the rate and magnitude of cellular Ca^{2+} cycling. As fish are ectotherms, and live where environmental conditions are variable, the Ca^{2+} cycling proteins, β-ARs, and myosin ATPase isoforms are directly impacted by changes in temperature or other environmental factors and serve as integrating regulatory mechanisms intrinsic to the myocyte that may be cardioprotective.

REFERENCES

Aho, E., Vornanen, M., 1998. Ca2+-ATPase activity and Ca2+ uptake by sarcoplasmic reticulum in fish heart: effects of thermal acclimation. J. Exp. Biol. 201, 525–532.

Aho, E., Vornanen, M., 1999. Contractile properties of atrial and ventricular myocardium of the heart of rainbow trout (Oncorhynchus mykiss): effects of thermal acclimation. J. Exp. Biol. 202, 2663–2677.

Alderman, S.L., Klaiman, J.M., Deck, C.A., Gillis, T.E., 2012. Effect of cold acclimation on troponin I isoform expression in striated muscle of rainbow trout. Am. J. Physiol. Regul. Integr. Comp. Physiol. 303, R168–R176.

Ask, J.A., 1983. Comparative aspects of adrenergic receptors in the hearts of lower vertebrates. Comp. Biochem. Physiol. A Comp. Physiol. 76, 543–552.

Axelsson, M., 1988. The importance of nervous and humoral mechanisms in the control of cardiac performance in the Atlantic cod Gadus morhua at rest and during non-exhaustive exercise. J. Exp. Biol. 137, 287–301.

Bass, A., Ošťádal, B., Pelouch, V., Vítek, V., 1973. Differences in weight parameters, myosin-ATPase activity and the enzyme pattern of energy supplying metabolism between the compact and spongious cardiac musculature of carp (Cyprinus carpio) and turtle (Testudo Horsfieldi). Pflugers Arch. 343, 65–77.

Bean, B.P., Nowycky, M.C., Tsien, R.W., 1984. [beta]-Adrenergic modulation of calcium channels in frog ventricular heart cells. Nature 307, 371–375.

Bers, D.M., 2001. Excitation-Contraction Coupling and Contractile Force. Kluwer Academic Publishers, Dordrecht, The Netherlands.

Bers, D.M., 2002. Cardiac excitation-contraction coupling. Nature 415, 198–205.

Bers, D.M., 2014. Cardiac sarcoplasmic reticulum calcium leak: basis and roles in cardiac dysfunction. Annu. Rev. Physiol. 76, 107–127.

Beyer, E., Goodenough, D., Paul, D., 1988. The Connexins, a Family of Related Gap Junction Proteins. Alan R. Liss, Inc., New York, pp. 165–175.

Birkedal, R., Shiels, H.A., 2007. High [Na+]i in cardiomyocytes from rainbow trout. Am. J. Physiol. 293, R861–R866.

Birkedal, R., Shiels, H.A., Vendelin, M., 2006. Three-dimensional mitochondrial arrangement in ventricular myocytes: from chaos to order. Am. J. Physiol. Cell Physiol. 291, C1148–C1158.

Birkedal, R., Christopher, J., Thistlethwaite, A., Shiels, H.A., 2009. Temperature acclimation has no effect on ryanodine receptor expression or subcellular localization in rainbow trout heart. J. Comp. Physiol. B 179, 961–969.

Bolamba, D., Patino, R., Yoshizaki, G., Thomas, P., 2003. Changes in homologous and heterologous gap junction contacts during maturation-inducing hormone-dependent meiotic resumption in ovarian follicles of Atlantic croaker. Gen. Comp. Endocrinol. 131, 291–295.

Borrmann, C.M., Grund, C., Kuhn, C., Hofmann, I., Pieperhoff, S., Franke, W.W., 2006. The area composita of adhering junctions connecting heart muscle cells of vertebrates. II. Colocalizations of desmosomal and fascia adhaerens molecules in the intercalated disk. Eur. J. Cell Biol. 85, 469–485.

Bovo, E., Dvornikov, A.V., Mazurek, S.R., de Tombe, P.P., Zima, A.V., 2013. Mechanisms of Ca handling in zebrafish ventricular myocytes. Pflugers Arch. 465, 1775–1784.

Bowler, K., Tirri, R., 1990. Temperature dependence of the heart isolated from the cold or warm acclimated perch (Perca fluviatilis). Comp. Biochem. Physiol. A 96, 177–180.

Boyett, M.R., Inada, S., Yoo, S., Li, J., Liu, J., Tellez, J., Greener, I.D., Honjo, H., Billeter, R., Lei, M., Zhang, H., Efimov, I.R., Dobrzynski, H., 2006. Connexins in the sinoatrial and atrioventricular nodes. Adv. Cardiol. 42, 175–197.

Breisch, E.A., White, F., Jones, H.M., Laurs, R.M., 1983. Ultrastructural morphometry of the myocardium of Thunnus alalunga. Cell Tissue Res. 233, 427–438.

Brette, F., Luxan, G., Cros, C., Dixey, H., Wilson, C., Shiels, H.A., 2008. Characterization of isolated ventricular myocytes from adult zebrafish (Danio rerio). Biochem. Biophys. Res. Commun. 374, 143–146.

Brodde, O.-E., Bruck, H., Leineweber, K., Seyfarth, T., 2001. Presence, distribution and physiological function of adrenergic and muscarinic receptor subtypes in the human heart. Basic Res. Cardiol. 96, 528–538.

Bruzzone, R., White, T.W., Paul, D.L., 1996. Connections with connexins: the molecular basis of direct intercellular signaling. Eur. J. Biochem. 238, 1–27.

Campbell, A.S., Johnstone, S.R., Baillie, G.S., Smith, G., 2014. β-Adrenergic modulation of myocardial conduction velocity: connexins vs. sodium current. J. Mol. Cell. Cardiol. 77, 147–154.

Cerra, M.C., Imbrogno, S., 2012. Phospholamban and cardiac function: a comparative perspective in vertebrates. Acta Physiol. 205, 9–25.

Chatterjee, B., Chin, A.J., Valdimarsson, G., Finis, C., Sonntag, J.M., Choi, B.Y., Tao, L., Balasubramanian, K., Bell, C., Krufka, A., Kozlowski, D.J., Johnson, R.G., Lo, C.W., 2005. Developmental regulation and expression of the zebrafish connexin43 gene. Dev. Dyn. 233, 890–906.

Cheng, H., Lederer, W.J., Cannell, M.B., 1993. Calcium sparks: elementary events underlying excitation-contraction coupling in heart muscle. Science 262, 740–744.

Cheng, S.H., Shakespeare, T., Mui, R., White, T.W., Valdimarsson, G., 2004. Connexin 48.5 is required for normal cardiovascular function and lens development in zebrafish embryos. J. Biol. Chem. 279, 36993–37003.

Chi, N.C., Bussen, M., Brand-Arzamendi, K., Ding, C., Olgin, J.E., Shaw, R.M., Martin, G.R., Stainier, D.Y.R., 2010. Cardiac conduction is required to preserve cardiac chamber morphology. Proc. Natl. Acad. Sci. U.S.A. 107, 14662–14667.

Christie, T.L., Mui, R., White, T.W., Valdimarsson, G., 2004. Molecular cloning, functional analysis, and RNA expression analysis of connexin45.6: a zebrafish cardiovascular connexin. Am. J. Physiol. Heart Circ. Physiol. 286, H1623-H1632.

Chugun, A., Taniguchi, K., Murayama, T., Uchide, T., Hara, Y., Temma, K., Ogawa, Y., Akera, T., 2003. Subcellular distribution of ryanodine receptors in the cardiac muscle of carp (Cyprinus carpio). Am. J. Physiol. Regul. Integr. Comp. Physiol. 285, R601–R609.

Churcott, C.S., Moyes, C.D., Bressler, B.H., Baldwin, K.M., Tibbits, G.F., 1994. Temperature and pH effects on Ca2+ sensitivity of cardiac myofibrils: a comparison of trout with mammals. Am. J. Physiol. 267, R62–R70.

Cobb, J.L.S., 1974. Gap junctions in the heart of teleost fish. Cell Tissue Res. 154, 131–134.

Costa, M.J., Olle, C.D., Kalinin, A.L., Rantin, F.T., 2004. Role of the sarcoplasmic reticulum in calcium dynamics of the ventricular myocardium of Lepidosiren paradoxa (Dipnoi) at different temperatures. J. Therm. Biol. 29, 81–89.

Cros, C., Salle, L., Warren, D.E., Shiels, H.A., Brette, F., 2014. The calcium stored in the sarcoplasmic reticulum acts as a safety mechanism in rainbow trout heart. Am. J. Physiol. Regul. Integr. Comp. Physiol. 307, R1493–R1501.

Di Maio, A., Block, B., 2008. Ultrastructure of the sarcoplasmic reticulum in cardiac myocytes from Pacific bluefin tuna. Cell Tissue Res. 334, 121–134.

Driedzic, W.R., Gesser, H., 1994. Energy-metabolism and contractility in ectothermic vertebrate hearts-hypoxia, acidosis, and low-temperature. Physiol. Rev. 74, 221–258.

Eliason, E.J., Clark, T.D., Hague, M.J., Hanson, L.M., Gallagher, Z.S., Jeffries, K.M., Gale, M.K., Patterson, D.A., Hinch, S.G., Farrell, A.P., 2011. Differences in thermal tolerance among sockeye salmon populations. Science 332, 109–112.

Farrell, A.P., 1991. From hagfish to tuna: a perspective on cardiac-function in fish. Physiol. Zool. 64, 1137–1164.

Farrell, A.P., Jones, D.R., 1992. The heart. In: Hoar, W.S., Randall, D.J., Farrell, A.P. (Eds.), Fish Physiology, Vol. XIIA. Academic Press, San Diego, pp. 1–88.

Farrell, A.P., Milligan, C.L., 1986. Myocardial intracellular pH in a perfused rainbow trout heart during extracellular acidosis in the presence and absence of adrenaline. J. Exp. Biol. 125, 347–359.

Farrell, A.P., Olson, K.R., 2000. Cardiac natriuretic peptides: a physiological lineage of cardioprotective hormones? Physiol. Biochem. Zool. 73, 1–11.

Farrell, A.P., Smith, F., 2017. Cardiac form, function and physiology. In: Gamperl, A.K., Gillis, T.E., Farrell, A.P., Brauner, C.J. (Eds.), Fish Physiology. In: The Cardiovascular System: Morphology, Control and Function, vol. 36A. Academic Press, San Diego, pp. 155–264.

Farrell, A.P., MacLeod, K.R., Driedzic, W.R., Wood, S., 1983. Cardiac performance in the in situ perfused fish heart during extracellular acidosis: interactive effects of adrenaline. J. Exp. Biol. 107, 415–429.

Farrell, A.P., MacLeod, K.R., Chancey, B., 1986. Intrinsic mechanical properties of the perfused rainbow trout heart and the effects of catecholamines and extracellular calcium under control and acidotic conditions. J. Exp. Biol. 125, 319–345.

Farrell, A.P., Jones, D.R., Hoar, W.S., Randall, D.J., 1992. The heart. In: Hoar, W.S., Randall, D.J., Farrell, A.P. (Eds.), The Cardiovascular System. Academic Press, San Diego, CA, pp. 1–88.

Farrell, A.P., Gamperl, A.K., Hicks, J.M.T., Shiels, H.A., Jain, K.E., 1996. Maximum cardiac performance of rainbow trout (Oncorhynchus mykiss) at temperatures approaching their upper lethal limit. J. Exp. Biol. 199, 663–672.

Farrell, A.P., Axelsson, M., Altimiras, J., Sandblom, E., Claireaux, G., 2007. Maximum cardiac performance and adrenergic sensitivity of the sea bass Dicentrarchus labrax at high temperatures. J. Exp. Biol. 210, 1216–1224.

Galfre, E., Pitt, S.J., Venturi, E., Sitsapesan, M., Zaccai, N.R., Tsaneva-Atanasova, K., O'Neill, S., Sitsapesan, R., 2012. FKBP12 activates the cardiac ryanodine receptor Ca2+-release channel and is antagonised by FKBP12.6. PLoS One 7. e31956

Galli, G.L.J., 2006. The role of the sarcoplasmic reticulum in the generation of high heart rates and blood pressures in reptiles. J. Exp. Biol. 209, 1956–1963.

Galli, G.L.J., Shiels, H.A., 2012. The sarcoplasmic reticulum in the vertebrate heart. In: Wang, T., Sedmera, D. (Eds.), Ontogeny and Phylogeny of the Vertebrate Heart. Advances in Experimental Medicine and Biology Springer Verlag, New York, pp. 103–124.

Galli, G.L., Taylor, E.W., Shiels, H.A., 2006. Calcium flux in turtle ventricular myocytes. Am. J. Physiol. Regul. Integr. Comp. Physiol. 291, R1781–R1789.

Galli, G.L.J., Shiels, H.A., Brill, R.W., 2009a. Temperature sensitivity of cardiac function in pelagic fishes with different vertical mobilities: yellowfin tuna (Thunnus albacares), bigeye tuna (Thunnus obesus), mahimahi (Coryphaena hippurus), and swordfish (Xiphias gladius). Physiol. Biochem. Zool. 82, 280–290.

Galli, G.L., Warren, D.E., Shiels, H.A., 2009b. Ca2+ cycling in cardiomyocytes from a high-performance reptile, the varanid lizard (Varanus exanthematicus). Am. J. Physiol. Regul. Integr. Comp. Physiol. 297, R1636–R1644.

Galli, G.L., Lipnick, M.S., Shiels, H.A., Block, B.A., 2011. Temperature effects on Ca2+ cycling in scombrid cardiomyocytes: a phylogenetic comparison. J. Exp. Biol. 214, 1068–1076.

Gamperl, A.K., Wilkinson, M., Boutilier, R.G., 1994. Beta-adrenoreceptors in the trout (Oncorhynchus mykiss) heart: characterization, quantification, and effects of repeated catecholamine exposure. Gen. Comp. Endocrinol. 95, 259–272.

Gamperl, A.K., Vijayan, M.M., Pereira, C., Farrell, A.P., 1998. Beta-receptors and stress protein 70 expression in hypoxic myocardium of rainbow trout and chinook salmon. Am. J. Physiol. 274, R428–R436.

Garofalo, F., Parisella, M.L., Amelio, D., Tota, B., Imbrogno, S., 2009. Phospholamban S-nitrosylation modulates Starling response in fish heart. Proc. R. Soc. B Biol. Sci. 276, 4043–4052. http://dx.doi.org/10.1098/rspb.2009.1189.

Gauthier, C., Leblais, V., Kobzik, L., Trochu, J.-N., Khandoudi, N., Bril, A., Balligand, J.-L., Le Marec, H., 1998. The negative inotropic effect of beta3-adrenoceptor stimulation is mediated by activation of a nitric oxide synthase pathway in human ventricle. J. Clin. Invest. 102, 1377.

Genge, C., Tibbits, G.F., Hove-Madsen, L., 2012. Functional and Structural Differences in Atria Versus Ventricles in Teleost Hearts. INTECH Open Access Publisher. https://www.intechopen.com/books/editor/new-advances-and-contributions-to-fish-biology.

Genge, C.E., Davidson, W.S., Tibbits, G.F., 2013. Adult teleost heart expresses two distinct troponin C paralogs: cardiac TnC and a novel and teleost-specific ssTnC in a chamber-and temperature-dependent manner. Physiol. Genomics 45, 866–875.

Gillis, T.E., 2012. Evolution of the regulatory control of the vertebrate heart: the role of the contractile proteins. In: Wang, D.S.T. (Ed.), Ontogeny and Phylogeny of the Vertebrate Heart. Springer, New York, pp. 125–145.

Gillis, T.E., Tibbits, G.F., 2002. Beating the cold: the functional evolution of troponin C in teleost fish. Comp. Biochem. Physiol. A Mol. Integr. Physiol. 132, 763–772.

Gillis, T.E., Marshall, C.R., Xue, X.H., Borgford, T.J., Tibbits, G.F., 2000. Ca2+ binding to cardiac troponin C: effects of temperature and pH on mammalian and salmonid isoforms. Am. J. Physiol. 279, R1707–R1715.

Gillis, T.E., Moyes, C.D., Tibbits, G.F., 2003. Sequence mutations in teleost cardiac troponin C that are permissive of high Ca2+ affinity of site II. Am. J. Physiol. Cell Physiol. 284, C1176–C1184.

Gillis, T.E., Liang, B., Chung, F., Tibbits, G.F., 2005. Increasing mammalian cardiomyocyte contractility with residues identified in trout troponin C. Physiol. Genomics 22, 1–7.

Goaillard, J.M., Vincent, P.V., Fischmeister, R., 2001. Simultaneous measurements of intracellular cAMP and L-type Ca2+ current in single frog ventricular myocytes. J. Physiol. 530, 79–91.

Gourdie, R.G., Dimmeler, S., Kohl, P., 2016. Novel therapeutic strategies targeting fibroblasts and fibrosis in heart disease. Nat. Rev. Drug Discov. 15, 620–638. advance online publication.

Graham, M.S., Farrell, A.P., 1989. The effect of temperature-acclimation and adrenaline on the performance of a perfused trout heart. Physiol. Zool. 62, 38–61.

Guderley, H., 2011. Mitochondria and temperature. In: Farrell, A. (Ed.), Encyclopedia of Fish Physiology: From Genome to Environment. vols. 1–3. Elsevier, pp. 1709–1716.

Harvey, R.D., Calaghan, S.C., 2012. Caveolae create local signalling domains through their distinct protein content, lipid profile and morphology. J. Mol. Cell. Cardiol. 52, 366–375.

Haverinen, J., Vornanen, M., 2009. Comparison of sarcoplasmic reticulum calcium content in atrial and ventricular myocytes of three fish species. Am. J. Physiol. Regul. Integr. Comp. Physiol. 297, R1180–R1187.

Haworth, T.E., Haverinen, J., Shiels, H.A., Vornanen, M., 2014. Electrical excitability of the heart in a Chondrostei fish, the Siberian sturgeon (Acipenser baerii). Am. J. Physiol. Regul. Integr. Comp. Physiol. 307 (9), R1157–R1166.

Hilton, Z., Clements, K.D., Hickey, A.J.R., 2010. Temperature sensitivity of cardiac mitochondria in intertidal and subtidal triplefin fishes. J. Comp. Physiol. B 180, 979–990.

Hove-Madsen, L., 1992. The influence of temperature on ryanodine sensitivity and the force-frequency relationship in the myocardium of rainbow trout. J. Exp. Biol. 167, 47–60.

Hove-Madsen, L., Tort, L., 1998. L-type Ca2+ current and excitation-contraction coupling in single atrial myocytes from rainbow trout. Am. J. Physiol. 275, 2061–2069.

Hove-Madsen, L., Tort, L., 2001. Characterization of the relationship between Na+-Ca2+ exchange rate and cytosolic calcium in trout cardiac myocytes. Pflugers Arch. 441, 701–708.

Hove-Madsen, L., Llach, A., Tort, L., 1999. Quantification of calcium release from the sarcoplasmic reticulum in rainbow trout atrial myocytes. Pflugers Arch. 438, 545–552.

Hove-Madsen, L., Llach, A., Tort, L., 2000. Na(+)/Ca(2+)-exchange activity regulates contraction and SR Ca(2+) content in rainbow trout atrial myocytes. Am. J. Physiol. Regul. Integr. Comp. Physiol. 279, 1856–1864.

Hove-Madsen, L., Llach, A., Tibbits, G.F., Tort, L., 2003. Triggering of sarcoplasmic reticulum Ca2+ release and contraction by reverse mode Na+/Ca2+ exchange in trout atrial myocytes. Am. J. Physiol. Regul. Integr. Comp. Physiol. 284, R1330–R1339.

Icardo, J.M., 2012. The Teleost Heart: A Morphological Approach. In: Sedmera, D., Wang. T. (Eds.), Ontogeny and Phylogeny of the Vertebrate Heart. New York, Springer, pp. 35–53.

Icardo, J.M., 2017. Heart morphology and anatomy. In: Gamperl, A.K., Gillis, T.E., Farrell, A.P., Brauner, C.J. (Eds.), Fish Physiology. In: The Cardiovascular System: Morphology, Control and Function, vol. 36A. Academic Press, San Diego, pp. 1–54.

Icardo, J.M., Colvee, E., Schorno, S., Lauriano, E.R., Fudge, D.S., Glover, C.N., Zaccone, G., 2016. Morphological analysis of the hagfish heart. I. The ventricle, the arterial connection and the ventral aorta. J. Morphol. 277, 326–340.

Iftikar, F.I., Hickey, A.J.R., 2013. Do mitochondria limit hot fish hearts? Understanding the role of mitochondrial function with heat stress in Notolabrus celidotus. PLoS One 8. e64120

Imbrogno, S., Cerra, M.C., 2017. Hormonal and autacoid control of cardiac function. In: Gamperl, A.K., Gillis, T.E., Farrell, A.P., Brauner, C.J. (Eds.), Fish Physiology. In: The Cardiovascular System: Morphology, Control and Function, vol. 36A. Academic Press, San Diego, pp. 265–315.

Imbrogno, S., Angelone, T., Adamo, C., Pulera, E., Tota, B., Cerra, M.C., 2006. Beta3-adrenoceptor in the eel (Anguilla anguilla) heart: negative inotropy and NO-cGMP-dependent mechanism. J. Exp. Biol. 209, 4966–4973.

Jensen, B., Wang, T., Christoffels, V.M., Moorman, A.F., 2013. Evolution and development of the building plan of the vertebrate heart. Biochim. Biophys. Acta 1833 (4), 783–794.

Jewett, P.H., Sommer, J.R., Johnson, E.A., 1971. Cardiac muscle. Its ultrastructure in the finch and hummingbird with special reference to the sarcoplasmic reticulum. J. Cell Biol. 49, 50–65.

Jeyakumar, L.H., Ballester, L., Cheng, D.S., McIntyre, J.O., Chang, P., Olivey, H.E., Rollins-Smith, L., Barnett, J.V., Murray, K., Xin, H.B., Fleischer, S., 2001. FKBP binding characteristics of cardiac microsomes from diverse vertebrates. Biochem. Biophys. Res. Commun. 281, 979–986.

Johnstone, S., Isakson, B., Locke, D., 2009. Biological and biophysical properties of vascular connexin channels. Int. Rev. Cell Mol. Biol. 278, 69–118.

Keen, J.E., Farrell, A.P., Tibbits, G.F., Brill, R.W., 1992. Cardiac physiology in tunas. 2. Effect of ryanodine, calcium, and adrenaline on force frequency relationships in atrial strips from skipjack tuna, Katsuwonus Pelamis. Can. J. Zool. 70, 1211–1217.

Keen, J.E., Viazon, D.M., Farrell, A.P., Tibbits, G.F., 1993. Thermal acclimation alters both adrenergic sensitivity and adrenoreceptor density in cardiac tissue of rainbow trout. J. Exp. Biol. 181, 27–47.

Keen, A.N., Fenna, A.J., McConnell, J.C., Sherratt, M.J., Gardner, P., Shiels, H.A., 2015. The dynamic nature of hypertrophic and fibrotic remodeling of the fish ventricle. Front. Physiol. 6, 427.

Kim, C.S., Davidoff, A.J., Maki, T.M., Doye, A.A., Gwathmey, J.K., 2000. Intracellular calcium and the relationship to contractility in an avian model of heart failure. J. Comp. Physiol. B 170, 295–306.

Kirkpatrick, K.P., Robertson, A.S., Klaiman, J.M., Gillis, T.E., 2011. The influence of trout cardiac troponin I and PKA phosphorylation on the $Ca2^{+}$ affinity of the cardiac troponin complex. J. Exp. Biol. 214 (12), 1981–1988.

Klaiman, J.M., Fenna, A.J., Shiels, H.A., Macri, J., Gillis, T.E., 2011. Cardiac remodeling in fish: strategies to maintain heart function during temperature change. PLoS One 6 (9):e24464.

Korajoki, H., Vornanen, M., 2009. Expression of calsequestrin in atrial and ventricular muscle of thermally acclimated rainbow trout. J. Exp. Biol. 212, 3403–3414.

Korajoki, H., Vornanen, M., 2012. Expression of SERCA and phospholamban in rainbow trout (Oncorhynchus mykiss) heart: comparison of atrial and ventricular tissue and effects of thermal acclimation. J. Exp. Biol. 215, 1162–1169.

Korajoki, H., Vornanen, M., 2014. Species- and chamber-specific responses of 12 kDa FK506-binding protein to temperature in fish heart. Fish Physiol. Biochem. 40, 539–549.

Krasnov, A., Teerijoki, H., Gorodilov, Y., Mölsä, H., 2003. Cloning of rainbow trout (*Oncorhynchus mykiss*) α-actin, myosin regulatory light chain genes and the 5′-flanking region ofα-tropomyosin. Functional assessment of promoters. J. Exp. Biol. 206, 601–608.

Kubly, K.L., Stecyk, J.A., 2015. Temperature-dependence of L-type Ca2+ current in ventricular cardiomyocytes of the Alaska blackfish (Dallia pectoralis). J. Comp. Physiol. B 185, 845–858.

Lafontant, P.J., Behzad, A.R., Brown, E., Landry, P., Hu, N., Burns, A.R., 2013. Cardiac myocyte diversity and a fibroblast network in the junctional region of the zebrafish heart revealed by transmission and serial block-face scanning electron microscopy. PLoS One 8, e72388.

Lague, S.L., Speers-Roesch, B., Richards, J.G., Farrell, A.P., 2012. Exceptional cardiac anoxia tolerance in tilapia (Oreochromis hybrid). J. Exp. Biol. 215, 1354–1365.

Landeira-Fernandez, A., Morrisette, J.M., Blank, J.M., Block, B.A., 2004. Temperature dependence of Ca2+-ATPase (SERCA2) in the ventricles of tuna and mackerel. Am. J. Physiol. 286, R398–R404.

Lefkowitz, R.J., Stadel, J.M., Caron, M.G., 1983. Adenylate cyclase-coupled beta-adrenergic receptors: structure and mechanisms of activation and desensitization. Annu. Rev. Biochem. 52, 159–186.

Llach, A., Tibbits, G.F., Sedarat, F., Tort, L., Hove-Madsen, L., 2001. Low temperature reduces Na+-Ca2+ exchange rate but not SR Ca2+ release in trout atrial myocytes. Biophys. J. 80, 585A.

Llach, A., Molina, C.E., Alvarez-Lacalle, E., Tort, L., Benítez, R., Hove-Madsen, L., 2011. Detection, properties, and frequency of local calcium release from the sarcoplasmic reticulum in teleost cardiomyocytes. PLoS One 6. e23708

Louch, W.E., Koivumaki, J.T., Tavi, P., 2015. Calcium signalling in developing cardiomyocytes: implications for model systems and disease. J. Physiol. 593, 1047–1063.

Lurman, G.J., Petersen, L.H., Gamperl, A.K., 2012. In situ cardiac performance of Atlantic cod (*Gadus morhua*) at cold temperatures: long-term acclimation, acute thermal challenge and the role of adrenaline. J. Exp. Biol. 215, 4006–4014.

Marshall, C.R., Pan, T.C., Le, H.D., Omelchenko, A., Hwang, P.P., Hryshko, L.V., Tibbits, G.F., 2005. cDNA cloning and expression of the cardiac Na+/Ca2+ exchanger from Mozambique tilapia (Oreochromis mossambicus) reveal a teleost membrane transporter with mammalian temperature dependence. J. Biol. Chem. 280, 28903–28911.

McGuigan, K., Phillips, P.C., Postlethwait, J.H., 2004. Evolution of sarcomeric myosin heavy chain genes: evidence from fish. Mol. Biol. Evol. 21, 1042–1056.

Meissner, G., 2010. Regulation of ryanodine receptor ion channels through posttranslational modifications. Curr. Top. Membr. 66, 91–113.

Mendonca, P.C., Genge, A.G., Deitch, E.J., Gamperl, A.K., 2007. Mechanisms responsible for the enhanced pumping capacity of the in situ winter flounder heart (Pseudopleuronectes americanus). Am. J. Physiol. Regul. Integr. Comp. Physiol. 293, R2112–R2119.

Midttun, B., 1983. Ultrastructure of the junctional region of the fish heart ventricle. Comp. Biochem. Physiol. A Physiol. 76, 471–474.

Milligan, C.L., 1991. Adrenergic stimulation of substrate utilization by cardiac myocytes isolated from rainbow trout. J. Exp. Biol. 159, 185–202.

Moss, R.L., Fitzsimons, D.P., 2002. Frank-Starling relationship: long on importance, short on mechanism. Circ. Res. 90, 11–13.

Negretti, N., Varro, A., Eisner, D.A., 1995. Estimate of net calcium fluxes and sarcoplasmic reticulum calcium content during systole in rat ventricular myocytes. J. Physiol. 486, 581–591.

Nemtsas, P., Wettwer, E., Christ, T., Weidinger, G., Ravens, U., 2010. Adult zebrafish heart as a model for human heart? An electrophysiological study. J. Mol. Cell. Cardiol. 48 (1), 161–171.

Nickerson, J.G., Dugan, S.G., Drouin, G., Moon, T.W., 2001. A putative 2-adrenoceptor from the rainbow trout (Oncorhynchus mykiss). Molecular characterization and pharmacology. Eur. J. Biochem. 268, 6465–6472.

Nickerson, J.G., Dugan, S.G., Drouin, G., Perry, S.F., Moon, T.W., 2003. Activity of the unique beta-adrenergic Na+/H+ exchanger in trout erythrocytes is controlled by a novel beta3-AR subtype. Am. J. Physiol. Regul. Integr. Comp. Physiol. 285, R526–R535.

Nihei, Y., Kobiyama, A., Ikeda, D., Ono, Y., Ohara, S., Cole, N.J., Johnston, I.A., Watabe, S., 2006. Molecular cloning and mRNA expression analysis of carp embryonic, slow and cardiac myosin heavy chain isoforms. J. Exp. Biol. 209, 188–198.

Nikolaev, V.O., Bünemann, M., Schmitteckert, E., Lohse, M.J., Engelhardt, S., 2006. Cyclic AMP imaging in adult cardiac myocytes reveals far-reaching β1-adrenergic but locally confined β2-adrenergic receptor–mediated signaling. Circ. Res. 99, 1084–1091.

Nilius, B., Carbone, E., 2014. Amazing T-type calcium channels: updating functional properties in health and disease. Pflügers Archiv. 466 (4), 623–626.

O'Brien, J., Valdivia, H.H., Block, B.A., 1995. Physiological differences between the alpha and beta ryanodine receptors of fish skeletal muscle. Biophys. J. 68, 471–482.

Ollivier, H., Marchant, J., Le Bayon, N., Servili, A., Claireaux, G., 2015. Calcium response of KCl-excited populations of ventricular myocytes from the European sea bass (Dicentrarchus labrax): a promising approach to integrate cell-to-cell heterogeneity in studying the cellular basis of fish cardiac performance. J. Comp. Physiol. B 185, 755–765.

Patrick, S.M., Hoskins, A.C., Kentish, J.C., White, E., Shiels, H.A., Cazorla, O., 2010. Enhanced length-dependent Ca2+ activation in fish cardiomyocytes permits a large operating range of sarcomere lengths. J. Mol. Cell. Cardiol. 48 (5), 917–924.

Petersen, L.H., Burleson, M.L., Huggett, D.B., 2015. Temperature and species-specific effects on ß3-adrenergic receptor cardiac regulation in two freshwater teleosts: channel catfish (Ictalurus punctatus) and common carp (Cyprinus carpio). Comp. Biochem. Physiol. A Mol. Integr. Physiol. 185, 132–141.

Pieperhoff, S., Bennett, W., Farrell, A.P., 2009. The intercellular organization of the two muscular systems in the adult salmonid heart, the compact and the spongy myocardium. J. Anat. 215, 536–547.

Polyakova, V., Hein, S., Kostin, S., Ziegelhoeffer, T., Schaper, J., 2004. Matrix metalloproteinases and their tissue inhibitors in pressure-overloaded human myocardium during heart failure progression. J. Am. Coll. Cardiol. 44, 1609–1618.

Richards, M.A., Clarke, J.D., Saravanan, P., Voigt, N., Dobrev, D., Eisner, D.A., Trafford, A.W., Dibb, K.M., 2011. Transverse tubules are a common feature in large mammalian atrial myocytes including human. Am. J. Physiol. Heart Circ. Physiol. 301, H1996–H2005.

Rocha, M.L., Rantin, F.T., Kalinin, A.L., 2007. Importance of the sarcoplasmic reticulum and adrenergic stimulation on the cardiac contractility of the neotropical teleost Synbranchus marmoratus under different thermal conditions. J. Comp. Physiol. B 177, 713–721.

Rodnick, K.J., Gesser, H., 2017. Cardiac energy metabolism. In: Gamperl, A.K., Gillis, T.E., Farrell, A.P., Brauner, C.J. (Eds.), Fish Physiology. In: The Cardiovascular System: Morphology, Control and Function, vol. 36A. Academic Press, San Diego, pp. 317–367.

Sanchez-Quintana, D., García-Martínez, V., Climent, V., Hurlé, J., 1996. Myocardial fiber and connective tissue architecture in the fish heart ventricle. J. Exp. Zool. 275, 112–124.

Santer, R.M., 1974. The organization of the sarcoplasmic reticulum in teleost ventricular myocardial cells. Cell Tissue Res. 151, 395–402.

Satoh, H., Delbridge, L.M., Blatter, L.A., Bers, D.M., 1996. Surface: volume relationship in cardiac myocytes studied with confocal microscopy and membrane capacitance measurements: species-dependence and developmental effects. Biophys. J. 70, 1494–1504.

Seebacher, F., Brand, M.D., Else, P.L., Guderley, H., Hulbert, A.J., Moyes, C.D., 2010. Plasticity of oxidative metabolism in variable climates: molecular mechanisms. Physiol. Biochem. Zool. 83, 721–732.

Severs, N.J., Bruce, A.F., Dupont, E., Rothery, S., 2008. Remodelling of gap junctions and connexin expression in diseased myocardium. Cardiovasc. Res. 80, 9–19.

Shaffer, J.F., Gillis, T.E., 2010. Evolution of the regulatory control of vertebrate striated muscle: the roles of troponin I and myosin binding protein-C. Physiol. Genomics 42, 406–419.

Shiels, H.A., Farrell, A.P., 1997. The effect of temperature and adrenaline on the relative importance of the sarcoplasmic reticulum in contributing Ca2+ to force development in isolated ventricular trabeculae from rainbow trout. J. Exp. Biol. 200, 1607–1621.

Shiels, H.A., Galli, G.L., 2014. The sarcoplasmic reticulum and the evolution of the vertebrate heart. Physiology 29 (6), 456–469.

Shiels, H.A., Sitsapesan, R., 2015. Is there something fishy about the regulation of the ryanodine receptor in the fish heart? Exp. Physiol. 100, 1412–1420.

Shiels, H.A., White, E., 2005. Temporal and spatial properties of cellular Ca2+ flux in trout ventricular myocytes. Am. J. Physiol. Regul. Integr. Comp. Physiol. 288, R1756–R1766.
Shiels, H.A., White, E., 2008a. The effect of mechanical stimulation on vertebrate hearts: a question of class. In: Kamkin, A., Kiseleva, I. (Eds.), Mechanosensitivity in Cells and Tissues. Mechanosensitive Ion Channels. Springer, New York, pp. 323–342.
Shiels, H.A., White, E., 2008b. The Frank-Starling mechanism in vertebrate cardiac myocytes. J. Exp. Biol. 211, 2005–2013.
Shiels, H.A., Freund, E.V., Farrell, A.P., Block, B.A., 1999. The sarcoplasmic reticulum plays a major role in isometric contraction in atrial muscle of yellowfin tuna. J. Exp. Biol. 202, 881–890.
Shiels, H.A., Vornanen, M., Farrell, A.P., 2000. Temperature-dependence of L-type Ca2+ channel current in atrial myocytes from rainbow trout. J. Exp. Biol. 203, 2771–2780.
Shiels, H.A., Vornanen, M., Farrell, A.P., 2002. Temperature dependence of cardiac sarcoplasmic reticulum function in rainbow trout myocytes. J. Exp. Biol. 205, 3631–3639.
Shiels, H.A., Vornanen, M., Farrell, A.P., 2003. Acute temperature change modulates the response of ICa to adrenergic stimulation in fish cardiomyocytes. Physiol. Biochem. Zool. 76, 816–824.
Shiels, H.A., Calaghan, S.C., White, E., 2006a. The cellular basis for enhanced volume-modulated cardiac output in fish hearts. J. Gen. Physiol. 128, 37–44.
Shiels, H.A., Paajanen, V., Vornanen, M., 2006b. Sarcolemmal ion currents and sarcoplasmic reticulum Ca2+ content in ventricular myocytes from the cold stenothermic fish, the burbot (Lota lota). J. Exp. Biol. 209, 3091–3100.
Shiels, H.A., Di Maio, A., Thompson, S., Block, B.A., 2011. Warm fish with cold hearts: thermal plasticity of excitation-contraction coupling in bluefin tuna. Proc. R. Soc. B Biol. Sci. 278, 18–27.
Shiels, H.A., Galli, G.L.J., Block, B.A., 2015. Cardiac function in an endothermic fish: cellular mechanisms for overcoming acute thermal challenges during diving. Proc. R. Soc. Lond. B Biol. Sci. 282. 20141989.
Standen, N.B., Gray, P.T., Whitaker, M.J. (Eds.), 1987. Microelectrode Techniques: The Plymouth Workshop Handbook. Company of Biologists, Cambridge.
Sultana, N., Nag, K., Hoshijima, K., Laird, D.W., Kawakami, A., Hirose, S., 2008. Zebrafish early cardiac connexin, Cx36.7/Ecx, regulates myofibril orientation and heart morphogenesis by establishing Nkx2.5 expression. Proc. Natl. Acad. Sci. U.S.A. 105, 4763–4768.
Sutko, J.L., Airey, J.A., 1996. Ryanodine receptor Ca2+ release channels: does diversity in form equal diversity in function? Physiol. Rev. 76, 1027–1071.
Tibbits, G.F., Moyes, C.D., Hove-Madsen, L., Hoar, W.S., Randall, D.J., Farrell, A.P., 1992. Excitation-contraction coupling in the teleost heart. In: Hoar, W.S., Randall, D.J., Farrell, A.P. (Eds.), The Cardiovascular System. Academic Press, San Diego, CA, pp. 267–304.
Tiitu, V., Vornanen, M., 2002. Morphology and fine structure of the heart of the burbot (Lota lota), a cold stenothermal fish. J. Fish Biol. 61, 106–121.
Trafford, A.W., Diaz, M.E., Eisner, D.A., 1999. A novel, rapid and reversible method to measure Ca buffering and time-course of total sarcoplasmic reticulum Ca content in cardiac ventricular myocytes. Pflugers Arch. 437, 501–503.
Valiunas, V., 2002. Biophysical properties of connexin-45 gap junction hemichannels studied in vertebrate cells. J. Gen. Physiol. 119, 147–164.
Valiunas, V., Weingart, R., 2000. Electrical properties of gap junction hemichannels identified in transfected HeLa cells. Pflugers Arch. 440, 366–379.
Valiunas, V., Weingart, R., Brink, P.R., 2000. Formation of heterotypic gap junction channels by connexins 40 and 43. Circ. Res. 86, E42–E49.

Venetucci, L.A., Trafford, A.W., Diaz, M.E., O'Neill, S.C., Eisner, D.A., 2006. Reducing ryanodine receptor open probability as a means to abolish spontaneous Ca2+ release and increase Ca2+ transient amplitude in adult ventricular myocytes. Circ. Res. 98, 1299–1305.

Venturi, E., Galfre, E., O'Brien, F., Pitt, S.J., Bellamy, S., Sessions, R.B., Sitsapesan, R., 2014. FKBP12.6 activates RyR1: investigating the amino acid residues critical for channel modulation. Biophys. J. 106, 824–833.

Vornanen, M., 1994. Seasonal and temperature-induced changes in myosin heavy-chain composition of crucian carp hearts. Am. J. Physiol. Regul. Integr. Comp. Physiol. 36, R1567–R1573.

Vornanen, M., 1997. Sarcolemmal Ca influx through L-type Ca channels in ventricular myocytes of a teleost fish. Am. J. Physiol. 41, R1432–R1440.

Vornanen, M., 1998. L-type Ca2+ current in fish cardiac myocytes: effects of thermal acclimation and beta-adrenergic stimulation. J. Exp. Biol. 201, 533–547.

Vornanen, M., 1999. Na+/Ca2(+) exchange current in ventricular myocytes of fish heart: contribution to sarcolemmal Ca2+ influx. J. Exp. Biol. 202, 1763–1775.

Vornanen, M., 2006. Temperature- and Ca 2+-dependence of [3H] ryanodine binding in the burbot (Lota lota L.) heart. Am. J. Physiol. Regul. Integr. Comp. Physiol. 290, R345–R351.

Vornanen, M., 2017. Electrical excitability of the fish heart and its autonomic regulation. In: Gamperl, A.K., Gillis, T.E., Farrell, A.P., Brauner, C.J. (Eds.), Fish Physiology. In: The Cardiovascular System: Morphology, Control and Function, vol. 36A. Academic Press, San Diego, pp. 99–153.

Vornanen, M., Haverinen, J., 2012. A significant role of sarcoplasmic reticulum in cardiac contraction of a basal vertebrate, the river lamprey (Lampetra fluviatilis). Acta Physiol. (Oxf.) 207, 269–279.

Vornanen, M., Haverinen, J., 2013. A significant role of sarcoplasmic reticulum in cardiac contraction of a basal vertebrate, the river lamprey (Lampetra fluviatilis). Acta Physiol. 207, 269–279.

Vornanen, M., Shiels, H.A., Farrell, A.P., 2002. Plasticity of excitation-contraction coupling in fish cardiac myocytes. Comp. Biochem. Physiol. A Mol. Integr. Physiol. 132, 827–846.

Vornanen, M., Hassinen, M., Koskinen, H., Krasnov, A., 2005a. Steady-state effects of temperature acclimation on the transcriptome of the rainbow trout heart. Am. J. Physiol. Regul. Integr. Comp. Physiol. 289 (4), R1177–R1184.

Vornanen, M., Haverinen, J., Hassinen, M., Koskinen, H., Krasnov, A., 2005b. Plasticity of cardiac function in thermally acclimated rainbow trout. Comp. Biochem. Physiol. A Mol. Integr. Physiol. 141, S355.

Wei, A.-C., Liu, T., O'Rourke, B., 2015. Alterations of mitochondrial calcium handling in heart failure. Circ. Res. 117, A389.

Ye Sheng, X., Qu, Y., Dan, P., Lin, E., Korthout, L., Bradford, A., Hove-Madsen, L., Sanatani, S., Tibbits, G.F., 2011. Isolation and characterization of atrioventricular nodal cells from neonate rabbit heart. Circ. Arrhythm. Electrophysiol. 4, 936–946.

Yue, D.T., 1987. Intracellular [Ca2+] related to rate of force development in twitch contraction of heart. Am. J. Physiol. 252, H760–H770.

3

ELECTRICAL EXCITABILITY OF THE FISH HEART AND ITS AUTONOMIC REGULATION

MATTI VORNANEN[1]

University of Eastern Finland, Joensuu, Finland
[1]Corresponding author: matti.vornanen@uef.fi

A key factor determining the pumping function of the fish heart is the regulation of electrical excitability of cardiac myocytes in the pacemaker region, the atrium and the ventricle. Cardiac excitability must be maintained at a level that ensures steady contractility of the heart under various physiological and environmental stressors, such as exercise and changes in temperature and oxygen availability. On the other hand, the heart must retain electrical stability and resist irregularities in the rhythm of the heart beat which could compromise cardiac contractility. This chapter reviews the current state of knowledge on electrical excitability of the fish heart, with a major emphasis on the recent

The Cardiovascular System: Morphology, Control and Function, Volume 36A
FISH PHYSIOLOGY

DOI: http://dx.doi.org/10.1016/bs.fp.2017.04.002

advances in our understanding of ion currents in fish cardiac myocytes, their molecular basis, and their regulation by the autonomic nervous system. As most fishes are ectotherms, the temperature dependence of cardiac excitability is of special interest. This is because it varies depending on the timescale of temperature changes: from acute, to seasonal, to the evolutionary as seen with adaptation to different thermal habitats. Further, the temperature dependence of cardiac electrophysiology has implications with regard to fishes living under current scenarios of climate change.

1. INTRODUCTION

The fish heart has four chambers: *sinus venosus*, atrium, ventricle, and *bulbus arteriosus* (*conus arteriosus* in elasmobranchs). In teleost fishes, the *sinus venosus* and *bulbus arteriosus* consist mainly of connective tissue and are non-contractile (Santer, 1985). The atrium and ventricle form the muscular pump of the fish heart and are comprised of electrically excitable and contractile cells, the cardiac myocytes (see Chapter 1, Volume 36A: Icardo, 2017 for more details on cardiac morphology). The autonomous rhythm of the heart is produced by the pacemaker cells of the nodal tissue at the border area between the *sinus venosus* and atrium. Blood returning from the body to the *sinus venosus* is propelled by the work of atrial and ventricular myocytes, through the *bulbus arteriosus* and the ventral aorta to the gills for oxygenation, and finally, to the systemic circulation.

Performance of the heart is basically the product of two parameters, heart rate (f_H) and the stroke volume (V_S). In exercising fish, changes in cardiac output are attained by increases in both V_S and f_H, while temperature-related changes in cardiac output are almost exclusively due to changes in f_H (Gollock et al., 2006; Randall, 1982; Steinhausen et al., 2008) (see Chapter 4, Volume 36A: Farrell and Smith, 2017; Chapter 4, Volume 36B: Eliason and Anttila, 2017). In most fishes f_H varies, in a temperature-dependent manner, from a few beats per minute (bpm) to a maximum heart rate of 70–120 bpm (Gollock et al., 2006; Lillywhite et al., 1999; Mendonca and Gamperl, 2010). However, in some species at high temperatures, maximum f_H can be much higher, up to 300 bpm as in zebrafish (*Danio*) species and *Bathygobius soporator* (Lin et al., 2014; Rantin et al., 1998; Sidhu et al., 2014; Vornanen and Hassinen, 2016).

Electrical excitability, and its modulation by extrinsic (environmental) and intrinsic (e.g., the autonomic nervous system) factors, is crucial for the adjustment of cardiac performance to meet physiological demands. A number of findings during the past 20 years have significantly advanced our understanding of cardiac excitation in fishes. This progress is based on the application of various research approaches at different levels of biological organization, including optical mapping of cardiac excitation with voltage-sensitive dyes (Hou et al., 2014; Lin et al., 2014), *in vivo* recording of electrocardiograms (ECGs) (Badr et al., 2016; Milan et al., 2006; Yu et al., 2012), patch-clamp measurement of ion currents (Ganim et al., 1998; Haverinen and Vornanen, 2009b; Hove-Madsen and Tort, 1998; Nemtsas et al., 2010; Paajanen and Vornanen, 2002; Vornanen, 1997), immunohistochemical detection of cardiac innervation (Newton et al., 2014; Stoyek et al., 2015), ligand binding characterization and quantification of adrenergic receptors (Gamperl et al., 1994), molecular studies of ion channel expression and function (Hassinen et al., 2011, 2015a), and genetic modification of the cardiac phenotype in model species (Arrenberg et al., 2010; Steele et al., 2009, 2011). This chapter reviews some of these achievements.

2. ELECTRICAL EXCITABILITY OF THE FISH HEART

Mechanical function of the vertebrate heart is preceded by electrical excitation of the sarcolemma (SL) cardiac myocytes [i.e., the cardiac action potential (AP)], which spreads throughout the heart and sets contraction in motion *via* increases in intracellular free Ca^{2+} concentration Ca^{2+} (Coraboeuf, 1978). The process through which electrical excitation of the SL is coupled to force production of the myofilaments *via* changes in Ca^{2+} is called excitation–contraction (E–C) coupling and involves a number of ion channels, ion pumps, and transporters and their regulation by second messenger systems (Bers, 2002; Fabiato, 1983; Tibbits et al., 1992; Vornanen et al., 2002b). The correct timing of sequential contractions of atrial and ventricular muscles is controlled by the conduction of electrical excitation through the heart (Sedmera et al., 2003).

Electrical excitability of cardiac myocytes is powered by electrochemical gradients of Na^+, K^+, and Ca^{2+} ions across the SL, created through the operation of SL Na^+, K^+ and Ca^{2+}-ATPases. The unequal distribution of cations and the voltage difference across the SL provide the driving force for ion entry or exit through ion selective channels, which are gated to open when membrane potential changes or external ligands bind to channel proteins (Bezanilla, 2005). The opening and closing of several ion channels result in the generation of chamber-specific APs which propagate from the site of origin through the heart and cause contraction of the atrial and ventricular chambers (Coraboeuf, 1978; Maltsev et al., 2006). The basic principles of electrical excitation, including the major cardiac ion currents, are the same in mammalian and fish hearts, but the molecular basis of ion currents and the temperature dependence of current densities and their kinetics can be markedly different (Vornanen, 2016; Vornanen and Hassinen, 2016).

The unequal distribution of ions across the SL determines the Nernst equilibrium potential for each ion species. For a typical teleost fish (e.g., rainbow trout, *Oncorhynchus mykiss*) the extracellular concentrations of Na^+, K^+, and Ca^{2+} ions are approximately 155, 4 and 2 mM, respectively (Houston and Koss, 1984). Assuming that intracellular concentrations are 13, 150, and 0.0001 mM for Na^+, K^+ and Ca^{2+} ions, respectively, Nernst equilibrium potentials of +55, −81, and +108 mV are obtained for Na^+, K^+ and Ca^{2+} at 14°C (Fig. 1). By opening specific channels in the cardiac SL for each ion species, inward currents are generated at the negative side of the equilibrium potential and outward currents at voltages more positive than the equilibrium potential. Considering that the normal voltage range of the cardiac AP is between −90 and +50 mV, physiologically meaningful Na^+ and Ca^{2+} currents are always inward and K^+ currents mainly outward (Fig. 1). These values apply for most freshwater and marine teleost fishes, but are not completely representative for marine species which are osmoconformers or poorly regulate the ion concentrations of their body fluids (e.g., hagfishes, sharks) (Prosser, 1973).

As ectotherms, the body temperature of fishes is directly determined by environmental temperature, with the exception of about 30 regionally

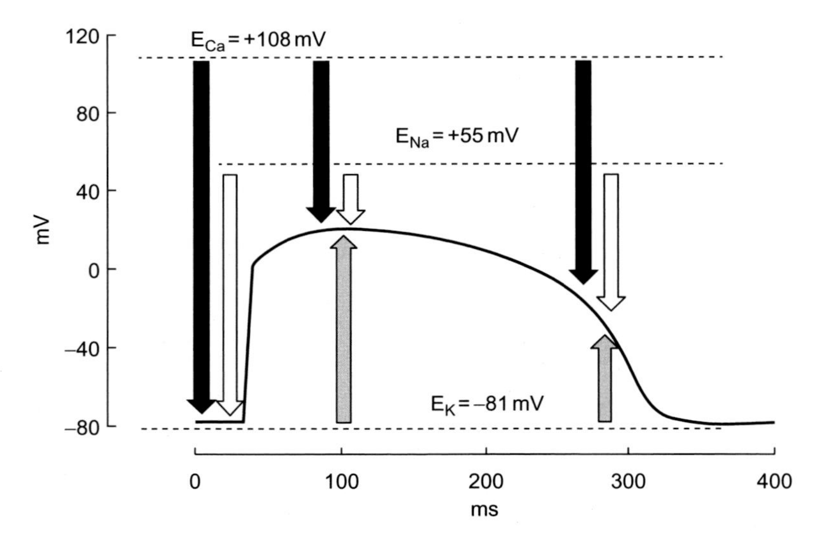

Fig. 1. Electrochemical gradients for Na^+, K^+ and Ca^{2+} ions across the cardiac sarcolemma in fishes. The *dashed lines* show Nernst equilibrium potentials for physiological concentrations of Na^+, K^+ and Ca^{2+} in extra- and intracellular fluid compartments of teleost fishes. The *arrows* indicate changes in electrochemical driving force during the cardiac action potential. *Downward arrow* = inward (depolarizing) current, *upward arrow* = outward (repolarizing) current.

endothermic species (Dickson and Graham, 2004). Increases and decreases in ambient temperature will be reflected as changes in the metabolic rate of the fish, and thus in f_H, the latter the main regulator of cardiac output under acute temperature changes (Cech et al., 1976; Gollock et al., 2006). The temperature sensitivity of f_H means that the electrical excitability of cardiac myocytes needs to be adjusted in a temperature- and rate-dependent manner. Electrical excitability, thus, needs to be sensitive to temperature changes to produce temperature-dependent acceleration and deceleration of f_H, and coordinated changes in the conduction rate of APs through the heart; i.e., the adjustments must be such that myocytes are sensitive enough to be excited, but are able to maintain electrical stability to prevent severe cardiac arrhythmias. A proper balance between electrical excitability and electrical stability, which are conflicting properties of the cardiac SL (Milstein et al., 2012; Varghese, 2016), should be achieved at all temperatures that the animal may encounter in its habitat. At the same time, the duration of the cardiac AP must adjust to the changes in f_H so that the balance between the durations of diastole and systole is maintained (Vornanen et al., 2014).

Electrical excitability of the cardiac myocyte is dependent on the function of Na^+-, K^+-, and Ca^{2+}-specific ion channels, integral membrane proteins, or protein assemblies of the cardiac SL. It has been reported that there are 232 and 311 genes in the human and puffer fish (*Fugu rubripes*) genomes, respectively, which encode for the pore-forming α-subunits of plasma membrane ion channels (Jegla et al., 2009). The number of ion channel genes expressed and the diversity of ion channel isoforms in the vertebrate heart are large when the ancillary subunits are taken into account; i.e., a large variety of channel assemblies with different biophysical properties can be produced. In fishes, the diversity is higher than in tetrapods due to the whole genome duplication in the teleost lineage, which may have allowed sub- and neofunctionalization between the gene paralogs (Hoekstra and Coyne, 2007). Each of the different functional parts of the heart (nodal, atrial, and ventricular tissues) has special electrophysiological characteristics, and therefore, different ion channel compositions.

3. CARDIAC ACTION POTENTIAL

The action potential of the vertebrate heart can be divided into five phases (Roden et al., 2002; Vornanen, 2016) (Fig. 2A). Phase 4 is the negative resting membrane potential (RMP) of atrial and ventricular myocytes in diastole. It is a stable level of −70 to −90 mV in unexcited myocytes and needs to be depolarized by a voltage wave from neighboring myocytes *via* gap junctions before myocyte contraction can occur. When RMP is depolarized to the

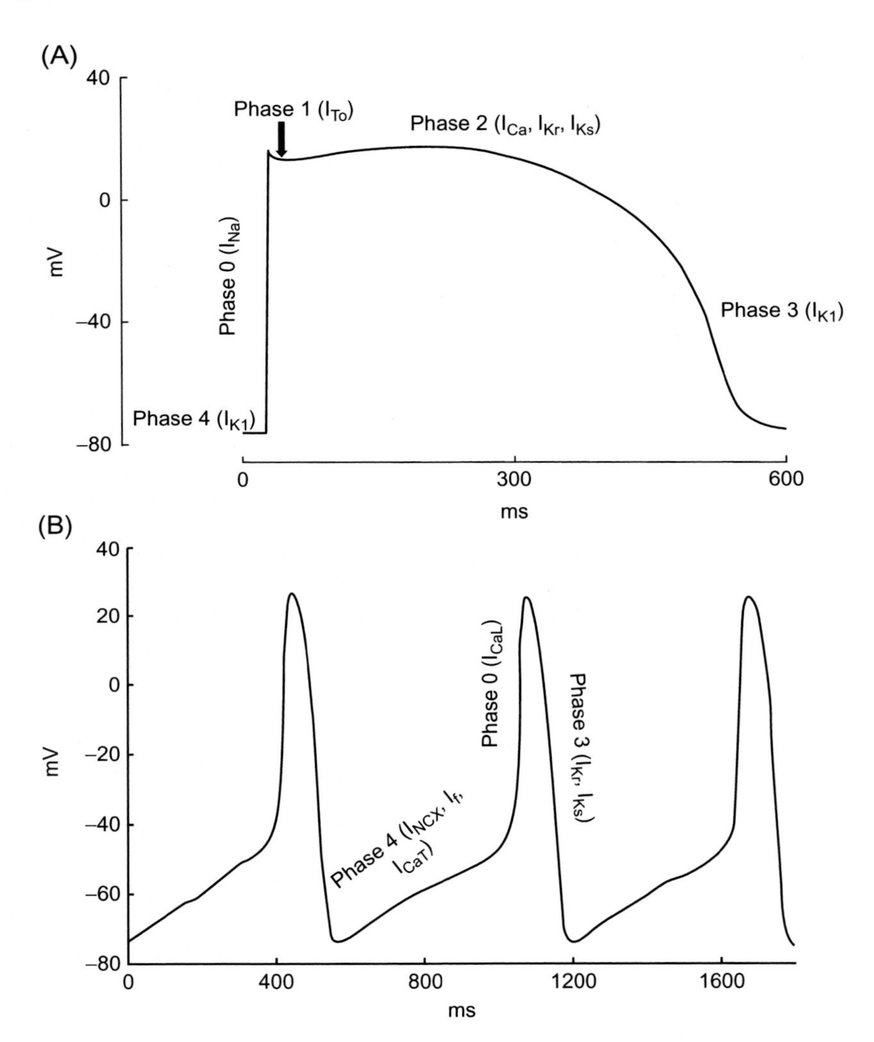

Fig. 2. (A) Five phases of the fish cardiac action potential (AP). Major ion currents active at each phase of the AP are shown in parentheses. (B) APs of an enzymatically isolated pacemaker cell of the brown trout (*Salmo trutta fario*) with five distinct phases. Main ion currents known to be present in vertebrate pacemaker cells are shown in parenthesis. I_{K1}, inward rectifier K^+ current; I_{Na}, Na^+ current; I_{To}, transient outward current; I_{CaL}, L-type Ca^{2+} current; I_{CaT}, T-type Ca^{2+} current; I_{NCX}, Na^+–Ca^{2+} exchange current; I_f, funny current; I_{Kr}, the rapid component of the delayed rectifier K^+ current; I_{Ks}, the slow component of the delayed rectifier K^+ current.

threshold level of AP firing, i.e., to the voltage where the density of inward Na^+ current exceeds the total density of outward K^+ currents, a fast upstroke or Phase 0 of the AP is generated. Within a few milliseconds, the membrane is depolarized from the threshold value (approx. −45 to −50 mV) to between +10 and +50 mV. The maximum value of the membrane potential above the zero voltage level is called the AP overshoot. The rapid Phase 1 repolarization, which is typical for many mammalian hearts, is small or completely absent in the hearts of all fish examined (Vornanen and Hassinen, 2016). This is probably the most prominent qualitative difference between fish and mammalian cardiac APs. In fish cardiac myocytes, the fast upstroke of the AP is followed by a long plateau phase, or Phase 2 of the cardiac AP, which is the basis for the prolonged cardiac contraction. Phase 2 gradually turns into Phase 3 (repolarization) and finally results in complete restoration of the negative RMP (Phase 4).

Principally, the same five phases of the cardiac AP are found in both ventricular and atrial APs of the fish heart. However, there is a striking difference between atrial and ventricular myocytes in AP duration (Haverinen and Vornanen, 2009b; Lin et al., 2014; Saito and Tenma, 1976). In comparison to the ventricular AP, the atrial AP is much shorter, which is a necessary adjustment to the brief and fast atrial contraction. In five teleost species, the ratio of atrial to ventricular durations (APD_{50}) was 0.3–0.5 at 4°C (Haverinen and Vornanen, 2009b). However, in a cyclostome species, the European river lamprey (*Lampetra fluviatilis*), the relative duration of the atrial AP is even shorter, with an atrioventricular APD_{50} ratio of only 0.22 (Haverinen et al., 2014). The short atrial AP and slow rate of AP propagation make the atrial repolarization visible as a Pt wave between the P wave (atrial depolarization) and QRS complex (ventricular depolarization) in the lamprey ECG. In the lampey ECG, atrial repolarization is not concealed under the much larger ventricular depolarization, as happens in the teleost ECG.

In contrast to the stable RMP of atrial and ventricular myocytes, the membrane potential of the pacemaker cells is continuously changing, which makes them fire spontaneous pacemaker APs (Irisawa, 1978) (Fig. 2B). Pacemaker APs are characterized by gradual and slow diastolic depolarization (Phase 4) toward the threshold voltage of the AP upstroke (phase 0) (Harper et al., 1995; Haverinen and Vornanen, 2007; Saito, 1969, 1973; Tessadori et al., 2012). The faster the rate of diastolic depolarization, the sooner the threshold for AP firing is reached, and the faster is f_H. Phases 1 and 2 are thought to be absent in the pacemaker AP, since repolarization starts immediately after the upstroke. The amplitude and duration of the pacemaker AP are smaller, and the rate of upstroke and repolarization slower, than those of atrial and ventricular APs. Transmission of pacemaker APs to the neighboring atrial myocytes depolarizes them, and triggers the wave of electrical excitation throughout the heart.

Increases in beating frequency reduce the duration and plateau height of the fish ventricular AP (Harwood et al., 2000), and similarly, acute changes in temperature cause profound changes in the shape of the cardiac AP, including those of atrial and ventricular myocytes as well as the pacemaker cells (Harper et al., 1995; Haverinen and Vornanen, 2007; Lin et al., 2014; Vornanen et al., 2014).

4. RHYTHM OF THE HEARTBEAT AND IMPULSE CONDUCTION

In the vertebrate heart, there is a small aggregate of special myocardial cells that spontaneously fire APs with a higher rate than secondary pacemakers and thereby determine the autonomous rhythm of the heart (Irisawa, 1978). The pacemaker of fish hearts was anatomically recognized over 100 years ago (Keith and Mackenzie, 1910), and functional separation of different cardiac compartments with ligatures has confirmed that the sinoatrial border zone is the primary pacemaker of the fish heart (von Skramlik, 1935). Since then, the nodal tissue has been described in several fish species as a ring-like structure at the border zone between the *sinus venosus* and atrium (Haverinen and Vornanen, 2007; Laurent, 1962; Saito, 1969; Santer and Cobb, 1972; Yamauchi et al., 1973; Yousaf et al., 2012). In the zebrafish (*Danio rerio*) heart, the primary pacemaker tissue can be identified by mRNA expression of the islet-1 transcription factor, and in zebrafish and goldfish (*Carassius auratus*) by an antibody against the pacemaker ion channel HCN4 (Newton et al., 2014; Stoyek et al., 2015; Tessadori et al., 2012). Sinoatrial pacemaker cells of the rainbow trout also contain natriuretic peptides with unknown function (Yousaf et al., 2012). The fish cardiac pacemaker tissue is densely innervated (Haverinen and Vornanen, 2007; Keith and Mackenzie, 1910; Yamauchi and Burnstock, 1968) and includes both cholinergic and adrenergic nerve endings (Newton et al., 2014; Zaccone et al., 2010) (see Chapter 4, Volume 36A: Farrell and Smith, 2017).

The first pacemaker APs in the fish heart were recorded by Jensen from the hearts of three hagfish species, and by Saito from the hearts of several teleost species (Jensen, 1965; Saito, 1969, 1973). Three morphologically distinct types of pacemaker cells have been isolated from the rainbow trout primary pacemaker tissue, each capable of generating pacemaker APs (Haverinen and Vornanen, 2007).

The impulse generated by the pacemaker must be conducted through the heart at spatially variable rates to prevent regurgitation of the blood on the way to the ventral aorta (Fig. 3). Although no morphologically recognizable conducting tissue has been found in fish hearts, functional studies suggest the

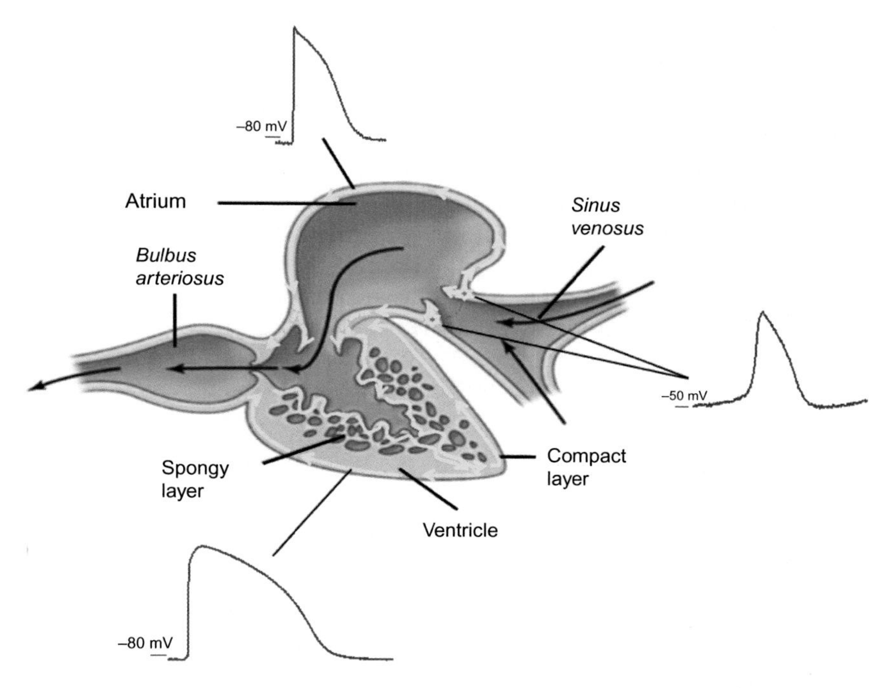

Fig. 3. Origin and propagation of electrical excitation in the fish heart. The site of origin is indicated by a *yellow star*, whereas the conduction of electrical excitation is indicated by the *yellow line*. Pacemaker cells basal to the sinoatrial valve produce pacemaker action potentials that proceed with variable rate through the atrium and ventricle. The *continuous line* marks fast impulse propagation in the atrial and ventricular walls, and the *dashed line* indicates deceleration of action potential spread at sinoatrial and atrioventricular junctions. Note that the ventricular excitation in this scheme is from apex to base, even though the excitation pattern may vary between fish species (i.e., occurring from base to apex or showing a spiral pattern).

presence of a conducting pathway that accelerates and decelerates the rate of impulse spread in an analogous manner to the conducting system of the mammalian heart (Chi et al., 2008; Sedmera et al., 2003). The AP proceeds slowly from the sinoatrial pacemaker to the atrium, and then quickly in the atrial wall to the atrioventricular canal, where the rate of impulse propagation slows down again to allow sufficient time for ventricular filling. The slow rate of impulse propagation in the atrioventricular canal is suggested to result from the circular arrangement of cardiomyocytes, but the slow rate of AP upstroke (Saito and Tenma, 1976) strongly suggests that SL ion channels in this area are of a unique composition. From the atrioventricular canal to the apex of the ventricle, the velocity of impulse conduction is again fast along the endocardial trabeculae of the heart (Chi et al., 2008; Icardo and Colvee, 2011; Sedmera et al., 2003). From the apex of the ventricle, the impulse quickly propagates epicardially toward the base of the ventricle, generating a forward movement of blood, i.e., ejection (Chi et al., 2008; Poon and Brand, 2013; Randall, 1968; Sedmera et al., 2003). It should be noted, however, that some studies suggest an opposite direction for ventricular activation, i.e., from the base of the ventricle caudally to the apex (Jensen et al., 2012; Noseda et al., 1963; Vaykshnorayte et al., 2011). In fact, ventricular activation may occur in a spiral manner from the base to the apex of the ventricle. It remains to be shown to what extent the opposite results are related to variations in methodology (e.g., voltage-sensitive dyes *vs* calcium indicators) or differences in the shape and morphology of the heart, the latter possibly depending on the relative proportion of spongy and compact muscle layers (Santer, 1985).

5. ION CURRENTS OF THE FISH HEART

5.1. Inward Currents

There are three physiologically important inward current systems in the vertebrate heart, the Na^+ current (I_{Na}), the Ca^{2+} current (I_{Ca}) and the hyperpolarization-activated "funny" current (I_f) have a variety of roles in pacemaker cells, conducting pathways and working cardiac myocytes of the atrium and ventricle.

5.1.1. Sodium Current (I_{Na})

The first current to be activated in the excitation of atrial and ventricular myocytes of the vertebrate heart is the inward Na^+ current (I_{Na}), which is due to the opening of voltage-gated Na^+ channels (Fozzard and Hanck, 1996). The threshold voltage of the AP, where Na^+ channels of the fish ventricular myocytes start progressive and explosive opening, is −46 to −49 mV in the cold-acclimated (4°C) rainbow trout (Haverinen and Vornanen, 2006). The rate of Na^+ channel opening is very fast, and once the threshold voltage is achieved all available Na^+ channels of the myocyte will open within a few ms. This produces the fast AP upstroke and prominent overshoot of the cardiac AP (Fig. 4). Most of the Na^+ channels will also spontaneously close by the mechanism of fast inactivation, although a small population of Na^+ channels may remain open and maintain a tiny sustained I_{Na} (Liu et al., 1992).

The rate of AP upstroke (depolarization) and the propagation velocity of the AP along the SL are determined by the density of I_{Na}. Therefore, factors that affect the density and kinetics of I_{Na} will also have an impact on the rate of impulse transmission through the heart, and this may have consequences on

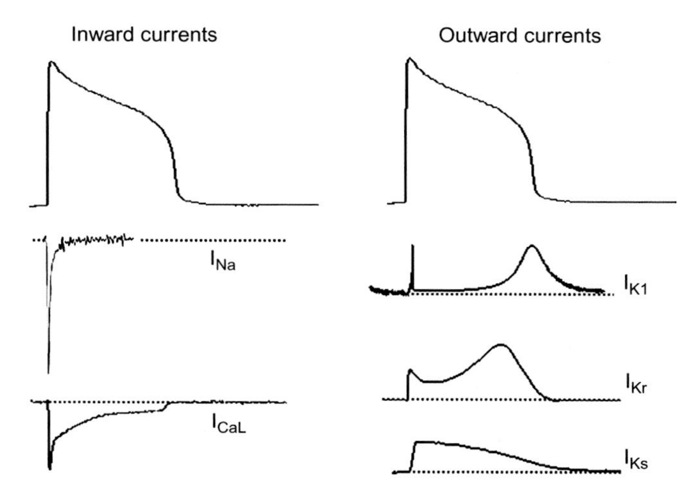

Fig. 4. Major inward and outward currents of fish atrial and ventricular myocytes. The figure depicts the time course of major inward and outward currents during a ventricular action potential. Inward currents: Na^+ current (I_{Na}) and L-type Ca^{2+} current (I_{CaL}). Outward currents: the inward rectifier K^+ current (I_{K1}), the rapid component of the delayed rectifier K^+ current (I_{Kr}), and the slow component of the delayed rectifier K^+ current (I_{Ks}).

the rate of myocardial contraction, the frequency of the heartbeat, and the electrical stability of the heart. However, not all Na^+ channels of the myocyte are available for opening, since during repetitive firing of APs, Na^+ channels get inactivated *via* the slow inactivation or do not open (closed-state inactivation). The available pool of functional Na^+ channels is dependent on RMP: the more negative the RMP the larger the pool of available Na^+ channels (Maltsev and Undrovinas, 1998). Therefore, K^+ currents that contribute to RMP will have an indirect effect on I_{Na} by regulating the number of available Na^+ channels. Depolarization of RMP due to the failure of K^+ channels can reduce the rate of AP propagation.

Acute temperature changes affect the density and kinetics of I_{Na} in fish hearts. With the lowering of temperature, the rates of opening and closing of Na^+ channels become slower and the density of I_{Na} decreases, as the probability of Na^+ channels being in an open state is reduced. I_{Na} density reduces with a temperature coefficient (Q_{10}) of 2.0–2.5 (Haverinen and Vornanen, 2004), and consequently the rate of AP upstroke and impulse conduction will change roughly in a similar manner to the beating rate of the heart (Q_{10} about 2). In this regard, it is notable that in the cold-acclimated (4°C) rainbow trout the density of ventricular I_{Na} is about 60% higher than in the warm-acclimated (18°C) trout (Haverinen and Vornanen, 2004). This is probably necessary to maintain a sufficient rate of AP propagation in the face of the compensatory increase in f_H, which occurs in rainbow trout in the chronic cold (Aho and Vornanen, 2001). In another teleost species, the crucian carp (*Carassius carassius* L.), the response of cardiac I_{Na} to temperature acclimation is opposite, i.e., a reduction of I_{Na} density (30%) in the cold (2°C) (Haverinen and Vornanen, 2004). This interspecific difference in response to thermal acclimation is probably related to the different activity patterns and lifestyles of these species. In contrast to the cold-active rainbow trout, which produce compensatory changes in f_H under chronic cold, crucian carp become inactive at near freezing temperatures. Crucian carp do not show a compensatory increase in f_H in the cold, but develop a sustained bradycardia in anoxic waters in the winter (Matikainen and Vornanen, 1992; Tikkanen et al., 2017). Therefore, the cold-induced depression of I_{Na} (Haverinen and Vornanen, 2006) is a physiologically meaningful response for this species. The molecular basis of these temperature-related variations in I_{Na} density is not known, but is likely to involve changes in the number of Na^+ channels, and possibly, shifts in the Na^+ channel isoform composition.

The density of I_{Na} does not differ between atrial and ventricular myocytes of rainbow trout and zebrafish hearts. However, the voltage dependences of steady-state inactivation and activation are slightly more negative for trout atrial I_{Na} than ventricular I_{Na} (Haverinen and Vornanen, 2006). Similarly, the atrial I_{Na} of the zebrafish shows a more negative voltage dependence of inactivation in comparison to ventricular I_{Na} (Warren et al., 2001). The relatively negative voltage dependence of I_{Na} activation, together with the small size of the inward rectifier K^+ current (I_{K1}) (see Section 5.2), is expected to make atrial myocytes easily excitable by the depolarization wave from the pacemaker cells.

Voltage-gated Na^+ channels are large multimers that consist of the pore-forming α-subunits and auxiliary β-subunits (Catterall and Waxman, 2005). In mammals, nine genes encode for the α-subunits of the voltage-gated Na^+ channels, while fishes have eight Na^+ channel genes. The evolutionary origin of these genes is somewhat different in fishes and tetrapod vertebrates (Widmark et al., 2011). The chordate predecessor of vertebrates had only one α-subunit, which after two rounds of whole genome duplication (1R, 2R) gave rise to four Na^+ channel α-subunits in the early vertebrate ancestor. In teleost fishes, a third round of whole genome duplication (3R) over 300 million years ago gave rise to the eight Na^+ channels genes, and all of them have persisted in fish genomes (Widmark et al., 2011). There seems to be substantial variation in cardiac Na^+ channel isoform composition between vertebrates and among fish species. In mammalian hearts, I_{Na} is mainly produced by $Na_v1.5$ α-subunits, while in rainbow trout the main cardiac isoform is $Na_v1.4$ and only small amounts of $Na_v1.5$ and $Na_v1.6$ are expressed. In crucian carp and zebrafish, both cyprinid species, $Na_v1.5$ is the main cardiac isoform, although they express some $Na_v1.4$ channels (Hassinen, personal communication; Vornanen et al., 2011). Irrespective of the Na^+ channel isoform, the fish cardiac I_{Na} is always tetrodotoxin sensitive, in contrast to the mammalian cardiac I_{Na} which is about 1000 times less sensitive to this marine toxin (Vornanen et al., 2011). The molecular basis of this difference is the substitution of a nonaromatic amino acid (serine, glycine) in position 401 with an aromatic amino acid (tyrosine, aspartate), but the significance of this striking difference in drug sensitivity for cardiac function, if any, remains unknown. Further, diversity of Na^+ channels is produced by β-subunits which are encoded by five genes in the zebrafish (Chopra et al., 2007).

5.1.2. Calcium Currents (I_{Ca})

Ca^{2+} currents provide for the voltage-dependent entry of extracellular Ca^{2+} into the cardiac myocyte, which is crucial for diastolic depolarization and the upstroke of the pacemaker AP and for the maintenance of the long AP plateau and activation of atrial and ventricular myocytes (McDonald et al., 1994). Two distinct families of Ca^{2+} channels are generally expressed in vertebrate cardiac myocytes: (1) T-type channels that are activated at more negative voltages and produce transient Ca^{2+} currents (I_{CaT}) (Bean, 1985; Nilius et al., 1985) and (2) L-type channels that generate long-lasting currents (I_{CaL}) at more depolarized voltages (Isenberg and Klöckner, 1980; Trautwein et al., 1975). T-type Ca^{2+} currents are blocked by low concentrations (20–300 μM) of Ni^{2+} and L-type Ca^{2+} currents can be inhibited by dihydropyridines (e.g., nifedipine) and phenylalkylamines (e.g., verapamil) (McDonald et al., 1994), even though none of these substances is completely specific for any type of Ca^{2+} channel.

5.1.2.1. L-type Ca^{2+} Current (I_{CaL}). The L-type Ca^{2+} current (I_{CaL}) is the major Ca^{2+} current of the fish heart and has been characterized in cardiac myocytes of several fish species. It is activated at about −40 mV, reaches peak magnitude in the range of AP plateau voltage (0 to +10 mV) and inactivates more slowly than I_{Na} (Brette et al., 2008; Coyne et al., 2000; Hove-Madsen and Tort, 1998; Nemtsas et al., 2010; Shiels et al., 2000; Vornanen, 1997, 1998). Therefore, it provides an important inward current for the maintenance of a long AP duration (Fig. 4). In fish hearts, Ca^{2+} influx through Ca^{2+} channels makes a significant direct contribution to the activation of contraction and may trigger Ca^{2+} release from the sarcoplasmic reticulum (SR), which further augments the force of contraction (Tibbits et al., 1991; Vornanen et al., 2002b).

The density of I_{CaL} in fish cardiac myocytes ranges between 2 and 10 pA/pF at room temperature (~20°C). The basal ventricular I_{Ca} appears to be bigger in the hearts of cyprinid species, crucian carp, and zebrafish, than, for example, in rainbow trout or burbot (*Lota lota*) hearts (Shiels et al., 2006; Vornanen, 1997, 1998; Zhang et al., 2011). However, in the presence of maximal β-adrenergic activation (10 μM isoprenaline) the density of I_{Ca} is similar in rainbow trout and crucian carp ventricular myocytes (Vornanen, 1998). This is explainable if the L-type Ca^{2+} channels of crucian carp myocytes are almost maximally phosphorylated under basal conditions, leaving little scope for upregulation by the β-adrenergic stimulation. Consistent with this assumption, the basal I_{Ca} of crucian carp myocytes is suppressed by carbacholine, a cholinergic agonist that antagonizes the β-adrenergic cascade (Vornanen et al., 2010). Although β-adrenergic effects on cardiac I_{CaL} have been studied in fish, the role of different β-adrenergic receptors and their signaling pathways in I_{CaL} regulation are poorly elucidated. Molecular studies have shown that β_1-, β_2-, and β_3-adrenergic receptors are expressed in teleost hearts, β_1- and β_2-receptors being the dominant isoforms and β_3-receptor the minor isoform (Giltrow et al., 2011; Nickerson et al., 2003; Steele et al., 2011; Wang et al., 2009). β_1-Adrenergic receptors increase cardiac contractility, while β_2- and β_3-receptors can either stimulate or inhibit cardiac function in fishes (Gamperl et al., 1994; Nickerson et al., 2002; Petersen et al., 2013, 2015; Steele et al., 2011). Studies on frogs indicate that both β_1- and β_2-receptors are capable of stimulating cAMP production and increasing I_{CaL} in cardiac myocytes, albeit in a spatially different manner; cAMP production by β_1-stimulation activates protein kinase A (PKA) globally throughout the cell, whereas the β_2-receptors activate PKA locally in the vicinity of L-type Ca^{2+} channels (Harvey and Hell, 2013). Whether similar differences exist in β-adrenergic regulation of fish cardiac myocytes remains to be shown. Considering the significant modification of β-adrenergic stimulation of the fish heart under acute and chronic temperature changes (Keen et al., 1993; Shiels et al., 2003), temperature-dependent regulation of fish cardiac I_{CaL} by various β-adrenergic receptor pathways needs to be studied more carefully.

Effects of acute temperature changes on the peak density of fish cardiac I_{CaL} are slightly weaker in comparison to the mammalian cardiac I_{CaL} (Kim et al., 2000; Shiels et al., 2000). Q_{10} values of the trout cardiac I_{CaL} are 1.8–2.2 between 7°C and 21°C, while temperature dependence of the mammalian cardiac I_{CaL} varies between 2.6 and 3.2. The modest temperature sensitivity of the fish cardiac I_{Ca} could moderate temperature-dependent changes in I_{CaL} when fish swim at different depths or cross the thermocline. Furthermore, temperature affects the rate of I_{Ca} inactivation, and thereby counteracts temperature-dependent changes in Ca^{2+} influx due to changes in the peak I_{Ca}

density. Furthermore, it has been shown that a physiological level of β-adrenergic tone (5 nM isoprenaline) attenuates temperature-dependent changes in I_{Ca} density in rainbow trout atrial myocytes (Shiels et al., 2003). Collectively, these findings suggest that Ca^{2+} influx through L-type Ca^{2+} channels is not highly sensitive to acute temperature changes. It should be noted, however, that in ventricular myocytes of the air-breathing Alaska black fish (*Dallia pectoralis*), the Q_{10} value of the peak I_{Ca} is as high as 8 between 5°C and 15°C (Kubly and Stecyk, 2015).

It is notable that the acclimation of fish (rainbow trout, crucian carp) to different temperatures does not induce any changes in ventricular I_{CaL} (Vornanen, 1998); i.e., I_{CaL} does not show similar thermal plasticity which is typical for Na^+ and K^+ currents. Therefore, any changes in the density or kinetics of I_{Ca} that might be associated with acute temperature changes are not compensated for by thermal acclimation in these species. However, in seasonally acclimatized crucian carp, the density of L-type Ca^{2+} channels and I_{CaL} are transiently elevated in early summer (May–July), while remaining constant in other months of the year (Vornanen and Paajanen, 2004). This ion channel remodeling may be associated with energy allocation to growth and reproduction, which occur in crucian carp from late May to mid-July (Holopainen et al., 1997).

In mammals, there are four α1-subunits of the L-type Ca^{2+} channel (α1S, α1C, α1D, α1F or $Ca_V1.1$–4), each of which forms functional channels as a multimeric protein complex that includes auxiliary β, α2δ, and γ subunits (Benitah et al., 2010). The α1-subunit incorporates the ion-conducting pore and defines the regulation of different Ca^{2+} channels by second messenger systems, drugs, and toxins (Catterall et al., 2005). The predominant cardiac isoform in ventricular and atrial myocytes of mammals is α1C ($Ca_V1.2$), while sinoatrial and atrioventricular nodal cells also express α1D ($Ca_V1.3$) channels. Due to their 10–15 mV more negative activation threshold, α1D channels are important for cardiac pacemaking and conduction (Striessnig et al., 2014). Until recently, the molecular basis of the fish cardiac I_{CaL} had not been examined. However, a lethal mutation of the $Ca_V1.2$ gene in the zebrafish abolishes cardiac I_{CaL}, suggesting that homologous genes may be involved in producing L-type Ca^{2+} channels in fish and mammalian ventricular myocytes (Rottbauer et al., 2001). It should be noted, however, that the diversity of Ca^{2+} channel α-subunits is probably much higher in fishes than mammals due to the whole genome duplication in teleosts (Hurley et al., 2007). For example, in Fugu (*F. rubripes*), 21 Ca^{2+} channel α-subunit isoforms have been found and as many as 16 of these appear to be expressed in the heart (Wong et al., 2006). In the crucian carp ventricle $Ca_V1.2$ and $Ca_V1.3$ L-type Ca^{2+} channel α-subunits are expressed (Tikkanen et al., 2017).

5.1.2.2. T-type Ca^{2+} Current (I_{CaT}). T-type Ca^{2+} current (I_{CaT}) has a peak current amplitude of around −30 mV, is inactivated by depolarized holding potentials (−40 mV), and is kinetically faster than I_{CaL} (Bean, 1985; Nilius et al., 1985). I_{CaT} is fairly resistant to nifedipine but can be blocked by Ni^{2+} (Lee et al., 1999). In fetal atria and ventricles of mammals, I_{CaT} is abundantly expressed, but almost disappears in these tissues during maturation to adulthood (Alvarez and Vassort, 1992; Maylie and Morad, 1995). In adult animals I_{CaT} is a functionally important current component in pacemaker tissue and cardiac conduction pathways of mammals, birds, and frogs (Mangoni et al., 2006). Two α-subunits of T-type Ca^{2+} channel proteins, α1G ($Ca_V3.1$) and α1H ($Ca_V3.2$), are functionally expressed in mammalian hearts (Vassort et al., 2006). These channels differ in regard to Ni^{2+} sensitivity: half-maximal inhibition occurs at concentrations of 167 and 5.7 μM for α1G and α1H, respectively (Lee et al., 1999).

In fishes, I_{CaT} has been recorded from embryonic (Rottbauer et al., 2001; Warren et al., 2001) and adult cardiac myocytes of the zebrafish (Nemtsas et al., 2010). Both atrial and ventricular myocytes of the adult zebrafish have a large I_{CaT}, although not quite the size of I_{CaL}. In zebrafish cardiac myocytes, I_{CaT} is half-maximally inhibited by 124 μM Ni^{2+}, suggesting that the current is produced by a homolog to the mammalian α1G channel (Nemtsas et al., 2010). Notably, in the Siberian sturgeon (*Acipenser baerii*), a Chondrostean species, I_{CaT} is the main type of atrial Ca^{2+} current. In this species the atrial I_{CaT} is almost 2.5 times as large as the atrial I_{CaL}. In ventricular myocytes of *A. baerii* I_{CaT} is only about 9% of the peak I_{CaL} (Haworth et al., 2014). It is not clear why cardiac I_{CaT} is so dominating in atrial myocytes and almost absent in the ventricle. Careful comparison of E–C coupling between sturgeon atrial and ventricular myocytes could shed light on the physiological significance of I_{CaT} in extranodal tissues. Densities of I_{CaT} and I_{CaL} are similar in ventricular myocytes of the shark (dogfish, *Squalus acanthias*) (Maylie and Morad, 1995). The significance of T-type Ca^{2+} channels in temperature-related modulation of fish cardiac pacemaking and impulse conduction deserves further study, and the molecular basis of fish cardiac I_{CaT} needs to be determined.

5.1.2.3. The Hyperpolarization-Activated Funny Current (I_f). Unlike atrial and ventricular myocytes, pacemaker cells of the vertebrate heart do not have a stable RMP, but show a diastolic depolarization that results in a spontaneous AP discharge (Huang, 1973; Saito, 1969). It should be noted that ionic and molecular mechanisms of fish cardiac pacemaking are still poorly known. Therefore, the following discussion is largely based on our knowledge about mammalian and frog pacemakers (Irisawa, 1978; Mangoni and Nargeot, 2008), with some relevant findings from fish studies. Two "clocks" have been recognized in the pacemaker cells of the vertebrate heart: "a membrane clock" produced by various ion channels and transporters of the SL and "a Ca^{2+} clock" generated by cyclic uptake and release of Ca^{2+} by the SR (DiFrancesco and Noble, 2012; Yaniv et al., 2011). The membrane clock involves a small sodium current (I_{Na}), various K^+ currents (I_{Kr}, I_{Ks}, I_{to}), two Ca^{2+} currents (I_{CaT}, I_{CaL}), Na^+–Ca^{2+} exchange current (I_{NCX}), Na^+ pump current (I_{NaK}), and funny current (I_f). The Ca^{2+} clock is based on spontaneous local Ca^{2+} releases from the SR *via* the ryanodine receptors (RyR), which generate a small inward current *via* the reverse mode of Na^+–Ca^{2+} exchange (I_{NCX}) and drive the membrane clock to the threshold of I_{Ca}-dependent APs. It should be noted that the membrane clock can work on its own, while the Ca^{2+} clock requires the operation of SL ion channels and is actually "a coupled clock" (Yaniv et al., 2015). Although both clocks appear to contribute to the discharge of pacemaker APs, controversy exists about their relative importance in the cardiac rhythm. Proponents of the membrane clock hypothesis regard the funny current, I_f, as the key operator in providing inward current for diastolic depolarization and mediating the effects of the autonomic nervous system on f_H. Proponents of the Ca^{2+} clock hypothesis think that local spontaneous releases of Ca^{2+} from the SR are necessary to push membrane potential to the threshold voltage of I_{Ca}, and to explain the effects of autonomic nervous system control on cardiac pacemaking.

I_f current is a time-dependent and non-specific Na^+/K^+ conductance activated upon hyperpolarization of the SL (DiFrancesco, 1993). Although the I_f channels are permeable to both Na^+ and K^+ ions, the physiological I_f (at voltages from positive to −70 mV) is an inward Na^+ current (Fig. 1). Although I_f is fairly specific for the nodal cells, it is sometimes present in atrial myocytes and even in ventricular myocytes of some mammalian species. I_f has been recorded from embryonic cardiac myocytes and atrial and ventricular myocytes of the adult zebrafish heart (Baker et al., 1997; Warren et al., 2001), but thus far, there are no I_f recordings from the primary pacemaker cells of the fish heart. The zebrafish I_f is activated at voltages more negative than −40 mV and is completely blocked by 2 mM Cs^+ (Baker et al., 1997). Homozygous mutants for the *slow mo* gene (encoding for a mitochondrial protein) cause slowing of the basal f_H in zebrafish and a reduction in the density of atrial and ventricular I_f (Warren et al., 2001). This suggests that I_f might be a significant component of the zebrafish cardiac pacemaker.

I_f of the mammalian sinoatrial node is produced by voltage-gated cyclic nucleotide dependent (HCN1–4) channels in tetrameric assemblies. The predominant isoform of the mammalian sinoatrial node is HCN4, although HCN1 and HCN2 are expressed in smaller amounts (Satoh, 2003). Little is known about the HCN channels of the fish heart. HCN4 channels have been localized in putative pacemaker tissues at the sinoatrial and atrioventricular border of zebrafish and goldfish (Newton et al., 2014; Stoyek et al., 2015;

Tessadori et al., 2012), and Wilson et al. found transcript expression for six HCN channel isoforms in the hagfish (*Eptatretus stoutii*) atrium and ventricle (Wilson et al., 2013). In the hagfish heart HCN3c was the most abundant isoform with minor contributions by HCN4 and HCN2.

5.2. Outward Potassium Currents

Physiological K^+ currents are outward currents, and therefore, repolarizing (Fig. 4). They maintain negative RMP, regulate AP duration, and repolarize membrane potential back to the resting level in atrial and ventricular myocytes. K^+ currents play a major role in producing different AP morphologies in atrial and ventricular myocytes and regulate excitability and refractoriness of cardiac contractility. K^+ currents are also expressed in the pacemaker and other nodal tissues. K^+-selective channels are the largest and most diverse group of ion channels (Gutman et al., 2003), which in different molecular assemblies enable great flexibility in cardiac AP phenotype under different physiological and environmental demands. In fishes, K^+ currents are strongly responsive to chronic temperature changes and, therefore, constitute a central mechanism in acclimation of cardiac excitability to seasonal temperature changes (Hassinen et al., 2008a,b, 2014; Vornanen et al., 2002a). There appears to also be phylogenetic differences in the expression of cardiac K^+ currents, possibly reflecting adaptation to different environments, and these have potential implications for the ability of the fish heart to accommodate to environmental changes (Haverinen and Vornanen, 2009b).

5.2.1. Voltage-Gated K^+ Currents (I_K)

A variety of voltage-gated and nonvoltage-gated inward rectifier K^+ channels are involved in the regulation of phase 1, 2, and 3 of the cardiac AP (Nerbonne and Kass, 2005). The long duration of the cardiac AP is produced by a balance between inward I_{Ca} and outward K^+ currents. The voltage-gated K^+ currents of the vertebrate heart are grouped into transient outward currents (I_{to}) and delayed rectifier currents (I_K). In mammalian hearts two distinct transient outward currents, referred to as I_{to1} and I_{to2}, are expressed and produce the fast phase 1 repolarization of the cardiac AP. I_{to1} is carried by K^+ ions and can be blocked by 4-amino pyridine (4-AP), whereas I_{to2} is a chloride current and insensitive to 4-AP (Nerbonne and Kass, 2005). In fish cardiac myocytes, the phase 1 repolarization is small or totally absent, and thus far, no I_{to} has been recorded in cardiac myocytes from any fish species (Alday et al., 2014; Nemtsas et al., 2010; Vornanen and Hassinen, 2016). Indeed, it has been suggested that I_{to} is a mammalian "invention." The cardiac delayed rectifier K^+ current has two major components, the rapid component (I_{Kr}) and the slow component (I_{Ks}), characterized by different rates of current activation.

5.2.1.1. The Rapid Component of the Delayed Rectifier (I_{Kr}). The rapid component of the delayed rectifier K^+ current (I_{Kr}) of the fish heart was first found in embryonic myocytes of the zebrafish (Warren et al., 2001). Electrophysiological properties of the fish I_{Kr} were subsequently characterized from atrial and ventricular myocytes of the adult rainbow trout (Hassinen et al., 2008a; Vornanen et al., 2002a). The presence of cardiac I_{Kr} has been documented in all fish species where it has been sought (Hassinen et al., 2008a; Haverinen and Vornanen, 2009b). This includes teleost species where it has been found in atrial, ventricular and pacemaker cells (Galli et al., 2009; Haverinen and Vornanen, 2007), as well as in cardiac myocytes of the Siperian sturgeon (*Acipenser baerii*), a Chondrostei fish, and the European river lamprey (*Lampetra fluviatilis*), a cyclostome species (Haverinen et al., 2014; Haworth et al., 2014). The density of I_{Kr} channels is generally higher in atrial (2–7 pA/pF, at 11°C) than ventricular (1–5 pA/pF) myocytes of the fish heart (Galli et al., 2009; Haverinen and Vornanen, 2009b; Vornanen et al., 2002a) (Fig. 5). The high density and the ubiquitous presence of I_{Kr} channels in fish hearts suggest that they have a particularly significant role in the regulation of AP duration and refractoriness of the fish heart, and are possibly involved in the pacemaker mechanism (Haverinen and Vornanen, 2007; Langheinrich et al., 2003; Stengl et al., 2003).

The central role of I_{Kr} in repolarization of fish cardiac AP is reflected in thermal responses of the I_{Kr} (Hassinen et al., 2008a; Vornanen et al., 2002a). In a study of six teleost species, I_{Kr} was upregulated by cold acclimation in all species except in the pike (*Esox lucius*) (Fig. 5) (Haverinen and Vornanen, 2009b). This response is not limited to fresh water teleosts, but has also been found in the partially endothermic bluefin tuna (*Thunnus thynnus*) (Galli et al., 2009) and in the White Sea navaga cod (*Eleginus navaga*) (Abramochkin and Vornanen, 2015; Hassinen et al., 2014). The increased density of the I_{Kr} channels in cold-acclimated fishes counteracts the cold-related increase in AP duration, and thereby, makes room for compensatory increases in f_H. Indeed, the density of I_{Kr} channels and f_H are closely correlated in cold-acclimated and warm-acclimated fishes (Vornanen, 2016). The cold-induced compensatory decrease in AP duration allows more rapid restitution of excitability and contractility and, therefore, improves force production at short diastolic intervals (Haverinen and Vornanen, 2009b).

Electrophysiological findings suggest that the cold-induced increase in the density of the I_{Kr} channels is achieved by doubling the number of I_{Kr} (Erg) channels in the myocyte SL without changes in activation and inactivation kinetics of the I_{Kr} (Hassinen et al., 2008a). *In situ* hybridization and quantitative PCR indicated only moderate increases in the expression of channel transcripts and cannot alone explain the doubling of the I_{Kr} current density in the trout heart. Therefore, temperature-related

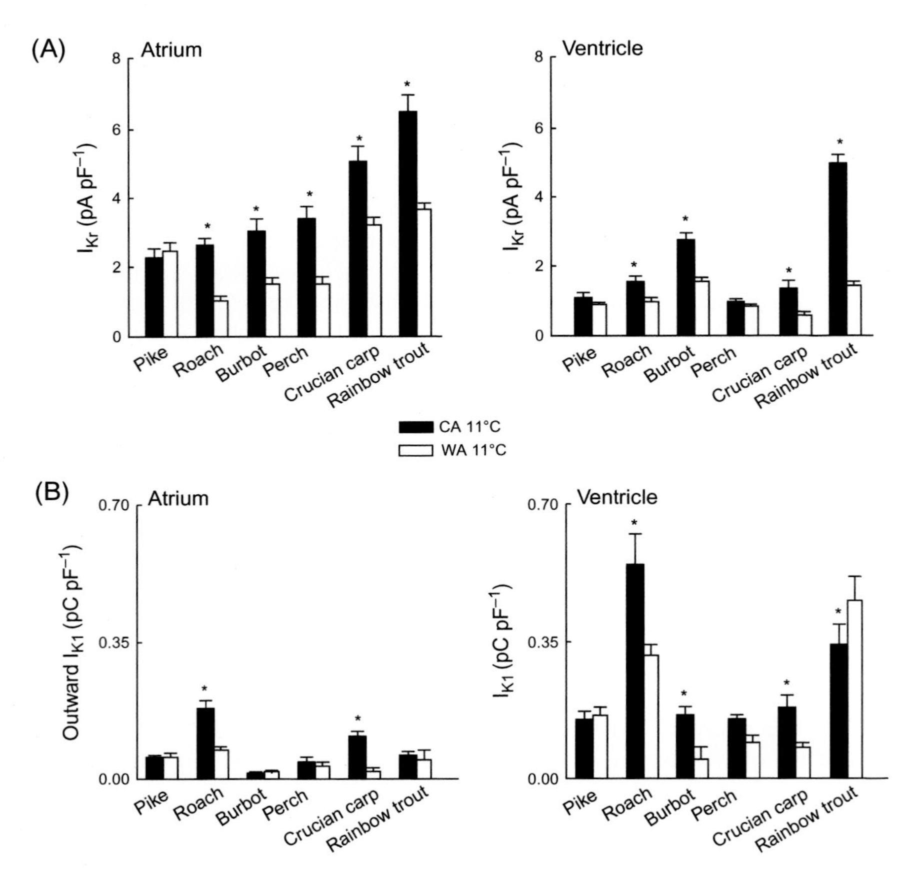

Fig. 5. The rapid component of the delayed rectifier current (I_{Kr}) and the background inward rectifier current (I_{K1}) of fish atrial and ventricular myocytes. (A) Current density of the I_{Kr} in atrium and ventricle of six teleost species. (B) Charge transfer by the outward I_{K1} in the atrium and ventricle of six teleost species. *Asterisk* (*) indicates statistically significant differences between cold-acclimated (4°C) and warm-acclimated (18°C) fishes at the experimental temperature of 11°C. Pike (*Esox Lucius*), roach (*Rutilus rutilus*), burbot (*Lota lota*), perch (*Perca fluviatilis*), crucian carp (*Carassius carassius*), and rainbow trout (*Oncorhynchus mykiss*). The results are from Haverinen, J., Vornanen, M., 2009. Responses of action potential and K^+ currents to temperature acclimation in fish hearts: phylogeny or thermal preferences? Physiol. Biochem. Zool. 82, 468–482.

expression of the I_{Kr} channel is probably modulated by other mechanisms in addition to increased transcript levels. These could include translation efficiency and temperature dependence of trafficking of the channels to the membrane.

I_{Kr} is generated by voltage-gated K^+ channels, encoded by three ERG-subfamily (ether-à-go-go-related gene) genes (ERG1–3 or KCNH2, KCNH6, KCNH7), which encode the respective α-subunits of the channel (Kv11.1–Kv11.3) (Warmke and Ganetzky, 1994). Functional ERG channels of the mammalian heart are homotetramers of the ERG1 (KCNH2) α-subunit which may assemble with accessory subunits MinK or MiRP1. In the zebrafish heart four ERG gene products (KCNH2A, KCNH2B, KCNH6, KCNH7) are expressed (Vornanen and Hassinen, 2016). Interestingly, the main isoform of the zebrafish heart is orthologous to the brain isoform of the mammalian channel (ERG2 or KCNH6). The zebrafish ortholog to the mammalian cardiac ERG1 is only weakly expressed in the zebrafish heart (Leong et al., 2010; Vornanen and Hassinen, 2016). The cardiac ERG channel has also been cloned from the rainbow trout heart, and this isoform is also orthologous to the mammalian ERG2 (KCNH6), although it was initially believed to be an ortholog of the mammalian ERG1 (Hassinen et al., 2008a). Similarly, the ERG channels found in crucian carp and navaga sturgeon hearts are produced by the ERG2 gene (Hassinen et al., 2014, 2015b). Collectively, these findings indicate that the cardiac I_{Kr} is produced by non-orthologous genes in fish and mammals.

The I_{Kr} appears to be a significant part of the pacemaker mechanism of the fish heart (Haverinen and Vornanen, 2007; Langheinrich et al., 2003; Stengl et al., 2003). In zebrafish embryos, 22 of the 23 known I_{Kr} blockers caused bradycardia, and knockdown of the ERG channels by antisense morpholinos dose-dependently reduced *in vivo* f_H (Langheinrich et al., 2003; Stengl et al., 2003). In rainbow trout, 50% inhibition of the cardiac I_{Kr} with 0.1 μM E-4031, a specific blocker of the ERG channels, depressed (at 11°C) intrinsic f_H *in vitro* by 47% and 15% in warm-acclimated and cold-acclimated trout, respectively (Haverinen and Vornanen, 2007). The smaller inhibition of f_H in the cold-acclimated trout is explained by the higher initial density of I_{Kr} channels in this acclimation group; after the half-maximal block the size of I_{Kr} is still similar to that in the warm-acclimated fish before the block. These findings suggest that the cold-induced increase in the density of the I_{Kr} channel contributes to the higher f_H of the cold-acclimated fish, and that f_H changes due to acute temperature increases are largely dependent on the I_{Kr}. In contrast, the cold-induced acceleration of f_H is largely due to the shorter duration of pacemaker AP in the cold-acclimated fish, i.e., earlier start of diastolic depolarization (Harper et al., 1995; Haverinen and Vornanen, 2007).

There are obvious differences in the contribution of I_{Kr} to cardiac excitability between mammalian and fish hearts. In mammalian hearts, loss-of-function mutations of the ERG channel or drug-related reductions in I_{Kr} channel density cause severe cardiac arrhythmias in the form of long QT syndrome. *In vivo* these are associated with a prolonged Q-T interval of the ECG, and at the single cell level these are due to prolonged duration of the cardiac AP. Partial reductions of the mammalian cardiac I_{Kr} can cause life-threatening ventricular tachyarrhythmia known as *torsades de pointes*. Although the cardiac AP duration and Q-T interval are also prolonged in fish hearts by reductions in I_{Kr} channel density, chaotic ventricular tachyarrhythmia has not been seen in zebrafish knocked downs for the cardiac ERG channel or under drug-induced block of the I_{Kr} (Arnaout et al., 2007; Langheinrich et al., 2003) (see Section 7.2).

5.2.1.2. The Slow Component of the Delayed Rectifier (I_{Ks}). The slow component of the delayed rectifier K^+ current (I_{Ks}) functionally differs from I_{Kr} in that it activates more slowly than I_{Kr} and does not completely deactivate during membrane depolarization. By virtue of these biophysical properties, I_{Ks} current increases throughout the plateau phase of the cardiac AP until the decreasing electrochemical driving force abolishes I_{Ks} during phase 3 of the AP (Tristani-Firouzi et al., 2001). I_{Ks} plays an important role in controlling the repolarizing phase of the cardiac AP in various mammalian species. In particular, I_{Ks} functions as a repolarization reserve that prevents excessive prolongation of cardiac AP, when β-adrenergic tone is high or when other K^+ currents (I_{Kr}, I_{K1}) fail, i.e., I_{Ks} is often activated in situations where additional repolarizing current is needed (Roden and Yang, 2005). Owing to the slow deactivation of I_{Ks}, the current increases (accumulates) at high pacing frequencies, and this is important in shortening AP duration at high f_Hs. Recruitment of I_{Ks} in various physiological and pathological states is a significant mechanism protecting against cardiac arrhythmias (Guo et al., 2012; Jost et al., 2005; Marx et al., 2002).

Thus far, I_{Ks} has been documented in only one teleost species, the crucian carp (Hassinen et al., 2011). In the crucian carp heart, the density of I_{Ks} current is considerably larger in atrial than ventricular myocytes. I_{Ks} has not been found in cardiac myocytes of the zebrafish (a cyprinid species like crucian carp) and a specific blocker of the I_{Ks}, chromanol 239B, did not have any effect on ventricular AP duration in the zebrafish (Alday et al., 2014; Nemtsas et al., 2010). However, mRNA of the gene (KCNQ1) encoding α-subunits of the I_{Ks} channel is expressed in the zebrafish heart (Alday et al., 2014; Wu et al., 2014). An alternative explanation for the apparent absence of I_{Ks} in zebrafish cardiac myocytes is that I_{Ks} is concealed by the presence of the large I_{Kr}, and appears only when I_{Kr} is blocked. Further studies are necessary to resolve the role of I_{Ks} in the heart of zebrafish and other fish species.

I_{Ks} channels are heteromultimeric proteins consisting of four identical $K_v7.1$ α-subunits (KCNQ1 gene) assembled with auxiliary MinK (KCNE1 gene) β-subunits. Though $K_v7.1$ alone can generate a K^+ current, association with MinK is required to recapitulate the biophysical properties of the endogenous mammalian cardiac I_{Ks} (Sanguinetti et al., 1996). When $K_v7.1$ is coexpressed with the ancillary MinK subunit, the behavior of the channel dramatically changes. MinK also affects drug binding, which is expressed as a higher sensitivity of the $K_v7.1$/MinK channels to chromanol 293B compared with the homotetrameric $K_v7.1$ channels. MinK is also necessary for β-adrenergic enhancement of the I_{Ks} and frequency-dependent modulation of the I_{Ks}. Finally, the $K_v7.1$/MinK channels and the native cardiac I_{Ks} channels display extremely slow activation kinetics and strong temperature dependence in comparison to the homotetrameric $K_v7.1$ channels. Biophysical properties of the crucian carp I_{Ks} are markedly different from those of the mammalian cardiac I_{Ks}, and probably represent evolutionary adaptation for function at the low body temperature of fishes. The I_{Ks} of crucian carp atrial myocytes is characterized by (i) a complete insensitivity to β-adrenergic stimulation, (ii) low sensitivity to chromanol 293B, (iii) frequency insensitivity, (iv) fast activation kinetics, and (v) fairly low temperature dependence (Hassinen et al., 2011). When the electrophysiological properties of the endogenous I_{Ks} of crucian carp atrial myocytes were compared with the currents produced by homotetrameric $K_v7.1$ channels and heteromeric Kv7.1/MinK channels in Chinese hamster ovary cells, it appeared that the endogenous I_{Ks} is similar to the current produced by homotetrameric Kv7.1 channels, and clearly different from Kv7.1/MinK coassemblies (Hassinen et al., 2011). These findings suggest that instead of the Kv7.1/MinK composition of mammalian hearts, the fish cardiac channels are mainly composed of $K_v7.1$ homotetramers. Consistent with the biophysical properties, transcript expression of KCNE1 in crucian carp atrial myocytes was less than 2% of the KCNQ1 expression, while a 50% expression level would be expected on the basis of 2:4 MinK/$K_v7.1$ stoichiometry.

One interesting question is why are homomeric Kv7.1 channels found in fish hearts instead of Kv7.1/MinK heteromers? One obvious reason is the rate of current activation. Heteromeric assemblies of Kv7.1 and MinK subunits are extremely slow to activate ($t_{0.5}=20.59$ s at 18°C) in comparison to homomeric $K_v7.1$ channels (0.67 s) and endogenous I_{Ks} of the atrial myocytes (0.92 s) (Hassinen et al., 2011) (Fig. 6). At the generally low body temperature of fishes, heteromeric channels would be far too slow to make any contribution to repolarization of the cardiac AP. The homomeric Kv7.1 assembly provides a simple molecular solution for thermal adaptation of ion channel function, since omission of the MinK subunit from the channel assembly makes the activation kinetics of the I_{Ks} much faster. In addition to fast activation kinetics, the homomeric assembly of the channel has other electrophysiological consequences. For example, the I_{Ks} of crucian carp atrial myocytes is insensitive to activation by an increase in pacing rate and enhancement of the cAMP-dependent signaling pathway (Hassinen et al., 2011). These properties of

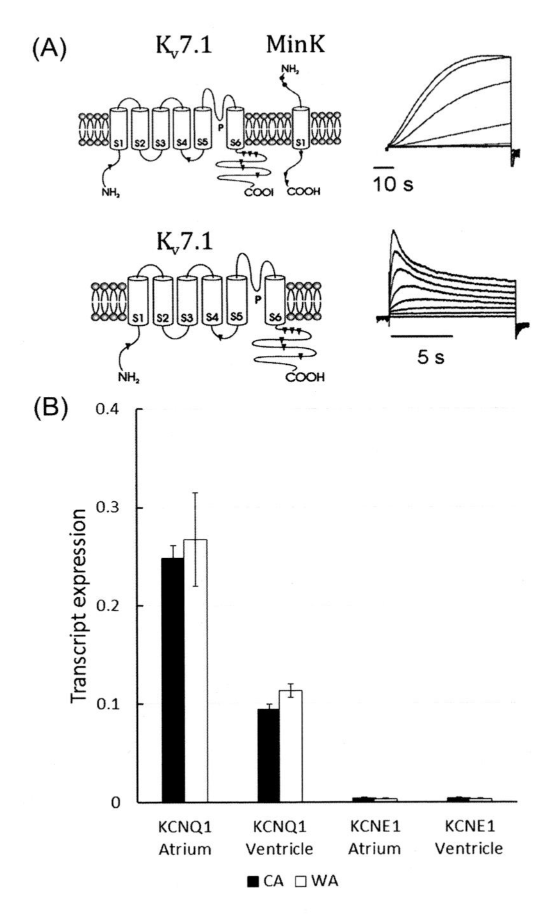

Fig. 6. Molecular composition of the slow component of the delayed rectifier K^+ current (I_{Ks}) has a strong effect on the activation kinetics of the I_{Ks}. Cloned crucian carp K_v7.1 (ccKCNQ1 gene product) channels expressed alone or in association with the accessory β-subunit MinK (KCNE1 gene product) in HEK cells. (A) The homotetrameric Kv7.1 channels show much faster activation in comparison to channels composed of both K_v7.1 and MinK subunits (note the difference in timescales). (B) The slow component of the I_{Ks} of the crucian carp heart is mainly produced by homotetramers of K_V7.1 proteins without the accessory β-subunit MinK, which is weakly expressed in the crucian carp heart. Thermal acclimation has no effect on ccCNQ1 and ccKCNE1 transcript expression. CA, cold acclimation (4°C); WA, warm-acclimation (18°C). The results are adapted from Hassinen, M., Laulaja, S., Paajanen, V., Haverinen, J., Vornanen, M., 2011. Thermal adaptation of the crucian carp (*Carassius carassius*) cardiac delayed rectifier current, I_{Ks}, by homomeric assembly of Kv7.1 subunits without MinK. Am. J. Physiol. 301, R255–R2665.

the I_{Ks} are fully consistent with the blunted frequency response of the crucian carp heart to β-adrenergic activation (Vornanen et al., 2010).

Acute cooling decreases the density of I_{Ks} channels, but there is no compensatory increase of I_{Ks} following prolonged cold exposure that would counteract the acute temperature effect. Neither transcript expression of KCNQ1 or KCNE1 genes (Fig. 6B) nor current density of the cardiac I_{Ks} displays any differences between cold-acclimated and warm-acclimated crucian carp (Hassinen et al., 2011). Therefore, the importance of the I_{Ks} in AP regulation is associated with acute temperature changes to accelerate/decelerate repolarization of the cardiac AP when temperature and f_H rise/fall. The presence of I_{Ks} could be particularly important for eurythermic fishes like crucian carp, which need a large repolarization reserve to regulate AP duration in widely varying thermal habitats. Analogous to the antiarrhythmic role of the mammalian I_{Ks} (Tristani-Firouzi et al., 2001), recruitment of the repolarization reserve of the I_{Ks} under acute temperature changes might be a significant mechanism protecting against cardiac arrhythmias in eurythermic fishes. Under basal conditions, contribution of the I_{Ks} to AP duration could be rather small (Hassinen et al., 2011), and only when the repolarization power of the I_{Kr} has been used does I_{Ks} come into play. Therefore, it would be interesting to know how widely cardiac I_{Ks} current is expressed among fish species, and whether it is more common in eurythermic than stenothermic species.

5.2.2. Inward Rectifier K^+ Currents (I_{Kir})

Inward rectifier K^+ currents of the vertebrate heart are important in maintaining negative RMP (Phase 4) and accelerating the rate of final Phase 3 repolarization (Hibino et al., 2010). Three separate entities of the inward rectifier current have been resolved in vertebrate cardiac myocytes: (1) the background inward rectifier current (I_{K1}); (2) the acetylcholine-activated inward rectifier current (I_{KACH}); and (3) the ATP-sensitive inward rectifier current (I_{KATP}). These currents are produced by pore-forming α-subunits of Kir2, Kir3, and Kir6 subfamilies, respectively (Ehrlich, 2008). They all have the same basic structure of two transmembrane-spanning segments separated by the pore loop, the latter forming conducting pores as homo- or heterotetramers (Fig. 7). All these current systems have been also found in fish hearts.

5.2.2.1. The Background Inward Rectifier Current (I_{K1}). The background inward rectifier K^+ (Kir) channels allow small K^+ efflux at membrane voltages positive to the equilibrium potential of K^+ ions, thus generating the physiologically important outward I_{K1}. At RMP, the inward rectifier channels are constitutively open, and depolarization of the membrane during AP increases the driving force for K^+. K^+ efflux is, however, impeded by voltage-dependent

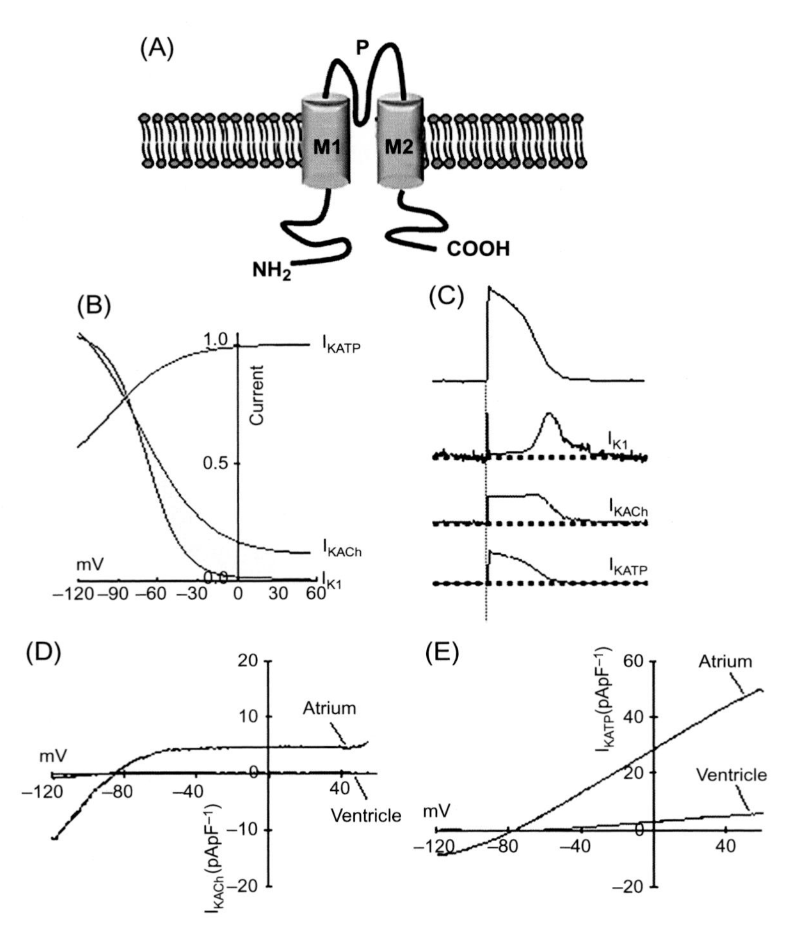

Fig. 7. Ligand-gated K^+ currents of the fish heart. (A) Schematic structure of the pore-forming Kir2, Kir3, and Kir6 inward rectifier channels. Kir channels are tetramers of proteins having two transmembrane domains (M1, M2) and the pore loop (P). (B) Inward rectification of the acetylcholine-activated inward rectifier (I_{KACH}) and of the ATP-sensitive inward rectifier (I_{KATP}) in comparison to the background inward rectifier current (I_{K1}). At the plateau voltage of the cardiac action potential (0–40 mV) I_{K1} is completely abolished, while I_{KACH} is still partly activated. I_{KATP} is fully activated at the AP plateau and, therefore, strongly repolarizing. (C) Relative conductance changes of I_{KACH} and I_{KATP} in comparison to the background I_{K1} in fish atrial myocytes. (D) Voltage dependence of I_{KACH}. I_{KACH} is present in atrial, but not in ventricular, myocytes of the sturgeon (*Acipenser baerii*) heart. (E) I_{KATP} is much larger in atrial than ventricular myocytes of the sturgeon heart. The results are from Haworth, T.E., Haverinen, J., Shiels, H.A., Vornanen, M., 2014. Electrical excitability of the heart in a Chondrostei fish, the Siberian sturgeon (*Acipenser baerii*). Am. J. Physiol. 307, R1157–R1166.

block of the channels by intracellular Mg^{2+} ions and polyamines, especially spermine (Hibino et al., 2010). The voltage-dependent block of the channels results in the typical current–voltage curve of the I_{K1}, which is characterized by strong deviation from the linear current–voltage relationship to an inward direction at depolarizing voltages. Physiologically, this means that at the AP plateau the outward I_{K1} is small, but increases when membrane potential starts to repolarize (due to decreasing I_{Ca} and increasing I_{Kr}). Due to its inward rectifying properties, I_{K1} helps to maintain the AP plateau and to increase the rate of the final Phase 3 repolarization. At the beginning of the AP, when the channels are not yet blocked by Mg^{2+} and polyamines, the conductance of I_{K1} is still high and generates a transient surge of outward I_{K1} which antagonizes the depolarizing I_{Na}. In cardiac myocytes K^+ currents are almost always outwardly directed. However, if some mechanism tends to hyperpolarize membrane potential, e.g., increased activity of the electrogenic Na^+-pump, the inward I_{K1} forces it to the equilibrium potential of K^+ ions. Therefore, I_{K1} has a stabilizing effect on membrane potential.

The cardiac myocytes of all the fish species studied have a well-defined I_{K1}, and the density of I_{K1} channels is generally much larger in ventricular than atrial myocytes (Haverinen and Vornanen, 2009b; Vornanen et al., 2002a) (Fig. 5). There are also substantial interspecies differences in the density of cardiac I_{K1} channels. The inward I_{K1} is particularly large in cyprinid species (roach and crucian carp) in comparison to rainbow trout, burbot, or perch (*Perca fluviatilis*) cardiac myocytes. In contrast, the outward I_{K1} is large in ventricular myocytes of rainbow trout and roach (Haverinen and Vornanen, 2009b). These species-specific differences in the density of I_{K1} channels probably stem from differences in inward rectifier channel composition between species and tissues.

The density of cardiac I_{K1} channels is modified by chronic temperature changes in some fish species (Galli et al., 2009; Haverinen and Vornanen, 2009b; Vornanen et al., 2002a). Comparison of cardiac I_{K1} in six teleost species showed that after cold acclimation the density of the I_{K1} channels increased in ventricular myocytes of roach, crucian carp, burbot, and perch, but not in those of the pike (*Esox lucius*) (Fig. 5). In atrial myocytes, a cold-induced increase was evident only in cyprinid species, the roach and crucian carp. These findings indicate that the density of I_{K1} channels is often increased with cold acclimation/acclimatization, in particular in the ventricular myocytes. The compensatory increase in repolarizing I_{K1} helps limit the duration of AP in the cold. The increase in the density of I_{K1} channels also has a stabilizing effect on membrane potential, which may decrease excitability in the cold, if it is not associated with increases in I_{Na}. Among the studied species, a clear exception is the rainbow trout, since acclimation to cold (4°C) depresses ventricular I_{K1} (Hassinen et al., 2007; Vornanen et al., 2002b).

This kind of response is expected to increase excitability in the cold. It remains to be shown whether this response occurs more generally in salmonid fishes and other cold-active teleosts, and if this is an adaptive response to cold or a limitation in the absence of relevant molecular mechanisms.

The I_{K1} current is produced by the inward rectifier K^+ channels of the subfamily-2, i.e., Kir2 channels. In mammalian cardiac myocytes three major gene products are expressed: Kir2.1, Kir2.2, and Kir2.3, with the Kir2.1 isoform being predominant (Hibino et al., 2010). In cyprinid species (zebrafish, crucian carp, and roach) Kir2.4 is the main cardiac isoform (Hassinen et al., 2015a), while in the rainbow trout heart Kir2.1 appears to be the prevailing channel (Hassinen et al., 2007). Kir2.2a and Kir2.2b are also expressed in fish hearts (Hassinen et al., 2007, 2008b, 2014). In the crucian carp heart, expression of Kir2.2a and Kir2.2b (initially named Kir2.5) channels is strongly temperature dependent. Fishes maintained in the cold (2°C) express more Kir2.2b channels, while fishes reared in warm (18°C) have more Kir2.2a channels. Such temperature-dependent expression suggests that these gene paralogs (Leong et al., 2014) may be under the regulation of separate promoters, i.e., after the duplication event they may have been afforded slightly different functions. Biophysically, Kir2.2a and Kir2.2b channels of the crucian carp are similar, but not identical. Physiologically, the most important difference exists in inward rectification, with Kir2.2b channels being stronger rectifiers, and therefore passing less outward current.

5.2.2.2. Acetylcholine-Activated Inward Rectifier Current (I_{KACH}). Under parasympathetic tone, acetylcholine (ACh) activates atrial I_{KACH} *via* muscarinic cholinergic receptors. Coupling of the type-2 muscarinic cholinergic receptors to the ligand-gated Kir channels occurs *via* the βγ-subunit of the pertussis toxin-sensitive G-proteins (Hibino et al., 2010). In addition to ACH, I_{KACH} can be triggered by adenosine *via* adenosine receptors, which are coupled to the inwardly rectifying channels *via* the same G-proteins (Belardinelli and Isenberg, 1983). I_{KACH} rectifies weakly in comparison to I_{K1}, and therefore, I_{KACH} has a strong effect on the duration of atrial AP, when parasympathetic tone is increased (Fig. 7). Among teleost fishes, I_{KACH} has been recorded from atrial myocytes of rainbow trout and crucian carp, but is absent in ventricular myocytes of these species (Abramochkin et al., 2014; Molina et al., 2007; Vornanen and Tuomennoro, 1999; Vornanen et al., 2010). Atrial myocytes of some fish species, e.g., rainbow trout, perch, and burbot, have a very small I_{K1} (Haverinen and Vornanen, 2009b; Vornanen et al., 2002a), which may not be sufficient to maintain a negative RMP. Considering that fish hearts usually have a resting cholinergic tone (Campbell et al., 2004; Holmgren, 1977), it is possible that I_{KACH} contributes to the negative RMP in fish atrial myocytes (Molina et al., 2007). Whether a constitutively active

I_{KACH}, i.e., current in the absence of agonist (Yeh et al., 2007), exists in the fish heart remains to be shown.

I_{KACH} significantly shortens the atrial AP with a consequent reduction in atrial force generation. In atrial muscle of the common carp (*Cyprinus carpio*), ACh can make the atrial tissue completely inexcitable and stop it in diastole (Abramochkin et al., 2008; Saito and Tenma, 1976). Interestingly, the effect of ACH on atrial excitability and contractility occurs at lower concentrations than its effect on HR (Lin et al., 1995; Vornanen et al., 2010), suggesting that the reduction in atrial contraction precedes bradycardia under a weak cholinergic tone.

In contrast to the other cardiac K^+ currents (I_{Kr} and I_{K1}), which are upregulated or do not change after exposure to chronic cold, I_{KACH} appears to be downregulated by cold acclimation. Adenosine (100 μM) stimulates atrial I_{KACH} in warm-acclimated (18°C) rainbow trout, but not in cold-acclimated (4°C) trout, suggesting that I_{KACH} signaling is weakened by chronic cold (Aho and Vornanen, 2002). In the winter-acclimatized (2°C) navaga, the atrial I_{KACH} is 26 times smaller than in the summer-acclimatized (12°C) fish (Abramochkin and Vornanen, 2016). These changes might be part of the mechanistic basis for the low cholinergic tone in some cold-acclimated fish hearts (Seibert, 1979; Sureau et al., 1989). However, in another teleost species, the crucian carp, seasonal temperature acclimation had only a minor effect on the I_{KACH} density of atrial myocytes (Vornanen et al., 2010).

In mammalian hearts I_{KACH} is produced by Kir3.1 and Kir3.4 channels in homo- and heterotetrameric compositions (Dobrzynski et al., 2001). Thus far, there are no studies on the molecular basis of the fish cardiac I_{KACH}.

5.2.2.3. ATP-Sensitive Potassium Current (I_{KATP}). Under normal non-stressful conditions K_{ATP} channels are closed and do not contribute to cardiac excitability. However, when intracellular ATP concentration suddenly drops and ADP concentrations rise, these nucleotide-gated K^+ channels quickly open. A repolarizing I_{KATP} shortens cardiac AP and consequently reduces SL Ca^{2+} influx and cardiac contractility. By this means, I_{KATP} provides short-term cardiac protection under ischemic insult and during hypoxia and anoxia (Grover and Garlid, 2000; Noma, 1983). K_{ATP} channels rectify very weakly, and therefore, pass a large outward current at the AP plateau (Fig. 7). Indeed, I_{KATP} is the strongest repolarizing current among cardiac inward rectifiers and is very effective in shortening the duration of the cardiac AP. The opening of as few as 1% of K_{ATP} channels can produce significant shortening of the cardiac AP. In mammals, I_{KATP} generally increases stress tolerance by providing a protective feedback mechanism in the cardiovascular system (Zingman et al., 2002). For example, knockout of K_{ATP} channels attenuates the exercise tolerance of mice due to diminished cardiac output with adrenergic stimulation.

Oxygen content in the water is much lower than in air. Because it is fairly common that oxygen availability becomes a limiting factor for animal performance in aquatic habitats, it might be assumed that I_{KATP} would play a more prominent role in hypoxic cardioprotection in fish than in air-breathing vertebrates. Surprisingly, the whole-cell density of I_{KATP} is much lower in rainbow trout and crucian carp cardiac myocytes than in cardiac myocytes of tetrapod vertebrates (Paajanen and Vornanen, 2002). Furthermore, I_{KATP} is much more difficult to induce in fish cardiac myocytes than in myocytes of tetrapod ectotherms. In ventricular myocytes of the warm-acclimated (18°C) rainbow trout, a hypoxia-sensitive species, I_{KATP} could not be induced at all, not even when the myocytes were perfused intracellularly with ATP-free solution and simultaneously exposed to inhibitors of both aerobic (0.1 mM $Na_2S_2O_3$) and anaerobic (5 mM jodoacetate) metabolism (Paajanen and Vornanen, 2002). Further, experiments on crucian carp showed that cardiac K_{ATP} channels are not activated under severe oxygen shortage or anoxia. The current–voltage relationship of total ventricular I_{Kir} was practically the same in carp exposed to severe hypoxia (O_2 <0.4 mg/L at 4°C) for 1–4 weeks and fishes maintained under normoxia; there was no increase in the outward K_{ir} current at 0 mV as should be the case if I_{KATP} was sustainably induced (Paajanen and Vornanen, 2003). The strongest evidence against activation of I_{KATP} in the anoxic crucian carp heart comes from long-term (up to 57 days at 2°C) *in vivo* recordings of ECGs, which showed a strong prolongation of the QT interval (i.e., the average duration of the ventricular AP) (Tikkanen et al., 2017). Experiments on crucian carp and rainbow trout hearts suggest that activation of I_{KATP} is not involved in anoxia protection in these fish species. In contrast, in goldfish maintained under moderately hypoxic conditions (2.6 mg/L O_2 for 7 days at 21°C), the activity of single K_{ATP} channels was increased and this resulted in a 16%–18% reduction in ventricular AP duration (Cameron et al., 2013). Furthermore, AP duration in normoxia-acclimated goldfish decreased by 15% when the hearts were exposed to moderate hypoxia *in vitro* (Chen et al., 2005). Nitric oxide was suggested as a physiological mediator for activation of I_{KATP} in the hypoxic goldfish (Cameron et al., 2003; Chen et al., 2005). Whether this difference in induction of I_{KATP} between crucian carp and goldfish is explained by the different acclimation/experimental temperatures (cold *vs* warm), the extent of oxygen limitation (anoxia *vs* moderate hypoxia), or represents a real species-specific difference, remains to be shown.

I_{KATP} is not easily induced in metabolically compromised crucian carp cardiac myocytes, but can be activated by acute increases in temperature. In 56% of whole-cell recordings and 33% of cell-attached single-channel recordings, ventricular I_{KATP} of the cold-acclimated crucian carp (4°C) was induced by acute increases in temperature between 13°C and 19°C (Paajanen and Vornanen, 2004). In the goldfish heart, the duration of ventricular AP in the cold-acclimated fish (7°C), but not in the warm-acclimated fish (21°C), was dependent on a glibenclamide-sensitive current (I_{KATP}) (Ganim et al., 1998). These differences were associated with higher activity (mean open time, open probability) of the K_{ATP} channels in the cold-acclimated goldfish. Collectively, the present data suggest that I_{KATP} could be involved in acute and chronic thermal responses of the fish heart. However, additional research is needed to resolve the physiological role of I_{KATP} in electrical excitability of the fish heart in hypoxia-sensitive and hypoxia-tolerant fish species.

K_{ATP} channels are heteromeric complexes of the pore-forming inward rectifier channels of the subfamily Kir6 (α-subunits) and the sulfonylurea receptor (SUR) β-subunits (Flagg and Nichols, 2011). Functional channels are octamers of four Kir6 proteins and four SUR proteins. The Kir6 subunits (Kir6.1 and Kir6.2) form the ion pore and determine the biophysical properties of the I_{KATP} and mediate the inhibitory effect of ATP. The SUR subunits (SUR1 and SUR2 with two splice variants SUR2A and SUR2B) sense intracellular levels of the nucleotides ATP and ADP, fine-tune the ATP sensitivity of the channel, and facilitate opening or closing of Kir6 channels. In the mouse heart, K_{ATP} channel composition differs between atrial and ventricular tissues. In atrial myocytes, the channels are predominantly formed by SUR1/Kir6.2, while in ventricular myocytes SUR2A/Kir6.2 channels are mainly expressed (Flagg et al., 2008). Transcripts of Kir6.1 and Kir6.2 channels and SUR2 β-subunits are expressed in goldfish and crucian carp hearts (Cameron et al., 2013; Tikkanen et al., 2017). Chronic hypoxia does not affect Kir6.1 transcripts, but causes a decrease in Kir6.2 transcripts and an increase in SUR2 transcripts of the goldfish ventricle (Cameron et al., 2013). Nonetheless, the significance of these changes in K_{ATP} channel function in hypoxic goldfish hearts remains to be explained. On the basis of transcript expression, K_{ATP} channels of the crucian carp ventricle are mainly comprised of Kir6.2/SUR2 channels. Notably, prolonged anoxia exposure (39–57 days) of winter-acclimatized fish does not affect Kir6.2/SUR2 transcript expression, but acclimatization to winter is associated with strong depression of those channel subunits (Tikkanen et al., 2017). These findings suggest K_{ATP} channels are not involved in anoxia acclimatization of the crucian carp heart.

6. EFFECTS OF AUTONOMIC NERVOUS CONTROL ON CARDIAC EXCITABILITY

The hearts of teleost fishes are innervated by both the inhibitory cholinergic and excitatory adrenergic components of the autonomous nervous system, although the cholinergic control of the heart is often stronger

(Axelsson et al., 1987; Farrell, 1984; Holmgren, 1977; Laurent et al., 1983; Randall, 1968). The most densely innervated area of the teleost heart is at the base of the sinoatrial valves and around the primary pacemaker tissue and sinoatrial nerve plexus. Atrial muscle is more densely innervated than ventricular muscle (Newton et al., 2014; Stoyek et al., 2015; Yamauchi and Burnstock, 1968; Zaccone et al., 2010).

In fish hearts, cholinergic activity is increased under oxygen shortage and causes a depression in f_H (i.e., a negative chronotropic effect or bradycardia) and a reduction of contractile force (i.e., a negative inotropic effect) in the atrium, but has little effect on ventricular contraction (Fritsche and Nilsson, 1990; Laurent et al., 1983; Steele et al., 2009). In mammalian hearts, cholinergic effects include a reduction in the rate of impulse conduction (i.e., a negative dromotropic effect). Adrenergic effects are important for fishes under conditions of exercise, acidosis, hypoxia, and acute changes in temperature (Aho and Vornanen, 2002; Farrell and Milligan, 1986; Keen et al., 1994; Mendonca and Gamperl, 2009; Shiels et al., 2003, 2015). Adrenaline and noradrenaline increase f_H (a positive chronotropic effect) and the force of cardiac contraction (a positive inotropic effect), and may increase the rate of impulse transmission over the heart (a positive dromotropic effect). Although chronotropic, inotropic, and dromotropic responses of the fish heart involve changes in excitability of the cardiac myocytes, relatively little is known about the effects of cholinergic and adrenergic agonists on ion channel function of the fish heart.

6.1. Cholinergic Regulation of Nodal Tissues and the Atrium

Despite its physiological significance, our knowledge of autonomic regulation of excitability of the fish cardiac pacemaker, and the associated ionic mechanisms, is modest. In the mammalian cardiac pacemaker, acceleration and deceleration of f_H by adrenergic and cholinergic agonists, respectively, are mediated by changes in the density of SL ion currents (I_f and I_{CaL}) channels and/or in the rate of Ca^{2+} cycling through the SR, even though the relative significance of different factors is disputed. Both clock mechanisms depend on either phosphorylation/dephosphorylation of the key molecular operators (L-type Ca^{2+} channels, RyRs, calmodulin) in a cAMP-dependent manner or direct stimulation of HCN channels by cAMP. Although HCN4 pacemaker channels have been located in the pacemaker region of goldfish (*Carassius auratus*) and zebrafish hearts (Newton et al., 2014; Tessadori et al., 2012), nothing is known about adrenergic or cholinergic responses of the fish cardiac I_f. Considering that the positive chronotropic effects of adrenergic stimulation in fish hearts are relatively weak, it can be anticipated that either the dependence of the fish pacemaker mechanism on I_f and I_{Ca} is minor, or the cAMP-dependent stimulation of these channels is weak. On the other hand, the contribution of the SR to cardiac E–C coupling in many fish species is fairly small (Bovo et al., 2013; Vornanen et al., 2002b; Zhang et al., 2011), possibly meaning that the role of SR Ca^{2+} release in cardiac pacemaking is also minor. However, enhancement of SR Ca^{2+} release by adrenaline has been suggested for zebrafish ventricular and trout atrial myocytes (Bovo et al., 2013; Cros et al., 2014). The strong negative chronotropic effect of cholinergic stimulation in the fish heart is probably mediated by the activation of the I_{KACH}, which is a major current system in fish atrial myocytes (Abramochkin and Vornanen, 2016; Molina et al., 2007; Vornanen et al., 2010), although I_{KACH} has not yet been measured from the fish pacemaker cells.

The velocity of cardiac impulse conduction is dependent on the rate of the AP upstroke (Phase 0), which is largely determined by the density of inward I_{Na} and I_{Ca}. The AP propagates more slowly in the nodal tissues than in atrial and ventricular muscles (Sedmera et al., 2003). For example, in the sinoatrial and atrioventricular canals of the carp (*Cyprinus carpio*), the rate of AP transmission is slowed down and can be completely blocked by vagal stimulation (Saito, 1973; Saito and Tenma, 1976; von Skramlik, 1935). In the mammalian sinoatrial node the expression of Na^+ channels is low and the slow rate of AP conduction is largely based on I_{Ca} (Satoh, 2003). Assuming a similar ion channel composition for fish nodal tissues, the slow rate of AP conduction in sinoatrial and atrioventricular regions of the fish heart could be due to a vagally mediated decrease in the density of I_{Ca} caused by cholinergic inhibition of the cAMP pathway. In atrial myocytes the density of I_{Na} is much higher than that of I_{Ca} (Haverinen and Vornanen, 2006; Vornanen, 1998), and therefore, I_{Na} practically determines the rate of impulse conduction in the atrium. In the fast conducting atrial tissue, vagal slowing of AP transmission and complete conduction block are possibly caused by the large I_{KACH} (Molina et al., 2007; Vornanen et al., 2010), which hyperpolarizes RMP and thereby short circuits the AP (Abramochkin et al., 2008).

6.2. Adrenergic Regulation

Fish ventricles do not have cholinergic control, but ventricular function is modulated by circulating catecholamines and those released from sympathetic nerves. Resting adrenergic tonus is low (or even negative) (Mendonca and Gamperl, 2009), but adrenergic effects increase under various stressors. Catecholamine-induced increases in f_H-although often modest in fish—are expected to shorten AP duration and allow enough time for diastolic filling of the heart. Counterintuitively, adrenaline induces prolongation of the ventricular AP in juvenile pink salmon (*Oncorhynchus gorbuscha*) and Pacific bluefin tuna (*Thunnus orientalis*) (Ballesta et al., 2012; Shiels et al., 2015). While this might improve intracellular Ca^{2+} management and contractility, the findings from rainbow trout indicate the opposite, i.e., a stronger negative inotropic frequency response in the presence of adrenaline (Hove-Madsen and Gesser, 1989). In two Antarctic fish species (*Chaenocephalus aceratus*, *Notothenia coriiceps*), adrenaline failed to improve the frequency response of the heart (Skov et al., 2009).

In mammals, the slow component of the delayed rectifier current, I_{Ks}, is considered to be the main current system that mediates the repolarizing effects of β-adrenergic stimulation on the duration of the cardiac AP. Prerequisites for the β-adrenergic stimulation of I_{Ks} are that the channel complex includes, in addition to the pore-forming α-subunit (K_v11.1), the β-subunit (MinK) and the A-kinase anchoring protein (AKAP) which binds the catalytic subunits of PKA and protein phosphatases. Furthermore, the α-subunit must have a phosphorylation site in the N-terminus (Ser-27) of the protein. Phosphorylation of Kv11.1 channels slows the inactivation of I_{Ks} and moves its voltage dependence of steady-state activation to more negative voltages, which appear as an increase in I_{Ks} channel density. Thus far, the presence of the fish cardiac I_{Ks} has been demonstrated only in crucian carp atrial myocytes (Hassinen et al., 2011). In this fish species, the delayed rectifier channel appears to consist of homotetramers of the K_v11.1 α-subunits without the MinK β-subunit. Moreover, the crucian carp K_v11.1 protein does not have the N-terminal phosphorylation site. Thus, the crucian carp cardiac I_{Ks} channel is lacking at least two of the prerequisites of β-adrenergic activation (the presence of AKAP has not been examined). Consistent with this, the crucian carp I_{Ks} is not stimulated by forskolin, an activator of adenylate cyclase. These findings suggest that the duration of fish cardiac APs cannot be regulated by I_{Ks} under β-adrenergic activation. However, electrophysiological and molecular studies of I_{Ks} in other fish species are needed to corroborate this hypothesis.

Shortening of the AP could be achieved by increases in the density of I_{Kr} and I_{K1}, which are the main repolarizing currents of the fish heart. ERG2 (I_{Kr}) channels of trout and zebrafish hearts have four consensus sites for PKA-dependent phosphorylation, but thus far, no data exist on β-adrenergic regulation of the fish cardiac I_{Kr} (Hassinen et al., 2008a). I_{K1} accelerates the final phase (Phase 3) of repolarization of the AP and thereby also reduces the duration of the vertebrate cardiac AP. Findings from the mammalian Kir2 channels indicate that the inward rectifier channels can be targets for β-adrenergic regulation. For example, PKA-dependent phosphorylation of the C-terminal site of cloned rat Kir2.1 channels, expressed in *Xenopus* oocytes or HEK239 cells, significantly increases the density of I_{K1} (Zhang et al., 2013). Nevertheless, in guinea-pig ventricular myocytes, I_{K1} is depressed by the β-adrenergic cascade (Koumi et al., 1995). Cardiac myocytes of different fish species mainly express Kir2.4, Kir2.1, Kir2.2a, and Kir2.2b channels, which all have the C-terminal consensus phosphorylation site (Hassinen et al., 2007, 2008b, 2015a). Thus, the I_{K1} of fish cardiac myocytes could be regulated by the β-adrenergic pathway. Experiments on adrenergic regulation of the fish cardiac I_{K1} are, however, lacking.

L-type I_{Ca} provides inward current for the maintenance of the AP plateau in cardiac myocytes. Therefore, any changes in the density and kinetics of I_{Ca} will have an immediate effect on the duration of the cardiac AP. β-Adrenergic activation stimulates I_{Ca} in fish atrial and ventricular myocytes, although the strength of the effect varies between species (Hove-Madsen and Tort, 1998; Shiels et al., 2003; Vornanen, 1997, 1998). If everything else remains constant, a cAMP- and PKA-dependent increase in I_{Ca} should prolong the duration of the cardiac AP. However, Ca^{2+} influx through L-type Ca^{2+} channels is under negative feedback control *via* Ca^{2+}-dependent inactivation of I_{Ca}: increased sarcolemmal Ca^{2+} influx accelerates inactivation of I_{Ca}, and faster closure of the Ca^{2+} channels counteracts prolongation of the cardiac AP. This autoregulation of I_{Ca} shortens fish cardiac AP under control conditions, and is probably also effective under β-adrenergic activation (Ballesta et al., 2012; Harwood et al., 2000). The current evidence shows, that in fish ventricular myocytes, adrenaline increases AP duration, probably *via* an increase in I_{CaL}, but it is not clear whether this improves cardiac contractility, because the response is associated with slowed or unchanged force restitution (Hove-Madsen and Gesser, 1989).

7. SIGNIFICANCE OF ION CHANNEL FUNCTION IN THERMAL TOLERANCE OF FISH HEARTS

7.1. Thermal Tolerance Limits of the Fish Heart in Comparison With Other Vertebrates

The body temperature of ectothermic fishes is variable, with some eurythermic species able to tolerate all temperatures between 0°C and 40°C (Beitinger and Bennett, 2000; Bennett and Beitinger, 1997; Rantin et al., 1998). Therefore, in the many eurythermic fish species, the temperature tolerance window of cardiac excitation appears to be as wide as in small hibernating mammals. At the other extreme, fishes of the Southern Ocean live under constant cold, and these cold stenotherms have a narrow temperature tolerance range and less capacity for thermal acclimation (Bilyk and Devries, 2011; Egginton and Campbell, 2016; Franklin et al., 2007; Somero and DeVries, 1967). Salmonid fishes (family *Salmonidae*) are mesothermic and tolerate temperatures from about 0°C to 22–28°C (Elliott and Elliott,

2010). The first signs of cardiac disturbance in the salmonid heart appear 2–4° C below the upper thermal tolerance limit of the fish (Heath and Hughes, 1973; Vornanen et al., 2014), thus giving a thermal tolerance window for undisturbed cardiac function of 18–24°C. Comparison of mammalian and piscine data suggests that in both vertebrate groups, temperature tolerance of cardiac function is highly variable between species reflecting the animals' adaptation to different habitat/body temperatures. Species with constant body temperature (Antarctic fishes, non-hibernating mammals) are more sensitive to temperature changes than species with largely variable body temperature (eurythermic fishes, hibernating mammals). Fishes living in temperate climates have to tolerate large seasonal temperature changes: their hearts have to function close to 0°C in winter, while summer temperatures may be 20–30°C higher. This raises an interesting question: how are temperature-induced cardiac arrhythmias—typical for non-hibernating mammals—avoided in fishes?

7.2. Are Fish Hearts Resistant to Temperature-Induced Arrhythmias?

Factors recognized as significant causes of cardiac arrhythmias in mammals include APs that are too long or too short, slow AP conduction, and intracellular Ca^{2+} loading (Nattel et al., 2007). Depression of I_{Na} density and reduced gap junction conduction between myocytes can slow impulse transmission in the heart and cause reentry-type arrhythmias. If APs are abnormally long, early afterdepolarizations during the AP plateau (Phase 2 or 3) may be provoked by reactivation of Ca^{2+} or Na^{+} currents in the voltage "window," where all Ca^{2+} and Na^{+} channels have not yet been inactivated and can be reactivated. Even minor increases in AP duration may lead to life-threatening chaotic ventricular tachyarrhythmia (*torsades des pointes*) in human hearts. In addition, early afterdepolarizations are promoted by spontaneous Ca^{2+} release from the SR that activates inward I_{NCX} *via* the reverse-mode operation of Na^{+}–Ca^{2+} exchange (Choi et al., 2002). APs that are too short can predispose the heart to delayed afterdepolarizations, which occur in early diastole (Phase 4), when spontaneous Ca^{2+} releases from the SR activate the inward I_{NCX}. Afterdepolarizations may depolarize membrane potential to the AP threshold and induce extra systoles (triggered activity).

Different types of arrhythmias also occur in fish hearts due to acute changes in temperature, and can be induced by experimentally generated changes in cardiac ion channel composition or caused by drug-induced changes in ion channel function (Table 1). Typical responses of the fish heart to high temperatures—approaching or exceeding the normal physiological tolerance limit of the fish—include missed beats, bradycardia, and bursts of rapid beating, and finally complete cessation of heartbeat (asystole)

Table 1

Type of cardiac arrhythmias in fish induced by high temperature, ion channel mutations, manipulation of ion channel expression, and cardioactive drugs

Species	Type of manipulation	Current(s) involved	Characterization of arrhythmia	Remarks	Reference
Danio rerio (embryo)	QT-prolonging drugs; MO toward zERG (AF532865); breakdance mutant of zERG	I_{Kr}, I_{Ca}	Bradycardia; atrioventricular block; ectopic beats; irregularities in heart rhythm	No tachyarrhythmia or *torsades de pointes*	Langheinrich et al. (2003)
Danio rerio (embryo)	Breakdance mutant of zERG	I_{Kr}	Atrioventricular block		Meder et al. (2011)
Danio rerio (embryo)	QT-prolonging drugs	I_{Kr}	Atrioventricular block; "arrhythmia"		Dhillon et al. (2013)
Danio rerio	Mutation of zERG (NM_212837); terfenadine (an ERG blocker)	I_{Kr}	Ventricular asystole (homozygous mutants); 2:1 atrioventricular block; bradycardia; ventricular asystole (heterozygous mutant + terfenadine)		Arnaout et al. (2007)
Danio rerio	QT-prolonging drugs (23 different molecules); MO toward zERG	I_{Kr}	Bradycardia; atrioventricular block		Milan et al. (2003)
Danio rerio (adult)	QT-prolonging drugs	I_{Kr}	No arrhythmias reported	QT prolongation	Milan et al. (2006)
Danio rerio (embryo)	QT-prolonging drugs (15 different molecules)	I_{Kr}	Bradycardia		Park et al. (2013)
Danio rerio, *D. albolineatus*, *D. choprae*	High temperature; acclimation temperature 25–27°C		Variability and depression of f_H	Anesthetized and atropinized fish; isoprenaline 7.8 μg/kg	Sidhu et al. (2014)
Oncorhynchus mykiss	Deltamethrin (a pyrethroid insecticide)	I_{Na}	Missing beats; increased variability of beat interval and contraction force		Haverinen and Vornanen (2014b)
Carassius carassius	Deltamethrin	I_{Na}	Missing beats; increased variability of beat interval and contraction force		Haverinen and Vornanen (2014a)

(*Continued*)

Table 1 (Continued)

Species	Type of manipulation	Current(s) involved	Characterization of arrhythmia	Remarks	Reference
Carassius auratus	High temperature; acclimation temperatures 12°C, 20°C, and 28°C		Missing QRS complexes; variable QRS amplitude; depression of f_H	Anesthetized and atropinized fish; isoprenaline 4 μg/kg	Ferreira et al. (2014)
Oncorhynchus mykiss	High temperature (24–25°C); acclimation temperature 15°C		Bradycardia; missing beats		Heath and Hughes (1973)
Oncorhynchus mykiss (juvenile)	High temperature (>24.9°C); acclimation temperature 10°C		Missing QRS complexes; missing P waves		Anttila et al. (2013)
Oncorhynchus tshawytscha	High temperature (>25°C); acclimation temperature 13–14°C		Bradycardia; missing beats		Clark et al. (2008)
Oncorhynchus kisutch (juvenile)	High temperature (>20–23°C); acclimation temperature 10°C		Missing QRS complexes; bradycardia	Anesthetized and atropinized fish; isoprenaline 4 μg/kg	Casselman et al. (2012)
Gadus morhua	High temperature (>18°C); acclimation temperature 10–11°C		Bradycardia; missed beats; bursts of rapid beating		Gollock et al. (2006)
Gadus morhua	High temperature (>16°C); acclimation temperature 10°C		"Arrhythmia"	Arrhythmia appeared as large variability in blood pressure	Lannig et al. (2004)
Salmo trutta fario	High temperature (>21°C); acclimation temperature 12°C	I_{Na}	Bradycardia; missing beats; bursts of rapid beating		Vornanen et al. (2014)
Clupea pallasii	Weathered crude oil		Bradycardia; "irregular arrhythmia"		Incardona et al. (2009)
Rutilus rutilus	High temperature; acclimation temperatures 4°C and 18°C		Missing QRS complexes; bradycardia; atrial tachycardia; asystole		Badr et al. (2016)

MO, antisense morpholinos; zERG, a zebrafish ortholog to human ERG channel (I_{Kr} channel).

(Anttila et al., 2013; Badr et al., 2016; Casselman et al., 2012; Ferreira et al., 2014; Verhille et al., 2013; Vornanen et al., 2014). Similarly, genetic modifications of cardiac ion channel composition or drug-induced changes in ion channel function often appear as atrioventricular block, missed beats, bradycardia, ectopic beats, and asystole (Table 1). Thus far, there are no data on early and delayed afterdepolarizations or chaotic tachyarrhythmias (similar to *torsade de pointes*) in fish hearts under any experimental manipulations.

Although cardiac APs shorten greatly with increasing temperature (Ballesta et al., 2012; Harper et al., 1995; Haverinen and Vornanen, 2009b; Talo and Tirri, 1991; Vornanen et al., 2002a), there are no data demonstrating that high temperatures cause delayed afterdepolarizations in any fish heart. The reason for this resistance against afterdepolarizations may reside in the special features of the E–C coupling of the fish heart. Generation of delayed afterdepolarizations requires spontaneous Ca^{2+} release from the SR, which subsequently activates Ca^{2+} efflux *via* Na^{+}–Ca^{2+} exchange (inward I_{NCX}) and membrane depolarization to the AP threshold (Kihara and Morgan, 1991). In many fish hearts, SR makes only a minor, or no, contribution to cardiac E–C coupling (Bovo et al., 2013; Rantin et al., 1998; Tibbits et al., 1991; Vornanen et al., 2002b; Zhang et al., 2011), probably due to the low Ca^{2+} sensitivity of the RyRs (Chugun et al., 1999; Vornanen, 2006) and high Ca^{2+} buffering capacity of the SR which makes spontaneous Ca^{2+} releases from the SR rare or nonexistent (Haverinen and Vornanen, 2009a; Hove-Madsen et al., 1998; Shiels and White, 2005; Vornanen, 2006). The relative independence of fish heart contraction from Ca^{2+}-induced Ca^{2+} release (CICR) probably protects the heart against delayed afterdepolarizations and triggered activity. However, there are some fish species that rely more on CICR in cardiac E–C coupling, and in several species the activity of the cardiac SR is increased after acclimation to cold temperatures (Vornanen et al., 2002b). For example, the hearts of some tuna species (Shiels et al., 2011), cold-acclimated salmonids (Aho and Vornanen, 1999; Keen et al., 1994) and the cold-adapted burbot (Tiitu and Vornanen, 2002), are more strongly dependent on SR activity for force generation and might be more prone to develop triggered activity at high temperatures. On the other hand, acute temperature shifts are much smaller in winter under the ice than in summer. Finally, eurythermic species like goldfish, crucian carp, and zebrafish rely more heavily on SL Ca^{2+} transport mechanism in cardiac E–C coupling (Tiitu and Vornanen, 2001; Vornanen, 1999; Zhang et al., 2011), and are probably less likely to suffer from triggered activity at high temperatures.

Acute lowering of temperature will increase the duration of the cardiac AP, which could potentially increase the likelihood for reactivation of Ca^{2+} or Na^{+} currents, and thus, provoke early afterdepolarizations and associated tachyarrhythmias in fish hearts. Even a slight prolongation (about 15%) of the ventricular AP due to mutations of cardiac K^{+} or Na^{+} channels (long QT syndromes) predisposes human hearts to the life-threatening ventricular arrhythmias in the form of *torsades des pointes* (Kaye et al., 2013). Evidently, a minor imbalance between depolarizing and repolarizing currents can destroy the normal rhythm of the mammalian heart. Considering that an acute decrease in temperature within the physiological temperature range of the fish (e.g., by crossing a thermocline) can double the duration of cardiac AP, the excitability of the fish cardiac myocyte appears to be resistant to early afterdepolarizations. In mammalian ventricular myocytes early afterdepolarizations can be abolished by reducing the size of the window for I_{CaL} (Madhvani et al., 2015) or by blocking the SR Ca^{2+} release with ryanodine (Volders et al., 1997). Thus, the minor role of SR Ca^{2+} cycling in fish cardiac excitation–contraction coupling may also be protective against the delayed afterdepolarizations. Interestingly, in crucian carp and rainbow trout ventricular myocytes, acclimation to cold (4°C) reduces the size of the window for I_{CaL} by about 50% (Vornanen, 1998), which may also reduce the likelihood of early afterdepolarizations during long APs in the cold. It should be also noted that exposure of zebrafish embryos to QT-prolonging drugs (e.g., ERG channel inhibitors) does not cause tachyarrhythmia or *torsades de pointes* like arrhythmias (Langheinrich et al., 2003), indicating that fish hearts are inherently more tolerant to severe ventricular arrhythmias. Reduced dispersion of repolarization across the ventricular wall is antiarrhythmic, whereas increased dispersion is proarrhythmic. Therefore, it is possible that small hearts, like that of the zebrafish, are less prone to cardiac arrhythmias (Nerbonne, 2004).

7.3. Is Ion Channel Function Involved in the Heat Tolerance of Fish Heart Function?

Although fish hearts may be tolerant to cardiac arrhythmias caused by afterdepolarizations, cardiac excitation is disturbed at temperatures that are close to the upper thermal tolerance limit of the fish (Badr et al., 2017). Indeed, several findings suggest that heart function could be a limiting factor for the upper thermal tolerance of fishes (Eliason et al., 2011; Gollock et al., 2006; Lannig et al., 2004), even though a number of counter-arguments have been recently presented (Brijs et al., 2015; Gräns et al., 2014). Usually the limits of thermal tolerance appear earlier at higher levels of biological organization in intact animals or functioning organs, and only later at the cellular and molecular level (Lagerspetz, 1987; Prosser and Nelson, 1981; Ushakov, 1964). This seems to apply to cardiac excitation as well (Badr et al., 2017). Looking from the systemic perspective, it is conceivable that heat-dependent deterioration of cardiac performance could have its origin outside the heart, for example, in inadequate myocardial oxygen supply (Farrell, 2002) and not directly caused by temperature sensitivity of cardiac mechanics. Here only the direct effects of temperature on electrical excitability of the fish heart, and their possible implications on heat tolerance of heart function, will be discussed.

Thermal disturbances of cardiac excitation must ultimately reflect temperature-related deterioration or suboptimal function of some cellular or molecular process, or a mismatch of linked processes. A molecular mechanism could become limiting for heart function at high temperatures without the rate of the reaction actually declining. A limitation could appear if the temperature-dependent increase in the rate of the reaction is not sufficiently fast (low Q_{10}) to support the demand of the higher-level function (high Q_{10}), or if coupled reactions supporting the same function have a mismatch in their temperature dependencies. Electrical excitation is a complicated sequence of events involving several closely linked and interacting molecular mechanisms, and imbalance in this system could become limiting for cardiac function at high temperatures. Impaired excitability could compromise fish heart function either by retarding temperature-related increases in f_H, disrupting impulse conduction or normal cardiac rhythm, or completely preventing AP generation. It has been repeatedly shown that f_H slows down at high temperatures and, therefore, results in a reduction in cardiac output (Table 1). This could be due to: (1) deterioration of ion channel function in the cardiac pacemaker with direct effects on AP frequency; (2) refractoriness of atrial and ventricular excitation impairing their response to pacemaker rate; or (3) slowing of impulse propagation in nodal tissues or in the working myocardium with consequent failure to transmit the pacemaker rate through the heart.

The heat sensitivity of cardiac ion currents has been tested in cardiac myocytes of the brown trout (Vornanen et al., 2014). f_H in brown trout *in vivo* starts to decline at 22.3°C, i.e., 2–3°C below the upper thermal tolerance limit of the fish and similar to the findings from other fish species (Gollock et al., 2006; Heath and Hughes, 1973). Most ion currents of brown trout atrial and ventricular myocytes tolerate much higher temperatures than f_H: I_{K1} shows no signs of deterioration at temperatures below 35°C, the density of I_{CaL} channels continues to increase up to 32°C, and I_{Kr} is observed to increase up to 28°C. A clear exception is I_{Na}, which starts to decline at 20.9°C, i.e., at lower temperature than *in vivo* f_H. I_{Na} appears to be the weak link in electrical excitation of the brown trout heart, and possibly a limiting factor for f_H, and hence cardiac output. I_{Na} is necessary for the generation of atrial and ventricular APs and determines the rate of impulse conduction through the heart. A reduction in I_{Na} will increase the threshold voltage for AP firing, and as a result, a stronger depolarization from the nodal tissues is needed to trigger APs in atrial and ventricular myocytes. In the intact heart, this could initially appear as a sporadic appearance of missed beats and then as a complete cessation of the heartbeat (asystole). This would be analogous to the sinoatrial or atrioventricular conduction block of mammalian hearts as an outcome from the loss of Na^+

channel function (Derangeon et al., 2012). Essential in this hypothesis of temperature-dependent deterioration of electrical excitation (TDEE) is the antagonism of I_{Na} and I_{K1} in the initiation of atrial and ventricular APs, and the different temperature sensitivities of those currents (Vornanen, 2016). An AP will fail if the outward I_{K1} overwhelms the inward I_{Na} (Varghese, 2016). This is likely to occur at high temperatures because I_{K1} is much more heat tolerant than I_{Na} and generates a fast surge of outward current at the beginning of atrial and ventricular APs. Another putative factor in addition to an elevated AP threshold - that could prevent excitation of atrial and ventricular myocytes - the reduced amplitude of the nodal APs (Haverinen et al., 2017; Vornanen, 2016). This trigger could become smaller at high temperatures due to the temperature-dependent increase in outward K^+ currents. Finally, temperature-dependent failure of the pacemaker mechanism itself could cause arrhythmias, missed beats, and cessation of the heartbeat. However, the recent findings on enzymatically isolated pacemaker cells of the brown trout heart show that the pacemaker cells can generate much higher AP frequencies (193 bpm) than are found in the intact animal (77 bpm) or in sinoatrial preparations *in vitro* (94 bpm) (Haverinen et al., 2017).

It is likely not accidental that I_{Na} is the most heat-sensitive current of the fish heart. The opening and closing of ion channel requires conformational changes of the channel (Bezanilla, 2005; Collins and Rojas, 1982), and ion channel proteins must be sufficiently flexible to undergo conformational changes at the low body temperature of the fish. The kinetics of Na^+ channel gating is very fast, resulting in an almost instantaneous opening of the channels upon small membrane depolarizations, followed by a large Na^+ influx and rapid inactivation during maintained depolarization (Patlak, 1991). The high catalytic activity of Na^+ channels requires high molecular flexibility, which probably comes with the trade-off of low thermal stability. Considering the implicit kinetic compromise, the high sensitivity of I_{Na} for thermal inactivation may reside in its high catalytic activity. Whether the low thermal tolerance is purely a property of ion channel proteins or is also contributed by the lipid environment of channels remains to be shown. The large variability in heat tolerance between different ion currents of the brown trout heart suggests that thermal properties of the bulk lipid membrane are not decisive. However, the contribution of the immediate lipid environment of the ion channel, the lipid annulus, remains to be studied (Powl et al., 2007).

8. SUMMARY

Pumping of the fish heart is dependent on the sequential and correct timing of excitation of atrial and ventricular cardiomyocytes. The shape and propagation rate of the cardiac AP is typical for each functional region of the heart. Excitability of cardiac myocytes is adjusted to fulfill cardiovascular demands under physiological and environmental stressors *via* activation of inward Na^+ and Ca^{2+} currents and outward K^+ currents. The major current systems of the fish heart are the same as in the endothermic vertebrates, but the molecular basis of ion currents is often different and more diverse in comparison to the mammalian cardiac ion currents. This is probably due to the adaptation of fish to a wide variety of environmental conditions.

Cardiac function is a key factor with regard to the thermal responses of fish and may set the thermal tolerance limits of fishes in our warming climate. Cardiac excitability in fishes is thermally flexible, enables acclimation of f_H and rhythm to seasonally changing temperatures, and provides some physiological plasticity when fishes face warming waters. Recent evidence indicates that K^+ currents are particularly malleable in adjusting the excitability of fish heart to different thermal environments and seasonal temperature changes, while Ca^{2+} currents are less flexible; but on the other hand they are intrinsically less dependent on acute temperature changes. Na^+ current is the most heat-sensitive ion current of the fish heart, and therefore, particularly important in regulating thermal tolerance of atrial and ventricular excitation. Indeed, the I_{Na}–I_{K1} antagonism in atrial and ventricular myocytes could be setting the upper limit of f_H, and thus, cardiac output. The largest gap in our knowledge with regard to the excitability of the fish heart exists in the generation and regulation of f_H and rhythm by the pacemaker cells. Despite recent progress, our knowledge of cardiac excitability in fishes is far from complete, and many novel and exciting things will be revealed by future studies.

REFERENCES

Abramochkin, D.V., Vornanen, M., 2015. Seasonal acclimatization of the cardiac potassium currents (I_{K1} and I_{Kr}) in an arctic marine teleost, the navaga cod (*Eleginus navaga*). J. Comp. Physiol. B 185, 883–890.

Abramochkin, D.V., Vornanen, M., 2016. Seasonal changes of cholinergic response in the atrium of arctic navaga cod (*Eleginus navaga*). J. Comp. Physiol. B 187, 329–338.

Abramochkin, D.V., Suris, M.A., Sukhova, G.S., Rozenshtraukh, L.V., 2008. Acetylcholine-induced suppression of electric activity of working myocardium of the cod atrium. Dokl. Biol. Sci. 419, 73–76.

Abramochkin, D.V., Tapilina, S.V., Vornanen, M., 2014. A new potassium ion current induced by stimulation of M2 cholinoreceptors in fish atrial myocytes. J. Exp. Biol. 217, 1745–1751.

Aho, E., Vornanen, M., 1999. Contractile properties of atrial and ventricular myocardium of the heart of rainbow trout *Oncorhynchus mykiss*: effects of thermal acclimation. J. Exp. Biol. 202, 2663–2677.

Aho, E., Vornanen, M., 2001. Cold-acclimation increases basal heart rate but decreases its thermal tolerance in rainbow trout (*Oncorhynchus mykiss*). J. Comp. Physiol. B 171, 173–179.

Aho, E., Vornanen, M., 2002. Effects of adenosine on the contractility of normoxic rainbow trout heart. J. Comp. Physiol. B 172, 217–225.

Alday, A., Alonso, H., Gallego, M., Urrutia, J., Letamendia, A., Callol, C., Casis, O., 2014. Ionic channels underlying the ventricular action potential in zebrafish embryo. Pharmacol. Res. 84, 26–31.

Alvarez, J.L., Vassort, G., 1992. Properties of the low threshold Ca current in single frog atrial cardiomyocytes. A comparison with the high threshold Ca current. J. Gen. Physiol. 100, 519–545.

Anttila, K., Casselman, M.T., Schulte, P.M., Farrell, A.P., 2013. Optimum temperature in juvenile salmonids: connecting subcellular indicators to tissue function and whole-organism thermal optimum. Physiol. Biochem. Zool. 86, 245–256.

Arnaout, R., Ferrer, T., Huisken, J., Spitzer, K., Stainier, D.Y., Tristani-Firouzi, M., Chi, N.C., 2007. Zebrafish model for human long QT syndrome. Proc. Natl. Acad. Sci. U.S.A. 104, 11316–11321.

Arrenberg, A.B., Stainier, D.Y., Baier, H., Huisken, J., 2010. Optogenetic control of cardiac function. Science 330, 971–974.

Axelsson, M., Ehrenström, F., Nilsson, S., 1987. Cholinergic and adrenergic influence on the teleost heart *in vivo*. Exp. Biol. 46, 179–186.

Badr, A., El-Sayed, M.F., Vornanen, M., 2016. Effects of seasonal acclimatization on temperature-dependence of cardiac excitability in the roach, *Rutilus rutilus*. J. Exp. Biol. 219, 1495–1504.

Badr, A., Hassinen, M., El-Sayed, M.F., Vornanen, M., 2017. Effects of seasonal acclimatization on action potentials and sarcolemmal K^+ currents in roach (*Rutilus rutilus*) cardiac myocytes. Comp. Biochem. Physiol. A, 205, 1–10.

Baker, K., Warren, K.S., Yellen, G., Fishman, M.C., 1997. Defective "pacemaker" current (I_h) in a zebrafish mutant with a slow heart rate. Proc. Natl. Acad. Sci. U.S.A. 94, 4554–4559.

Ballesta, S., Hanson, L.M., Farrell, A.P., 2012. The effect of adrenaline on the temperature dependency of cardiac action potentials in pink salmon *Oncorhynchus gorbuscha*. J. Fish Biol. 80, 876–885.

Bean, B.P., 1985. Two kinds of calcium channels in canine atrial cells. J. Gen. Physiol. 86, 1–30.

Beitinger, T.L., Bennett, W.A., 2000. Quantification of the role of acclimation temperature in temperature tolerance of fishes. Environ. Biol. Fishes 58, 277–288.

Belardinelli, L., Isenberg, G., 1983. Actions of adenosine and isoproterenol on isolated mammalian ventricular myocytes. Circ. Res. 53, 287–297.

Benitah, J.P., Alvarez, J.L., Gomez, A.M., 2010. L-type Ca^{2+} current in ventricular cardiomyocytes. J. Mol. Cell. Cardiol. 48, 26–36.

Bennett, W.A., Beitinger, T.L., 1997. Temperature tolerance of the sheepshead minnow *Cyprinodon variegatus*. Copeia (1), 77–87.

Bers, D.M., 2002. Cardiac excitation-contraction coupling. Nature 415, 198–205.

Bezanilla, F., 2005. Voltage-gated ion channels. IEEE Trans. Nanobioscience 4, 34–48.

Bilyk, K.T., Devries, A.L., 2011. Heat tolerance and its plasticity in Antarctic fishes. Comp. Biochem. Physiol. A 158, 382–390.

Bovo, E., Dvornikov, A.V., Mazurek, S.R., de Tombe, P.P., Zima, A.V., 2013. Mechanisms of Ca^{2+} handling in zebrafish ventricular myocytes. Pflugers Arch. 465, 1775–1784.

Brette, F., Luxan, G., Cros, C., Dixey, H., Wilson, C., Shiels, H.A., 2008. Characterization of isolated ventricular myocytes from adult zebrafish (*Danio rerio*). Biochem. Biophys. Res. Commun. 374, 143–146.

Brijs, J., Jutfelt, F., Clark, T.D., Gräns, A., Ekström, A., Sandblom, E., 2015. Experimental manipulations of tissue oxygen supply do not affect warming tolerance of European perch. J. Exp. Biol. 218, 2448–2454.

Cameron, J.S., Hoffmann, K.E., Zia, C., Hemmett, H.M., Kronsteiner, A., Lee, C.M., 2003. A role for nitric oxide in hypoxia-induced activation of cardiac K_{ATP} channels in goldfish (*Carassius auratus*). J. Exp. Biol. 206, 4057–4065.

Cameron, J.S., DeWitt, J.P., Ngo, T.T., Yajnik, T., Chan, S., Chung, E., Kang, E., 2013. Cardiac K_{ATP} channel alterations associated with acclimation to hypoxia in goldfish (*Carassius auratus* L.). Comp. Biochem. Physiol. A 164, 554–564.

Campbell, H.A., Taylor, E.W., Egginton, S., 2004. The use of power spectral analysis to determine cardiorespiratory control in the short-horned sculpin *Myoxocephalus scorpius*. J. Exp. Biol. 207, 1969–1976.

Casselman, M.T., Anttila, K., Farrell, A.P., 2012. Using maximum heart rate as a rapid screening tool to determine optimum temperature for aerobic scope in Pacific salmon *Oncorhynchus* spp. J. Fish Biol. 80, 358–377.

Catterall, W.A., Waxman, S.G., 2005. International Union of Pharmacology. XLVII. Nomenclature and structure-function relationship of voltage-gated sodium channels. Pharmacol. Rev. 57, 397–409.

Catterall, W.A., Perez-Reyes, E., Snutch, T.P., Striessnig, J., 2005. International Union of Pharmacology. XLVIII. Nomenclature and structure-function relationships of voltage-gated calcium channels. Pharmacol. Rev. 57, 411–425.

Cech Jr., J.J., Bridges, D.W., Rowell, D.M., Balzer, P.J., 1976. Cardiovascular responses of winter flounder, *Pseudopleuronectes americanus* (Walbaum), to acute temperature increase. Can. J. Zool. 54, 1383–1388.

Chen, J., Zhu, J.X., Wilson, I., Cameron, J.S., 2005. Cardioprotective effects of K_{ATP} channel activation during hypoxia in goldfish *Carassius auratus*. J. Exp. Biol. 208, 2765–2772.

Chi, N.C., Shaw, R.M., Jungblut, B., Huisken, J., Ferrer, T., Arnaout, R., Scott, I., Beis, D., Xiao, T., Baier, H., et al., 2008. Genetic and physiologic dissection of the vertebrate cardiac conduction system. PLoS Biol. 6, e109.

Choi, B.R., Burton, F., Salama, G., 2002. Cytosolic Ca^{2+} triggers early afterdepolarizations and Torsade de Pointes in rabbit hearts with type 2 long QT syndrome. J. Physiol. 543, 615–631.

Chopra, S.S., Watanabe, H., Zhong, T.P., Roden, D.M., 2007. Molecular cloning and analysis of zebrafish voltage-gated sodium channel beta subunit genes: implications for the evolution of electrical signaling in vertebrates. BMC Evol. Biol. 7, 113.

Chugun, A., Oyamada, T., Temma, K., Hara, Y., Kondo, H., 1999. Intracellular Ca^{2+} storage sites in the carp heart: comparison with the rat heart. Comp. Biochem. Physiol. A 123A, 61–67.

Clark, T.D., Sandblom, E., Cox, G.K., Hinch, S.G., Farrell, A.P., 2008. Circulatory limits to oxygen supply during an acute temperature increase in the Chinook salmon (Oncorhynchus tshawytscha). Am. J. Physiol. Regul. Integr. Comp. Physiol. 295 (5), R1631–R1639.

Collins, C.A., Rojas, E., 1982. Temperature dependence of the sodium channel gating kinetics in the node of Ranvier. Q. J. Exp. Physiol. 67, 41–55.

Coraboeuf, E., 1978. Ionic basis of electrical activity in cardiac tissues. Am. J. Physiol. 234, H101–H116.

Coyne, M.D., Kim, C.S., Cameron, J.S., Gwathmey, J.K., 2000. Effects of temperature and calcium availability on ventricular myocardium from rainbow trout. Am. J. Physiol. 278, R1535–R1544.

Cros, C., Salle, L., Warren, D.E., Shiels, H.A., Brette, F., 2014. The calcium stored in the sarcoplasmic reticulum acts as a safety mechanism in rainbow trout heart. Am. J. Physiol. 307, R1493–R1501.

Derangeon, M., Montnach, J., Baro, I., Charpentier, F., 2012. Mouse models of SCN5A-related cardiac arrhythmias. Front. Physiol. 3, 210.

Dhillon, S.S., Dóró, É., Magyary, I., Egginton, S., Sík, A., Müller, F., 2013. Optimisation of embryonic and larval ECG measurement in zebrafish for quantifying the effect of QT prolonging drugs. PLoS One 8 (4), e60552.

Dickson, K.A., Graham, J.B., 2004. Evolution and consequences of endothermy in fishes. Physiol. Biochem. Zool. 77, 998–1018.

DiFrancesco, D., 1993. Pacemaker mechanisms in cardiac tissue. Annu. Rev. Physiol. 55, 455–472.

DiFrancesco, D., Noble, D., 2012. The funny current has a major pacemaking role in the sinus node. Heart Rhythm 9, 299–301.

Dobrzynski, H., Marples, D.D.R., Musa, H., Yamanushi, T.T., Hendersonxyl, Z., Takagishi, Y., Honjo, H., Kodama, I., Boyett, M.R., 2001. Distribution of the muscarinic K^+ channel proteins Kir3.1 and Kir3.4 in the ventricle, atrium, and sinoatrial node of heart. J. Histochem. Cytochem. 49, 1221–1234.

Egginton, S., Campbell, H.A., 2016. Cardiorespiratory responses in an Antarctic fish suggest limited capacity for thermal acclimation. J. Exp. Biol. 219, 1283–1286.

Ehrlich, J.R., 2008. Inward rectifier potassium currents as a target for atrial fibrillation therapy. J. Cardiovasc. Pharmacol. 52, 129–135.

Eliason, E.J., Anttila, K., 2017. Temperature and the cardiovascular system. In: Gamperl, A.K., Gillis, T.E., Farrell, A.P., Brauner, C.J. (Eds.), Fish Physiology. In: The Cardiovascular System: Development, Plasticity and Physiological Responses, vol. 36B. Academic Press, San Diego (In press).

Eliason, E.J., Clark, T.D., Hague, M.J., Hanson, L.M., Gallagher, Z.S., Jeffries, K.M., Gale, M.K., Patterson, D.A., Hinch, S.G., Farrell, A.P., 2011. Differences in thermal tolerance among sockeye salmon populations. Science 332 (6025), 109–112.

Elliott, J.M., Elliott, J.A., 2010. Temperature requirements of Atlantic salmon *Salmo salar*, brown trout *Salmo trutta* and Arctic charr *Salvelinus alpinus*: predicting the effects of climate change. J. Fish Biol. 77, 1793–1817.

Fabiato, A., 1983. Calcium-induced release of calcium from the cardiac sarcoplasmic reticulum. Am. J. Physiol. 245, C1–C14.

Farrell, A.P., 1984. A review of cardiac performance in the teleost heart: intrinsic and humoral regulation. Can. J. Zool. 62, 523–536.

Farrell, A.P., 2002. Cardiorespiratory performance in salmonids during exercise at high temperature: insights into cardiovascular design limitations in fishes. Comp. Biochem. Physiol. A Mol. Integr. Physiol. 132 (4), 797–810.

Farrell, A.P., Smith, F., 2017. Cardiac form, function and physiology. In: Gamperl, A.K., Gillis, T.E., Farrell, A.P., Brauner, C.J. (Eds.), Fish Physiology. In: The Cardiovascular System: Morphology, Control and Function, vol. 36A. Academic Press, San Diego, pp. 155–264.

Farrell, A.P., Milligan, C.L., 1986. Myocardial intracellular pH in a perfused raibow trout heart during extracellular acidosis in the presence and absence of adrenaline. J. Exp. Biol. 125, 347–359.

Ferreira, E.O., Anttila, K., Farrell, A.P., 2014. Thermal optima and tolerance in the eurythermic goldfish (*Carassius auratus*): relationships between whole-animal aerobic capacity and maximum heart rate. Physiol. Biochem. Zool. 87, 599–611.

Flagg, T.P., Nichols, C.G., 2011. "Cardiac K_{ATP}": a family of ion channels. Circulation 4, 796–798.

Flagg, T.P., Kurata, H.T., Masia, R., Caputa, G., Magnuson, M.A., Lefer, D.J., Coetzee, W.A., Nichols, C.G., 2008. Differential structure of atrial and ventricular KATP: atrial KATP channels require SUR1. Circ. Res. 103, 1458–1465.

Fozzard, H.A., Hanck, D.A., 1996. Structure and function of voltage-dependent sodium channels: comparison of brain II and cardiac isoforms. Physiol. Rev. 76, 887–926.

Franklin, C.E., Davison, W., Seebacher, F., 2007. Antarctic fish can compensate for rising temperatures: thermal acclimation of cardiac performance in *Pagothenia borchgrevinki*. J. Exp. Biol. 210, 3068–3074.

Fritsche, R., Nilsson, S., 1990. Autonomic nervous control of blood pressure and heart rate during hypoxia in the cod, *Gadus morhua*. J. Comp. Physiol. B 160, 287–292.

Galli, G.L., Lipnick, M.S., Block, B.A., 2009. Effect of thermal acclimation on action potentials and sarcolemmal K^+ channels from Pacific bluefin tuna cardiomyocytes. Am. J. Physiol. 297, R502–R509.

Gamperl, A.K., Wilkinson, M., Boutlier, R.G., 1994. β-Adrenoceptors in the trout (*Oncorhynchus mykiss*) heart: characterization, quantification, and effects of repeated catcholamine exposure. Gen. Comp. Endocrinol. 95, 259–272.

Ganim, R.B., Peckol, E.L., Larkin, J., Ruchhoef, M.L., Cameron, J.S., 1998. ATP-sensitive K^+ channels in cardiac muscle from cold-acclimated goldfish: characterization and altered response to ATP. Comp. Biochem. Physiol. A 119A, 395–401.

Giltrow, E., Eccles, P.D., Hutchinson, T.H., Sumpter, J.P., Rand-Weaver, M., 2011. Characterisation and expression of β1-, β2- and β3-adrenergic receptors in the fathead minnow (*Pimephales promelas*). Gen. Comp. Endocrinol. 173, 483–490.

Gollock, M.J., Currie, S., Petersen, L.H., Gamperl, A.K., 2006. Cardiovascular and haematological responses of Atlantic cod (*Gadus morhua*) to acute temperature increase. J. Exp. Biol. 209, 2961–2970.

Gräns, A., Jutfelt, F., Sandblom, E., Jonsson, E., Wiklander, K., Seth, H., Olsson, C., Dupont, S., Ortega-Martinez, O., Einarsdottir, I., et al., 2014. Aerobic scope fails to explain the detrimental effects on growth resulting from warming and elevated CO_2 in Atlantic halibut. J. Exp. Biol. 217, 711–717.

Grover, G.J., Garlid, K.D., 2000. ATP-sensitive potassium channels: a review of their cardioprotective pharmacology. J. Mol. Cell. Cardiol. 32, 677–695.

Guo, X., Gao, X., Wang, Y., Peng, L., Zhu, Y., Wang, S., 2012. I_{Ks} protects from ventricular arrhythmia during cardiac ischemia and reperfusion in rabbits by preserving the repolarization reserve. PLoS One 7, e31545.

Gutman, G.A., Chandy, K.G., Adelman, J.P., Aiyar, J., Bayliss, D.A., Clapham, D.E., Covarriubias, M., Desir, G.V., Furuichi, K., Ganetzky, B., et al., 2003. International Union of Pharmacology. XLI. Compendium of voltage-gated ion channels: potassium channels. Pharmacol. Rev. 55, 583–586.

Harper, A.A., Newton, I.P., Watt, P.W., 1995. The effect of temperature on spontaneous action potential discharge of the isolated sinus venosus from winter and summer plaice (*Pleuronectes platessa*). J. Exp. Biol. 198, 137–140.

Harvey, R.D., Hell, J.W., 2013. CaV1.2 signaling complexes in the heart. J. Mol. Cell. Cardiol. 58, 143–152.

Harwood, C.L., Howarth, F.C., Altringham, J.D., White, E., 2000. Rate-dependent changes in cell shortening, intracellular Ca^{2+} levels and membrane potential in single, isolated rainbow trout (*Oncorhynchus mykiss*) ventricular myocytes. J. Exp. Biol. 203, 493–504.

Hassinen, M., Paajanen, V., Haverinen, J., Eronen, H., Vornanen, M., 2007. Cloning and expression of cardiac Kir2.1 and Kir2.2 channels in thermally acclimated rainbow trout. Am. J. Physiol. 292, R2328–R2339.

Hassinen, M., Haverinen, J., Vornanen, M., 2008a. Electrophysiological properties and expression of the delayed rectifier potassium (ERG) channels in the heart of thermally acclimated rainbow trout. Am. J. Physiol. 295, R297–R308.

Hassinen, M., Paajanen, V., Vornanen, M., 2008b. A novel inwardly rectifying K^+ channel, Kir2.5, is upregulated under chronic cold stress in fish cardiac myocytes. J. Exp. Biol. 211, 2162–2171.

Hassinen, M., Laulaja, S., Paajanen, V., Haverinen, J., Vornanen, M., 2011. Thermal adaptation of the crucian carp (*Carassius carassius*) cardiac delayed rectifier current, I_{Ks}, by homomeric assembly of Kv7.1 subunits without MinK. Am. J. Physiol. 301, R255–R2665.

Hassinen, M., Abramochkin, D.V., Vornanen, M., 2014. Seasonal acclimatization of the cardiac action potential in the Arctic navaga (*Eleginus navaga*, Gadidae). J. Comp. Physiol. B 184, 319–327.

Hassinen, M., Haverinen, J., Hardy, M.E., Shiels, H.A., Vornanen, M., 2015a. Inward rectifier potassium current (I_{K1}) and Kir2 composition of the zebrafish (*Danio rerio*) heart. Pflugers Arch. 467, 2437–2446.

Hassinen, M., Haverinen, J., Vornanen, M., 2015b. Molecular basis and drug sensitivity of the delayed rectifier (I_{Kr}) in the fish heart. Comp. Biochem. Physiol. C 176–177, 44–51.

Haverinen, J., Vornanen, M., 2004. Temperature acclimation modifies Na^+ current in fish cardiac myocytes. J. Exp. Biol. 207, 2823–2833.

Haverinen, J., Vornanen, M., 2006. Significance of Na^+ current in the excitability of atrial and ventricular myocardium of the fish heart. J. Exp. Biol. 209, 549–557.

Haverinen, J., Vornanen, M., 2007. Temperature acclimation modifies sinoatrial pacemaker mechanism of the rainbow trout heart. Am. J. Physiol. 292, R1023–R1032.

Haverinen, J., Vornanen, M., 2009a. Comparison of sarcoplasmic reticulum calcium content in atrial and ventricular myocytes of three fish species. Am. J. Physiol. 297, R1180–R1187.

Haverinen, J., Vornanen, M., 2009b. Responses of action potential and K^+ currents to temperature acclimation in fish hearts: phylogeny or thermal preferences? Physiol. Biochem. Zool. 82, 468–482.

Haverinen, J., Vornanen, M., 2014a. Deltamethrin is cardiotoxic to the fish (crucian carp, *Carassius carassius*) heart. Pestic. Biochem. Phys. 159, 1–9.

Haverinen, J., Vornanen, M., 2014b. Effects of deltamethrin on excitability and contractility of the rainbow trout (*Oncorhynchus mykiss*) heart. Comp. Biochem. Physiol. C 159, 1–9.

Haverinen, J., Egginton, S., Vornanen, M., 2014. Electrical excitation of the heart in a basal vertebrate, the European river lamprey (*Lampetra fluviatilis*). Physiol. Biochem. Zool. 87, 817–828.

Haverinen, J., Abramochkin, D.V., Kamkin, A., Vornanen, M., 2017. The maximum heart rate in brown trout (*Salmo trutta fario*) is not limited by firing rate of pacemaker cells. Am. J. Physiol. 312, R165–R171. ajpregu.00403.2016.

Haworth, T.E., Haverinen, J., Shiels, H.A., Vornanen, M., 2014. Electrical excitability of the heart in a Chondrostei fish, the Siberian sturgeon (*Acipenser baerii*). Am. J. Physiol. 307, R1157–R1166.

Heath, A.G., Hughes, G.M., 1973. Cardiovascular and respiratory changes during heat stress in rainbow trout (*Salmo gairdneri*). J. Exp. Biol. 59, 323–338.

Hibino, H., Inanobe, A., Furutani, K., Murakami, S., Findlay, I., Kurachi, Y., 2010. Inwardly rectifying potassium channels: their structure, function and physiological role. Physiol. Rev. 90, 291–366.

Hoekstra, H.E., Coyne, J.A., 2007. The locus of evolution: evo devo and the genetics of adaptation. Evolution 61, 995–1016.

Holmgren, S., 1977. Regulation of the heart of a teleost, *Gadus morhua*, by autonomic nerves and circulating catecholamines. Acta Physiol. Scand. 99, 62–74.

Holopainen, I.J., Tonn, W.M., Paszkowski, C.A., 1997. Tales of two fish: the dichotomous biology of crucian carp (*Carassius carassius* (L.)) in northern Europe. Ann. Zool. Fenn. 28, 1–22.

Hou, J.H., Kralj, J.M., Douglass, A.D., Engert, F., Cohen, A.E., 2014. Simultaneous mapping of membrane voltage and calcium in zebrafish heart in vivo reveals chamber-specific developmental transitions in ionic currents. Front. Physiol. 5, 344.

Houston, A.H., Koss, T.F., 1984. Plasma and red cell ionic composition in rainbow trout exposed to progressive temperature increases. J. Exp. Biol. 110, 53–67.

Hove-Madsen, L., Gesser, H., 1989. Force frequency relation in the myocardium of rainbow trout. Effects of K^+ and adrenaline. J. Comp. Physiol. B 159, 61–69.

Hove-Madsen, L., Tort, L., 1998. L-type Ca^{2+} current and excitation-contraction coupling in single atrial myocytes from rainbow trout. Am. J. Physiol. 275, R2061–R2069.

Hove-Madsen, L., Llach, A., Tort, L., 1998. Quantification of Ca^{2+} uptake in the sarcoplasmic reticulum of trout ventricular myocytes. Am. J. Physiol. 275, R2070–R2080.

Huang, T.F., 1973. The action potential of the myocardial cells of the golden carp. Jpn. J. Physiol. 23, 529–540.

Hurley, I.A., Mueller, R.L., Dunn, K.A., Schmidt, E.J., Friedman, M., Ho, R.K., Prince, V.E., Yang, Z., Thomas, M.G., Coates, M.I., 2007. A new time-scale for ray-finned fish evolution. Proc. R. Soc. B Biol. Sci. 274, 489–498.

Icardo, J.M., 2017. Heart morphology and anatomy. In: Gamperl, A.K., Gillis, T.E., Farrell, A.P., Brauner, C.J. (Eds.), Fish Physiology. In: The Cardiovascular System: Morphology, Control and Function, vol. 36A. Academic Press, San Diego, pp. 1–54.

Icardo, J.M., Colvee, E., 2011. The atrioventricular region of the teleost heart. A distinct heart segment. Anat. Rec. 294, 236–242.

Incardona, J.P., Carls, M.G., Day, H.L., Sloan, C.A., Bolton, J.L., Collier, T.K., Scholz, N.L., 2009. Cardiac arrhythmia is the primary response of embryonic Pacific herring (*Clupea pallasi*) exposed to crude oil during weathering. Environ. Sci. Technol. 43, 201–207.

Irisawa, H., 1978. Comparative physiology of the cardiac pacemaker mechanism. Physiol. Rev. 58, 461–498.

Isenberg, G., Klöckner, U., 1980. Glycocalyx is not required for slow inward calcium current in isolated rat heart myocytes. Nature 284, 358–360.

Jegla, T.J., Zmasek, C.M., Batalov, S., Nayak, S.K., 2009. Evolution of the human ion channel set. Comb. Chem. High Throughput Screen. 12, 2–23.

Jensen, D., 1965. The aneural heart of the hagfish. Ann. N. Y. Acad. Sci. 127, 443–458.

Jensen, B., Boukens, B.J.D., Postma, A.V., Gunst, Q.D., van den Hoff, M.J., Moorman, A.F.M., Wang, T., Christoffels, V.M., 2012. Identifying the evolutionary building blocks of the cardiac conduction system. PLoS One 7, e44231.

Jost, N., Virag, L., Bitay, M., Takacs, J., Lengyel, C., Biliczki, P., Nagy, Z., Bogats, G., Lathrop, D.A., Papp, J.G., et al., 2005. Restricting excessive cardiac action potential and QT prolongation: a vital role for I_{Ks} in human ventricular muscle. Circulation 112, 1392–1399.

Kaye, A.D., Volpi-Abadie, J., Bensler, J.M., Kaye, A.M., Diaz, J.H., 2013. QT interval abnormalities: risk factors and perioperative management in long QT syndromes and Torsades de Pointes. J. Anesth. 27, 575–587.

Keen, J.E., Vianzon, D.M., Farrell, A.P., Tibbits, G.F., 1993. Thermal acclimation alters both adrenergic sensitivity and adrenoceptor density in cardiac tissue of rainbow trout. J. Exp. Biol. 181, 27–47.

Keen, J.E., Vianzon, D.M., Farrell, A.P., Tibbits, G.F., 1994. Effect of temperature and temperature acclimation on the ryanodine sensitivity of the trout myocardium. J. Comp. Physiol. B 164, 438–443.

Keith, A., Mackenzie, I., 1910. Recent researches on the anatomy of the heart. Lancet 1, 101–103.

Kihara, Y., Morgan, J.P., 1991. Intracellular calcium and ventricular fibrillation. Studies in the aequorin-loaded isovolumic ferret heart. Circ. Res. 68, 1378–1389.

Kim, C.S., Coyne, M.D., Gwathmey, J.K., 2000. Voltage-dependent calcium channels in ventricular cells of rainbow trout: effect of temperature changes in vitro. Am. J. Physiol. 278, R1524–R1534.

Koumi, S.I., Wasserstrom, J.A., Ten Eick, R.E., 1995. Beta-adrenergic and cholinergic modulation of the inwardly rectifying K^+ current in guinea-pig ventricular myocytes. J. Physiol. 486 (3), 647–659.

Kubly, K.L., Stecyk, J.A., 2015. Temperature-dependence of L-type Ca^{2+} current in ventricular cardiomyocytes of the Alaska blackfish (*Dallia pectoralis*). J. Comp. Physiol. B 185, 845–858.

Lagerspetz, K.Y.H., 1987. Temperature effects on different organization levels in animals. Symp. Soc. Exp. Biol. 41, 429–449.

Langheinrich, U., Vacun, G., Wagner, T., 2003. Zebrafish embryos express an orthologue of HERG and are sensitive toward a range of QT-prolonging drugs inducing severe arrhythmia. Toxicol. Appl. Pharmacol. 193, 370–382.

Lannig, G., Bock, C., Sartoris, F.J., Pörtner, H.O., 2004. Oxygen limitation of thermal tolerance in cod, Gadus morhua L., studied by magnetic resonance imaging and on-line venous oxygen monitoring. Am. J. Physiol. Regul. Integr. Comp. Physiol. 287 (4), R902–R910.

Laurent, P., 1962. Contribution A l'etude morphologique et physiologique de l'innervation du ceur des telosteens. Arch. Anat. Micr. Morphol. 51, 337–458.

Laurent, P., Holmgren, S., Nilsson, S., 1983. Nervous and humoral control of the fish heart: structure and function. Comp. Biochem. Physiol. A 76A, 525–542.

Lee, J.H., Gomora, J.C., Cribbs, L.L., Perez-Reyes, E., 1999. Nickel block of three cloned T-type calcium channels: low concentrations selectively block α1H. Biophys. J. 77, 3034–3042.

Leong, I.U.S., Skinner, J.R., Shelling, A.N., Love, D.R., 2010. Identification and expression analysis of kcnh2 genes in the zebrafish. Biochem. Biophys. Res. Commun. 396, 817–824.

Leong, I.U.S., Skinner, J.R., Shelling, A.N., Love, D.R., 2014. Expression of a mutant kcnj2 gene transcript in zebrafish. ISRN Mol. Biol. 324839, 1–14.

Lillywhite, H.B., Zippel, K.C., Farrell, A.P., 1999. Resting and maximal heart rates in ectothermic vertebrates. Comp. Biochem. Physiol. A 124A, 369–382.

Lin, T.C., Hsieh, J.C., Lin, C.I., 1995. Electromechanical effects of acetylcholine on the atrial tissues of the cultured tilapia (*Oreochromis nilotica* x *O. aureus*). Fish Physiol. Biochem. 14, 449–457.

Lin, E., Ribeiro, A., Ding, W., Hove-Madsen, L., Sarunic, M.V., Beg, M.F., Tibbits, G.F., 2014. Optical mapping of the electrical activity of isolated adult zebrafish hearts: acute effects of temperature. Am. J. Physiol. 306, R823–R836.

Liu, Y.M., DeFelice, L.J., Mazzanti, M., 1992. Na channels that remain open throughout the cardiac action potential plateau. Biophys. J. 63, 654–662.

Madhvani, R.V., Angelini, M., Xie, Y., Pantazis, A., Suriany, S., Borgstrom, N.P., Garfinkel, A., Qu, Z., Weiss, J.N., Olcese, R., 2015. Targeting the late component of the cardiac L-type Ca^{2+} current to suppress early afterdepolarizations. J. Gen. Physiol. 145, 395–404.

Maltsev, V.A., Undrovinas, A.I., 1998. Relationship between steady-state activation and availability of cardiac sodium channel: evidence of uncoupling. Cell. Mol. Life Sci. 54, 148–151.

Maltsev, V.A., Vinogradova, T.M., Lakatta, E.G., 2006. The emergence of a general theory of the initiation and strength of the heartbeat. J. Pharmacol. Sci. 100, 338–369.

Mangoni, M.E., Nargeot, J., 2008. Genesis and regulation of the heart automaticity. Physiol. Rev. 88, 919–982.

Mangoni, M.E., Couette, B., Marger, L., Bourinet, E., Striessnig, J., Nargeot, J., 2006. Voltage-dependent calcium channels and cardiac pacemaker activity: from ionic currents to genes. Prog. Biophys. Mol. Biol. 90, 38–63.

Marx, S.O., Kurokawa, J., Reiken, S., Motoike, H., D'Armiento, J., Marks, A.R., Kass, R.S., 2002. Requirement of a macromolecular signaling complex for beta adrenergic receptor modulation of the KCNQ1-KCNE1 potassium channel. Science 295, 496–499.

Matikainen, N., Vornanen, M., 1992. Effect of season and temperature acclimation on the function of crucian carp (*Carassius carassius*) heart. J. Exp. Biol. 167, 203–220.

Maylie, J.G., Morad, M., 1995. Evaluation of T-type and L-type Ca^{2+} currents in shark ventricular myocytes. Am. J. Physiol. 269, H1695–H1703.

McDonald, T.F., Pelzer, S., Trautwein, W., Pelzer, D.J., 1994. Regulation and modulation of calcium channels in cardiac, skeletal, and smooth muscle cells. Physiol. Rev. 74, 365–507.

Meder, B., Scholz, E.P., Hassel, D., Wolff, C., Just, S., Berger, I.M., Patzel, E., Karle, C., Katus, H.A., Rottbauer, W., 2011. Reconstitution of defective protein trafficking rescues Long-QT syndrome in zebrafish. Biochem. Biophys. Res. Commun. 408, 218–224.

Mendonca, P.C., Gamperl, A.K., 2009. Nervous and humoral control of cardiac performance in the winter flounder (*Pleuronectes americanus*). J. Exp. Biol. 212, 934–944.

Mendonca, P.C., Gamperl, A.K., 2010. The effects of acute changes in temperature and oxygen availability on cardiac performance in winter flounder (*Pseudopleuronectes americanus*). Comp. Biochem. Physiol. A 155, 245–252.

Milan, D.J., Peterson, T.A., Ruskin, J.N., Peterson, R.T., MacRae, C.A., 2003. Drugs that induce repolarization abnormalities cause bradycardia in zebrafish. Circulation 107 (10), 1355–1358.

Milan, D.J., Jones, I.L., Ellinor, P.T., MacRae, C.A., 2006. *In vivo* recording of adult zebrafish electrocardiogram and assessment of drug-induced QT prolongation. Am. J. Physiol. 291, H269–H273.

Milstein, M.L., Musa, H., Balbuena, D.P., Anumonwo, J.M.B., Auerbach, D.S., Furspan, P.B., Hou, L., Hu, B., Schumacher, S.M., Vaidyanathan, R., et al., 2012. Dynamic reciprocity of sodium and potassium channel expression in a macromolecular complex controls cardiac excitability and arrhythmia. Proc. Natl. Acad. Sci. U.S.A. 109, E2134–E2143.

Molina, C.E., Gesser, H., Llach, A., Tort, L., Hove-Madsen, L., 2007. Modulation of membrane potential by an acetylcholine-activated potassium current in trout atrial myocytes. Am. J. Physiol. 292, R388–R395.

Nattel, S., Maguy, A., Le Bouter, S., Yeh, Y.H., 2007. Arrhythmogenic ion-channel remodeling in the heart: heart failure, myocardial infarction, and atrial fibrillation. Physiol. Rev. 87, 425–456.

Nemtsas, P., Wettwer, E., Christ, T., Weidinger, G., Ravens, U., 2010. Adult zebrafish heart as a model for human heart? An electrophysiological study. J. Mol. Cell. Cardiol. 48, 161–171.

Nerbonne, J.M., 2004. Studying cardiac arrhythmias in the mouse—a reasonable model for probing mechanisms? Trends Cardiovasc. Med. 14, 83–93.

Nerbonne, J.M., Kass, R.S., 2005. Molecular physiology of cardiac repolarization. Physiol. Rev. 85, 1205–1253.

Newton, C.M., Stoyek, M.R., Croll, R.P., Smith, F.M., 2014. Regional innervation of the heart in the goldfish, *Carassius auratus*: a confocal microscopy study. J. Comp. Neurol. 522, 456–478.

Nickerson, J.G., Dugan, S.G., Drouin, G., Moon, T.W., 2002. A putative β2-adrenoceptor from the rainbow trout (*Oncorhynchus mykiss*). Molecular characterization and pharmacology. Eur. J. Biochem. 268, 6465–6472.

Nickerson, J.G., Dugan, S.G., Drouin, G., Perry, S.F., Moon, T.W., 2003. Activity of the uinque β-adrenergic Na^+/H^+ exchanger in trout erythrocytes is controlled by a novel β3-AR subtype. Am. J. Physiol. 285, R526–R535.

Nilius, B., Hess, P., Lansman, J.B., Tsien, R.W., 1985. A novel type of cardiac calcium channel in ventricular cells. Nature 316, 443–446.

Noma, A., 1983. ATP-regulated K^+ channels in cardiac muscle. Nature 305, 147–148.

Noseda, V., Chiesa, F., Marchetti, R., 1963. Vectocardiogram of *Anguilla anguilla* L. and *Pleurodeles waltlii* Mich. Nature 197, 816–818.

Paajanen, V., Vornanen, M., 2002. The induction of an ATP-sensitive K^+ current in cardiac myocytes of air- and water-breathing vertebrates. Pflugers Arch. 444, 760–770.

Paajanen, V., Vornanen, M., 2003. Effects of chronic hypoxia on inward rectifier K^+ current (I_{K1}) in ventricular myocytes of crucian carp (*Carassius carassius*) heart. J. Membr. Biol. 194, 119–127.

Paajanen, V., Vornanen, M., 2004. Regulation of action potential duration under acute heat stress by $I_{K,ATP}$ and I_{K1} in fish cardiac myocytes. Am. J. Physiol. 286, R405–R415.

Park, M.J., Lee, K.R., Shin, D.S., Chun, H.S., Kim, C.H., Ahn, S.H., Bae, M.A., 2013. Predicted drug-induced bradycardia related cardio toxicity using a zebrafish in vivo model is highly correlated with results from in vitro tests. Toxicol. Lett. 216 (1), 9–15.

Patlak, J., 1991. Molecular kinetics of voltage-dependent Na^+ channels. Physiol. Rev. 71, 1047–1080.

Petersen, L.H., Needham, S.L., Burleson, M.L., Overturf, M.D., Huggett, D.B., 2013. Involvement of β3-adrenergic receptors in in vivo cardiovascular regulation in rainbow trout (*Oncorhynchus mykiss*). Comp. Biochem. Physiol. A 164, 291–300.

Petersen, L.H., Burleson, M., Huggett, D., 2015. Temperature and species-specific effects on ß3-adrenergic receptor cardiac regulation in two freshwater teleosts: channel catfish (*Ictalurus punctatus*) and common carp (*Cyprinus carpio*). Comp. Biochem. Physiol. A 185, 132–141.

Poon, K.L., Brand, T., 2013. The zebrafish model system in cardiovascular research: a tiny fish with mighty prospects. Glob. Cardiol. Sci. Pract. 2013, 9–28.

Powl, A.M., East, J.M., Lee, A.G., 2007. Different effects of lipid chain length on the two sides of a membrane and the lipid annulus of MscL. Biophys. J. 93, 113–122.

Prosser, C.L., 1973. Comparative Animal Physiology. Saunders, Philadelphia/London/Toronto.

Prosser, C.L., Nelson, D., 1981. The role of nervous systems in temperature adaptation of poikilotherms. Annu. Rev. Physiol. 43, 281–300.

Randall, D.J., 1968. Functional morphology of the heart in fishes. Am. Zool. 8, 179–189.

Randall, D.J., 1982. The control of respiration and circulation in fish during exercise and hypoxia. J. Exp. Biol. 100, 275–288.

Rantin, F.T., Gesser, H., Kalinin, A.L., Querra, C.D.R., De Freitas, J.C., Driedzic, W.R., 1998. Heart performance, Ca^{2+} regulation and energy metabolism at high temperatures in *Bathygobius soporator*, a tropical marine teleost. J. Therm. Biol. 23, 31–39.

Roden, D.M., Yang, T., 2005. Protecting the heart against arrhythmias: potassium current physiology and repolarization reserve. Circulation 112, 1376–1378.

Roden, D.M., Balser, J.R., George Jr., A.L., Anderson, M.E., 2002. Cardiac ion channels. Annu. Rev. Physiol. 64, 431–475.

Rottbauer, W., Baker, K., Wo, Z.G., Mohideen, M.A.P.K., Cantiello, H.F., Fishman, M.C., 2001. Growth and function of the embryonic heart depend upon the cardiac-specific L-type calcium channel α1 subunit. Dev. Cell 1, 265–275.

Saito, T., 1969. Electrophysiological studies on the pacemaker of several fish hearts. Zool. Mag. 78, 291–296.

Saito, T., 1973. Effects of vagal stimulation on the pacemaker action potentials of carp heart. Comp. Biochem. Physiol. A 44A, 191–199.

Saito, T., Tenma, K., 1976. Effects of left and right vagal stimulation on excitation and conduction of the carp heart (*Cyprinus carpio*). Comp. Biochem. Physiol. B 111, 39–53.

Sanguinetti, M.C., Curran, M.E., Zou, A., Shen, J., Specter, P.S., Atkinson, D.L., Keating, M.T., 1996. Coassembly of KVLQT1 and minK (IsK) proteins to form cardiac I_{Ks} potassium channel. Nature 384, 80–83.

Santer, R.M., 1985. Morphology and innervation of the fish heart. Adv. Anat. Embryol. Cell Biol. 89, 1–102.

Santer, R.M., Cobb, J.L.S., 1972. The fine structure of the heart of the teleost, *Pleuronectes platessa* L. Z. Zellforsch. Mikrosk. Anat. 131, 1–14.

Satoh, H., 2003. Sino-atrial nodal cells of mammalian hearts: ionic currents and gene expression of pacemaker ionic channels. J. Smooth Muscle Res. 39, 175–193.

Sedmera, D., Reckova, M., deAlmeida, A., Sedmerova, M., Biermann, M., Volejnik, J., Sarre, A., Raddatz, E., McCarthy, R.A., Gourdie, R.G., et al., 2003. Functional and morphological evidence for a ventricular conduction system in zebrafish and *Xenopus* hearts. Am. J. Physiol. 284, H1152–H1160.

Seibert, H., 1979. Thermal adaptation of heart rate and its parasympathetic control in the European eel *Anguilla anguilla*. Comp. Biochem. Physiol. 64C, 275–278.

Shiels, H.A., White, E., 2005. Temporal and spatial properties of cellular Ca^{2+} flux in trout ventricular myocytes. Am. J. Physiol. 288, R1756–R1766.

Shiels, H.A., Vornanen, M., Farrell, A.P., 2000. Temperature-dependence of L-type Ca^{2+} channel current in atrial myocytes from rainbow trout. J. Exp. Biol. 203, 2771–2880.

Shiels, H.A., Vornanen, M., Farrell, A.P., 2003. Acute temperature change modulates the response of I_{Ca} to adrenergic stimulation in fish cardiomyocytes. Physiol. Biochem. Zool. 76, 816–824.

Shiels, H.A., Paajanen, V., Vornanen, M., 2006. Sarcolemmal ion currents and sarcoplasmic reticulum Ca^{2+} content in ventricular myocytes from the cold stenothermic fish, the burbot (*Lota lota*). J. Exp. Biol. 209, 3091–3100.

Shiels, H.A., Di Maio, A., Thompson, S., Block, B.A., 2011. Warm fish with cold hearts: thermal plasticity of excitation-contraction coupling in bluefin tuna. Proc. R. Soc. Lond. B 278, 18–27.

Shiels, H.A., Galli, G.L., Block, B.A., 2015. Cardiac function in an endothermic fish: cellular mechanisms for overcoming acute thermal challenges during diving. Proc. R. Soc. Lond. B 282, 20141989.

Sidhu, R., Anttila, K., Farrell, A.P., 2014. Upper thermal tolerance of closely related *Danio* species. J. Fish Biol. 84, 982–995.

Skov, P.V., Bushnell, P.G., Tirsgaard, B., Steffensen, J.F., 2009. The role of adrenaline as a modulator of cardiac performance in two Antarctic fishes. Polar Biol. 32, 215–223.

Somero, G.N., DeVries, A.L., 1967. Temperature tolerance of some Antarctic fishes. Science 156, 257–258.

Steele, S.L., Lo, K.H., Li, V.W., Cheng, S.H., Ekker, M., Perry, S.F., 2009. Loss of M2 muscarinic receptor function inhibits development of hypoxic bradycardia and alters cardiac β-adrenergic sensitivity in larval zebrafish (*Danio rerio*). Am. J. Physiol. 297, R412–R420.

Steele, S.L., Yang, X., Debiais-Thibaud, M., Schwerte, T., Pelster, B., Ekker, M., Tiberi, M., Perry, S.F., 2011. *In vivo* and *in vitro* assessment of cardiac β-adrenergic receptors in larval zebrafish (*Danio rerio*). J. Exp. Biol. 214, 1445–1457.

Steinhausen, M.F., Sandblom, E., Eliason, E.J., Verhille, C., Farrell, A.P., 2008. The effect of acute temperature increases on the cardiorespiratory performance of resting and swimming sockeye salmon (*Oncorhynchus nerka*). J. Exp. Biol. 211, 3915–3926.

Stengl, M., Volders, P.G.A., Thomsen, M.B., Spätjens, R.L.H.M.G., Sipido, K.R., Vos, M.A., 2003. Accumulation of slowly activating delayed rectifier potassium current (IKs) in canine ventricular myocytes. J. Physiol. 551, 777–786.

Stoyek, M.R., Croll, R.P., Smith, F.M., 2015. Intrinsic and extrinsic innervation of the heart in zebrafish (*Danio rerio*). J. Comp. Neurol. 523, 1683–1700.

Striessnig, J., Pinggera, A., Kaur, G., Bock, G., Tuluc, P., 2014. L-type Ca^{2+} channels in heart and brain. Wiley Interdiscip. Rev. Membr. Transp. Signal. 3, 15–38.

Sureau, D., Lagardere, J.P., Pennec, J.P., 1989. Heart rate and its cholinergic control in the sole (*Solea vulgaris*), acclimatized to different temperatures. Comp. Biochem. Physiol. A 92A, 49–51.

Talo, A., Tirri, R., 1991. Temperature acclimation of the perch (*Perca fluviatilis* L.): changes in duration of cardiac action potential. J. Therm. Biol. 16, 89–92.

Tessadori, F., van Weerd, J.H., Burkhard, S.B., Verkerk, A.O., de Pater, E., Boukens, B.J.D., Vink, A., Christoffels, V.M., Bakkers, J., 2012. Identification and functional characterization of cardiac pacemaker cells in zebrafish. PLoS One 7, e47644.

Tibbits, G.F., Hove-Madsen, L., Bers, D.M., 1991. Calcium transport and the regulation of cardiac contractility in teleosts—a comparison with higher vertebrates. Can. J. Zool. 69, 2014–2019.

Tibbits, G.F., Moyes, C.D., Hove-Madsen, L., 1992. Excitation-Contraction Coupling in the Teleost Heart. In: Hoar, W.S., Randall, D.J., Farrell, A.P. (Eds.), Fish Physiology. In: The Cardiovascular System, vol. 12. Academic Press, Inc., San Diego, pp. 267–304. Part A.

Tiitu, V., Vornanen, M., 2001. Cold adaptation suppresses the contractility of both atrial and ventricular muscle of the crucian carp (*Carassius carassius* L.) heart. J. Fish Biol. 59, 141–156.

Tiitu, V., Vornanen, M., 2002. Regulation of cardiac contractility in a cold stenothermal fish, the burbot *Lota lota* L. J. Exp. Biol. 205, 1597–1606.

Tikkanen, E., Haverinen, J., Egginton, S., Hassinen, M., Vornanen, M., 2017. Effects of prolonged anoxia on electrical activity of the heart in crucian carp (*Carassius carassius*). J. Exp. Biol., 220 (3), 1–10.

Trautwein, W., McDonald, T.F., Tripathi, O.N., 1975. Calcium conductance and tension in mammalian ventricular muscle. Pflugers Arch. 354, 55–74.

Tristani-Firouzi, M., Chen, J., Mitcheson, J.S., Sanguinetti, M.C., 2001. Molecular biology of K^+ channels and their role in cardiac arrhythmias. Am. J. Med. 110, 59.

Ushakov, B., 1964. Thermostability of cells and proteins of poikilotherms and its significance in speciation. Physiol. Rev. 44, 518–560.

Varghese, A., 2016. Reciprocal modulation of I_{K1}–I_{Na} extends excitability in cardiac ventricular cells. Front. Physiol. 7, 542.

Vassort, G., Talavera, K., Alvarez, J.L., 2006. Role of T-type Ca^{2+} channels in the heart. Cell Calcium 40, 205–220.

Vaykshnorayte, M.A., Azarov, J.E., Tsvetkova, A.S., Vityazev, V.A., Ovechkin, A.O., Shmakov, D.N., 2011. The contribution of ventricular apicobasal and transmural repolarization patterns to the development of the T wave body surface potentials in frogs (*Rana temporaria*) and pike (*Esox lucius*). Comp. Biochem. Physiol. A 159, 39–45.

Verhille, C., Anttila, K., Farrell, A.P., 2013. A heart to heart on temperature: impaired temperature tolerance of triploid rainbow trout (*Oncorhynchus mykiss*) due to early onset of cardiac arrhythmia. Comp. Biochem. Physiol. A 164, 653–657.

Volders, P.G., Kulcsar, A., Vos, M.A., Sipido, K.R., Wellens, H.J., Lazzara, R., Szabo, B., 1997. Similarities between early and delayed afterdepolarizations induced by isoproterenol in canine ventricular myocytes. Cardiovasc. Res. 34, 348–359.

von Skramlik, E., 1935. Über den Kreislauf bei den Fischen. Ergeb. Biol. 11, 1–130.

Vornanen, M., 1997. Sarcolemmal Ca influx through L-type Ca channels in ventricular myocytes of a teleost fish. Am. J. Physiol. 272, R1432–R1440.

Vornanen, M., 1998. L-type Ca current in fish cardiac myocytes: effects of thermal acclimation and β-adrenergic stimulation. J. Exp. Biol. 201, 533–547.

Vornanen, M., 1999. Na^{+}-Ca^{2+} exchange current in ventricular myocytes of fish heart: contribution to sarcolemmal Ca^{2+} influx. J. Exp. Biol. 202, 1763–1775.

Vornanen, M., 2006. Temperature and Ca^{2+} dependence of [^{3}H]ryanodine binding in the burbot (*Lota lota* L.) heart. Am. J. Physiol. 290, R345–R351.

Vornanen, M., 2016. The temperature-dependence of electrical excitability of fish heart. J. Exp. Biol. 219, 1941–1952.

Vornanen, M., Hassinen, M., 2016. Zebrafish heart as a model for human cardiac electrophysiology. Channels 10, 101–110.

Vornanen, M., Paajanen, V., 2004. Seasonality of dihydropyridine receptor binding in the heart of an anoxia-tolerant vertebrate, the crucian carp (*Carassius carassius* L.). Am. J. Physiol. 287, R1263–R1269.

Vornanen, M., Tuomennoro, J., 1999. Effects of acute anoxia on heart function in crucian carp (*Carassius carassius* L.) heart: importance of cholinergic and purinergic control. Am. J. Physiol. 277, R465–R475.

Vornanen, M., Ryökkynen, A., Nurmi, A., 2002a. Temperature-dependent expression of sarcolemmal K^{+} currents in rainbow trout atrial and ventricular myocytes. Am. J. Physiol. 282, R1191–R1199.

Vornanen, M., Shiels, H.A., Farrell, A.P., 2002b. Plasticity of excitation-contraction coupling in fish cardiac myocytes. Comp. Biochem. Physiol. A 132, 827–846.

Vornanen, M., Hälinen, M., Haverinen, J., 2010. Sinoatrial tissue of crucian carp heart has only negative contractile responses to autonomic agonists. BMC Physiol. 10, 10. http://dx.doi.org/10.1186/1472-6793-10-10.

Vornanen, M., Hassinen, M., Haverinen, J., 2011. Tetrodotoxin sensitivity of the vertebrate cardiac Na^{+} current. Mar. Drugs 9, 2409–2422.

Vornanen, M., Haverinen, J., Egginton, S., 2014. Acute heat tolerance of cardiac excitation in the brown trout (*Salmo trutta fario*). J. Exp. Biol. 217, 299–309.

Wang, Z., Nishimura, Y., Shimada, Y., Umemoto, N., Hirano, M., Zang, L., Oka, T., Sakamoto, C., Kuroyanagi, J., Tanaka, T., 2009. Zebrafish β-adrenergic receptor mRNA expression and control of pigmentation. Gene 446, 18–27.

Warmke, J.W., Ganetzky, B., 1994. A family of potassium channel genes related to eag in Drosophila and mammals. Proc. Natl. Acad. Sci. U.S.A. 91, 3438–3442.

Warren, K.S., Baker, K., Fishman, M.C., 2001. The slow mo mutation reduces pacemaker current and heart rate in adult zebrafish. Am. J. Physiol. 281, H1711–H1719.

Widmark, J., Sundstrom, G., Ocampo Daza, D., Larhammar, D., 2011. Differential evolution of voltage-gated sodium channels in tetrapods and teleost fishes. Mol. Biol. Evol. 28, 859–871.

Wilson, C.M., Stecyk, J.A., Couturier, C.S., Nilsson, G.E., Farrell, A.P., 2013. Phylogeny and effects of anoxia on hyperpolarization-activated cyclic nucleotide-gated channel gene expression in the heart of a primitive chordate, the Pacific hagfish (*Eptatretus stoutii*). J. Exp. Biol. 216, 4462–4472.

Wong, E., Yu, W.P., Yap, W.H., Venkatesh, B., Soong, T.W., 2006. Comparative genomics of the human and Fugu voltage-gated calcium channel α1-subunit gene family reveals greater diversity in Fugu. Gene 366, 117–127.

Wu, C., Sharma, K., Laster, K., Hersi, M., Torres, C., Lukas, T.J., Moore, E.J., 2014. Kcnq1-5 (Kv7.1-5) potassium channel expression in the adult zebrafish. BMC Physiol. 14, 1.

Yamauchi, A., Burnstock, G., 1968. An electronmicroscopic study on the innervation of the trout heart. J. Comp. Neurol. 132, 567–588.

Yamauchi, A., Fujimaki, Y., Yokota, R., 1973. Fine structural studies of the sino-auricular nodal tissue in the heart of a teleost fish, Misgurnus, with particular reference to the cardiac internuncial cell. Am. J. Anat. 138, 407–430.

Yaniv, Y., Maltsev, V.A., Escobar, A.L., Spurgeon, H.A., Ziman, B.D., Stern, M.D., Lakatta, E.G., 2011. Beat-to-beat Ca^{2+}-dependent regulation of sinoatrial nodal pacemaker cell rate and rhythm. J. Mol. Cell. Cardiol. 51, 902–905.

Yaniv, Y., Lakatta, E.G., Maltsev, V.A., 2015. From two competing oscillators to one coupled-clock pacemaker cell system. Front. Physiol. 6, 28.

Yeh, Y.H., Ehrlich, J.R., Qi, X., Hebert, T.E., Chartier, D., Nattel, S., 2007. Adrenergic control of a constitutively active acetylcholine-regulated potassium current in canine atrial cardiomyocytes. Cardiovasc. Res. 74, 406–415.

Yousaf, M.N., Amin, A.B., Koppang, E.O., Vuolteenaho, O., Powell, M.D., 2012. Localization of natriuretic peptides in the cardiac pacemaker of Atlantic salmon (*Salmo salar* L.). Acta Histochem. 114, 819–826.

Yu, F., Zhao, Y., Gu, J., Quigley, K.L., Chi, N.C., Tai, Y.C., Hsiai, T.K., 2012. Flexible microelectrode arrays to interface epicardial electrical signals with intracardial calcium transients in zebrafish hearts. Biomed. Microdevices 14, 357–366.

Zaccone, G., Mauceri, A., Maisano, M., Giannetto, A., Parrino, V., Fasulo, S., 2010. Postganglionic nerve cell bodies and neurotransmitter localization in the teleost heart. Acta Histochem. 112, 328–336.

Zhang, P.C., Llach, A., Sheng, X.Y., Hove-Madsen, L., Tibbits, G.F., 2011. Calcium handling in zebrafish ventricular myocytes. Am. J. Physiol. 300, R56–R66.

Zhang, L., Liu, Q., Liu, C., Zhai, X., Feng, Q., Xu, R., Cui, X., Zhao, Z., Cao, J., Wu, B., 2013. Zacopride selectively activates the Kir2.1 channel via a PKA signaling pathway in rat cardiomyocytes. Sci. China Ser. C 56, 788–796.

Zingman, L.V., Hodgson, D.M., Bast, P.H., Kane, G.C., Perez-Terzic, C., Gumina, R.J., Pucar, D., Bienengraeber, M., Dzeja, P.P., Miki, T., et al., 2002. Kir6.2 is required for adaptation to stress. Proc. Natl. Acad. Sci. U.S.A. 99, 13278–13283.

4

CARDIAC FORM, FUNCTION AND PHYSIOLOGY

ANTHONY P. FARRELL[*,1]
FRANK SMITH[†]

[*]University of British Columbia, Vancouver, BC, Canada
[†]Dalhousie University, Halifax, NS, Canada
[1]Corresponding author: tony.farrell@ubc.ca

Herein, we consider the heart as an organ whose task is to supply blood flow to the tissues. A brief section on the form and function of the heart sets the stage for more detailed sections on cardiac physiology where we emphasize how cardiac output varies among species and with activity, and how heart rate and cardiac stroke volume are controlled. In separate sections, we consider the coronary circulation and its control.

The Cardiovascular System: Morphology, Control and Function, Volume 36A
FISH PHYSIOLOGY

DOI: http://dx.doi.org/10.1016/bs.fp.2017.07.001

1. INTRODUCTION

Our intent is to provide a comprehensive account of the form, function and physiology of the fish heart, but without undue replication of the excellent literature that already exists. An earlier review in the *Fish Physiology* series (Farrell and Jones, 1992) summarized and integrated the available literature up to 25 years ago. Also, many of the earlier reviews, monograms, and perspectives still have currency today (e.g., Brill and Bushnell, 1991; Butler, 1986; Butler and Metcalfe, 1988; Davie and Farrell, 1991a; Farrell, 1984, 1985, 1991; Forster et al., 1991; Johansen and Burggren, 1980; Johansen and Gesser, 1986; Laurent et al., 1983; Nilsson, 1983; Santer, 1985; Satchell, 1991; Tota, 1983, 1989; Wood and Perry, 1985). More recent reviews have covered specialized areas such as primitive fishes (Farrell, 2007a), cardiac plasticity (Gamperl and Farrell, 2004), temperature effects (Farrell, 2002; Farrell et al., 2009), anoxia tolerance (Farrell and Stecyk, 2007) and digestion (Seth et al., 2011), as well as general fish cardiovascular physiology (e.g., Gamperl and Shiels, 2013; Olson and Farrell, 2006). Brill and Lai (2016) also recently contrasted the cardiovascular systems of elasmobranch and teleost fishes. As a result, we focus on substantive and recent advances in cardiac function and physiology. Proper appreciation of these topics, however, requires an understanding of cardiac form, anatomy and morphology, as well as cardiomyocyte structure, which are all detailed elsewhere in this volume (Chapter 1, Volume 36A: Icardo, 2017; Chapter 2, Volume 36A: Shiels, 2017). Therefore, we briefly cover cardiac form to better place cardiac function and physiology in perspective. As with any review, the general statements we make are likely to have exceptions given that cardiac physiology has been studied in only a few hundred of the more than 30,000 extant fish species. Thus, we note important exceptions, where known, as well as any key knowledge gaps that would be fruitful avenues for future work.

2. CARDIAC FORM AND FUNCTION

The fish heart is an organ with three main chambers (sinus venosus, atrium and ventricle), an outflow tract (OFT; comprised of both the conus and bulbus arteriosus) that connects to the ventral aorta, and the atrioventricular (AV) segment (see Chapter 1, Volume 36A: Icardo, 2017). All these structures are contained within a pericardial cavity (an exception occurs in hagfishes; Farrell, 2007a; Icardo et al., 2016), and work together to deliver venous blood to the gills where it can be oxygenated, and then to the systemic circulation; see

Chapter 7 (Volume 36A: Sandblom and Gräns, 2017) for details on vascular anatomy and physiology.

2.1. Sinus Venosus

The sinus venosus collects venous blood (from the large bilateral *ducti Cuvier*, hepatic veins, anterior jugular veins, and the secondary circulation) and delivers it to the atrium. The sinus venosus is comprised primarily of connective tissues, but can also contain smooth muscle, cardiac muscle and autonomic nerves in varying amounts (see Chapter 1, Volume 36A: Icardo, 2017). For example, the sinus venosus in the common eel (*Anguilla anguilla*) has a complete layer of cardiac muscle; in the plaice (*Pleuronectes platessa*) it has sparse cardiac muscle; in the goldfish (*Carassius auratus*) and common carp (*Cyprinus carpio*) only smooth muscle elements (Yamauchi, 1980) are present; whereas in loach (*Misgurnus anguillicaudatus*), brown trout (*Salmo trutta*), and zebrafish (*Zebra danio* = *Danio rerio*) it is virtually amuscular.

The volume of the sinus venosus varies considerably among species. In some teleost fishes, the internal volume equals that of the atrium, but in elasmobranch fishes it is typically smaller. Whether the sinus venosus can functionally vary its volume needs to be examined experimentally because this ability may be important when fish vary cardiac stroke volume (V_S), especially if the volume of the pericardial cavity is fixed (see Section 5). The vascular smooth muscle in the neighboring *ducti Cuvier* contracts when it is mechanically stimulated during surgery (A.P. Farrell, personal observation), and thus, the vascular smooth muscle of the sinus venosus may have a similar capability.

The role of the cardiac muscle in the sinus venosus is less clear. Some fish species have rhythmic electrical activity associated with the sinus venosus as part of the cardiac cycle (see Section 3). However, any rhythmic contractions that result are unlikely to propel a significant quantity of blood into the atrium because there are no valves between the sinus venosus and the *ducti Cuvier* to prevent retrograde blood flow should sinus blood pressure exceed central venous blood pressure (P_{CV}). Thus, filling of the atrium with blood during diastole depends primarily on energy inherent to the venous blood, namely its kinetic and potential energy; the latter being P_{CV}. The contracting heart can also transfer energy from ventricular contraction to assist in cardiac filling through a *vis-a-fronte* mechanism (see Section 5).

The other main function of the sinus venosus (and the atrial tissue immediately adjacent to the sinoatrial junction of some species) is to produce the rhythmic, intrinsic, heartbeat. This is the role of a sparse arrangement of specialized cardiac pacemaker cells, whose locations and physiology are detailed elsewhere in this volume (see Chapter 1, Volume 36A: Icardo, 2017;

Chapter 3, Volume 36A: Vornanen, 2017). Older reviews provide the histological and neurophysiological evidence for the existence of pacemaker tissue in fishes (Laurent, 1962; Laurent et al., 1983; Santer, 1985; Satchell, 1991; Yamauchi, 1980). Furthermore, recent studies have shown that functional pacemaker cells are located either in the sinus venosus, or in a ring of specialized myocardial cells (nodal tissue) located at the base of the sinoatrial ostium in crucian carp (Vornanen et al., 2010) and cyprinids (zebrafish: Tessadori et al., 2012; Stoyek et al., 2015, 2016; goldfish: Newton et al., 2014). Thus, the electrical signal produced by the pacemaker cells does not have to travel far to initiate atrial contraction.

2.2. Atrium

A single atrial chamber is found in all water-breathing fishes. The atrial wall is comprised primarily of a thin layer of cardiac muscle that is easily distended by the low blood pressures of the venous circulation (Forster and Farrell, 1994). Internal muscular trabeculae are present in the atrium of some species. Atrial mass is 8%–25% of ventricular mass and only 0.01%–0.03% of body mass (Table 1). When the atrial myocardium contracts, a sinoatrial valve prevents retrograde blood flow and the kinetic energy inherent to the venous blood increases, and this aids in filling the ventricle.

The primary function of the atrium is to move venous blood into the ventricle, even though the energy inherent in the venous blood is ample to fill the ventricle during atrial diastole. Echo-Doppler imaging in a variety of anesthetized or physically restrained teleost and elasmobranch fishes shows biphasic filling of the ventricle; the first surge of blood (termed an E-wave) is associated with atrial diastole and results from the pressure gradient between the venous system and ventricle, whereas the second surge of blood (termed an A-wave) is a result of energy imparted to the blood by atrial contraction. Measurements of the relative contribution of atrial contraction to ventricular filling vary considerably (see Section 5), and this may be partly due to the response of atrial contraction to various inotropic agents, such as catecholamines (adrenaline and noradrenaline). The atrium also typically receives autonomic innervation (see Section 4; Chapter 1, Volume 36A: Icardo, 2017). However, the extent that inotropic stimulation contributes to atrial contraction, and thus filling the ventricle, has not been quantified.

Beyond its contractile function, the atrium is also an important endocrine tissue. It releases nitric oxide and atrial natriuretic factor, which affect the heart, vascular resistance and kidney function (Cousins and Farrell, 1996; Cousins et al., 1997; Olson and Farrell, 2006; Takei and McCormick, 2013; Chapter 5, Volume, 36A: Imbrogno and Cerra, 2017). The endothelial lining of cardiac chambers has attracted considerable attention in this regard.

Table 1
Ventricular morphometrics in selected fishes

	Relative ventricular mass (%)	Compact myocardium (%)	Relative atrial mass (%)	Source
Cyclostomes				
Eptatretus cirrhatus (New Zealand hagfish)	0.096	None	0.039	P.S. Davie (personal communication)
Eptatretus cirrhatus (New Zealand hagfish)	0.088	None	0.031	Forster et al. (1992)
Eptatretus stoutii (Pacific hagfish)	0.1			Cox et al. (2011)
Elasmobranchs				
Chimaera monstrosa		5.0		Santer and Greer Walker (1980)
Charcharodon carcharias (great white shark)		36.0		Emery et al. (1985)
Isurus oxyrinchus (shortfin mako shark)	0.14	41.5	0.028	P.S. Davie (personal communication)
Isurus oxyrinchus (shortfin mako shark)	0.11			Gregory et al. (2004)
Isurus oxyrinchus (shortfin mako shark)		38		Cox (2015)
Galeocerdo cuvieri (tiger shark)		19.8		Emery et al. (1985)
Prionance glauca (blue shark)		16.7		Emery et al. (1985)
Prionance glauca (blue shark)	0.096			Gregory et al. (2004)
Squalus acanthias (spiny dogfish)		24.5		Santer and Greer Walker (1980)
Squalus acanthias (spiny dogfish)	0.086	22.1		Davie and Farrell (1991b)
Squalus acanthias (spiny dogfish)	0.084	9.0		Farrell et al. (2007a)
Squalus acanthias (spiny dogfish)		10		Cox (2015)
Squatina squatina (monkfish)		34.5		Santer and Greer Walker (1980)
Triakis semifasciata (leopard shark)	0.072			Cox (2015)
Aptychotrema rostrata (epaulette shark)	0.066			Speers-Roesch et al. (2012b)
Caracharhinus plumbeus (sandbar shark)	0.07	15.0		Cox (2015)
Heterodontus francisci (horn shark)	0.086			Gregory et al. (2004)
Raja clavata (thornback ray)		35.8		Santer and Greer Walker (1980)
Raja hyperborea (Arctic skate)		17.3		Santer and Greer Walker (1980)
Dasyatis violacea (pelagic ray)	0.054			Gregory et al. (2004)
Hemiscyllium oscellatum (shovelnose ray)	0.05			Speers-Roesch et al. (2012b)

(*Continued*)

Table 1 (Continued)

	Relative ventricular mass (%)	Compact myocardium (%)	Relative atrial mass (%)	Source
Teleosts with Only Spongiosa				
Chaenocephalus aceratus (hemoglobin-free icefish)	0.30	None		Holeton (1975)
Chionodraco humatus (hemoglobin-free icefish)	0.39	None		Tota et al. (1991)
Pagothenia borchgrevinki (red-blooded Antarctic fish)	0.16	None	0.042	Axelsson et al. (1992)
Pagothenia bernacchi (red-blooded Antarctic fish)	0.11	None	0.050	Axelsson et al. (1992)
Hemitripterus americanus (sea raven)	0.07	None		Farrell et al. (1985)
Chelidonichthys kumu (gurnard)	0.05	None	0.017	P.S. Davie (personal communication)
Gadus morhua (Atlantic cod)	0.08	None		Axelsson and Nilsson (1986)
Oreochromis hybrid (tilapia)	0.034	None		Lague et al. (2012)
Acipenser transmontanus (white sturgeon)	0.044			Gregory et al. (2004)
Dicentrarchus labrax 18°C	0.1			Farrell et al. (2007b)
Dicentrarchus labrax 22°C	0.078			Farrell et al. (2007b)
Pleuronectes platessa (plaice)	0.035	None		Santer et al. (1983)
P. stellatus (starry flounder)	0.04	None		Watters and Smith (1973)
P. americanus (winter flounder)	0.05	None		Joaquim et al. (2004) and Cech et al. (1976)
Coryphaenoides rupestris (deep-sea fish)	0.06	None		Greer Walker et al. (1985)
Pterygoplichthys pardalis (*armoured catfish*)	0.028	None		Hanson et al. (2009)
Teleosts with Compacta				
Katsuwonus pelamis (skipjack tuna)	0.4	65.6	0.061	Farrell et al. (1992)
Thunnus albacores (yellowfin tuna)	0.29	54.5	0.056	Farrell et al. (1992)
Thunnus thynnus (northern bluefin tuna)	0.31	39.1		Poupa et al. (1981) and Santer and Greer Walker (1980)
Thunnus maccoyii (southern bluefin tuna)	0.29	48.5	0.053	Davie (1987)
Thunnus obesus (bigeye tuna)		73.6		Santer and Greer Walker (1980)
Makaira nigricans (Pacific blue marlin)	0.087	48.0	0.013	Davie (1987)
Scomber scombrus (mackerel)	0.18	43.0		Santer and Greer Walker (1980)
Engraulis encrasicolus (anchovy)		61.7		Santer and Greer Walker (1980)
Clupea harengus (herring)		24.8		Santer and Greer Walker (1980)
Seriola grandis (kingfish or yellowtail)	0.11	42.3	0.030	Davie (1987)
Megalops cyprinoides (juvenile tarpon)	0.066	20		Farrell et al. (2007a, b)
Megalops cyprinoides (adult tarpon)	0.066	60		Farrell et al. (2007a, b)

Cyprinus carpio (common carp)		37.0		Bass et al. (1973)
Anguilla (European eel)		34.0		Santer and Greer Walker (1980)
Anguilla dieffenbachii (long finned eel)	0.03–0.10	40.9	0.007–0.013	Davie et al. (1992)
Conger (conger eel)		16.0		Santer and Greer Walker (1980)
Salmo salar (Atlantic salmon)	0.076		0.026	Penney et al. (2014)
Salvelinus alpinus (Arctic char)	0.11		0.022	Penney et al. (2014)
Oncorhynchus tschawytscha (chinook salmon)	0.101			Gallaugher et al. (2001)
Oncorhynchus tschawytscha (chinook salmon; trained)	0.114			Gallaugher et al. (2001)
Oncorhynchus gorbuscha (pink salmon) male	0.166			Clark et al. (2011)
Oncorhynchus gorbuscha (pink salmon) female	0.152			Clark et al. (2011)
Oncorhynchus nerka (sockeye salmon; Early Stuart) male	0.13	42.0		Sandblom et al. (2009b)
Oncorhynchus nerka (sockeye salmon; Early Stuart) female	0.11	44.0		Sandblom et al. (2009b)
Oncorhynchus nerka (sockeye salmon; Harrison) female	0.11	36.0		Eliason et al. (2011)
Oncorhynchus nerka (sockeye salmon; Chilko) female	0.15	38.0		Eliason et al. (2011)
Oncorhynchus nerka (sockeye salmon; Early Stuart) female	0.14	45.0		Eliason et al. (2011)
Oncorhynchus mykiss (rainbow trout) 2 yo; 5°C	0.11	38.7	~0.017	Farrell et al. (1988c)
Oncorhynchus mykiss (rainbow trout) 2 yo; 10°C	0.077	44.7		Farrell et al. (1988c)
Oncorhynchus mykiss (rainbow trout) 2 yo; 15°C	0.078	44.8		Farrell et al. (1988c)
Oncorhynchus mykiss (rainbow trout) 3 yo; 5°C	0.13	34.1		Farrell et al. (1988c)
Oncorhynchus mykiss (rainbow trout) 3 yo; 10°C	0.081	39.3		Farrell et al. (1988c)
Oncorhynchus mykiss (rainbow trout) 3 yo; 15°C	0.11	47		Farrell et al. (1988c)
Oncorhynchus mykiss (rainbow trout) male 4°C	0.13			Klaiman et al. (2011)
Oncorhynchus mykiss (rainbow trout) male 12°C	0.1			Klaiman et al. (2011)
Oncorhynchus mykiss (rainbow trout) male 17°C	0.85			Klaiman et al. (2011)
Oncorhynchus mykiss (rainbow trout) 17.6°C; Hct=35%	0.086	32.2		Simonot and Farrell (2007)
Oncorhynchus mykiss (rainbow trout) 17.6°C; Hct=24%	0.126	40.5		Simonot and Farrell (2007)
Oncorhynchus mykiss (rainbow trout) 6.5°C; Hct=39%	0.087	29.4		Simonot and Farrell (2007)
Oncorhynchus mykiss (rainbow trout) 6.5°C; Hct=16%	0.118	37.0		Simonot and Farrell (2007)

Hct=hematocrit.
yo=years old.

The color of the fish atrium (and ventricle) varies considerably, from pale brown to deep red depending on the expression of cardiac myoglobin. The very active tunas appear to have the highest cardiac myoglobin concentrations among fishes (Poupa et al., 1981), whereas sluggish species such as ocean pout (*Macrozoarces americanus*) express very low levels of cardiac myoglobin (Driedzic, 1983). An extreme situation is the loss of the genes for producing myoglobin, which appears to have occurred independently several times among the Notothenoid fishes that inhabit frigid Antarctic waters (Sidell and O'Brien, 2006). Their hearts appear pale, with only their mitochondria providing a light brownish color (Fig. 1). Certain fishes within this group have

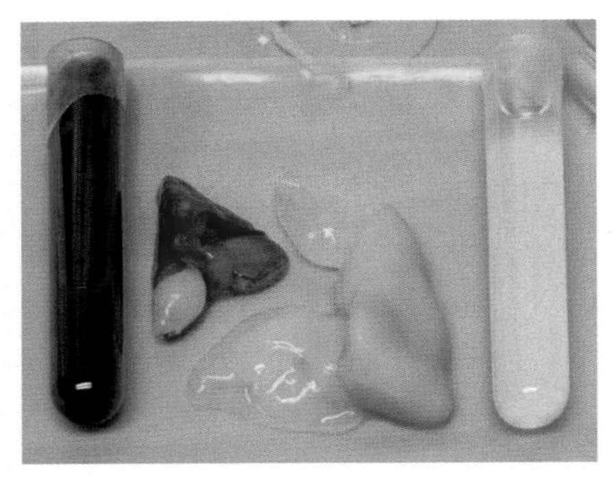

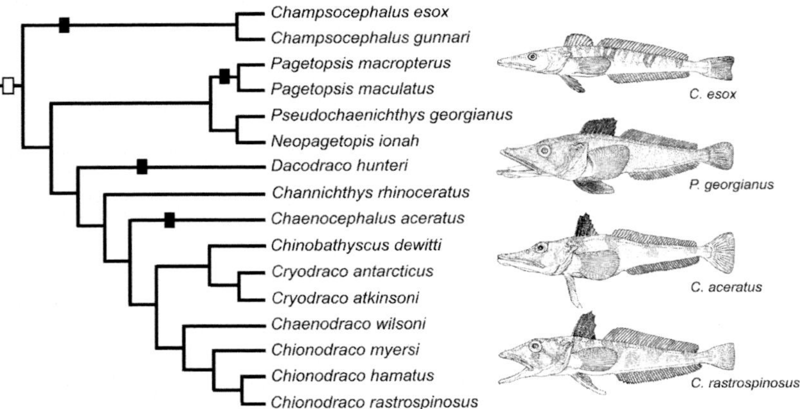

Fig. 1. Example of the cardiomegaly of the Type I fish heart (right side of photograph, *upper panel*) that results from mutations which prevent the production of hemoglobin (the blood in the tube is *white*) and myoglobin (the heart is *pale*) in *Chaenocephalus aceratus*; as compared to another Antarctic teleost (*Notothenia coriiceps*, *left side*) lacking these mutations. The phylogeny for hemoglobin-free Antarctic teleosts (*lower panel*) indicates that the mutations involving the myoglobin-free and hemoglobin-free condition represent several independent events, as suggested by Sidell and O'Brien (2006). Upper panel photo credit: Stuart Eggington. Lower panel photo credit: Kristin O'Brien.

also lost the genes necessary for hemoglobin production, which renders the blood white due to an absence of red blood cells (Fig. 1). The functional consequences of these two missing proteins are being intensely studied. Myoglobin has a much higher oxygen affinity than hemoglobin (its tetrameric cousin) and enhances both oxygen content and oxygen diffusion within cardiac cells. This myocardial oxygen store is thought to be particularly valuable when fish become hypoxemic, an event that can occur during aquatic hypoxia, exhaustive exercise, and at extremely warm temperatures when venous blood oxygen content decreases (Driedzic, 1983; Ekstrom et al., 2016; Eliason et al., 2013a; Farrell and Clutterham, 2003).

2.3. Ventricle

Ventricular size is a primary determinant of both V_s and ventral aortic blood pressure (P_{VA}). Relative ventricular mass (RVM, ventricular mass as a percentage of fish mass) varies about 10-fold among fish species [e.g., from as low as 0.035% of body mass in winter flounder (*Pseudopleuronectes americanus*) to as high as 0.38% of body mass in skipjack tuna, *Katsuwonus pelamis*; Table 1]. Active teleost fishes have a larger RVM than active elasmobranch fishes, but benthic elasmobranch fishes have a larger RVM than benthic teleost fishes (Poupa and Lindstrom, 1983). Also, endothermic sharks have a larger RVM than ectothermic sharks (Emery et al., 1985).

Although differences in RVM explain some of the species-specific variability in P_{VA} and cardiac output ($\dot{Q}$) (Tables 1 and 2), they do not fully account for the nearly 10-fold difference in P_{VA} between tunas and hagfishes, which represents the known extreme among fishes (Note: the reported peak systolic blood pressure in the ventral aorta of zebrafish of 0.033 kPa reported by Hu et al. (2001) is unbelievably low, even by hagfish standards). The RVM and $\dot{Q}$ of tunas are similar to those of Antarctic fishes without hemoglobin, but their P_{VA} values differ considerably (Table 2) because ventricular wall thickness and end-diastolic volume (and therefore V_s) are very different. The cardiac mass of fishes usually scales in direct proportion to body mass (Poupa and Lindstrom, 1983), as it does in other vertebrates. Two known exceptions are bluefin tuna (*Thunnus thynnus*) and the great white shark (*Carcharodon carcharias*) whose ventricular mass scales positively with body mass (Brill and Lai, 2016).

At the species level, RVM is surprisingly plastic (Gamperl and Farrell, 2004; Chapter 3, Volume 36B: Gillis and Johnson, 2017). For example, steelhead trout (*Oncorhynchus mykiss*) that are anadromous have a larger RVM than rainbow trout (*O. mykiss*) that inhabit lakes (Graham and Farrell, 1992); and populations of sockeye salmon (*Oncorhynchus nerka*) with a larger RVM swim further up the Fraser River to spawn (Table 1; Eliason et al., 2011,

Table 2

Direct measurements of cardiac output and related variables in fishes

Species	Mass (kg)	Acclimation temp. (°C)	State	$\dot{Q}$ ($mL\,min^{-1}\,kg^{-1}$)	f_H (min^{-1})	V_s (mL)	P_{VA} (kPa)	CPO ($mW\,g^{-1}$)	Total Peripheral resistance ($Pa\,min\,kg\,mL^{-1}$)	P_{DA} (kPa)	Systemic vascular resistance ($Pa\,min\,kg\,mL^{-1}$)	P_{CV} (kPa)	Source
Cyclostomes													
Myxine glutinosa	0.05–0.07	10–11	Routine (+AD)	8.7 (25.0)	22.3 (25.0)	0.41 (1.0)	1.04 (1.54)		119 (60)	0.77 (0.90)	36 (84)		Axelsson et al. (1990)
Eptatretus cirrhatus	1.4	17	Routine	15.8	24.9	0.67	1.6		101	1.30	84		Forster et al. (1992)
Eptatretus stouti	0.16	10	Routine (anoxia recovery)	12.3 (25)	10.4 (17.0)	1.3 (2.6)	0.9		70				Cox et al. (2010)
Elasmobranchs													
Carcharhinus plumbeus	1.6	26	Routine	17.5	42	0.40		2.2					Cox et al. (2017, 2016b)
Isurus oxyrinchus			Anesthetized	47.4	21	2.3							Lai et al. (1997)
Raja rhina		10		21.2–23.3									Satchell et al. (1970)
Triakis semifasciata		14–24	Routine	33.1									Lai et al. (1989a)
Teleosts													
Antarctic Species													
Chaenocephalus aceratus		1–2		20–30									Hemmingsen et al. (1972)
Chaenocephalus aceratus		0.5–2		66									Hemmingsen and Douglas (1977)
Pseudochaenichthyes		0.5–2		50–87									Hemmingsen and Douglas (1977)
Pagothenia bernacchii	0.05	0	Routine (+AD)	17.6 (24.6)	10.5 (20.0)	1.56 (1.86)	3.09 (4.00)	0.82 (1.48)	197		89		Axelsson et al. (1992)
Temperate Species													
Ptychocheilus oregonensis			Routine	18.7									
				56									

Gadus morhua		10	Routine	29.1							Petersson and Nilsson (1980)
		9–10		17–26							Jones et al. (1974)
		10–11		17.3							Axelsson and Nilsson (1986)
		10–12		19.2							Axelsson (1988)
Ophiodon elongatus		13	Routine	5.9							Stevens et al. (1972)
	3.8 – 6.5	10		10.9							Farrell (1982)
	3.8 – 6.5	10	Routine	11.2	29.4	0.38					Farrell (1982)
Hemitripterus americanus		7	Routine	10.8							Farrell (1986)
		10.5		14.6							Farrell (1986)
		10–12		18.8							Axelsson et al. (1989)
Pseudopleuronectes americanus	0.46	8	Routine	10	25.3	0.41			3.03	320	Mendonça and Gamperl (2010)
Anguilla		8.5–10.5		12.2							Hughes et al. (1982)
		15		11.5							Peyraud-Waitzenegger and Soulier (1989)
		18–20		16.6–25.7							Peyraud-Waitzenegger and Soulier (1989)
Anguilla australis		15.5–18.5		9.1							Hipkins et al. (1986)
		16–20		6.2–10.2							Hipkins and Smith (1983)
		16–20		10.4							Hipkins (1985)
Oncorhynchus mykiss	0.1 – 0.7		Routine	36.7			31.2	0.16	25.4	0.69	Wood and Shelton (1980)
	0.5–0.8	5	Routine	19.0	43.0	0.44			2.2	115	Gamperl et al. (1994a)
	3.3–3.5	8–10	Routine	39.4	45.3	0.87			4.8	133	Axelsson and Farrell (1993)
		13–16		34							Wood (1974)
		13–16		31.2–36.7							Wood and Shelton (1980)

(*Continued*)

Table 2 (Continued)

				Total								
Species	Mass (kg)	Acclimation temp. (°C)	State	$\dot{Q}$ (mL min^{-1} kg^{-1})	f_H (min^{-1})	V_s (mL)	P_{VA} (kPa)	CPO (mW g^{-1})	Peripheral resistance (Pa min kg mL^{-1})	P_{DA} (kPa)	Systemic vascular resistance (Pa min kg mL^{-1})	P_{CV} (kPa)	Source
	1.05	17	Routine	18.3	57.0	0.37							Simonot and Farrell (2007)
			Anemic (9.4% Hct)	21.7	47.9	0.45							
			Anemic (3.6% Hct)	33.3	53.2	0.62							
Tropical species													
Thunnus alalunga	7.9–9.9	16.5	Anesthetized	29.4	90	2.8							Lai et al. (1987)
Thunnus alalunga	7.8 – 11		Routine	29.4			84.2		1.23	48.1	1.64		Lai et al. (1987)
Thunnus albacares	1.0 – 2.0		Routine	61.9			84.4		0.64	44.9	0.73		Jones et al. (1993)
Katsuwanus pelamis		24–26		132.3									Bushnell (1988)
Air-Breathing Fishes													
Protopterus aethiopicus		18		20									Johansen et al. (1968)
Electrophorus electricus		28–30		40–70									Johansen et al. (1968)
Hoplerythrinus unitaeniatus		26–30		27.6–32.2									Farrell (1978)
Megalops cyprinoides	0.49	27	Normoxia	14	35	0.49							Clark et al. (2007)
			~1 bl s^{-1}	46	58	1							
			Hypoxia (2 kPa O_2)	43	53	1.02							
			Hypoxic swim	64	76	1.07							
Megalops cyprinoides	1.21	27	Normoxia	18	63	0.28							Clark et al. (2007)
			~1 bl s^{-1}	17	61	0.28							
			Hypoxia, 2 kPa O_2	35	77	0.45							
			Hypoxic swim	17	69	0.25							

Maximum Swimming Values												
Elasmobranchs												
Triakis semifasciata[c]	1.93	14–24	Routine	33.1	51.3	0.77	48					Lai et al. (1989a)
			0.3–0.7 $bl\,s^{-1}$	56.2	55.1	1.02	58					
			Post-exercise	60.4	49.5	1.22	55					
Isurus oxyrinchus	3.8–12.3	20–22	0.45 $bl\,s^{-1}$		52		7.5			4.1		Lai et al. (1997)
Teleosts												
Pagothenia barchgrevinki	0.064	0	Routine	29.6	11.3	2.16	3.60	1.05	159			Axelsson et al. (1992)
			1 $bl\,s^{-1}$	51.8	21.0	2.16	3.80	1.93	73			
Gadus morhua			Routine	12.0	28.3	0.45						Webber et al. (1998)
			Max swim	35.0	40.0	0.88						
Gadus morhua	0.4–0.8	10–11	Routine	17.3	43.2	0.39	36.8		0.74	24.0	1.39	Axelsson and Nilsson (1986)
			2–3 $bl\,s^{-1}$	25.4	51.2	0.49	46.5		0.65	30.0	1.18	
Gadus morhua	0.35–0.73	10–12	Routine	19.2	30.5	0.49						Axelsson (1988)
			2–3 $bl\,s^{-1}$	30.2	42.7	0.61						
Hemitripterus americanus	0.67–1.40	10–12	Routine	18.8	37.3	0.51	28.5		0.28	23.3	1.24	Axelsson et al. (1989)
			30 $cm\,s^{-1}$	30.9	49.1	0.64	35.3		0.29	26.3	0.85	
Catostomus macrocheilus			Routine	9.6	27.0	0.36						Kolok et al. (1993)
			Max swim	37.0	47.0	0.78						
Catostomus macrocheilus	0.66	5	Routine	12.0	21.0	0.58						Kolok et al. (1993)
			Max swim	18.0	25.0	0.70						
		10	Routine	10.0	26.0	0.35						
			Max swim	37.0	46.0	0.78						
		16	Routine	18.0	40.0	0.43						
			Max swim	44.0	72.0	0.60						
Pleuronectes americanus	0.57	4	Routine	9.8	20.5	0.50						Joaquim et al. (2004)
			Max swim	25.4	31.6	0.80						
	0.68	10	Routine	15.5	34.3	0.47						
			Max swim	39.2	52.4	0.74						

(*Continued*)

Table 2 (Continued)

				Total									
Species	Mass (kg)	Acclimation temp. (°C)	State	$\dot{Q}$ (mL min^{-1} kg^{-1})	f_H (min^{-1})	V_s (mL)	P_{VA} (kPa)	CPO (mW g^{-1})	Peripheral resistance (Pa min kg mL^{-1})	P_{DA} (kPa)	Systemic vascular resistance (Pa min kg mL^{-1})	P_{CV} (kPa)	Source
Ptychocheilus oregonensis	0.12–0.23	5	Routine	13.3	17.0	0.84							Kolok and Farrell (1994)
			Max swim	33.0	30.0	1.10							
	0.18–0.39	16	Routine	19.1	46.0	0.45							
			Max swim	52.5	72.0	0.73							
		10–12	Routine	18.7	43.0	0.45							
			Max swim	56.0	73.0	0.77							
Dicentrarchus labrax			Routine	100%	80.0	100%				3.6	100%	0.11	Sandblom et al. (2005)
			1 bl s^{-1}	115%	88.0	100%				3.6	100%	0.12	
			2 bl s^{-1}	138%	103.0	100%				3.5	195%	0.16	
Anguilla australis	0.62		Routine	11.3			38.6		1.34	23.5	2.08		Davie and Forster (1980)
			15 cm s^{-1}	11.3			39.0		1.36	23.6	2.09		
			24 cm s^{-1}	10.3			46.8		2.42	21.8	2.11		
Salmo trutta (3N)	0.29–0.61	14	Routine	100%	46.0	100%							Altimiras et al. (2002)
			Max swim	230%	92.0	125%							
			Routine	100%	66.0	100%							
			Max swim	200%	93.0	140%							
Oncorhynchus mykiss			Routine	22.6	43.2	0.58							Thorarensen et al. (1996a, b)
			Max swim	48.7	66.9	0.73							
Oncorhynchus mykiss	0.4–1.02	10	Routine	26.6	48.4	0.58				3.29	126		Thorarensen et al. (1996a, b)
			U_{crit}	48.7	66.9	0.73				4.12	90		
Oncorhynchus mykiss	0.9 – 1.5		Routine	17.6			38.8		0.44	31.0	1.76		Kiceniuk and Jones (1977)
			44 cm s^{-1}	28.4			40.2		0.36	30.0	1.06		
			63 cm s^{-1}	34.8			48.7		0.45	30.0	0.95		
			73 cm s^{-1}	42.9			52.2		0.43	33.7	0.79		

O. mykiss (good swimmer)	1.0–1.2	16	Routine	36	82	0.44	3.9	0.120	Claireaux et al. (2005)
			U_{crit}	68	98	0.77	4.2	0.064	
(poor swimmer)		16	Routine	28	80	0.44	3.7	0.135	
			U_{crit}	47	92	0.66	4.4	0.087	
O. tshawytscha (untrained)	0.34–0.39	9	Routine	35.8	57.0	0.63	3.2	0.095	Gallaugher et al. (2001)
			U_{crit}	65.6	63.0	1.04	4.0	0.071	
(trained)			Routine	34.0	53.0	0.63	3.5	0.112	
			U_{crit}	65.0	64.0	0.97	3.6	0.058	
O. gorbuscha	1.6	17	Max swim	154.0	115.0	1.34			Clark et al. (2011)
O. nerka (Early Stuart)	2.36	15–20	Routine	34.8	70.1	0.49			Eliason et al. (2013a, b, c)
			Max swim	100.3	93.1	1.08			
O. nerka (Quesnel)	2.53	15–20	Routine	34.7	61.0	0.57			Eliason et al. (2013a, b, c)
			Max swim	101.9	94.0	1.09			
O. nerka (Chilko)	2.33	15–20	Routine	34.8	67.0	0.53			Eliason et al. (2013a, b, c)
			Max swim	100.5	94.4	1.11			
O. nerka	2.2–2.9	15	Routine	25.3	65.0	0.38			Steinhausen et al. (2008)
			Max swim	58.0	81.0	0.69			
	2.2–2.9	21–24	Routine	43.0	100.0	0.40			
			Max swim	67.8	104.0	0.63			
		15	1 h recovery	33.8	82.5	0.41			
Gadus morhua	0.55–0.68	10	Routine	23.1	32.9	0.73			Petersen and Gamperl (2010)
			Normoxic swim	44.5	46.3	0.99			
			Hypoxic swim (8–9 kPa O_2)	34.6	38.8	0.99			
			Hypoxia acclimated	17.0	35.2	0.49			
			Normoxic swim	34.2	50.1	0.74			
			Hypoxic swim (8–9 kPa O_2)	31.3	48.5	0.71			

(Continued)

Table 2 (Continued)

				Total									
Species	Mass (kg)	Acclimation temp. (°C)	State	$\dot{Q}$ (mL min^{-1} kg^{-1})	f_H (min^{-1})	V_s (mL)	P_{VA} (kPa)	CPO (mW g^{-1})	Peripheral resistance (Pa min kg mL^{-1})	P_{DA} (kPa)	Systemic vascular resistance (Pa min kg mL^{-1})	P_{CV} (kPa)	Source
Temperature Acclimation													
Cyprinus carpio	0.79–1.23	6	Routine	9.4	7.1	1.22							Stecyk and Farrell (2002)
		10	Routine	18.7	12.0	1.58							
		15	Routine	15.3	16.7	0.77							
Acute Warming													
Squalus acanthias	1.89	10	Normoxia	100%	19	1.00				1.7	100%	−0.08	Sandblom et al. (2009b)
			Warming to 16°C	143%	30	0.80				1.9	80%	−0.08	
Gadus morhua	1.13	10	Routine	21.5	36.3	0.60							Gollock et al. (2006)
			Warming to CT_{max}	52.6	71.8	0.76							
Salmo salar	0.62	10	Routine	100%	52.0	100%							Penney et al. (2014)
		26.5	Warming to 26.5°C—CT_{max}	221%	134.0	111%							
Salvelinus alpinus	0.75	10	Routine	100%	68.0	100%							Penney et al. (2014)
		23.7	Warming to 23.7°C—CT_{max}	217%	115.0	126%							
Oncorhynchus mykiss	0.39–0.41	14–15	Routine	100%	60.0	100%				3.57	100%		Gamperl et al. (2011)
			Warming to CT_{max}	210%	125.0	100%				4.30	50%		
O. mykiss		10	Routine	100%	72.0	100%							Keen and Gamperl (2012)
			Warming to 23°C	189%	116.0	127%							
O. tshawytscha	2.1–5.4	13	Routine	28.5	58.5	0.49				5.20	182	0.03	Clark et al. (2008)
			Warming to 17°C	38.0	75.0	0.51				5.40	142	0.03	
			Warming to 21°C	45.0	92.0	0.50				5.60	124	0.03	
			Warming to 25°C	56.3	105.3	0.53				5.60	98	0.19	

O. nerka	2.2–2.9	15	Routine	25.3	65.2	0.38							Steinhausen et al. (2008)
			Warming to 21–24°C	43.0	100.0	0.40							
	2.2–2.9	15	Swimming 70% U_{crit}	58.0	81.4	0.69							
			Warming to 21–24°C	67.8	104.5	0.63							
			Recovery at 15°C	33.8	82.0	0.41							
O. gorbuscha	1.6	7	Max swimming	80	55.0	1.70							Clark et al. (2011)
			Warming to 21°C	140	110.0	1.27							
			Acute warming to 25°C	115	120.0	0.96							
Hypoxia													
Cyclostomes													
Myrine glutinosa	0.55–0.91	10–11	Routine	8.7					0.23	5.8	0.66		Axelsson et al. (1990)
			Hypoxia (2 kPa O_2)	8.7			9.4		0.30	6.8	0.78		
Eptatretus stoutii	0.16	10	Normoxia	12.3	10.0	1.3	0.90	0.26					Cox et al. (2010)
			Anoxia	8.5	4.0	2.5	0.95	0.25					
Elasmobranchs													
Squalus acanthias	1.89	10	Normoxia	100% (110)	19 (27)	100% (110)				1.7	100%	−0.06	Sandblom et al. (2009a, b)
			Hypoxia—6.9 kPa O_2 (+ atropine)	86% (107)	19 (27)	84% (107)				1.8		−0.04	
			Hypoxia-2.5 kPa O_2 (+ atropine)	54% (84)	15 (23)	69% (66)				1.6		−0.02	
Hemiscyllium oscellatum	1.3	28	Normoxia (16 kPa O_2)	44.0	60.0	0.72		5.9		3.1	0.072		Speers-Roesch et al. (2012a)
			~0.1 kPa O_2	18.0	20.0	0.92		1.4		1.8	0.108		
Aptychotrema rostrata	1.5	28	Normoxia (16 kPa O_2)	39.00	56.0	0.70		4.0		3.0	0.078		Speers-Roesch et al. (2012a, b)
			1.6 kPa O_2	20.00	22.0	0.95		1.2		1.7	0.08		

(*Continued*)

Table 2 (Continued)

									Total				
Species	Mass (kg)	Acclimation temp. (°C)	State	$\dot{Q}$ (mL min^{-1} kg^{-1})	f_H (min^{-1})	V_s (mL)	P_{VA} (kPa)	CPO (mW g^{-1})	Peripheral resistance (Pa min kg mL^{-1})	P_{DA} (kPa)	Systemic vascular resistance (Pa min kg mL^{-1})	P_{CV} (kPa)	Source
Teleosts													
Cyprinus carpio	0.79–1.23	5	Normoxia	4.5	9.0	0.52	2.62	0.294	0.663				Stecyk and Farrell (2006)
			Severe hypoxia	2.3	3.3	0.92	1.63	0.095	0.80				
		10	Normoxia	9.5	16.4	0.61	3.12	0.71	0.342				
			Severe hypoxia	3.0	4.5	0.70	2.30	0.167	0.806				
		15	Normoxia	11.3	24.0	0.57	3.81	0.855	0.356				
			Severe hypoxia	4.1	5.8	0.82	2.79	0.266	0.811				
Oncorhynchus mykiss	0.9–1.5	9	Routine	18.0	64.0	0.29				2.8	156		Gamperl et al. (1994a)
			Hypoxia (12 kPa O_2)	21.1	60.0	0.36				2.8	156		
Gadus morhua	0.4 – 1.3		Routine	19.2			36.8		0.70	23.3	1.21		Fritsche and Nilsson (1990)
			Hypoxia (4.6 kPa O_2)	19.2			47.3		0.51	37.5	1.95		
Ophiodon elongatus	3.8 – 6.5	10	Routine	11.2			38.0		0.87	28.3	2.53		Farrell (1982)
			Hypoxia (10 kPa O_2)	9.9			39.2		0.91	30.2	3.05		
			Hypoxia (4.6 kPa O_2)	7.7			33.8		1.29	23.9	3.10		
Anguilla	0.51		Routine	11.8			37.9		1.12	25.0	2.17		Peyraud-Waitzenegger and Soulier (1989)
			Hypoxia (5.2 kPa O_2)	7.8			32.3		1.88	17.6	2.26		

Anguilla japonica	0.3 – 0.6		Routine	11.0			24.2		0.66	16.9	1.54	Chan (1986)
			Hypoxia (10.5 kPa O_2)	9.5			23.3		0.62	17.4	1.83	
			Hypoxia (5.2 kPa O_2)	4.8			14.3		0.97	9.7	2.05	
Oreochromis hybrid	0.71	22	Normoxia	12.0	35.0	0.30	3.1	1.25	0.30			Speers-Roesch et al. (2010)
			8 h at ~5 kPa O_2	6.0	18.0	0.30	3.1	0.70	0.55			
			1 h recovery	21.0	49.0	0.42	2.2	1.85	0.10			
Katsuwonus pelamis	1.6	24–26	Routine	132.3			87.3		0.36	40.2	0.30	Bushnell (1988)
			Hypoxia (17 kPa O_2)	132.3			87.3		0.36	40.2	0.30	
			Hypoxia (12 kPa O_2)	105.3			87.3		0.49	36.2	0.34	
			Hypoxia (6.5 kPa O_2)	75.0			87.3		0.63	40.2	0.54	
Thunnus albacares	1.4	24–26	Routine	115			89.7		0.49	32.6	0.28	Bushnell and Brill (1991)
			Hypoxia (17 kPa O_2)	115			89.7		0.50	32.6	0.28	
			Hypoxia (12 kPa O_2)	115			89.7		0.50	32.6	0.28	
			Hypoxia (6.5 kPa O_2)	74.1			89.7		0.77	32.6	0.44	
Digestion												
Oncorhynchus mykiss	0.90	11	Routine	17.5	54	0.31						Eliason and Farrell (2014)
			Peak post-prandial	32.2	73	0.60						

$\dot{Q}$, cardiac output; f_H, heart rate; V_s, stroke volume; P_{VA}, ventral aortic pressure; *CPO*, cardiac power output; P_{DA}, dorsal aortic pressure; P_{CV}, central venous pressure; $+AD$=adrenaline injection, peak response; U_{crit}, critical swimming speed; *Hct*, hematocrit.

100% represents the routine $\dot{Q}$ and V_s, with the treatment expressed relative to this amount (measured with a Doppler flow probe).

3N indicates triploid fish.

2013c). RVM also increases by up to 70% in salmonids in response to cold acclimation, experimental anaemia, exercise training and male sexual maturation (Davison, 1989; Farrell, 1987; Farrell et al., 1988a, 1990; Goolish, 1987; Graham and Farrell, 1989; Holeton, 1970; Johnston et al., 1983; Klaiman et al., 2011; Simonot and Farrell, 2007, 2009; Tota et al., 1991; Tsukuda et al., 1985; Chapter 3, Volume 36B: Gillis and Johnson, 2017); although some studies report no change in RVM with cold acclimation. A change in RVM can directly impact cardiac performance through potential effects on V_s and P_{VA}, and potentially the cardiac response to P_{CV}.

The ventricle is comprised almost entirely of myocardial cells that have a variable architectural arrangement among fish species (Farrell and Jones, 1992; Santer, 1985; Chapter 1, Volume 36A: Icardo, 2017). The ventricle in all fish embryos has an entirely trabecular architecture (i.e., spongiosa or spongy myocardium). However, during development, when the ventricular wall is thickening, the dividing cardiomyocytes can follow one of two developmental trajectories (see Chapter 2, Volume 36B: Burggren et al., 2017). In about half of extant fish species, the ventricle retains its trabecular architecture, resulting in an adult ventricle that is composed entirely of spongy myocardium. The alternative trajectory involves condensation of cardiomyocytes on the outside of the ventricle, which forms an outer compact layer (compacta or compact myocardium), a layer that thickens with age (Farrell et al., 1988b) and encases the spongy myocardium. There are no examples of adult fish hearts with 100% compacta. Rather, the proportion of compacta ranges from 5% of ventricular mass in Chimaerid elasmobranch fishes to 40%–60% in active swimmers that ram-ventilate such as scombrids, clupeids, and salmonids (Table 1). The mechanisms controlling the start and end of myocardial compaction during fish growth are unknown.

Tota (1989) proposed four major morphological categories (Types I–IV) of adult fish hearts on the basis of the presence of compact myocardium in the ventricle and overall cardiac vascularity (see Chapter 1, Volume 36A: Icardo, 2017). The main distinguishing features of these categories were summarized by Farrell and Jones (1992).

Type I. The ventricle is composed of only trabecular myocardium, and there is no coronary circulation (Farrell et al., 2012; Tota, 1989). Instead, there is diffusional exchange of oxygen and carbon dioxide between the venous blood and the trabecular myocardium. In these hearts, blood supply is effectively the entire $\dot{Q}$, but venous blood has a much lower partial pressure of oxygen (PO_2) than arterial blood. Type I hearts are the most common because all cyclostomes and about half of the adult teleost species have Type I hearts.

Type II. The ventricle is partially composed of a vascularized compact myocardium. While the coronary circulation provides an arterial blood supply to the compact myocardium (see Section 2.5), the trabecular myocardium has no coronary vessels. Type II hearts are present in many active and hypoxia-tolerant teleosts (Fig. 2). Many teleost species have some ventricular compact myocardium with an associated coronary circulation.

Types III and IV. The ventricle has vascularized compact myocardium and vascularized trabecular myocardium (Fig. 2), with some of the atrial trabeculae also being vascularized. Elasmobranch fishes with <30% compact myocardium are categorized as Type III, whereas tunas (and any remaining elasmobranchs yet to be discovered) with >40% compact myocardium are currently categorized as Type IV (Davie and Farrell, 1991a; Tota, 1989). Yet, the value of making a distinction between Type III and IV hearts is now questionable because: (1) the percentage of compact myocardium is plastic within a species; (2) all compact ventricular myocardium is likely functionally similar, independent of the percentage (see Section 6); and (3) the reported estimate of compact myocardium depends on the methodology used. For example, peeling the compact layer from the trabecular layer (or vice versa) to compare their relative total masses (Farrell et al., 2007a) requires a decision as to where the compact layer ends and the trabecular layer begins, and the two layers are not always easy to separate. Many of the earliest measurements used histological images (ventricular cross-sections). While this approach has benefited from image analysis software that was not available to earlier investigators, the thickness of the compact layer is heterogeneous around the ventricle, which means that multiple cross-sections are needed to provide reliable estimates. Also, even with modern image analysis software, determining a cross-sectional area of a highly branched trabecular structure is difficult unless the ventricle is fully distended prior to fixation. Comparisons of the percentage compact myocardium between species (Table 1) must acknowledge these concerns. Alternatively, some studies now simply measure a change in the linear thickness of the compact myocardium at a common location on the ventricular wall (e.g., Anttila et al., 2014b, 2015; Klaiman et al., 2011; Keen et al., 2015; Chapter 3, Volume 36B: Gillis and Johnson, 2017).

The compact myocardium has up to three different fiber arrangements (depending to some extent on the percentage of compact myocardium), which are described in detail in Chapter 1 (Volume 36A: Icardo, 2017). The basal architectural arrangement appears to be longitudinal looped fibers, which when contracted, would reduce the longitudinal and transverse diameters of the ventricle (Sanchez-Quintana and Hurle, 1987). Adding an inner fiber layer (either as a circular arrangement as in the sac-like elasmobranch ventricle, or as coils around the vertices as in the pyramidal tuna ventricle) may provide a mechanical advantage to develop a higher P_{VA}. Contracting trabeculae of the spongy myocardium, in contrast, tend to pull the walls of the ventricle together to better expel blood from the ventricle and lower end-systolic volume.

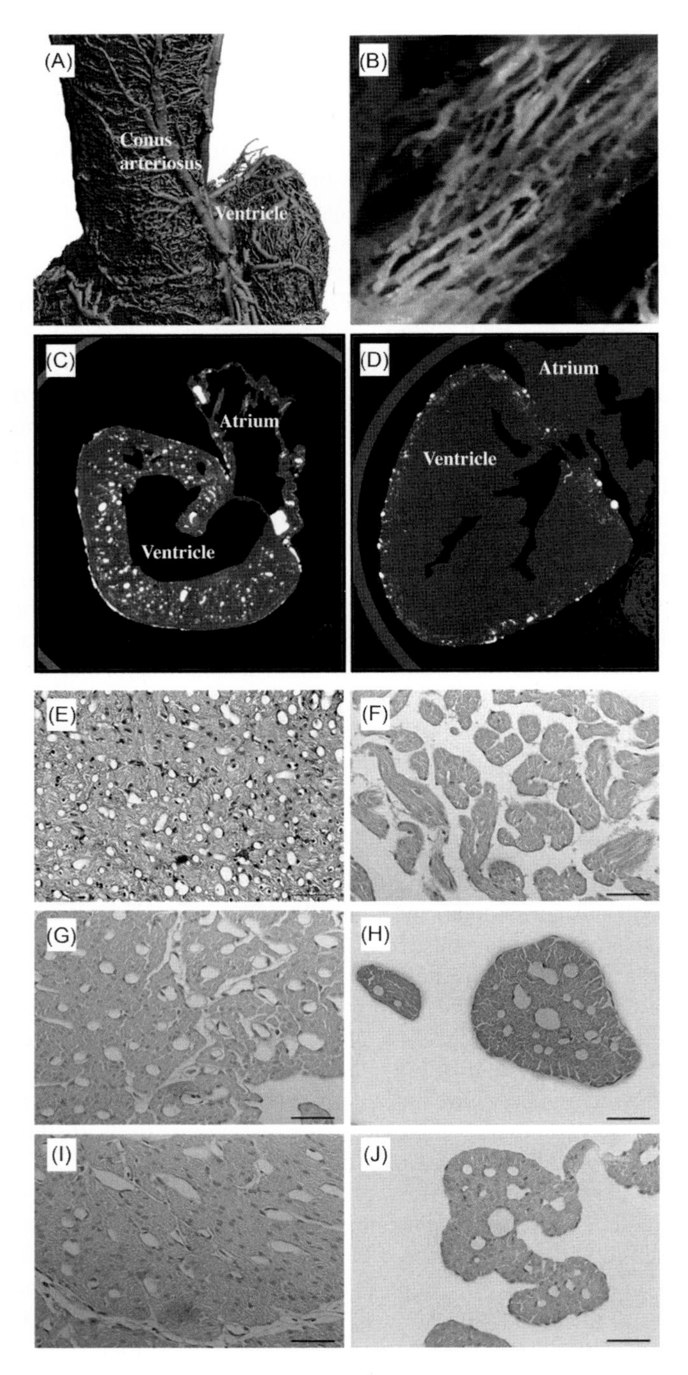

Fig. 2. A montage of images that illustrate the disparate arrangements of the coronary circulation in dogfish (Type III) and rainbow trout (Type II) hearts. (A) A vascular cast of the large coronary vessels in the outer layer of the dogfish shark conus arteriosus and ventricle. This aspect shows the coalescing of coronary veins on the conus as they drain toward the ventricle and onwards.

(B) A vascular cast of the extensive network of coronary microvessels that is typical for cardiac trabeculae in the dogfish shark ventricle. (C) A micro-X-ray image of vascular casting material showing large vessels (*white*) throughout a cross-section of the ventricle and atrium of the dogfish shark. (D) A similar micro-X-ray image of a cross-section through the ventricle and atrium of the rainbow trout. Note the confinement of the vascular casting material (*white*) to large vessels in the compact myocardium. (E–J) Standard histological cross-sections of various cardiac tissues in which the heart was perfusion-fixed after first washing out the red blood cells from the coronary circulation with perfusate; cross-sections of blood vessels therefore appear clear. (E) The compact ventricular myocardium of rainbow trout. (F) A ventricular trabeculae of rainbow trout, showing the absence of microvessels. (G) The ventricular compact myocardium of dogfish shark. Note the slightly larger diameter of the microvessels. (H) Two ventricular trabeculae from the dogfish shark. (I) Conal myocardium of dogfish shark. (J) Atrial trabeculum of dogfish shark. Scale bar in cross-sections = 50 μm. Credit: Cox, 2015. The functional significance and evolution of the coronary circulation in sharks. Dissertation, University of British Columbia; Cox, G.K., Kennedy, G.E., Farrell, A.P., 2016a. Morphological arrangement of the coronary vasculature in a shark (*Squalus suckleі*) and a teleost (*Oncorhynchus mykiss*). J. Morphol. 277, 896–905.

Burggren et al. (2014) detail the potential benefits of a complete or partial retention of a trabecular heart. In brief, a spongy ventricle is presumably more compliant and easier to fill when it is not wrapped with compact myocardium, and this has implications for the P_{CV} needed to fill the heart (see Section 5). As we note above, trabeculae that extend across the ventricle will pull the ventricular wall inwards when they contract, whereas contraction of compact myocardium compresses or stiffens the ventricle wall—depending on fiber orientation. These differences will influence cardiac pumping mechanics, end-systolic volume, and mechanical efficiency. Teleost fishes without compacta often have a smaller RVM and a lower P_{VA}. However, the very large heart (composed of spongy myocardium) of the hemoglobin-free Antarctic fishes is an obvious exception because they have a low P_{VA} (Table 2). Ventricular trabeculae have different cross-sectional shapes and areas, but the relative benefits are not entirely clear. For example, elasmobranch fishes have tubular trabeculae with diameters exceeding 200 μm (Cox, 2015; Cox et al., 2017; Santer, 1985), which may be why they require an extensive coronary circulation. Also, the low venous oxygen content in elasmobranch fishes may have triggered the need for a coronary circulation in both the spongy as well as the compact myocardium (Brill and Lai, 2016). In contrast, the trabeculae of salmonids are more sheet-like (i.e., very thin in one-dimension; Pieperhoff et al., 2009) and are not vascularized. Although trabecular cross-sectional area determines its contractile force and the diffusion distance, the P_{VA} developed is the sum of all contractile activity and is higher in the salmonid heart than in hearts of elasmobranch fishes (Table 2).

2.4. Cardiac Outflow Tract

Chapter 1 (Volume 36A: Icardo, 2017) and Grimes and Kirby (2009) provide details on the anatomy of the OFT, as well as the evolution and terminology of the OFT. In addition to past studies of extant species, sophisticated image analysis of fossil specimens has provided new insights into the phylogeny of the OFT (Maldanis et al., 2016). We only emphasize the functional aspects of the cardiac OFT in the following section.

The OFT of hagfishes lies outside the pericardium (Farrell, 2007a) and its relative stiffness causes diastolic blood flow in the ventral aorta to routinely fall to zero following ventricular contraction (Cox et al., 2010). In contrast, the relative content of elastin, collagen and vascular smooth muscle in the bulbus arteriosus of teleost fishes tunes the compliance of the OFT such that it matches the operational range for P_{VA}. Indeed, a pressure–volume loop of an excised bulbus arteriosus can accurately predict what mean P_{VA} would be *in vivo* (Jones et al., 2005). As a result, a considerable portion of the work associated with each ventricular contraction is stored in the elastic walls of the bulbus arteriosus, reducing peak systolic P_{VA} and ventricular work, and smoothing out blood velocity in the ventral aorta (e.g., Braun et al., 2003a, b). Reducing pulse pressure and flow pulsatility in the gill secondary lamellar vessels that are downstream may promote gas exchange, as well as protect the physical integrity of the secondary lamellar capillary sheet. Thus, unless there is a substantial bradycardia, blood flow in the ventral aorta of teleosts is continuous during diastole, and mean P_{VA} is usually estimated from diastolic pressure plus 50% of the pulse pressure, rather than the 33% of pulse pressure as used in mammalian physiology. The vascular smooth muscle of the bulbus arteriosus receives autonomic innervation and the muscle responds to vasoactive compounds (Farrell and Jones, 1992; Newton et al., 2014; Stoyek et al., 2015). This allows for modulation of the elastic properties of the bulbus arteriosus, perhaps in a functional manner, with benefits that remain to be described.

In all elasmobranch fishes, and in several other fish taxa, the conus arteriosus comprises the majority of the OFT (see Chapter 1, Volume 36A: Icardo, 2017). In these taxa, the conus arteriosus, like the bulbus arteriosus, contains elastin, collagen and vascular smooth muscle, but has a more uniform diameter as well as an outer layer of compact cardiac muscle over some or all of its length. The compact myocardium of the conus arteriosus, with its coronary blood supply, is currently a distinguishing feature between Type III and Type IV hearts. In fact, the amount of compact myocardium wrapping the conus is about equal to that wrapping the ventricle in the spiny dogfish (*Squalus acanthias*) (Farrell et al., 2007a). The conal myocardium contracts rhythmically in conjunction with each ventricular contraction, as indicated by the presence of a B-wave following the T-wave in an electrocardiogram (ECG) recording (see Section 3). Conal contraction is thought to assist, or smooth out, ventral aortic blood flow. Furthermore, it may improve the contact between opposing endothelial cusps on the inside of the conus that act as valves to prevent regurgitation of blood during ventricular diastole.

2.5. Coronary Circulation

Unlike the trabecular myocardium of Type I hearts, which relies on the venous blood to provide its nutrition and oxygen and unload its wastes, the bulk of the compact myocardium has no direct contact with the venous blood. Indeed, myocytes in the compact layer are generally too far from the luminal blood for rapid diffusion within a single cardiac contraction–relaxation cycle. Therefore, the compact myocardium must have a dedicated coronary circulation.

The origins and routings of coronary arteries have been described previously (e.g., Davie and Farrell, 1991a; DeAndres et al., 1990; Foxon, 1950; Grant and Regnier, 1926; Halpern and May, 1958; O'Donogue and Abbot, 1928; Parker, 1886; Parker and Davis, 1899; Tota, 1978) and are summarized in Chapter 1 (Volume 36A: Icardo, 2017). Regardless of whether the coronary supply route has an anterior or posterior origin (see Chapter 7, Volume 36A: Sandblom and Gräns, 2017), the origin is always on the efferent side of the branchial circulation, and so the driving blood pressure is equivalent to that in the dorsal aorta (P_{DA}). Consequently, coronary blood pressure in fishes is about two-thirds of the arterial blood pressure generated by the heart (i.e., P_{VA}), which is not the case for any other vertebrate group.

The coronary circulation of fish provides a well-oxygenated arterial blood supply to the myocardium of the ventricle, conus arteriosus and atrium (depending on type of heart; Fig. 2). Studies that quantify coronary microvasculature density are surprisingly sparse given its functional importance (Davie and Farrell, 1991a; Tota, 1989; Tota et al., 1983). Microvascular density in the ventricular myocardium of rainbow trout ranges from 770 to 1200 microvessels mm^{-2} (these estimates are based on slightly different methodologies: Egginton and Cordiner, 1997; Cox et al., 2017). In contrast, the compact myocardium of the ventricle and conus arteriosus of spiny dogfish has a microvascular density of 230 and 200 microvessels mm^{-2}, respectively, whereas those in ventricular (380 microvessels mm^{-2}) and atrial trabeculae (350 microvessels mm^{-2}) (Cox et al., 2017) are higher. Cox (2015) provides evidence that the trabecular myocardium has an even higher microvascular density than the compact myocardium in other elasmobranch species. This surprising finding challenges the necessity to separate Types III and IV hearts based on an arbitrary percentage of compact myocardium. Therefore, we suggest that all fish hearts with vascularization of both trabecular and compact myocardium should be classified as Type III (and not just elasmobranch fishes). We also note that defining a true coronary capillary is difficult in spiny dogfish because most vessels are thin walled, have a large range in diameter (Fig. 3), and are clearly suitable for diffusional exchange. Hence, we report microvascular density rather than capillary density.

The diffusion distance between a capillary and the most distant mitochondrion sets the maximum rate of oxygen consumption that can be supported by the system. Microvessels are usually more numerous and closely spaced in aerobic tissue. For example, capillarity of aerobic fibers in fish red skeletal muscle is about 10 times greater than that in white muscle fibers, even though the latter have a larger diameter (Egginton and Cordiner, 1997). This is because the white muscle is largely glycolytic. The average linear distance between microvessels is 12.6 μm in the compact myocardium of rainbow trout (Egginton and Cordiner, 1997) and slightly greater (15.4–17.3 μm) in compact and trabecular myocardia of spiny dogfish (Fig. 3). Microvessel diameter is also slightly larger in spiny dogfish than in rainbow trout (10.0–13.2 μm vs 8.8 μm; Fig. 3), and this appears to reflect the difference in red blood cell diameter between these two species (Cox et al., 2017).

In spiny dogfish, trabeculae with the smallest diameters are independent of the coronary circulation because no coronary vessels were found in atrial trabeculae with a cross-sectional area of $<2700\,\mu m^2$ (radius = 33 μm) and ventricular trabeculae $<3500\,\mu m^2$ (radius = 41 μm) in the study by Cox et al. (2017). However, in wider trabeculae only a small outer annulus of myocardium lacks coronary vessels. Thus, diffusional oxygen exchange between luminal venous blood and trabeculae may be reduced for the elasmobranch heart when compared with a Type I teleost heart, especially when venous oxygen content is reduced during aquatic hypoxia and activity (Farrell, 2002, 2007a) (see Section 3).

We summarize the phylogenetic associations for the coronary circulation in Fig. 4. The evolutionary loss of the coronary circulation to the entire heart (Type I teleost hearts) and to trabeculae (Type II teleost hearts) requires an explanation because it is now apparent that the trabeculae of the spongy myocardium became well vascularized before the teleost lineage emerged, and perhaps more so than even the compact myocardium of the conus and ventricle. New questions similarly arise with regards to the coronary organization in basal fish groups such as gars and lungfishes. In these groups, coronary arteries are clearly visible on the OFT and within the conal myocardium, but the ventricle is comprised of only spongy myocardium (Farrell, 2007a; Farrell et al., 2012).

Clearly, the absence of compact myocardium should no longer lead to the conclusion that a functional coronary circulation is absent from the ventricle. Indeed, if fishes are discovered without compact ventricular myocardium and only highly vascularized trabeculae, then a new cardiac classification will be needed for these hearts. Likewise, a vessel on the surface of the ventricle is not

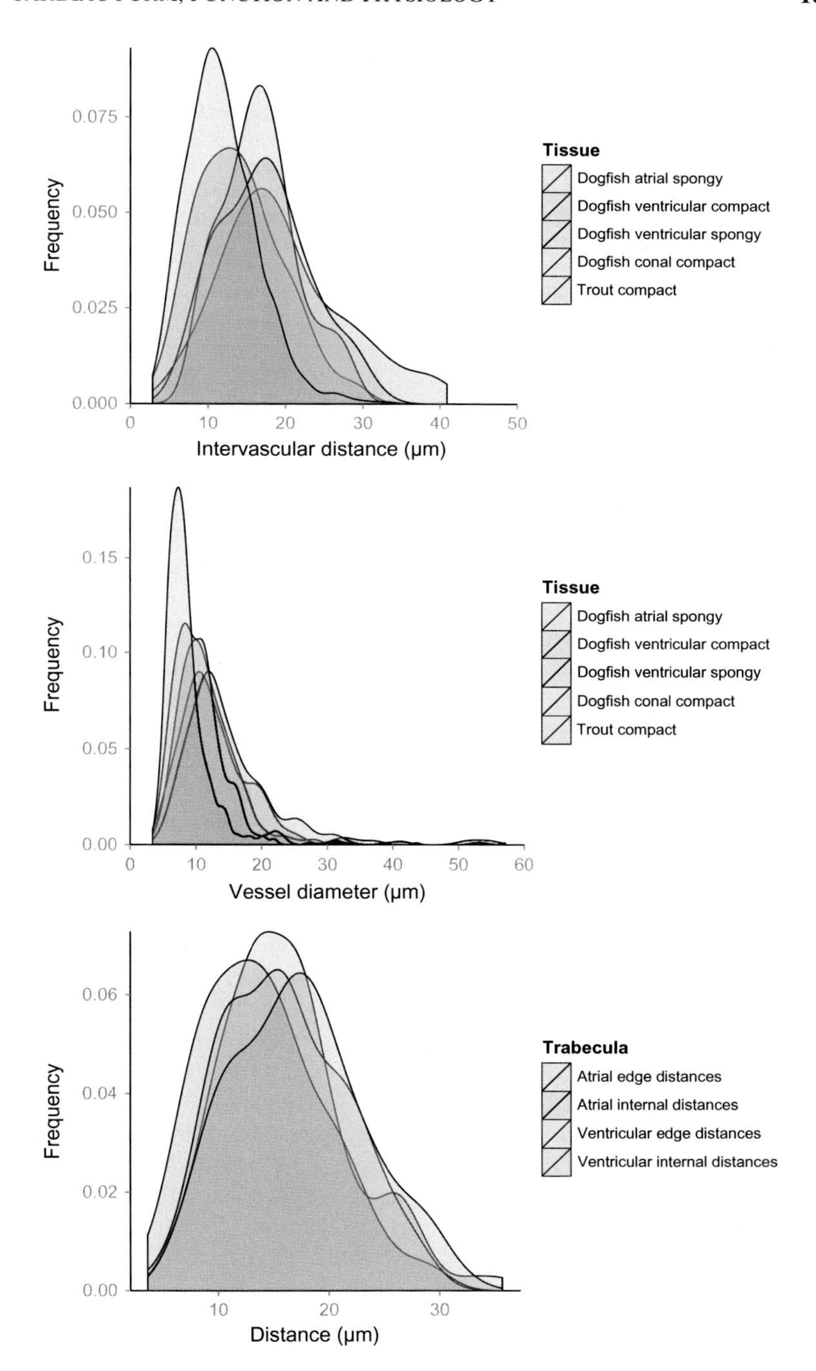

Fig. 3. See legend on next page.

Fig. 3. Frequency distributions for coronary microvascular characteristics from different regions of the dogfish shark and rainbow trout heart as determined from the tissue cross-sections shown in Fig. 2 using image analysis software. Note the similarity in intervascular distance (minimum distance between two neighboring vessels) and vessel diameter regardless of the type of cardiac tissue and species; with the exception of the greater homogeneity in the rainbow trout compact myocardium. The *lower panel* compares the minimum distance between microvessels (internal distance) with distances from the vessels to the edge of that tissue (edge distance). Credit: Cox, 2015. The functional significance and evolution of the coronary circulation in sharks. Dissertation, University of British Columbia; Cox, G.K., Kennedy, G.E., Farrell, A.P., 2016a. Morphological arrangement of the coronary vasculature in a shark (*Squalus sucklei*) and a teleost (*Oncorhynchus mykiss*). J. Morphol. 277, 896–905.

---

necessarily an indicator of a functional coronary circulation, because superficial arteries do not always penetrate the myocardium (Farrell et al., 2012; Chapter 1, Volume 36A: Icardo, 2017). Lastly, the role of the conal myocardium in driving the evolution of the coronary circulation is unknown. Given that chimerid fishes have conal myocardium, and only 5%–7% compact myocardium in the ventricle, it is possible that the evolution of the compact myocardium of the conus arteriosus drove the penetration of coronary arteries into the ventricle, and that ventricular compaction followed (Farrell et al., 2012).

Finally, a recent discovery has provided new insights into the conundrum of why salmonids have no coronary supply to the trabecular myocardium despite having a higher myocardial oxygen demand than elasmobranch fishes (Table 2). Admittedly, the sheet-like shape of the salmonid trabeculae (Pieperhoff et al., 2009) maximizes the surface area to volume ratio for diffusion between the myocardium and venous blood for trabeculae without coronary microvessels (Type II hearts). Nevertheless, the PO_2 of blood contained in the lumen of the fish heart is still routinely low and decreases during activity and environmental hypoxia. Recent work shows that plasma-accessible carbonic anhydrase (PACA) is located on the endocardium of the atrium of rainbow trout (Alderman et al., 2016) and suggests that it could be very important in aiding oxygen supply to cardiac trabeculae. Indeed, PACA is postulated to facilitate oxygen unloading from red blood cells in the red skeletal muscle of rainbow trout (McKenzie et al., 2004), perhaps doubling oxygen delivery (Rummer et al., 2013). By rapidly hydrating carbon dioxide released from the heart, PACA would locally reduce venous blood pH, and trigger a rapid unloading of oxygen from hemoglobin through the Root effect (in fish species whose blood displays this phenomenon) and a local increase in venous P_VO_2 within the heart (see Chapter 1, Volume 36B: Brauner and Harter, 2017). Further work on this mechanism clearly has the potential to provide new insights into the evolution of the coronary circulation and of tissue oxygen delivery in fishes.

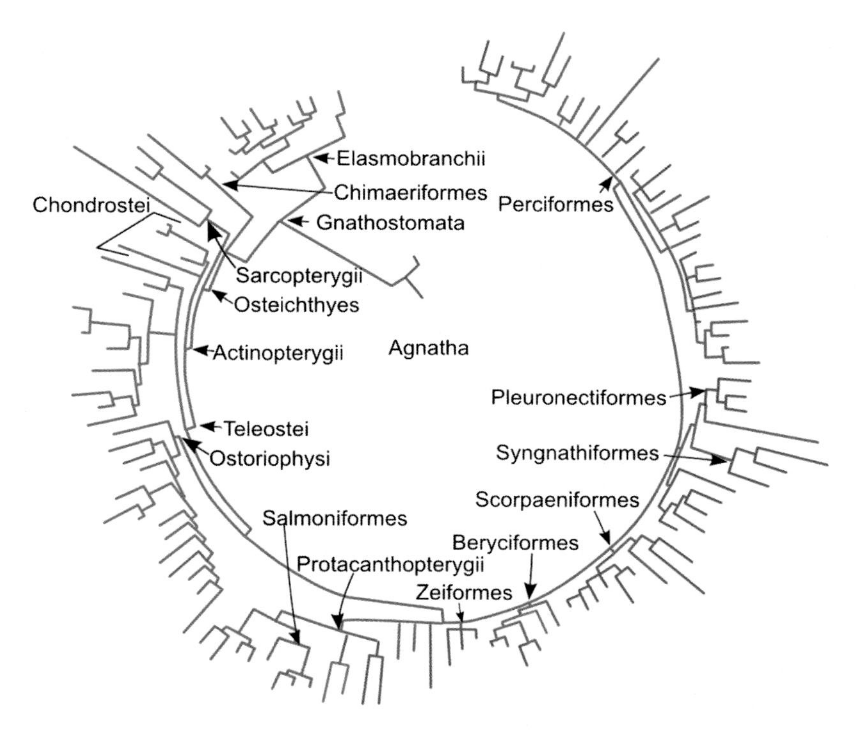

Fig. 4. A molecular tree for 123 fish species from different fish families, which illustrates the presence (*red*) or absence (*blue*) of a coronary artery. This figure is based on the inspection of preserved fish specimens. The phylogeny was constructed based on molecular sequences available on GenBank using concatenation of cytochrome *b* and 16S sequences consisting of 400 bp. The concatenated sequences were aligned using ClustalX with default parameters, and then imported to PHYML (http://atgc.lirmm.fr/phyml) for maximum likelihood analyses. MEGA software was also used to produce phylogenies using the Neighbor Joining, Maximum Parsimony and Minimum Evolution methods, which yielded similar results which are not shown here. Credit: Farrell, A.P., Farrell, N.D., Jourdan, H., Cox, G.K., 2012. A perspective on the evolution of the coronary circulation in fishes and the transition to terrestrial life. In: Sedmera, D., Wang, T. (Eds.), Ontogeny and Phylogeny of the Vertebrate Heart. Springer, New York. 75–102.

3. CARDIAC PHYSIOLOGY

3.1. The Cardiac Cycle

Fig. 5 shows representative examples of the electrical events, blood pressures, and blood flows during a cardiac cycle of a cyclostome, an elasmobranch, and a teleost. The electrical events (ECG) show many similarities with mammalian hearts. Indeed, the ECG terminology used for fishes is

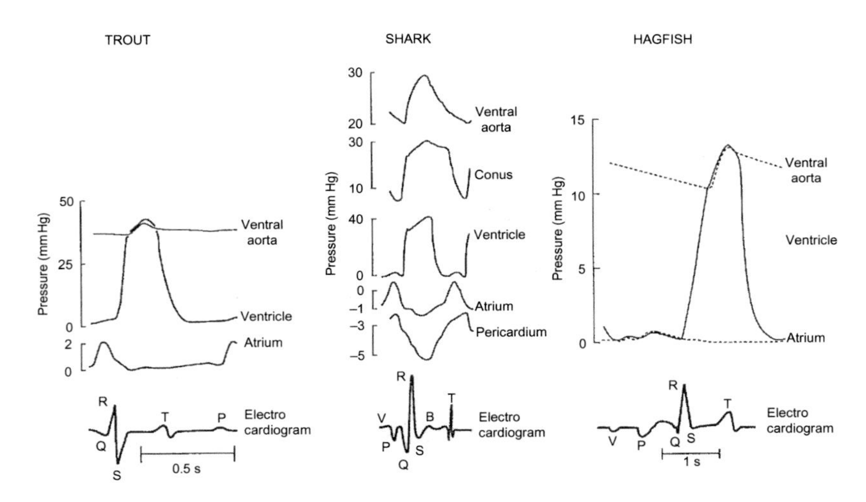

Fig. 5. Cardiac cycles for rainbow trout, dogfish shark and hagfish showing blood pressures and electrocardiograms, as presented in Farrell and Jones (1992). Material was adapted from Randall (1970) and Farrell (1991) for the trout heart, from Satchell (1971) for the shark heart, and from Davie et al. (1987) for the hagfish heart.

essentially the same as that for mammals: the P-wave indicates the start of atrial contraction; the QRS complex the start of ventricular contraction; and the T-wave the start of ventricular relaxation. The P-wave is small compared with the QRS-wave simply because the atrium in fishes has considerably less mass than the ventricle. Nevertheless, atrial contraction assists ventricular filling during ventricular diastole (Fig. 5) and provides appropriate adjustments in ventricular end-diastolic volume (i.e., to provide the "final kick" to ventricular filling; Korajoki and Vornanen, 2012).

Atrial and ventricular contractions must be sequentially coordinated. The P–Q interval represents the electrical delay through the atrioventricular canal that allows atrial contraction to complete ventricular filling before the QRS-wave begins. As a result, the QRS-wave masks the electrical activity associated with atrial relaxation. The time delay between the P-wave and the QRS complex is a consequence of the specialized electrophysiology of the cells in the atrioventricular canal in this region (Chapter 3, Volume 36A: Vornanen, 2017). Atrioventricular conduction in some fish can be compromised, however, when temperature is either too low (Graham and Farrell, 1989; Peyraud-Waitzenegger et al., 1980) or too high (Anttila et al., 2014a, b; Badr et al., 2016; Casselman et al., 2012; Drost et al., 2014, 2016). Under these circumstances, the QRS complex and the associated ventricular beat may be missing (Fig. 6), which then reveals the electrical activity associated with atrial

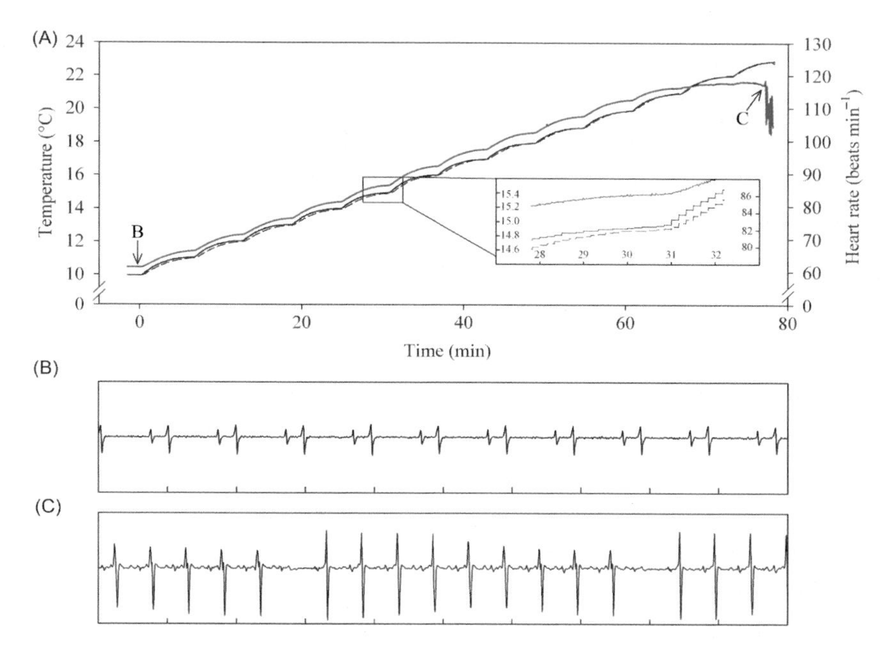

Fig. 6. An example of the effect of acute warming on the maximum heart rate of a coho salmon (Panel A). Also shown are the ECG for fish at their acclimation temperature (10–11°C; B) and an example of the missing QRS-wave in the ECG when the upper thermal limit is reached (Panel C). Credit: Casselman, M.T., Anttila, K., Farrell, A.P., 2011. Using maximum heart rate as a rapid screening tool to determine optimum temperature for aerobic scope in salmon Oncorhynchus spp. J. Fish Biol. 80, 358–377.

relaxation (the Pr-wave), as sometimes seen in hagfishes (Fig. 5). Chapter 3 (Volume 36A: Vornanen, 2017) proposes a mechanistic basis for the disruption of electrical conduction at thermal extremes.

In mammalian hearts, specialized (i.e., fast-conducting) cardiac fibers speed up the electrical propagation over the ventricle to ensure a syncytial contraction. Fast-conducting fibers have not been identified in fish (Chapter 1, Volume 36A: Icardo, 2017), but the spread of the ventricular contraction is clearly coordinated by some means. Studies of the electrical properties, and the distribution of ventricular gap junctions of cardiomyocytes, could be important for filling this knowledge gap.

The ventricle is responsible for pumping blood through the vascular system. Ventricular contraction increases luminal blood pressure, which first closes the atrioventricular valve and then ejects blood into the OFT once diastolic P_{VA} is exceeded (Fig. 5). Ventricular ejection fraction can approach 100% in fishes, i.e., end-systolic volume is approximately zero (Franklin and Davie, 1992). As a result, increases in V_s in fishes are primarily determined by increases in end-diastolic volume. The low end-systolic volume may be related to the trabecular architecture (Burggren et al., 2014; see Section 2.3), but this idea will be difficult to test experimentally.

A fish ECG can contain additional electrical waves. As noted earlier, the B-wave during the S–T interval indicates the start of contraction of conal myocardium in elasmobranch fishes (Fig. 5). Also, a V-wave that precedes the P-wave in hagfish indicates rhythmic electrical activity generated by the sinus venosus (Fig. 5). Which cells in the sinus venosus (pacemaker cells or others) produce this V-wave remains unclear. A V-wave is also evident in the ECG of the common eel (*A. anguilla*), but a weak V-wave could be easily missed in the noisy ECG signals that are typically recorded in fish immersed in water.

3.2. Cardiac Output

The primary task of the heart is to meet tissue requirements for oxygen and metabolic substrate delivery. Consequently, $\dot{Q}$ changes whenever oxygen uptake ($\dot{M}O_2$) changes as a result of either increased activity (i.e., swimming), the digestion and assimilation of nutrients, or an environmental stressor (e.g., an increase in temperature or hypoxia). This interdependence is easy to appreciate from the Fick equation, which relates $\dot{M}O_2$ to internal oxygen convection:

$$\dot{M}O_2 = \dot{Q} \times (C_aO_2 - C_vO_2) \quad (1)$$

where C_aO_2 is arterial oxygen content and C_vO_2 is venous oxygen content; and $(C_aO_2 - C_vO_2)$ is termed tissue oxygen extraction.

In addition, $\dot{Q}$ and tissue oxygen extraction both typically increase whenever $\dot{M}O_2$ increases. However, an increase in internal oxygen convection is primarily determined by $\dot{Q}$ and C_vO_2, (i.e., increased flow and increased tissue oxygen extraction), because the arterial blood is generally saturated with oxygen when leaving the gills. Arterial oxygen transport is the product of $\dot{Q}$ and C_aO_2 (Chapter 1, Volume 36B: Brauner and Harter, 2017). Furthermore, $\dot{Q}$ is set primarily by tissue oxygen demand, and so maximum arterial oxygen transport is linearly related to maximum $\dot{M}O_2$ across a number of fish species (Farrell, 2002; Gallaugher et al., 2001).

$\dot{Q}$ is altered by changing either V_s or heart rate (f_H; in beats min^{-1}), or some combination thereof. The heart also generates the P_{VA}, which is the potential energy that overcomes peripheral vascular resistance and powers blood flow. Total peripheral vascular resistance is the sum of the branchial vascular resistance (R_{gill}) and the systemic vascular resistance (R_{sys}); and about one-third of P_{VA} is lost to R_{gill} before oxygenated blood enters the systemic circulation (Bushnell et al., 1992). P_{VA} is centrally controlled through baroreceptor-mediated reflex changes in vascular resistance, likely involving baroreceptors located in the gills (Bushnell et al., 1992; Farrell and Jones, 1992).

Continuous direct measurements of $\dot{Q}$ and its pulsatility require that a cuff-type flow probe be placed around the ventral aorta (and this allows V_s and f_H to be derived from the pulsatile ventral aortic blood flow trace). The injection of some dyes or thermal indicators, and the subsequent downstream measurement of their concentrations/levels (i.e., dye and thermal dilution techniques; Brill and Bushnell, 1989; Bushnell and Brill, 1991) provides periodical measurements of an average $\dot{Q}$. Nevertheless, direct *in vivo* measurements of $\dot{Q}$ in fishes are not particularly numerous (Table 2), largely because it is not simple to apply the measurement technique to fishes (Farrell, 1991; Lillywhite et al., 1999).

3.2.1. Measuring Cardiac Function in Fishes

This section briefly describes how cardiac function is measured in fishes and highlights the technical challenges alluded to above. The techniques used to study cardiac function in fishes are largely borrowed from mammalian physiology, but their application to fish presents several problems beyond the fact that water and electrical devices are a poor mix. Chief among these are that unanesthetized fish normally must be placed into a restricted physical space, which limits activity and may create a stress that impacts cardiovascular function, and thus, the values reported (Note: the use of telemetry can avoid this constraint; see Gräns et al., 2009, 2010).

The earliest estimates of $\dot{Q}$ in fishes used the Fick equation. This indirect estimate of $\dot{Q}$ requires simultaneous measurements of $\dot{M}O_2$ and tissue oxygen extraction. Unfortunately, $\dot{Q}$ and V_s are overestimated by the Fick equation whenever the rate of oxygen removal from the water (i.e., the measured $\dot{M}O_2$) is not exactly the same as internal oxygen convection to the tissues. Indeed, skin and gill tissues are known to directly remove oxygen from the water for local consumption. Therefore, this oxygen never enters the circulatory system, but is part of the $\dot{M}O_2$ measurement [see Farrell et al. (2014) for a recent summary of the relevant literature and concerns with using this technique]. Consequently, the accuracy of any Fick estimate of $\dot{Q}$ should be confirmed with a direct measurement. This is especially true for flatfishes such as the winter flounder where this error is a particular concern due to high rates of transcutaneous O_2 uptake (Joaquim et al., 2004).

Dilution techniques provide estimates of mean $\dot{Q}$, averaged over the period of the measurement, which is a valuable technique for steady-state conditions. This technique requires two cannulation sites—one for injection (usually in the ventral aorta) and one for downstream measurement (usually the dorsal aorta). While the imposition of the gills is problematic for thermal dilution, it is not for dye dilution. However, repeated blood withdrawals to measure dye concentrations are problematic with small fish because of the limited total blood volume.

The gold standard for directly and continuously measuring blood flow is an ultrasonic cuff-type flow probe connected to a Transonic® flowmeter because they report absolute blood flow (and provide various cuff shapes and diameters). These probes have replaced the older cuff-type electromagnetic flow probes that needed an *in vivo* zero flow signal as well as an *in vitro* flow calibration. Nevertheless, ultrasonic flow probes are expensive and have other limitations. They are bulky and cannot be accommodated inside some fish species and especially small fish. The delicate electrical leads will break easily if a fish struggles, limiting work on very large fish. Also, the maximum length of the probe lead decreases with cuff diameter, again prohibiting work on small fish without greatly limiting fish movement. Lastly, ultrasonic flow probes come factory-calibrated to a particular temperature and so *in vitro* calibrations are needed for different test temperatures (e.g., Farrell et al., 2013; Powell and Gamperl, 2016). A cuff-type Doppler-flow probe, which measures ventral aortic blood velocity (rather than flow), circumvents some of the above limitations. Doppler probe leads are considerably lighter, stronger and longer; and the probe cuff is soft, flexible, less bulky and considerably cheaper (i.e., about 1/20th the cost). However, absolute blood flow cannot be reported unless a very careful *in situ* calibration is performed (e.g., Axelsson et al., 1992; Bushnell and Brill, 1991; Clark et al., 2007). To circumvent the need to calibrate these probes, many studies just report a response as a relative change in $\dot{Q}$ (e.g., Franklin et al., 2013; Gamperl et al., 2011; Gold and Farrell, 2015; Sandblom et al., 2005). Regardless, the Doppler signal must be focused to maximize signal strength and the probe must fit tightly around the blood vessel to ensure a constant vessel diameter (otherwise the relationship between blood velocity and blood flow is not constant).

Vascular anatomy differs considerably among fish species making surgical access to the ventral aorta difficult, if not impossible, in some species. The ventral aortic flow probe should be located outside of the pericardium if at all possible, but this has not always been the case. In teleosts with Type I ventricles [e.g., red Irish lord (*Hemitripterus americanus*), lingcod (*Ophiodon elongatus*), and shorthorn sculpin (*Myoxocephalus scorpius*)], the ventral aorta is easy to approach from the ventral surface of the isthmus by teasing apart overlaying muscle groups using blunt dissection and avoiding (or ligating) the hypobranchial vessels. In contrast, this ventral access route generates considerable tissue and vascular damage in eels and salmonids because muscle groups cannot be easily separated. Instead, the salmonid operculum can be retracted and a minor lateral incision made in the isthmus, followed by some blunt dissection to place probes on the ventral aorta (e.g., Farrell and Steffensen, 1987; Steffensen and Farrell, 1998). Extra care is also needed to avoid the coronary artery, when present.

In elasmobranch fishes, a more difficult anatomical problem exists because the two posterior pairs of afferent branchial arteries arise immediately after the conus arteriosus passes through the anterior wall of the pericardial cavity, leaving no room to place a cuff on the ventral aorta. Furthermore, total $\dot{Q}$ cannot be measured directly by placing a cuff on the conus arteriosus because this would compress the conal myocardium and its coronary vessels, in addition to breaking the integrity of the pericardial cavity. While it is easy to locate a bulky flow on the ventral aorta between the anterior and posterior afferent branchial arteries, this placement yields only a partial measurement of total $\dot{Q}$, which must be converted to total cardiac output by assuming equal flow distribution to the gill arches (e.g., Cox et al., 2016b; Satchell et al., 1970; Short et al., 1977). Radiographic imaging of the proportion of blood flow going to each set of gill arches in the blue shark (*Prionace glauca*) and leopard shark (*Triakis semifasciata*) supports this assumption (Lai et al., 1989b). Dye dilution techniques would be another way of validating this assumption.

Even though it is easier to obtain f_H from an ECG recording or arterial blood pressure (ventral or dorsal aortic) than to measure $\dot{Q}$, changes in f_H are not always a reliable quantitative or qualitative predictor of changes in $\dot{Q}$ in fishes. This is because V_s can vary appreciably and sometimes in the opposite direction to f_H (e.g., Gamperl and Driedzic, 2009; Thorarensen et al., 1996b). Nonetheless, in some situations, changes in f_H largely reflect the change in $\dot{Q}$ [e.g., during an acute change in temperature (Ekstrom et al., 2016; Eliason et al., 2011, 2013a, b, c; Farrell, 2009; Farrell et al., 2009; Sandblom and Axelsson, 2006; Chapter 4, Volume 36B: Eliason and Antilla, 2017) and during the digestion and assimilation of a meal (Eliason et al., 2008; Eliason and Farrell, 2014; see below).

Maximum $\dot{Q}$ is usually measured during a prolonged swimming test because aerobic swimming requires a sustained (and presumably maximal) $\dot{M}O_2$ (Farrell, 1991; Jones and Randall, 1978; Kiceniuk and Jones, 1977). Yet, only a small number of studies have directly measured $\dot{Q}$ during maximum prolonged swimming activity (Table 2), and even fewer have simultaneously measured $\dot{M}O_2$ to obtain comprehensive information on the cardiorespiratory response (e.g., Eliason et al., 2011, 2013a, b, c; Petersen and Gamperl, 2010; Steinhausen et al., 2008). For fishes that are reluctant to sustain a high swimming velocity in a flume, anaemia is a potential experimental approach to assess maximum $\dot{Q}$ because anemic fish will maximally increase $\dot{Q}$ to maintain $\dot{M}O_2$ provided the decrease in hematocrit (Gold and Farrell, 2015) and C_aO_2 are sufficiently large. Interestingly, the relative contributions of f_H and V_s to the response to anemia vary considerably among studies and species (Gold and Farrell, 2015). Increasing temperature to increase $\dot{M}O_2$ and obtain maximum $\dot{Q}$ could be a viable alternative because $\dot{Q}$ measured in Atlantic cod during a critical thermal maximum (CT_{max}) test was similar to that measured during swimming (Powell and Gamperl, 2016).

Working, perfused, heart preparations also yield reliable estimates of maximum $\dot{Q}$ (Table 3). For example, maximum $\dot{Q}$ measured *in situ* ranges from 33 to 74 $mL\,min^{-1}\,kg^{-1}$ for rainbow trout, and is 23 $mL\,min^{-1}\,kg^{-1}$ for sea raven (Farrell et al., 1983, 1985, 1991) and 39 $mL\,min^{-1}\,kg^{-1}$ in spiny dogfish (Davie and Franklin, 1992). However, *in vitro* and *in situ* values can exceed those measured *in vivo* (Table 2), likely because the perfused heart continuously receives fresh, well-oxygenated perfusate, whereas the heart is exposed to venous blood that has a progressively lower PO_2 when a fish exercises (i.e., [C_vO_2] decreases and wastes accumulate, as described in Section 3.2.5). For example, the maximum *in vitro* $\dot{Q}$ for long-finned eel (*Anguilla dieffenbachii*; 22 $mL\,min^{-1}\,kg^{-1}$) is about twice that observed *in vivo* (Davie et al., 1992). *In vitro* and *in situ* perfused heart studies also allow for the direct study of how hormones such as adrenaline (Fig. 7), temperature, oxygen levels, pH (Fig. 8) and other variables affect the maximum performance of the heart, including cardiac work (see Sections 3.2.3–3.2.6).

Cardiac work, also termed cardiac power output (CPO), is estimated from the product of $\dot{Q}$ and P_{VA} (Tables 2 and 3). While ignoring P_{CV} in this estimate may overestimate CPO in some situations, P_{CV} can be subambient in some fishes, which could lead to an underestimation of CPO in these species. Nonetheless, the error is likely no more than ~5% as P_{CV} is quite minimal as compared to P_{VA} (see Chapter 7, Volume 36A: Sandblom and Gräns, 2017). Because P_{DA} is easier to measure than P_{VA}, a useful approximation to estimate CPO is $P_{VA} = 1.5 \times P_{DA}$ because about one-third of P_{VA} is lost across the gill circulation (Farrell and Jones, 1992). Perfused heart preparations allow for a more accurate estimate of CPO because cardiac filling pressure is easily measured and controlled. Furthermore, isolated heart preparations develop maximum CPO when cardiac filling pressure (Fig. 9A) and the output resistance are raised (Fig. 9B).

CPO is usually normalized to ventricular mass (Tables 2 and 3). All the same, total CPO per kg of fish mass can be used whenever RVM differs among

Table 3
Maximum performance of perfused working heart preparations

Species	Acclimation/Test temperature (°C)	Perfusate conditions	$\dot{Q}$ ($mL^{-1} min^{-1} kg^{-1}$)	f_H (min^{-1})	V_s ($mL\,kg^{-1}$)	CPO ($mW\,g^{-1}$)	Source
Normoxia							
Myxine glutinosa	10–11	Air, 3 nM NAD and 1 nM AD	29.4	23	1.3	0.50	Johnsson and Axelsson (1996)
Oncorhynchus mykiss	5	50 nM AD	33	39	0.85	2.58	Graham and Farrell (1989)
	8	O_2, 5 nM AD	50	52	0.96	6.1	Keen and Farrell (1994)
	10	O_2, 5 nM AD (& Na lactate)	41 (46)	60 (65)	0.68 (0.70)	6.4 (6.5)	Milligan and Farrell (1991)
	15	50 nM AD	45	54	0.82	4.58	Graham and Farrell (1989)
	18	O_2, 5 nM AD	63	79	0.80	8.8	Keen and Farrell (1994)
Good swimmers	16	O_2, 5 nM AD (1000 nM AD)	48 (56)	89 (98)	0.54 (0.58)	6.3	Claireaux et al. (2005)
Poor swimmers	16	O_2, 5 nM AD (1000 nM AD)	42 (46)	87 (101)	0.49 (0.46)	5.4	Claireaux et al. (2005)
Untrained	10	O_2, 10 nM AD (1000 nM AD)	52 (63)	64 (73)	0.90 (0.98)	5.2 (6.9)	Farrell et al. (1991)
Trained	10	O_2, 10 nM AD (1000 nM AD)	65 (74)	67 (78)	1.05 (1.05)	6.7 (8.7)	
Pericardectomy	10	O_2, 10 nM AD (open pericardium)	53 (48)	50 (44)	1.05 (1.10)	6.22 (5.11)	Farrell et al. (1988a)
Thunnus albacares	25	O_2, pH 7.7, 100 nM AD	108	123	0.87	7.58	Farrell et al. (1991)
Katsuwonus pelamis	25	O_2, pH 7.7, 100 nM AD	85	139	0.68	2.55[a]	Farrell et al. (1991)

(Continued)

Table 3 (Continued)

Species	Acclimation/Test temperature (°C)	Perfusate conditions	$\dot{Q}$ ($mL^{-1} min^{-1} kg^{-1}$)	f_H (min^{-1})	V_s ($mL\,kg^{-1}$)	CPO ($mW\,g^{-1}$)	Source
Temperature (+ adrenaline)							
Salmo trutta (triploid)	14	O_2, 10 nM AD (1000 nM AD)	77 (87)	78 (87)	1.03 (1.03)	9.5 (11.4)	Mercier et al. (2002)
	18	O_2, 10 nM AD (1000 nM AD)	92 (119)	82 (93)	1.22	8.6 (12.1)	
Dicentrarchus laborax	18	O_2, 5 nM AD (50 nM AD)	91 (101)	66 (67)	1.38 (1.51)	11.4 (13.2)	Farrell et al. (2007a)
	22	O_2, 5 nM AD (50 nM AD)	99 (109)	94 (92)	1.05 (1.18)	15 (16.7)	
Oncorhynchus mykiss	15	O_2, 10 nM AD (1000 nM AD)	66 (76)	70 (82)	0.99 (1.02)	8.00 (9.3)	Farrell et al. (1996)
	18	O_2, 10 nM AD (1000 nM AD)	78 (76)	79 (88)	0.97 (0.91)	9.3 (8.6)	
	22	O_2, 10 nM AD (1000 nM AD)	71 (78)	84 (102)	0.81 (0.77)	7.2 (8.0)	
Myoxocephalus scorpius	1	O_2, 5 nM AD (100 nM AD)	54 (57)	28 (34)	1.92 (1.68)	2.61 (2.54)	Farrell et al. (2013)
	1	Tested at 6°C (100 nM AD)	56 (66)	40 (51)	1.4 (1.30)	2.72 (3.40)	
Myoxocephalus scorpius	6	O_2, 5 nM AD (100 nM)	53 (56)	41 (47)	2.12 (1.45)	2.35 (2.20)	Farrell et al. (2013)
	6	Tested at 1°C (100 nM AD)	49 (49)	37 (38)	1.32 (0.30)	2.67 (2.73)	
Hemitripterus americanus	3	Air, 1 nM AD	11	23	0.46	1.1	Graham and Farrell (1985)
	3	Tested at 13°C	13	45	0.29	2.1	
Hemitripterus americanus	13	Air, 1 nM AD	23	46	0.50	3.1	Graham and Farrell (1985)
	13	Tested at 4°C	14	34	0.41	1.7	
Acidosis (+ adrenaline)							
Oncorhynchus mykiss	10	O_2, pH 7.9	38	58	0.78	5.3	Farrell et al. (1986)
		pH 7.4 (10 nM AD)	35 (47)	53 (58)	0.65 (0.65)	4.8 (6.0)	

Oncorhynchus mykiss	10		O_2, pH 7.9, 1 nM AD (50 nM AD)	31 (39)		47 (59)	0.68 (0.67)	4.2 (5.7)	Farrell et al. (1988b)
			O_2, pH 7.4, 1 nM AD (50 nM)	22 (42)		35 (42)	0.71 (0.75)	3.2 (5.7)	
Oncorhynchus mykiss	10		Air, pH 7.9, 2.5 mM K^+ (5 nM AD)	51				5.9	Hanson et al. (2006)
			Air, pH 7.5, 5.0 mM K^+ (5000 nM AD)	35		−25%	−10%	3.8	
Oncorhynchus mykiss	18		Air, pH 7.9, 2.5 mM K^+ (5 nM AD)	54		90	0.60	5.9	Hanson and Farrell (2007)
			Air, pH 7.5, 5.0 mM K^+ (5000 nM AD)	35		81	0.43	3.8	
Hypoxic effects									
Anguilla dieffenbachii	10		O_2, 10 nM AD	22		37	0.60	3.40	Davie et al. (1992)
			Hypoxia (0.13 kPa O_2)	9.7		37	0.27	2.05	
Oreochromis hybrid	22		O_2 + 5 nM AD	39		58	0.67	5.0	Lague et al. (2012)
			N_2 + 5 nM AD	30		50	0.60	3.4	
			N_2 + acidosis (pH 7.3)	21		45	0.47	2.4	
			Na cyanide + N_2	18		38	0.47	1.7	
Oncorhynchus mykiss	10		Normoxia (0.89 kPa O_2), paced	30		43	0.7	4.4	Farrell et al. (1989)
			Hypoxia (0.61 kPa O_2), paced	22		43	0.51	3.4	
			Hypoxia (0.34 kPa O_2), paced	6.2		43	45	0.1	

AD and NAD = adrenaline and noradrenaline concentration in the perfusate, respectively.

[a]These preparations were not stable and this was the best performance of a failing heart preparation.

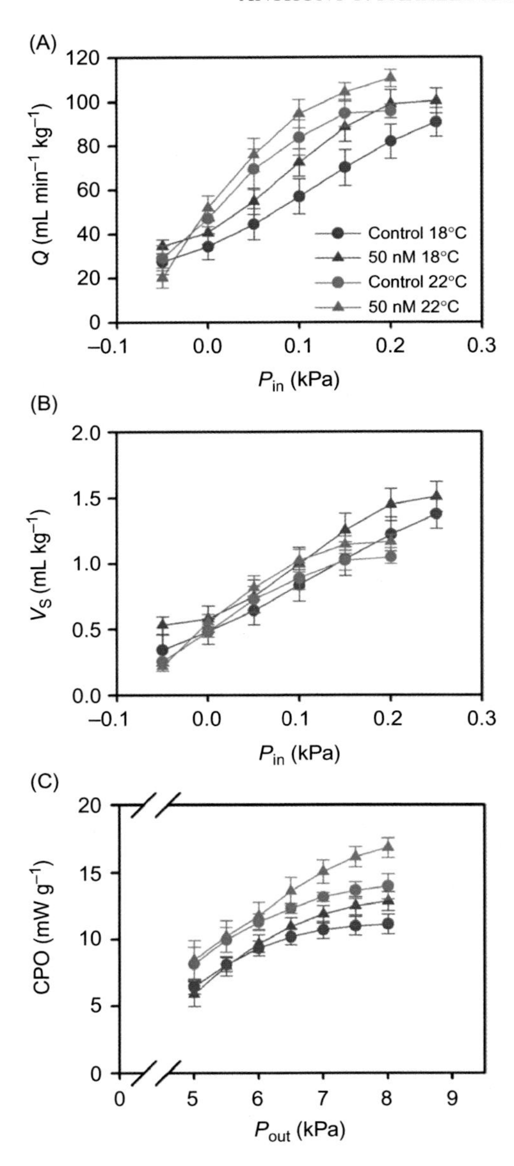

Fig. 7. An example of Frank–Starling curves for working, perfused, hearts from sea bass acclimated to either 18°C (*purple lines*) or 22°C (*blue lines*) to illustrate the effect of cardiac filling pressure on cardiac output ($\dot{Q}$, panel A), cardiac stroke volume (V_s, panel B), and cardiac power output (CPO, panel C). Note that increasing the adrenaline concentration from a tonic level (control = 5 nM) to a level that produces maximum stimulation (50 nM) increases maximum cardiac performance along with the sensitivity of the heart to filling pressure. This response is similar to that observed with the 4°C increase in temperature. Farrell, A.P., Axelsson, M., Altimiras, J., Sandblom, E., Claireaux, G., 2007b. Maximum cardiac performance and adrenergic sensitivity of the sea bass *Dicentrarchus labrax* at high temperatures. J. Exp. Biol. 210, 1216–1224.

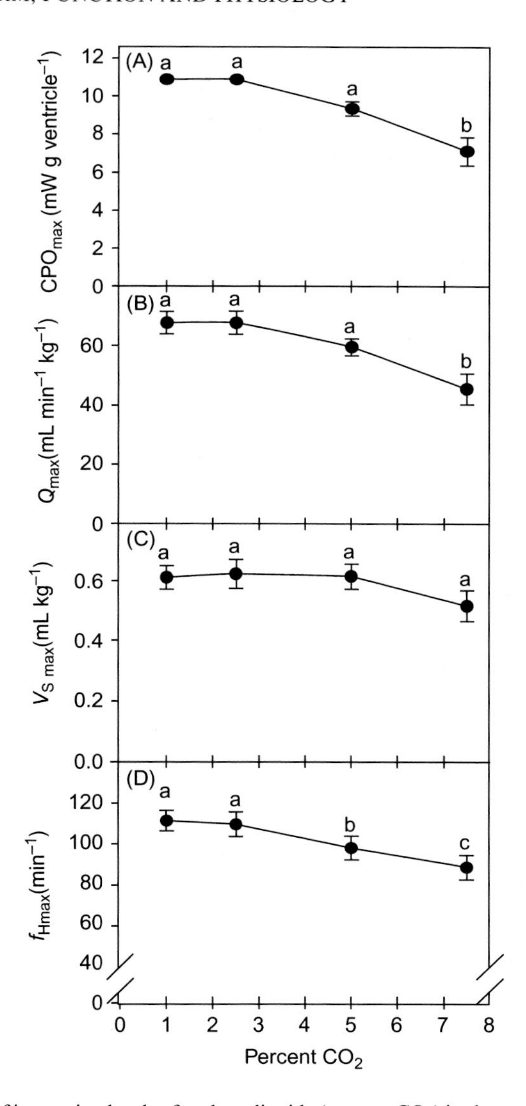

Fig. 8. The effect of increasing levels of carbon dioxide (percent CO_2) in the perfusate on the maximum cardiac power output (CPO_{max}, panel A), cardiac output ($\dot{Q}_{max}$, panel B), stroke volume (V_{smax}, panel C), and heart rate (f_{Hmax}; panel D) of a working perfused armored catfish heart. The data illustrate the resilience of cardiac pumping capacity to extreme extracellular acidosis. Symbols without a letter in common are statistically different ($P < 0.05$). Credit: Hanson, L.M., Baker, D.W., Kuchel, L.J., Farrell, A.P., Val, A.L., Brauner, C.J., 2009. Intrinsic mechanical properties of the perfused armoured catfish heart with special reference to the effects of hypercapnic acidosis on maximum cardiac performance. J. Exp. Biol. 212, 1270–1276.

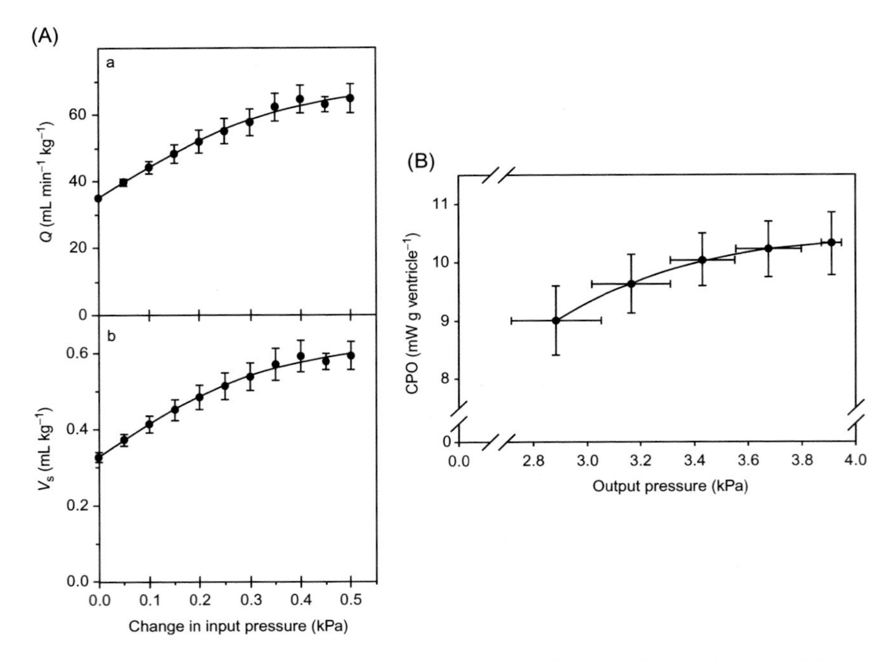

Fig. 9. (A) An example of the Frank–Starling curve for a perfused, working, armored catfish heart preparation. Data illustrate the sensitivity of cardiac output (panel a) and stroke volume (panel b) to increasing input pressure (i.e., cardiac filling pressure). (B) The effect of increasing output pressure on cardiac power output of the working armored catfish heart. In this case, filling pressure is set to generate maximum cardiac output. Cardiac power output increases and maximum cardiac output is maintained as output pressure is increased, until the maximum cardiac power output is reached. Credit: Hanson, L.M., Baker, D.W., Kuchel, L.J., Farrell, A.P., Val, A.L., Brauner, C.J., 2009. Intrinsic mechanical properties of the perfused armoured catfish heart with special reference to the effects of hypercapnic acidosis on maximum cardiac performance. J. Exp. Biol. 212, 1270–1276.

species or with environmental conditions (see Section 1). For example, the CPO of an Antarctic icefish normalized to ventricular mass is low compared with that of a temperate rainbow trout (Table 3), but the total work of the heart from a 1 kg icefish (6–8 $mW\,kg^{-1}$ body mass) is similar to that of a rainbow trout heart because the ventricular mass of the icefish is considerably larger. In contrast, a 1 kg tuna has a similar-sized ventricle as an icefish, but the heart produces $>20\,mW\,kg^{-1}$ body mass, albeit at a considerably warmer temperature.

If the oxygen content of the perfusate entering and leaving an isolated heart are measured, estimates of myocardial oxygen uptake and mechanical efficiency are also possible. External work is the primary determinant of myocardial oxygen demand, and thus, linear relationships exist between CPO and myocardial oxygen uptake for a number of fish species (Davie and Franklin, 1992; Farrell and Milligan, 1986; Farrell et al., 1985; Forster, 1991; Forster et al., 1992; Graham and Farrell, 1990; Houlihan et al., 1988; Lague et al., 2012). A reasonable approximation of myocardial oxygen uptake (across a range of species) is $0.3\,\mu L\ O_2\ s^{-1}\,mW^{-1}\,g^{-1}$ ventricular mass (Davie and Farrell, 1991a; Farrell and Jones, 1992). Such measurements also allow the cost of cardiac pumping to be estimated and these values range from 0.5% to 5.0% of whole animal $\dot{M}O_2$ (Farrell and Jones, 1992). Finally, estimates of routine cardiac adenosine triphosphate (ATP) turnover rates range from 150 to 300 nmol ATP $s^{-1}\,g^{-1}$ ventricular mass under normoxic conditions (Table 4).

Mechanical efficiency of the heart can be calculated from measured values of myocardial oxygen consumption and CPO. Several perfused, working heart studies show that mechanical efficiency is not constant. For example, it increases in rainbow trout from a low of about 12%–15% when the heart is working near routine values of CPO (0.7–1.0 $mW\,g^{-1}$) to >20% at maximum CPO (Davie and Franklin, 1992; Graham and Farrell, 1989). Interestingly, pressure loading the heart (Fig. 9B) increases myocardial oxygen extraction, whereas volume loading (Fig. 9A) decreases this parameter (Farrell et al., 1985). Maximum adrenergic stimulation alone increases mechanical efficiency of the rainbow trout heart from 11% to 16%. In contrast, acidosis decreases mechanical efficiency to 7%, but maximum adrenergic stimulation of the acidotic heart restores mechanical efficiency to 10% (Farrell and Milligan, 1986). Unfortunately, if the heart even partially uses glycolytic (anaerobic) metabolism to support its mechanical work, mechanical efficiency can be grossly overestimated. For example, perfusate with an oxygen partial pressure of 5.5 kPa increases the apparent mechanical efficiency of the sea raven heart from 15% to 20% to >30%, and a value of 65% is reached at 3.0 kPa (Farrell et al., 1985). A proper estimate of mechanical efficiency of a heart requires the measurement of heat production, but this has only been used for a non-working anoxic hagfish heart (Gillis et al., 2015).

3.2.2. Species Differences in Routine Cardiac Output

Cardiac output is usually normalized to body mass ($mL\,min^{-1}\,kg^{-1}$). This is because studying intraspecific allometry for $\dot{Q}$ is difficult. $\dot{Q}$ cannot be measured in fishes across a large range of body masses (Clark and Farrell, 2011), even though various allometric relationships exist for resting $\dot{M}O_2$ in fishes.

Differences in temperature must always be factored in when comparisons of $\dot{Q}$ are made between species. For example, the 75-fold difference in the routine $\dot{M}O_2$ between the New Zealand hagfish (*Eptatretus cirrhatus*) at 10°C and

Table 4

Anoxic cardiac ATP turnover rates based on working perfused heart preparations and isometrically contracting ventricular strips

Species	Temperature and condition (°C)		CPO ($mW\,g^{-1}$)	ATP turnover rate ($nmol\,s^{-1}\,g^{-1}$)	Source
Working hearts					
Hagfish (*Eptatretus cirrhatus*)	18	Anoxic (N_2)	0.28	41	Forster (1991)
Hagfish (*Myxine glutinosa*)	5	Na cyanide		0.27	Hansen and Sidell (1983)
Tilapia (*Oreochromis* hybrid)	22	Normoxic	5	340	Lague et al. (2012)
		Anoxic (N_2) and Na cyanide	1.75	113	
		Anoxic (N_2) and acidosis	2.3	120	
		Anoxic (N_2)	3.34	172	
Rainbow trout (*Oncorhynchus mykiss*)	16	Normoxic	2.55	153	Arthur et al. (1992)
		Anoxic (N_2)	0.54	72	
Rainbow trout (*Oncorhynchus mykiss*)	5	Anoxic (N_2)		33	Overgaard et al. (2004)
	10	Anoxic (N_2)		45	
	15	Anoxic (N_2)		68	
Crucian Carp (*Carassius carassius*)	8	Anoxic (N_2)	1.13	79	Stecyk, Stenslokken, Hanson, Farrell, and Nilsson (unpublished results)
Turtle	5	Normoxic	0.38	40	Farrell et al. (1994) and Arthur et al. (1997)
		Anoxic (N_2)	0.17	11	
	15	Normoxic	1.42	128	
		Anoxic (N_2)	0.76	73	
Turtle	24	Normoxic	0.88	192	Reeves (1963)
		Anoxic (N_2)	0.69	175	
Rat	37	Normoxic	7.25	490	Neely et al. (1973, 1975)
Isometrically contracting ventricular tissue					
Rainbow trout (*Oncorhynchus mykiss*)	15	Normoxic		300	Overgaard and Gesser (2004)
		Anoxic (N_2)		83	
Turtle	20	Normoxic		300	
		Anoxic (N_2)		83	

skipjack tuna at 26°C corresponds to a 15-fold difference in $\dot{Q}$ and a five-fold difference in C_aO_2—C_vO_2 (Table 2). Nevertheless, most of the seven-fold difference in f_H is most likely a direct result of the large temperature difference. Typically, f_H doubles with a 10°C increase in temperature (see Section 3.2.4). Even so, fishes at ~10°C with an active life style have a higher routine $\dot{Q}$ (e.g., 17 mL min^{-1} kg^{-1} in rainbow trout) than those with a demersal lifestyle (e.g., around 10 mL min^{-1} kg^{-1} in winter flounder, sea raven, and lingcod) (Table 2). Some of the highest routine $\dot{Q}$ values reported in fishes are for tunas; routine $\dot{Q}$ in skipjack tuna is 132 mL min^{-1} kg^{-1} at 26°C (Bushnell et al., 1992). Nonetheless, the highest measured $\dot{Q}$ is that of Notothenioids that inhabit the frigid waters of Antarctica. At 2°C, their heart can pump almost 200 mL min^{-1} kg^{-1} and their maximum V_s (up to 10 mL kg^{-1}) is unsurpassed by any vertebrate (Table 2).

3.2.3. Responses of Cardiac Output to Swimming

Prolonged swimming (i.e., at least 10–20 min at a constant velocity) increases $\dot{Q}$ to maximum levels (Table 2). Figs. 10 and 11 illustrate the cardiovascular changes that are typical for an incremental swimming speed test.

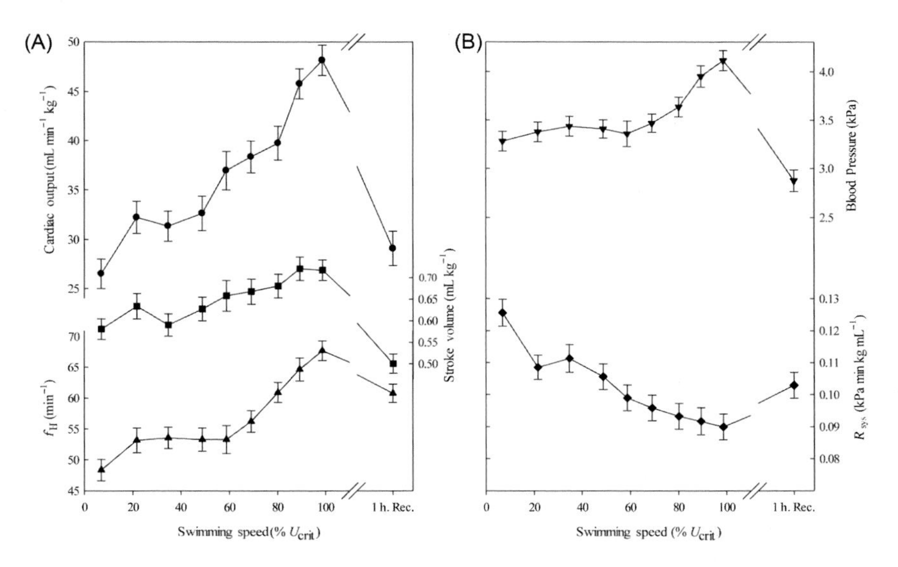

Fig. 10. An example of the cardiovascular changes recorded in rainbow trout during an incremental prolonged swimming test to U_{crit}, and during the first hour of recovery. Mean blood pressure was measured in the dorsal aorta; R_{sys} = systemic vascular resistance. Credit: Thorarensen, H., Gallaugher, P.E., Farrell, A.P., 1996a. Cardiac output in swimming rainbow trout, *Oncorhynchus mykiss*, acclimated to seawater. Physiol. Zool. 69, 139–153.

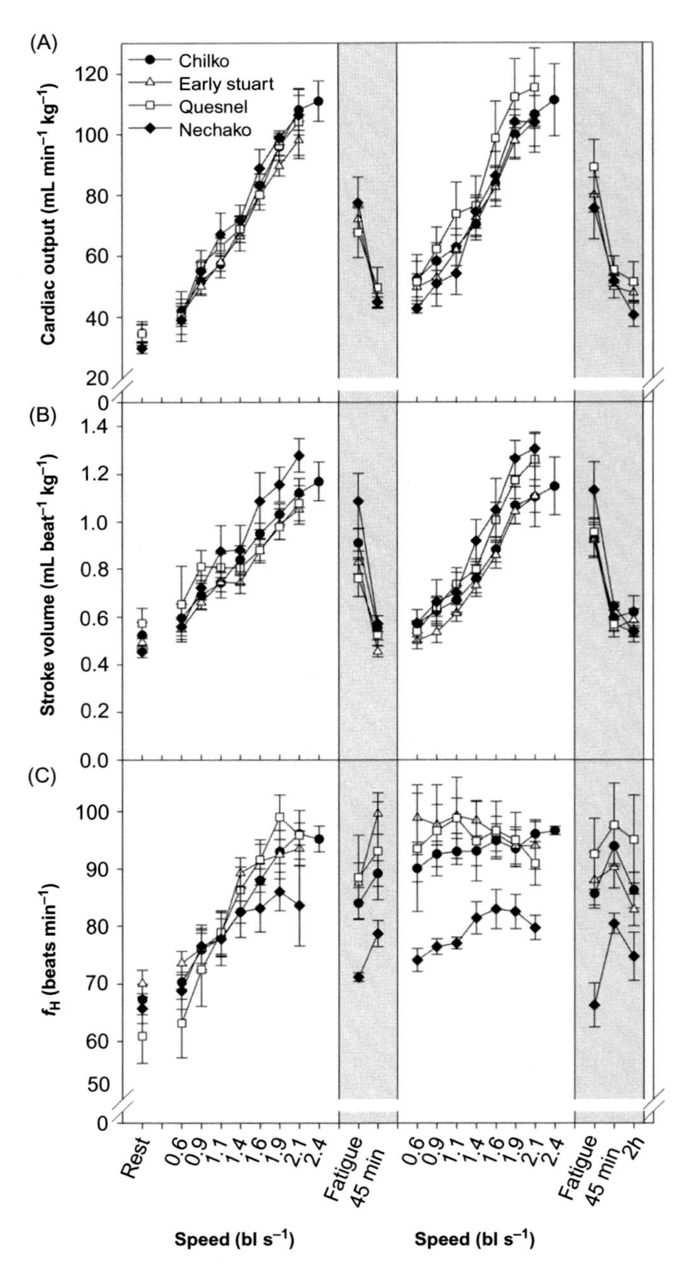

Fig. 11. Changes in cardiac output (A), stroke volume (B), and heart rate (C) with swimming speed during two consecutive prolonged swimming challenges to U_{crit} (and during their subsequent recovery periods) for four populations of sockeye salmon (Early Stuart; Chilko; Quesnel;

Nechako). There were no significant differences in cardiac output or stroke volume among populations, but heart rate was significantly lower in Nechako sockeye salmon compared to some of the other populations during the first three speeds of the second swim. Note that even though heart rate does not fully recover between swimming challenges, maximum cardiac output is reached. Credit: Eliason, E.J., Clark, T.D., Hinch, S.G., Farrell, A.P., 2013a. Cardiorespiratory collapse at high temperature in swimming adult sockeye salmon. Conserv. Physiol. 1, cot008.

The increases in $\dot{Q}$ with increases in swimming speed vary considerably across species: 60% for sea raven; 70% for the large-spotted dogfish (*Scyliorhinus stellaris*) and for the leopard shark; 50%–250% for Atlantic cod (*Gadus morhua*); 250%–270% for northern squawfish (*Ptychocheilus oregonensis*); 50%–370% for large-scale sucker (*Catostomus macrocheilus*); 90%–300% for rainbow trout; 290% for sockeye salmon (*O. nerka*); and 0% for the eel (*Anguilla australis schmidtii*). Although some of the variability within a species is related to temperature, it may be that the testing apparatus did not always induce maximum $\dot{Q}$, that the fish did not have a sufficient recovery period or were stressed prior to the initiation of the swimming test (i.e., resting $\dot{Q}$ was elevated), or that many experiments are performed on wild fishes that have never experienced a swimming flume before. We note, however, that the contribution of V_s and f_H to the increase in $\dot{Q}$ is also highly variable among species (Table 2). Some species only increase f_H while others largely increase V_s. In contrast to mammals, birds, reptiles and amphibians that routinely and primarily modulate f_H during aerobic exercise, fishes often modulate both V_s and f_H.

When a fish either sprints, undergoes spontaneous bursts of swimming, or uses a burst-and-coast swimming gait in a prolonged swimming test (usually near fatigue), it relies primarily on anaerobic glycolysis. Thus, the brief decreases in $\dot{Q}$, f_H and arterial blood pressure sometimes seen at the outset of the spontaneous activity (Farrell, 1982; Stevens et al., 1972) are perhaps not surprising even though repeated spontaneous activity does increase $\dot{Q}$ (Axelsson and Farrell, 1993). Because these glycolysis-powered activities last just a few seconds or less, the oxygen cost of the activity is delayed, which is termed the "excessive post-exercise oxygen consumption" (Eliason et al., 2013a, b, c; Lee et al., 2003a, b; Scarabello et al., 1991; Wagner et al., 2005, 2006). Therefore, after activities such as burst and prolonged exercise, manual chasing to exhaustion and hypoxic exposure, recovery is invariably associated with elevated $\dot{M}O_2$, $\dot{Q}$ and f_H, and reduced R_{sys} (Figs. 10 and 11; Eliason et al., 2013a, b, c; Steinhausen et al., 2008). In salmonids, $\dot{Q}$ returns to a routine level within an hour following a swimming speed test, whereas recovery of routine f_H takes longer (Figs. 8 and 9; Eliason et al., 2013b; Gallaugher et al., 2001; Thorarensen et al., 1996a). All the same, previously fatigued salmonids can repeat their maximum prolonged swimming performance (Farrell et al., 1998; Jain et al., 1998; Jain and Farrell, 2003; Wagner et al., 2005, 2006) and produce the same increase in $\dot{Q}$ (Eliason et al., 2013b) without fully recovering either $\dot{M}O_2$ or routine f_H; this implies deferment of some of the delayed cost of the first swim while performing the second swim.

When swimming at high speeds for prolonged periods (i.e., near critical swimming speed, U_{crit}) and during recovery from burst swimming, the heart faces the additional challenge of having to pump at near maximal levels at the same time that metabolic wastes have built up in venous blood (i.e., it is acidotic through a build up of H^+ and carbon dioxide, as well as hyperkalemic; see Section 3.2.5). Under these adverse conditions, the protective effect of catecholamines on maximum cardiac function may become important (Section 3.2.5). Indeed, rainbow trout injected with the β-adrenergic agonist propranolol would not exercise and those β-blocked immediately after strenuous exercise had a higher mortality than sham-injected fish (van Dijk and Wood, 1988).

3.2.4. Response of Cardiac Output to Changes in Temperature

Temperature is the ecological master factor for fishes, but no single species can tolerate the entire temperature range exploited by fishes as a whole (from −2°C in Antarctica to +42°C in Lake Magadi, Kenya; Farrell, 2009). As a result, the response of $\dot{Q}$ to a change in temperature very much depends on both the species and temperature range being considered. In fact, while many two-temperature comparisons exist (e.g., Table 2), these may not tell the whole story because the thermal relationship may be more complex than a straight line and a single Q_{10} value (ratio of rate functions over a 10°C temperature difference) (Farrell, 2007c). Temperature exposure experiments can be either acute (over minutes or hours) or chronic (over several days, weeks, or months). While both types of thermal change are important for fishes in temperate regions, thermal stratification is particularly common in polar and tropical regions. In this chapter, we only deal with some general points that are directly pertinent to subsequent sections because other chapters in this book deal with the cardiac (Chapter 4; Volume 36B: Eliason and Antilla, 2017), electrophysiological (Chapter 3, Volume 36A: Vornanen, 2017), and circulatory (Chapter 7, Volume 36A: Sandblom and Gräns, 2017) responses to temperature, as others have done previously (e.g., Farrell, 2007b, 2016).

Acute warming exponentially increases physiological and biochemical rates toward maximum values. Likewise, both routine f_H and $\dot{Q}$ increase in parallel up to maximum values during acute warming (Fig. 12), a response observed in almost all cardiovascular studies with fishes (e.g., Clark et al., 2008, 2011; Ekstrom et al., 2016; Eliason et al., 2011, 2013a, b, c; Gollock

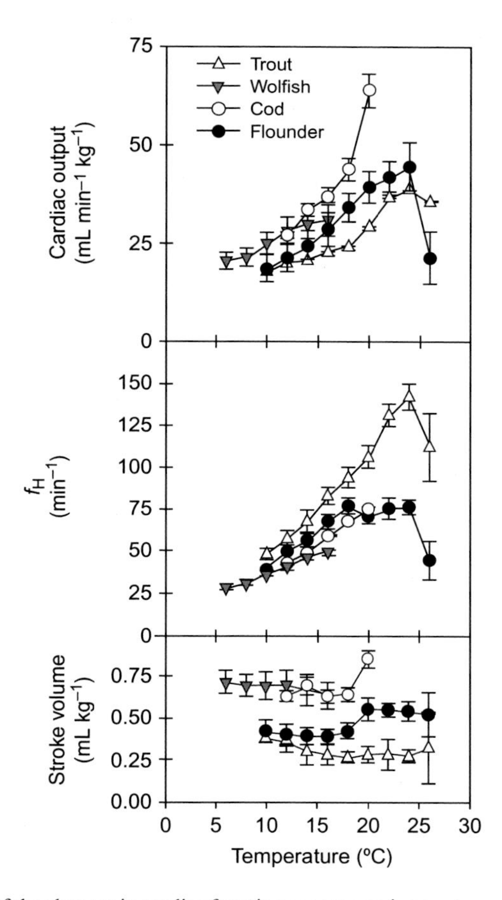

Fig. 12. Examples of the changes in cardiac function accompanying acute warming in several species of teleost fishes. These figures illustrate two common features: the thermal dependence of the rate functions (heart rate and cardiac output), but not of the capacity function (cardiac stroke volume); and the plateau for both heart rate and cardiac output. Credit: A. Kurt Gamperl, personal communication; Farrell, A.P., Eliason, E.J., Sandblom, E., Clark, T.D., 2009. Fish cardiorespiratory physiology in an era of climate change. Can. J. Zool. 87, 835–851.

et al., 2006; Steinhausen et al., 2008). Because a fish increases f_H when it swims, a swimming fish attains peak values of $\dot{Q}$ and f_H at a lower temperature than a resting fish during acute warming. Curiously, acute warming does not change routine V_s appreciably, despite the presence of excess capacity (although V_s may increase modestly near the peak f_H). Similarly, if a salmonid elevates its V_s by swimming, V_s remains at this elevated level during acute warming (Clark et al., 2008; Eliason et al., 2011, 2013a, b, c; Steinhausen et al., 2008). The question as to why resting fishes do not utilize the excess capacity to increase routine V_s during acute warming was addressed in steelhead trout (*O. mykiss*) that were acutely warmed and treated with the pharmacological agent zatebradine to limit increases in f_H (Keen and Gamperl, 2012). Zatebrabine blocks members of the hyperpolarization-activated cyclic nucleotide-gated (HCN) channel family, which help set the intrinsic pacemaker rate, even in hagfish (Wilson and Farrell, 2013; Wilson et al., 2013). In steelhead trout, zatebradine reduced routine f_H, but routine $\dot{Q}$ was maintained by elevating V_s. Moreover, these fish greatly increased V_s and elevated $\dot{Q}$ to similar levels as measured in sham-injected controls during acute warming. Thus, while steelhead trout have the potential to increase V_s during warming to raise Q, they do not under normal conditions. Both the regulation of venous return to the heart (Chapter 7, Volume 36A: Sandblom and Gräns, 2017) and chemoreception (Ekstrom et al., 2016) have been implicated in these cardiovascular responses. For example, Ekstrom et al. (2016) suggest that an adequate cardiac oxygen supply is critical to elevating V_s in resting fish at high temperatures. All the same, they suggest that a cardiac oxygen limitation does not explain the bradycardia and arrhythmia observed prior to CT_{max} being reached. A better understanding of these cardiac responses is a fertile area for further research.

The following section illustrates the general point that thermal compensation allows the heart to maintain it's functional capacity when the fish is exposed to a new temperature (i.e., cardiac plasticity). This will be done by highlighting three important changes to cardiac function during thermal acclimation. A large number of additional fundamental changes associated with thermal acclimation are dealt with elsewhere in these volumes (Chapter 3, Voume 36A: Vornanen, 2017; Chapter 4, Volume 36B: Eliason and Antilla, 2017).

First, thermal acclimation typically resets the intrinsic pacemaker rate (see Farrell, 1991, 2009). For example, by reducing the intrinsic pacemaker rate (Table 3), peak f_H (f_{Hmax}) is attained at a higher temperature after warm acclimation than if the fish were acutely warmed. In addition, it is possible to increase the scope for f_H with warm acclimation by also increasing f_{Hmax}. Although the changes that occur in atrial and ventricular cardiomyocytes with temperature acclimation are quite well understood (Chapter 3, Volume 36A: Vornanen, 2017), our understanding of how pacemaker cell function is altered by temperature acclimation is limited.

Second, the monitoring of routine f_H alone during thermal acclimation may not reveal changes at the level of the pacemaker because thermal acclimation also alters autonomic modulation of these cells (see Section 4). The pattern of autonomic outflow from the central nervous system (CNS) sets the relative contributions of adrenergic and cholinergic tone to the heart,

Table 5

Routine and intrinsic heart rates in different teleost species, and the estimated adrenergic and cholinergic tone on their heart[a]

Species	Temperature (°C)	Routine f_H (min^{-1})	Intrinsic f_H (min^{-1})	Adrenergic tone (%)	Cholinergic tone (%)	References
Gymnodraco acuticeps	−1.0	17.4	18.4	20	24	Axelsson et al. (2000a)
Pagothenia borchgrevinki	0.0	11.3	23.3	4	112	Axelsson et al. (1992)
Trematomus bernacchii	0.0	10.5	21.7	22	130	Axelsson et al. (1992)
Oncorhynchus mykiss	7.0	42.2	43.5	23	22	Gamperl et al. (1995)
Pleuronectes americanus	8.0	28	40.9	−12	26	Mendonça and Gamperl (2010)
Myoxocephalus scorpius	10.0	48.3	42.3	20	8	Axelsson et al. (1987)
Gadus morhua	10.0	37.2	39.8	13	21	Altimiras et al. (1997)
Gadus morhua	10.5	30.5	36.6	17	39	Axelsson (1988)
Hemitripterus americanus	11.0	37.6	33.0	29	23	Axelsson et al. (1989)
Polachius pollachius	11.5	46.0	40.0	25	13	Axelsson et al. (1987)
Labrus mixtus	11.5	52.0	50.0	14	12	Axelsson et al. (1987)
Labrus bergylta	11.5	41.0	49.0	14	36	Axelsson et al. (1987)
Ciliata mustela	11.5	67.0	58.0	23	9	Axelsson et al. (1987)
Raniceps raninus	11.5	31.0	28.0	22	8	Axelsson et al. (1987)
Zoarces viviparous	11.5	60.0	38.0	40	7	Axelsson et al. (1987)
Paranotothenia angustata	12.0	—	47.3	35	15	Egginton et al. (2001)
Oncorhynchus nerka (male)	12.0	43.0	50.0	24	46	Sandblom et al. (2009b)
Oncorhynchus nerka (female)	12.0	52.0	52.0	24	28	Sandblom et al. (2009b)
Oncorhynchus kisutch	12.0	31.8	30.0	38	34	Axelsson and Farrell (1993)
Sparus aurata	16.0	63.4	73.8	39	17	Altimiras et al. (1997)
Labrus bergylta	20.0	84.5	88.3	21	30	Altimiras et al. (1997)
Carassius auratus	22.5	36.0	57.0	18	97	Cameron (1979)
Thunnus	25.0	75.5	116.0	4	58	Keen et al. (1995)
Katsuwonus pelamis	25.0	79.4	183.0	6	131	Keen et al. (1995)

See Altimiras et al. (1997) for the calculation of adrenergic and cholinergic tone.

[a]Adapted from Axelsson, M., 2005. In: Polar Fishes, Farrell, A.P., Steffensen, J.F. (Ed.). Academic Press, New York, 239–280. Intrinsic heart rate is defined as heart rate after a complete blockade of muscarinic (+ atropine) and β-adrenoceptors (+ sotalol or propranolol).

and hence, routine f_H (Table 5; Section 4.1), and thermal acclimation can alter both of these contributions. In rainbow trout, for example, cardiac vagal tone decreases with warm acclimation, while cardiac β-adrenergic tone increases (Wood et al., 1979). Similarly, the effect of adrenergic stimulation can change with thermal acclimation. For example, the Q_{10} values (fish acclimated to 5 and 15°C) for intrinsic f_H and maximum $\dot{Q}$ of a perfused rainbow trout heart decrease from 2.20 and 1.72, respectively, without adrenergic stimulation to 1.38 and 1.34, respectively, with maximal adrenergic stimulation (Table 3; Graham and Farrell, 1989). Thus, a warm-acclimated rainbow trout heart is relatively less responsive to β-adrenergic stimulation (Farrell et al., 1996; Keen et al., 1993), which may reflect a reduction in the density of β-adrenoceptors on the cell membrane of the cardiomyocytes (Eliason et al., 2011; Gamperl et al., 1994c). Such changes, however, will also decrease the level of cardiac protection afforded by adrenergic stimulation (Section 3.2.5).

Third, thermal acclimation can also alter RVM, and this has an impact on cardiac function. Although the Q_{10} comparing maximum CPO expressed as g^{-1} ventricular mass for rainbow trout acclimated to 5 and 15°C is 1.78, warm-acclimated fish have a 50% smaller heart, which produces a much lower Q_{10} (1.34) when CPO is expressed per kg of body mass (Graham and Farrell, 1989). An area of research that is in need of attention is the study of cardiac adaptations. While fishes are broadly classified as stenotherms or eurytherms, cardiac adaptations that might restrict a particular fish species or population to its thermal range are poorly understood. However, intraspecific comparisons are providing some insight. For example, hatchery rainbow trout reared at 10°C in British Columbia, Canada, have a peak f_H at 20°C, with cardiac arrhythmias developing in half the fish by 21°C and by 24°C in the remainder (Verhille et al., 2013). Correspondingly, perfused heart preparations from 22°C-acclimated rainbow trout had a high failure rate (40%) and poorer maximum CPO (23% lower) compared with 18°C-acclimated fish where there were no failures when tested at their respective temperatures (Table 3; Farrell et al., 1996). In contrast, rainbow trout imported into Western Australia from California (United States), and then hatchery-selected over decades for high temperature tolerance, attain f_{Hmax} at about 24°C when acclimated to 15°C, and a peak maximum $\dot{M}O_2$ at 20°C, with 29%–45% of this peak $\dot{M}O_2$ being retained at 25°C (Chen et al., 2015). These findings are consistent with a population of rainbow trout from Tuolumne River in central California being able to maintain maximum $\dot{M}O_2$ up to 24°C (Verhille et al., 2016). Understanding the genetic and biochemical basis for such intraspecific differences in thermal performance of the cardiorespiratory system will be important for predicting the effects of global warming.

3.2.5. Response of Cardiac Output to Hypoxia and Anoxia

Aquatic hypoxia occurs for a variety of reasons and varies in intensity and duration (Diaz and Breitburg, 2009). Although most fishes maintain routine $\dot{Q}$ and f_H at a water $PO_2 > 7$ kPa (Farrell and Jones, 1992), most tunas are hypoxia-sensitive and reduce both f_H and $\dot{Q}$ at a water $PO_2 > 9$ kPa (Bushnell and Brill, 1991; Bushnell et al., 1990). Conversely, a few exceptional fishes do something that is impossible for the majority of vertebrates, their hearts tolerate severe hypoxia and prolonged anoxia (e.g., Farrell and Stecyk, 2007). The cardiac responses of various fishes to hypoxia and anoxia are presented in Table 2.

The cardiorespiratory responses ultimately triggered by aquatic hypoxia are presumably aimed at offsetting arterial hypoxemia and maintaining routine $\dot{M}O_2$ (Farrell and Richards, 2009). These cardiorespiratory responses have different water PO_2 thresholds that show species specificity (Gamperl and Driedzic, 2009), which is not surprising given that the PO_2 for the half saturation of whole blood (P_{50}) can vary from as low as 0.2 kPa to higher than 6.0 kPa among fishes (Chapter 1, Volume 36B: Brauner and Harter, 2017); even within a given family of fishes P_{50} can vary considerably (~2-fold for tidepool sculpins; Mandic et al., 2009). A good example of this species difference is the epaulette shark (*Hemiscyllium oscellatum*), which is hypoxia tolerant and reduces f_H and $\dot{Q}$ at a lower water PO_2 compared with the shovelnose ray (*Aptychotrema rostrata*); another Great Barrier Reef elasmobranch, but with a higher P_{50} (Speers-Roesch et al., 2012a; Stensløkken et al., 2004). Surprisingly, the relationship between hypoxic bradycardia and blood P_{50} has yet to be studied systematically.

Severe aquatic hypoxia ultimately depresses routine cardiac activity in almost all fish (Chapter 5, Volume 36B: Stecyk, 2017), although maximum cardiac pumping is impaired at modest levels of hypoxia (8–9 kPa O_2). For example, modest hypoxia attenuates the increases in $\dot{Q}$ and f_H (but not V_s) of Atlantic cod during prolonged swimming (Petersen and Gamperl, 2010). There can be a direct effect of hypoxia on cardiac function, as shown with the perfused rainbow trout heart where routine $\dot{Q}$ can no longer be maintained when perfusate PO_2 is <5.5 kPa (Farrell et al., 1988a, b). In contrast, a long-finned eel heart stills generates 80% of maximum $\dot{Q}$ with a perfusate PO_2 of 0.2 kPa (Davie et al., 1992). Nevertheless, fish usually impose a vagally mediated bradycardia well before the myocardial tissue loses its pumping capacity. For example, the common eel decreases both routine f_H and $\dot{Q}$ when water PO_2 is <5.5 kPa (Chan, 1986), but the maximum isometric force of myocardial strips is far more resilient to hypoxia (Gesser et al., 1982). Likewise, blocking the hypoxic bradycardia with atropine increases $\dot{Q}$ in spiny dogfish, but not V_s, again suggesting that cardiac activity is actively downregulated during hypoxia (Sandblom et al., 2009a). Indeed, moderate hypoxia (water PO_2 = 6.9 kPa) decreases V_s before f_H in spiny dogfish. Although this decreases $\dot{Q}$, R_{sys} increases to maintain P_{DA} (Sandblom et al., 2009a; Table 2). A more severe hypoxia (water PO_2 = 2.5 kPa) also triggers bradycardia and a further decrease in V_s and $\dot{Q}$ (by 46%), but once more R_{sys} increases to maintain P_{DA}.

Bradycardia seems a counterintuitive response to maintain $\dot{M}O_2$ during hypoxia given that CaO_2 is decreasing. Therefore, considerable speculation exists as to the potential benefits of hypoxic bradycardia (Farrell, 2007a; Farrell and Stecyk, 2007; Gamperl and Driedzic, 2009; McKenzie et al., 2009). Possible benefits include: a longer period of diastole, which favors CBF as well as diffusional exchange between the venous blood and cardiac trabeculae without blood vessels; a reduced speed of myocardial contraction, which lowers oxygen consumption; a longer cardiac filling time, which favors a higher V_s; and a reduction of blood flow pulsatility, which improves oxygen loading at the gills.

Farrell and Stecyk (2007) suggest that CPO is a primary consideration in understanding cardiac responses to severe hypoxia because the heart has a maximum capacity to produce ATP without oxygen. The maximum glycolytic ATP turnover rate of fish hearts can be measured under anoxic conditions using perfused heart or ventricular strip preparations. Maximum glycolytic ATP turnover rates in fish range from 33 to 110 nmol ATP s^{-1} g^{-1} ventricular mass, but depend on the level of myocardial work and temperature (Table 4). The glycolytic ATP turnover rate for the isometrically contracting ventricular myocardium is surprisingly similar for the hypoxia-sensitive rainbow trout and the anoxia-tolerant freshwater turtle (Table 4). Farrell and Stecyk (2007) suggest that a rainbow trout heart at 15°C has a maximum glycolytic capacity of ~0.7 mW g^{-1}, with an ATP turnover rate of ~70 nmol ATP s^{-1} g^{-1} (Table 4). The maximum glycolytic ATP turnover rate appears to have a Q_{10} of around 2.0 (Table 4) and a study on tilapia hearts (Lague et al., 2012) is consistent with this idea. Tilapia (*Oreochromis* hybrid) hearts at 22°C generate 1.7 mW g^{-1} and consume ~110 ATP s^{-1} g^{-1} after poisoning with cyanide. When gassed with pure nitrogen, the perfused tilapia heart performs even better (3.4 mW g^{-1} and ~170 ATP s^{-1} g^{-1}) (Table 2), suggesting that even a small amount of oxygen can be very helpful for this hypoxia-tolerant heart, even though the combination of extracellular acidosis and nitrogen gas reduces this benefit (maximum CPO is 2.4 mW g^{-1} under these conditions; ~120 ATP s^{-1} g^{-1}; Lague et al., 2012).

When routine CPO is at, or below, the heart's maximum glycolytic capacity, a fish theoretically has no need to downregulate cardiac activity during hypoxia. However, teleost and elasmobranch fishes appear to downregulate

CPO during severe hypoxia. As we noted above, the anoxic tilapia heart can support ~1.7 mW g^{-1} at 22°C. But, at a water PO_2 of just ~1 kPa, tilapia hearts at 22°C are clearly in a glycolytic state (plasma lactate increases to 17 mM) and the cardiac muscle tissue is acidotic, but they stabilize cardiac [ATP] for 8 h by halving routine f_H and Q and lowering CPO to 0.7 mW g^{-1}, while maintaining V_s and P_{VA} (Speers-Roesch et al., 2010). Species-specific differences in hypoxia tolerance can be reflected in the water PO_2 when a critical CPO is reached. For example, the epaulette shark and shovelnose ray substantially depress f_H, $\dot{Q}$, V_s, and P_{DA} by similar percentages during progressive hypoxia at 28°C (Speers-Roesch et al., 2012b; Stensløkken et al., 2004; Table 2). The epaulette shark, however, reaches a CPO of 1.6 mW g^{-1} (down from 5.8 mW g^{-1} and close to the maximum glycolytic CPO) when water PO_2 reaches 0.5 kPa, whereas the more hypoxia-sensitive shovelnose ray reaches a CPO of 1.1 mW g^{-1} (down from 3.9 mW g^{-1}) at a water PO_2 of 2 kPa (Table 2).

In the common carp, the minimal level of CPO seen with oxygen deprivation also depends on the acclimation temperature (0.29 mW g^{-1} at 5°C, 0.71 mW g^{-1} at 10°C, and 0.86 mW g^{-1} at 15°C), even though the relative decrease (from 72% to 87% for CPO, $\dot{Q}$, and f_H) is similar at all acclimation temperatures (Stecyk and Farrell, 2002; Table 2). Interestingly, acclimation temperature also increases the sympathetic- and parasympathetic-dependent downregulation of f_H, the primary driver for reductions in $\dot{Q}$ and CPO (Stecyk and Farrell, 2006). Acclimation temperature consequently influences hypoxic survival time, even in the hypoxia-tolerant crucian carp (Stensløkken et al., 2004). For example, common carp survive near anoxic conditions (< 0.3 mg O_2 L^{-1}) for 24 h when acclimated to 5°C, but for just 6 and 3 h when acclimated to 10 and 15°C, respectively (Stecyk and Farrell, 2002, 2006). Although the suppression of cardiac activity to a level lower than the heart's maximum glycolytic capacity clearly helps survival, cardiac tolerance of hypoxia in fish is determined more by their routine CPO, and to what level it needs to be depressed, rather than large species-specific differences in maximum glycolytic capacity (Farrell and Stecyk, 2007) provided temperature is considered. What sets the level of blood flow in metabolically depressed cardiac states is unknown.

Exactly when a heart will become functionally hypoxic requires knowledge of more than just the maximum glycolytic capacity and remains an intriguing and unanswered question. The presence of a coronary circulation is important because Type II and III (IV) hearts have an arterial oxygen supply to at least a portion of the myocardium. CBF certainly increases in hypoxic fish (Section 6), and fish hearts perform better *in vitro* under both normoxic and hypoxic conditions when the coronary circulation is perfused (Agnisola et al., 2003; Davie et al., 1992; Farrell et al., 1992). Theoretically, Type III (IV) hearts should fare better in hypoxia than Type II hearts because a much greater portion of the heart is supported by CBF. Nevertheless, tilapia have a Type I heart and, as described above, they tolerate severe hypoxia extremely well.

To what degree hypoxia tolerance reflects the venous PO_2 and when the myocardium becomes hypoxic is unknown. Most fish species have either a Type I or a Type II heart, where the spongy myocardium is supported entirely by the venous blood supply. As a result, the spongy myocardium likely has a threshold venous PO_2, perhaps in the range of 0.05–0.12 kPa (Davie and Farrell, 1991a). This is because the total amount of oxygen in venous blood far surpasses the needs of the heart (Farrell, 1987, 1991). Furthermore, when a fish exercises, or the ambient temperature increases, venous PO_2 decreases because the other tissues increase oxygen extraction. This, in turn, makes this cardiac oxygen supply route precarious when compared with a well-oxygenated coronary blood supply (Farrell, 2009). [Note: The increase in $\dot{Q}$ during swimming will increase the total venous oxygen delivered to the heart and compensate for the increase in cardiac $\dot{M}O_2$, which means that the lower C_VO_2 is not problematic; Farrell and Clutterham, 2003]. Implanting optical oxygen measurement probes in the *ductus Cuvier* to continuously monitor venous PO_2 (Ekstrom et al., 2016; Farrell and Clutterham, 2003) has provided insight into the concept of a threshold venous PO_2, although more thorough testing is needed. For example, cold-acclimated rainbow trout progressively decrease venous PO_2 to a threshold of ~1.2 kPa at 80% of U_{crit} (Farrell and Clutterham, 2003), a speed at which their swimming gait changes. A burst-and-coast swimming gait typically results in a temporary decrease in venous PO_2 below this threshold. Warm acclimation alters the venous oxygen threshold in rainbow trout, increasing the venous PO_2 threshold to 2.2 kPa, as might be expected with a right shift in the blood oxygen equilibrium curve (i.e., a decrease in blood oxygen affinity) and a similar degree of tissue oxygen extraction. In fact, warming European perch causes a decline in venous PO_2 below a threshold level, which is associated with declines in V_s, $\dot{Q}$, and CPO, but not f_H, before the fish reaches its CT_{max} (Ekstrom et al., 2016). However, a new line of thinking may be needed with regards to cardiac oxygen delivery. PACA on the cardiac endothelium may locally elevate PO_2 (Section 1; Alderman et al., 2016), and thus, ultimately determine the heart's PO_2 threshold (Fig. 13).

While most fishes tolerate hypoxia to varying degrees before decreasing $\dot{Q}$, a few species are exceptional because they maintain $\dot{Q}$ under nearly anoxic conditions. This appears possible because their routine CPO is low enough to be supplied by glycolytic ATP production alone (Farrell and Stecyk, 2007). For example, Pacific hagfish (*Eptatretus stouti*) maintain routine $\dot{Q}$ under anoxia at 10°C for more than 30 h with CPO at just $0.25\,mW\,g^{-1}$; although routine f_H is halved and V_s is elevated to compensate (Fig. 14; Table 2).

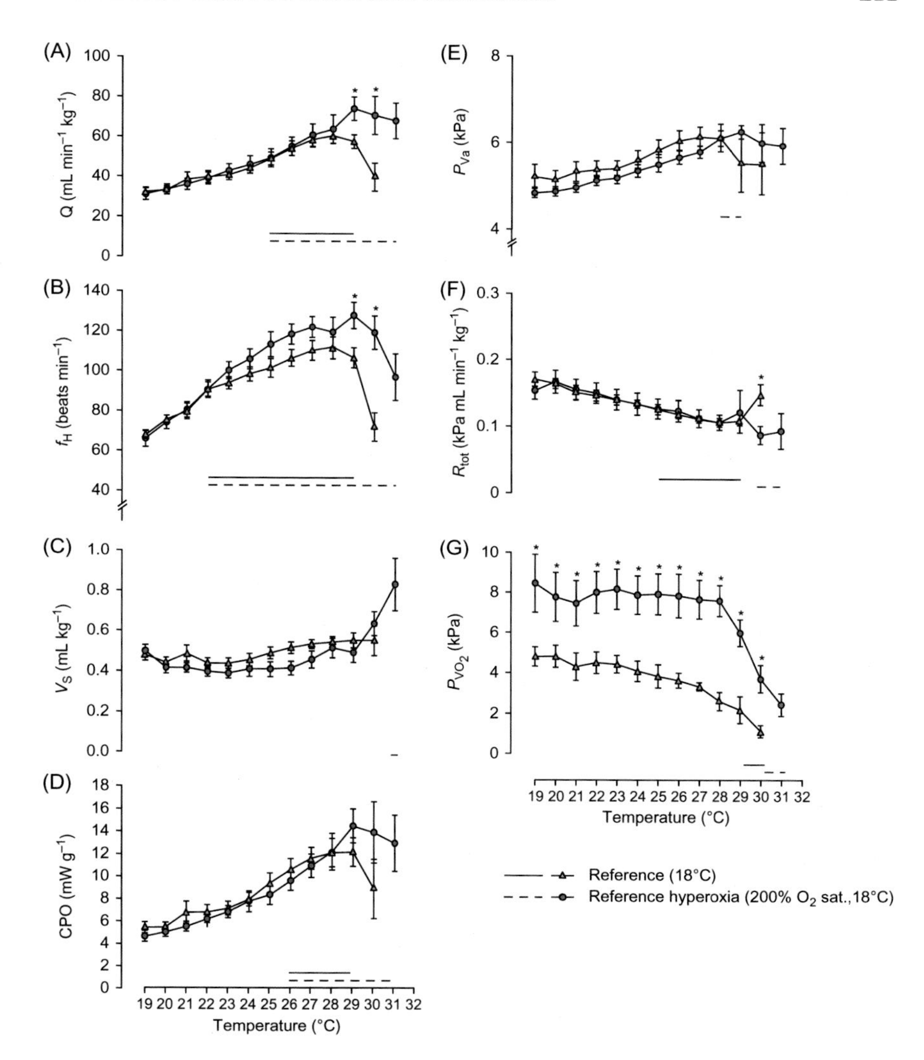

Fig. 13. The changes in routine cardiovascular parameters in European perch (*Perca fluviatilis*) during an incremental increase in water temperature up to their critical thermal maximum (CT_{max}) under normoxic (*blue lines*) or hyperoxic (*red lines*) conditions. The normoxic cardiac performance (cardiac output—$\dot{Q}$; heart rate—f_H; and cardiac power output—CPO) begins to peak as venous oxygen partial pressure (P_VO_2) falls. Hyperoxia, however, delays these declines until a higher temperature is reached. This is presumably by elevating P_VO_2, $\dot{Q}$ and V_s. P_{VA} = ventral aortic blood pressure; R_{tot} = total peripheral vascular resistance. Reprinted with permission and adapted from Ekstrom, A., Brijs, J., Clark, T.D., Grans, A., Jutfelt, F., Sandblom, E., 2016. Cardiac oxygen limitation during an acute lethal thermal challenge in the European perch: effects of chronic environmental warming and experimental hyperoxia. Am. J. Physiol. 311, R440–R449.

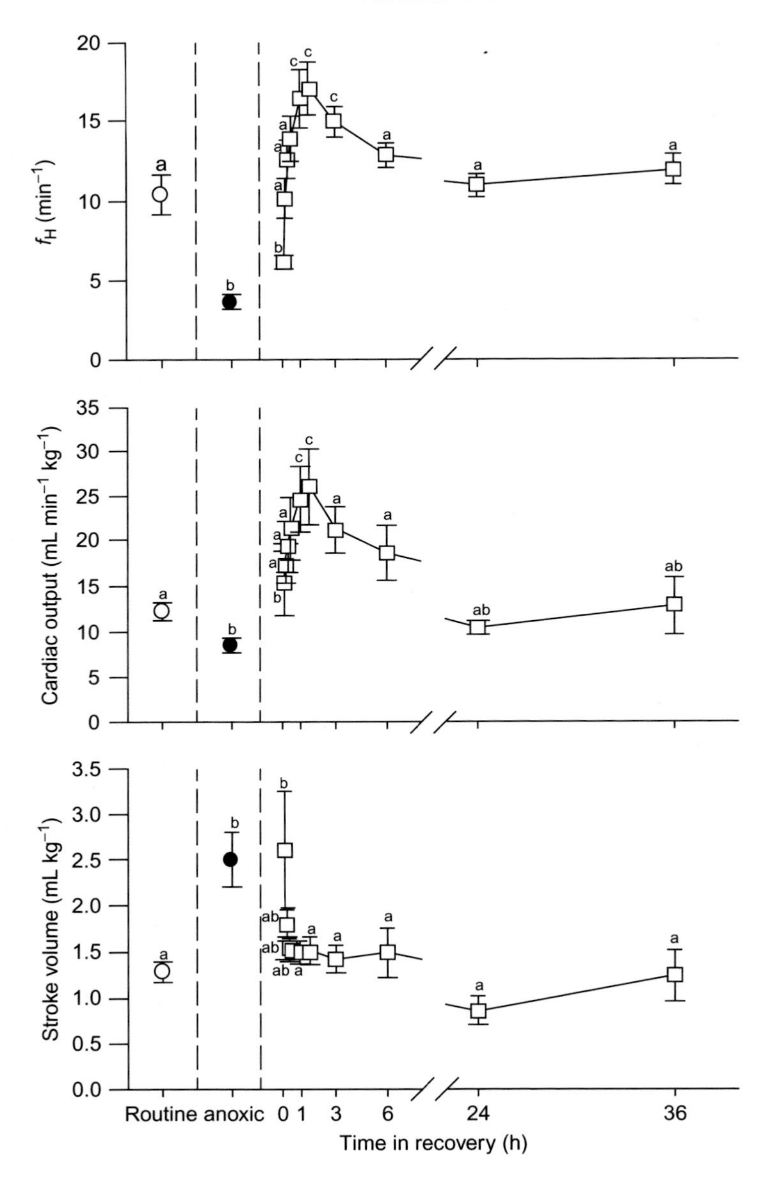

Fig. 14. Simultaneously recorded cardiovascular variables from hagfish before, at the end of a 36-h exposure to anoxia, and during recovery in normoxia (starting at time zero). Note that cardiac output is largely maintained despite the strong anoxic bradycardia, but that there are rapid increases in heart rate and cardiac output within the first hour of recovery. Statistical differences are indicated by *dissimilar letters*. Credit: Cox, G.K., Sandblom, E., Farrell, A.P., 2010. Cardiac responses to anoxia in the Pacific hagfish, *Eptatretus stoutii*. J. Exp. Biol. 213, 3692–3698.

The Atlantic hagfish (*Myxine glutinosa*) also maintains $\dot{Q}$ and f_H under severe hypoxia (water $PO_2 \sim 1$–2 kPa) with CPO at 0.4 mW g^{-1} (Axelsson et al., 1990). The New Zealand hagfish, however, increases $\dot{Q}$ and CPO (from 0.4 to 0.7 mW g^{-1}) during severe hypoxia (water $PO_2 \sim 4$ kPa) (Forster et al., 1992). Finally, the anoxia-tolerant crucian carp (*Carassius carassius*), which converts lactate to alcohol to prevent acidosis, maintains f_H, $\dot{Q}$ and CPO (0.4 mW g^{-1}) nearly unchanged after 5 days of anoxic exposure at 9°C (Stecyk et al., 2004; Chapter 5, Volume 36B: Stecyk, 2017).

3.2.6. Response of Cardiac Output to Extracellular Acidosis

Extracellular acidosis in fish is inevitable with strenuous or exhaustive exercise, extremely warm temperatures, or the hypoxemia produced by environmental hypoxia. This is because anaerobic glycolysis is activated either because internal oxygen delivery cannot keep up with tissue oxygen demand, and/or there is a need for a faster supply of ATP (glycolysis generates ATP faster than oxidative phosphorylation, while depleting metabolic substrates more rapidly). As a result, lactate and H^+ build up in the blood (i.e., there is a metabolic acidosis). Furthermore, any increase in oxidative phosphorylation will increase carbon dioxide levels in the blood leaving the tissues (i.e., there is also a respiratory acidosis). Thus, exhaustive exercise in fish can decrease blood pH by as much as 0.5 pH units (see Graham et al., 1982; Milligan and Wood, 1986; Ruben and Bennett, 1981; Wood et al., 1977) through a mixed metabolic and respiratory acidosis. Environmental hypercapnia also lowers blood pH simply because excess carbon dioxide enters the fish and full or partial metabolic compensation of the acidotic state takes time.

Myocardial acidosis is not inevitable during the extracellular acidosis associated with exhaustive activity. Exhaustive exercise causes ventricular intracellular pH to increase slightly in sea raven, remain unchanged in rainbow trout, and decrease slightly in starry flounder (*Platichthys stellatus*) (Milligan and Wood, 1986; Wood and Milligan, 1987). Although the intracellular pH of perfused rainbow trout (Milligan and Farrell, 1991) and white sturgeon (*Acipenser transmontanus*) (Baker et al., 2011) hearts tracks hypercapnic acidosis, the heart of the armored catfish (*Pterygoplichthys pardalis*) does not (Hanson et al., 2009). Mechanistic explanations for these differences in myocardial intracellular pH regulation among fishes remain unclear (Milligan and Farrell, 1991).

Although fish hearts are often required to pump maximally under these adverse extracellular conditions, extracellular acidosis can directly impair cardiac pumping through direct pH and/or CO_2 effects on the myocardium. The direct negative inotropic and chronotropic effects of hypercapnic acidosis are well established (Driedzic and Gesser, 1994; Farrell, 1984; Farrell and Jones,

1992; Gesser, 1985; Gesser and Poupa, 1983; Poupa and Johansen, 1975; Poupa et al., 1978). For example, an extreme extracellular pH (6.8–7.1) decreases maximum isometric force of the ventricular myocardium by as much as 55% in rainbow trout, European perch, Atlantic cod (*G. morhua*), plaice, common eel and winter flounder (Gesser, 1985; Gesser and Poupa, 1983; Gesser et al., 1982). A less extreme, and more physiological extracellular acidosis (pH 7.4), also reduces the maximum $\dot{Q}$ and CPO of perfused hearts, but only by <20% as compared to severe acidosis (pH 6.8–7.0) which halves maximum CPO (Table 3). Like hypoxia tolerance, some fish species are better at tolerating the severe extracellular acidosis associated with hypercapnia. For example, maximum $\dot{Q}$ and CPO decrease by only 12%–14% at pH 7.2 in armored catfish (Hanson et al., 2009). White sturgeon are even more tolerant; a decrease in blood pH to 6.8 during exposure to extreme hypercapnia only decreases maximum $\dot{Q}$ and CPO by 25% (Baker et al., 2011).

Fish defend maximum cardiac pumping capacity during hypercapnic acidosis through a variety of mechanisms. Winter flounder use an as yet undiscovered intrinsic mechanism to recover maximum isometric tension over time during exposure to hypercapnic (i.e., respiratory) acidosis, but not during lactacidosis (i.e., metabolic acidosis) (Gesser, 1985). The heart of the air-breathing catfish (*Pangasianodon hypophthalmus*) can also recover during lactacidosis (Joyce et al., 2015). Increasing extracellular concentrations of calcium, HCO_3^-, and catecholamines all improve performance under hypercapnic conditions, likely by improving the intracellular calcium available for cardiac muscle contraction (Driedzic and Gesser, 1994; Gesser, 1985; Gesser and Jorgensen, 1982; Gesser and Poupa, 1983; Gesser et al., 1982; Hanson and Farrell, 2007; Hanson et al., 2006; Milligan and Farrell, 1991). Intracellular pH regulation of cardiomyocytes appears to be important for cardiac protection in the armored catfish (Hanson et al., 2009). Among these potential protective mechanisms, β-adrenergic stimulation is perhaps the most universally important one because changes in extracellular calcium and HCO_3^- are not normally large enough to have a major impact, whereas catecholamines are released into the blood when fish are stressed with extracellular acidosis. Indeed, adrenergic stimulation can fully rescue maximum $\dot{Q}$ and CPO of white sturgeon, even at extreme values for extracellular pH (pH 6.8; Baker et al., 2011). Adrenergic stimulation also plays a very important role by rescuing maximum cardiac performance of the rainbow trout heart. At 10°C, 5 mM potassium decreases maximum $\dot{Q}$ and CPO by 30% (Fig. 15), largely by slowing f_H. Even though 7.5 mM potassium depolarizes the heart and stops cardiac contraction, a physiological concentration of adrenaline protects the myocardium from this fatal effect. Acidosis exacerbates the effect of 5 mM potassium, and decreases maximum $\dot{Q}$ to 33% and maximum CPO to 45%

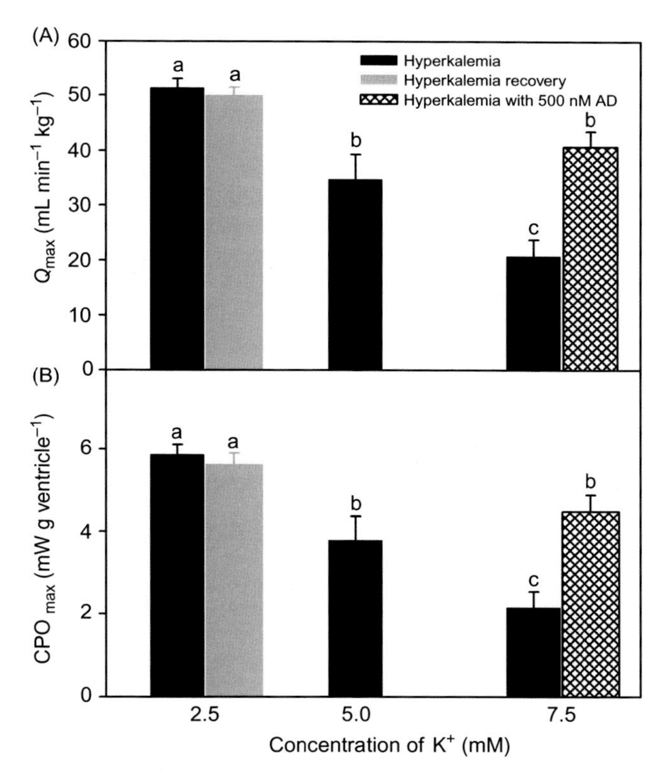

Fig. 15. Effects of hyperkalemia on the perfused rainbow trout heart. Data illustrate how adrenergic stimulation can rescue maximum cardiac performance under hyperkalemia, a condition that otherwise would eventually reduce cardiac function to zero. Credit: Hanson, L.M., Obradovich, S., Mouniargi, J., Farrell, A.P., 2006. The role of adrenergic stimulation in maintaining maximum cardiac performance in rainbow trout (*Oncorhynchus mykiss*) during hypoxia, hyperkalemia and acidosis at 10°C. J. Exp. Biol. 209, 2442–2451.

of normal levels through a further decrease in f_H. Again, adrenaline protects the heart from this debilitating effect. On the other hand, while the combination of 7.5 mM potassium and acidosis always reduced cardiac function to zero, adrenaline provided partial protection (Fig. 15).

The rainbow trout heart also shows that hypoxia, acidosis and high potassium have interactive effects. For example, maximum $\dot{Q}$ and CPO typically decline with a perfusate PO_2 of 3.3 kPa and hearts fail with a perfusate PO_2 below 2.0 kPa. Acidosis with 5 mM potassium increases the hypoxic threshold to 5.0 kPa, but again adrenaline can protect maximum performance down to a hypoxic threshold of 2.0 kPa. Thus, the hypoxic threshold for maximum performance of the heart is very dependent on the prevailing extracellular conditions, and adrenergic stimulation is critical to lowering this threshold. The interactive effects of extracellular acidosis and hypoxia with temperature and hyperkalemia adds even more complexity. Studies in perfused rainbow trout hearts provide insight into these interactions (Hanson and Farrell, 2007; Hanson et al., 2006). The extracellular conditions observed during exhaustive exercise can be simulated in the perfusate of working heart preparations [5 mM or 7.5 mM potassium; tonic (5 nM) or elevated (500 nM) adrenaline; and acidosis (pH 7.5)]. In general, the effects of combined extracellular acidosis, hyperkalemia and hypoxia exacerbate the individual effects of each of these factors, and their negative impacts are further exacerbated at higher temperatures. For example, the protection afforded by adrenergic stimulation during an acidotic and hyperkalemic challenge is greatly reduced when rainbow trout are acclimated to 18°C; all hearts fail completely at a hypoxic threshold of 5.1 kPa (some hearts fail even at 6.7 kPa). Consequently, the hypoxic threshold of the acidotic and hyperakalemic heart is also higher than at 10°C. This difference may reflect a change in the density of β-adrenoceptors (Farrell et al., 1996; Keen et al., 1993; Chapter 4, Volume 36B: Eliason and Antilla, 2017).

A mix of an unusually high temperature and hypoxia enhances the negative effects of hypercapnic acidosis on the rainbow trout heart (Hanson and Farrell, 2007; Hanson et al., 2006). In fact, the protective effect of β-adrenergic stimulation is essentially lost at 18°C in the acidotic and hypoxic perfused rainbow trout heart (Hanson and Farrell, 2007). This diminished protection could also be due to a downregulation of myocardial cell membrane β-adrenoceptors at high temperature, even though hypoxia may increase the density of these receptors in the spongy myocardium of Chinook salmon (*Oncorhynchus tshawytscha*) (Gamperl et al., 1998). In contrast to rainbow trout, Chilko sockeye salmon have a very high basal level of myocardial cell membrane β-adrenoceptors and upregulate these receptors at warmer temperatures (Eliason et al., 2011).

3.2.7. Responses of Cardiac Output to Feeding

Farrell et al. (2001) and Seth et al. (2011) have provided comprehensive accounts of the vascular anatomy of the gut circulation, the cardiac response to feeding, and the control of gut blood flow (GBF). Since very little has been added on this topic, we provide an update of what was previously reported. Seth et al. (2011) note the importance of using vascular casting techniques to advance our understanding of the diversity of arterial supply routes to the gastrointestinal tract (e.g., Thorarensen et al., 1991). Knowing these arterial supply routes is critical for an accurate estimate of GBF and the proportion of $\dot{Q}$ reaching the gut. For example, the teleost fish gut is primarily supplied by the coeliomesenteric artery that originates from the dorsal aorta proximate to the confluence of the efferent branchial arteries; it branches into a major intestinal artery (also called the mesenteric artery) and a minor gastrointestinal artery. In elasmobranch fishes, several arteries branch off the dorsal aorta toward the gut, with the coeliomesenteric artery being the major one. Total GBF has yet to be measured in fishes because some of the minor arteries are more difficult to access for flow probe placement. However, the underestimate of total GBF is likely minimal because only the smaller vessels have been missed. Under routine conditions, GBF is reported as 10%–40% of $\dot{Q}$, which is a proportion similar to that found in mammals (Seth et al., 2011).

To support the digestive process, fish increase $\dot{Q}$ to variable levels above routine $\dot{Q}$ (Altimiras et al., 2008; Axelsson and Fritsche, 1991; Axelsson et al., 1989, 2000b, 2002; Dupont-Prinent et al., 2009; Gräns et al., 2009; Seth et al., 2008). GBF between feeding sessions ranges from 3 to 18 $\mathrm{mL\,min^{-1}\,kg^{-1}}$ but during and post-feeding this can increase by as much as 71%–156%, with the peak response occurring within a few hours (Altimiras et al., 2008 Axelsson and Fritsche, 1991; Axelsson et al., 1989, 2000a, b, 2002; Crocker et al., 2000; Dupont-Prinent et al., 2009; Eliason et al., 2008; Gräns et al., 2009; Seth et al., 2008; Thorarensen and Farrell, 2006; Thorarensen et al., 1993). Still missing are systematic studies on the effect of meal size, meal quality and temperature on the cardiovascular responses to feeding. The postprandial increase in $\dot{Q}$ is almost exclusively mediated by an increase in f_H (Eliason et al., 2008).

The post-prandial increase in GBF occurs through the combined increase in $\dot{Q}$ and a 20%–52% increase in the proportion of $\dot{Q}$ reaching the gut (Seth et al., 2011). Also a routine α-adrenergic tonus exists in the gut circulation (Axelsson et al., 1989, 2000; Crocker et al., 2000; Seth and Axelsson, 2010). Therefore, given that postprandial changes in P_{DA} are minor, vasodilation of gut blood vessels to increase GBF likely involves an α-adrenergic mechanism (Seth et al., 2011). Nevertheless, other vascular control mechanisms in the gut include: substance P, neurokinin A, vasoactive intestinal polypeptide, and scyliorhinins, as well as chemical and mechanical stimuli (Seth et al., 2011; Chapter 7, Volume 36A; Sandblom and Gräns, 2017). The interaction of these various vasoactive mechanisms is unknown.

The gut appears to be a low priority circulation in unfed fish. Spontaneous bursts of exercise in white sturgeon greatly increase gut vascular resistance and halve GBF, even though $\dot{Q}$ and P_{DA} increase appreciably (Crocker et al., 2000). Similarly, the increase in $\dot{Q}$ associated with swimming in unfed fish is not matched by an increase in GBF (Axelsson and Fritsche, 1991; Thorarensen et al., 1993). In fact, the reduction in GBF during swimming has been linearly related to the exercise-induced increase in $\dot{M}O_2$ (Thorarensen et al., 1993). Similarly, fed fish decrease GBF during swimming (Dupont-Prinent et al., 2009). These results support the more general idea that must fish prioritize blood flow distribution during multiple activities because the capacity to increase $\dot{Q}$ is limited relative to the maximum flow capacity of all the systemic vascular beds combined.

An important technical advance for digestive physiology is the development of a cannulation technique for the hepatic portal vein in salmonids (Eliason et al., 2007; Karlsson et al., 2006). This vessel carries blood from the gut to the liver. Consequently, it is now possible to sequentially sample blood and follow what nutrients have been taken up by the gut during digestion without sacrificing the fish. This will provide greater insight into the absorptive processes during feeding.

4. HEART RATE AND ITS CONTROL

4.1. Intrinsic Heart Rate

All craniate hearts have a rhythmic, myogenic, heartbeat that is generated by specialized cardiac pacemaker cells (see Chapter 3, Volume 36A: Vornanen, 2017) and sets the intrinsic f_H. This intrinsic f_H can be measured in isolated beating heart preparations, in perfused heart preparations, or *in vivo* provided the primary modulatory mechanisms (adrenergic and cholinergic) are blocked pharmacologically (e.g., Keen et al., 1995). The atrium and ventricle themselves have slower intrinsic rates of spontaneous depolarizations, with the intrinsic atrial rhythm faster than that of the ventricle. Furthermore, myocardial strips from the atrium can be electrically stimulated to a higher frequency than can ventricular strips (Shiels et al., 2002: Chapter 2, Volume 36A: Shiels, 2017).

Intrinsic f_H in fishes is reported to be as high as 250 bpm in zebrafish during early development (Chapter 2, Volume 36B: Burggren et al., 2017), but f_H is rarely higher than about 150 bpm in most adult fish species examined to date (Farrell, 1991; Farrell and Jones, 1992; Tables 2 and 3). Beyond temperature effects, the mechanistic basis for the differences in intrinsic f_H among fish species is largely unexplored. Indeed, there is a clear need for further study on the physiology of pacemaker cells in fishes because the available information (see Chapter 3, Volume 36A: Vornanen, 2017) is sparse. A significant challenge will be finding a fish species where the pacemaker cells can be easily identified and isolated.

Temperature is a primary modulator of intrinsic f_H, and acute warming approximately doubles f_H with each 10°C increase (as predicted by Arrhenius over a century ago; Farrell, 2016). Nevertheless, the sinus and atrial rhythms can ultimately outpace the ventricular rhythm during acute warming, and this leads to missed ventricular contractions at supraphysiological temperatures (Farrell, 2016; Vornanen, 2016). In anesthetized fish, these cardiac arrhythmias appear during acute warming to its CT_{max} (Chapter 4; Volume 36B: Eliason and Antilla, 2017), and Chinook salmon and European perch all develop an irregular f_H as they near their upper thermal tolerance limit (Clark et al., 2008; Ekstrom et al., 2016; Eliason et al., 2013a). At very elevated temperatures, however, fish can apparently circumvent these arrhythmias by slowing the pacemaker rate with increased parasympathetic outflow from the CNS (as is described below). In fact, sockeye salmon unexpectedly decrease, rather than increase, f_H when they swim at a supraoptimal temperature (Eliason et al., 2013a). Also, the role of adrenergic stimulation in protecting against these arrhythmias should be studied.

4.1.1. Regulation of the Intrinsic Heart Rate

Routine f_H *in vivo* is rarely the intrinsic f_H because of simultaneous vagal and β-adrenergic modulation of the intrinsic pacemaker rate; these two are the most important neural and humoral control systems in fish. Their relative roles in altering the intrinsic pacemaker rate and setting routine f_H *in vitro* are studied by comparing f_H after blocking parasympathetic slowing of the intrinsic heartbeat with an injection of atropine (an antagonist of muscarinic acetylcholine receptors presumed to be associated with fish cardiac pacemaker cells, Section 4.2.4), and/or the injection of a β-adrenergic receptor blocker (e.g., sotalol, propranolol or timolol) which prevents catecholamine (i.e., noradrenaline or adrenaline)-mediated acceleration of the heart (humoral and neural, if the heart has sympathetic innervation). Vagal tone usually predominates over adrenergic tone in most fish (Sandblom and Axelsson, 2011; Table 5) and routine f_H is invariably lower than intrinsic f_H. A rapid means to increase f_H and $\dot{Q}$ is thus the removal of vagal tone, and the increase in f_H will be substantial if routine vagal tone is high. Vagal tone is very variable in resting fishes and appears to decrease with acute warming and warm acclimation (Table 5).

While sympathetic cardiac innervation is apparently entirely absent in cartilaginous fishes, its extent remains controversial in some bony fishes (Brill and Lai, 2016; Nilsson, 1983). Nevertheless, cardiac stimulation by circulating catecholamines is possible to some degree in all fishes. In fact, the heart is the first organ to be stimulated whenever noradrenaline and adrenaline are released from chromaffin tissue into the venous circulation, and cardiac stimulation by catecholamines plays a crucial role in achieving maximum inotropic and chronotropic cardiac performance, and protecting it when an alteration of extracellular conditions could greatly impair function (see Section 3).

Adrenergic modulation of intrinsic f_H involves β-adrenoceptors that increase f_H, and α-adrenoceptors that reduce it. In mammals, β-adrenoceptor-mediated acceleration of f_H occurs *via* coupling with a G_s-protein and stimulation of a transmembrane-bound adenylyl cyclase (TMAC) to produce cyclic adenosine monophosphate (cAMP), which binds to the tail region of HCN channels on the pacemaker cells to increase their open probability. Increasing the current through the HCN channels shortens the time taken for the pacemaker voltage to reach the threshold required for an action potential. A similar transduction mechanism likely exists in fish (Chapter 3, Volume 36A: Vornanen, 2017). Forskolin, a TMAC agonist that increases intracellular cAMP levels (mimicking β-adrenergic stimulation) increases intrinsic f_H of the isolated hagfish heart (Wilson et al., 2016). Nevertheless, the absolute change in f_H produced by adrenergic stimulation is very small in elasmobranch fishes, and not particularly large in some teleost fishes. In contrast, a β-adrenergic tonus is very important in setting routine f_H in hagfishes because β-adrenergic blockers slow f_H both *in vivo* and *in vitro* considerably (Axelsson et al., 1990; Fänge and Östlund, 1954). In fact, nadolol (a non-specific β-adrenergic antagonist) almost halves the f_H of isolated *Eptatretus stoutii* hearts (from 13.4 to 8.6 bpm; Wilson et al., 2013, 2016), which supports a long-standing notion that endogenous catecholamines released from intracardiac stores (Augustinsson et al., 1956; Greene, 1902; Jensen, 1961, 1965) set adrenergic tonus in this taxa (Farrell, 2007b). Furthermore, this routine paracrine-mediated β-adrenergic tonus may be maximal because routine f_H is notoriously unresponsive to further catecholamine stimulation (Axelsson et al., 1990; Fänge and Östlund, 1954; Forster et al., 1992).

Anoxia completely inhibits the β-adrenergic tonus in isolated *E. stouti* hearts, which can be restored by forskolin (Wilson et al., 2016). Yet, when *E. stouti* recover from anoxia, they increase f_H (Fig. 14; Cox et al., 2010) beyond the level produced by adrenergic stimulation. A mechanistic explanation for this tachycardia during recovery from anoxia has recently emerged as an equivalent tachycardia occurs when isolated hearts are exposed to increased extracellular bicarbonate levels during anoxia (Wilson et al., 2016). Although 2-h of anoxia reversibly decreased spontaneous f_H to 5.1 bpm, HCO_3^- produced a dose-dependent stimulation of f_H (to 16.9 bpm with 20 mM HCO_3^- and to 22.4 bpm with 40 mM HCO_3^-), such that f_H became 4.5-times higher than f_H during anoxia, and reached a level even higher than the *in vivo* tachycardia observed following anoxia (Cox et al., 2010).

This HCO_3-mediated modulation of intrinsic f_H in *E. stouti* is an apparently novel mechanism among vertebrates, as it is mediated by a soluble adenlyl cyclase (sAC), which like TMAC, supplies cAMP to modulate the HCN channels of hagfish. KH7, a sAC inhibitor, blocks the f_H response to

HCO_3^- and immunolabeling shows that sAC is distributed throughout atrial and ventricular myocardial cells (Wilson et al., 2016). Furthermore, neither HCO_3^- nor KH7 affect f_H in hagfish under normoxic conditions. Therefore, the sAC-mediated tachycardia during recovery from anoxia is replaced by a reactivation of TMAC-mediated tonic control of f_H once O_2 is available. This response may be due to catecholamine synthesis, given that 1 and 10 μM forskolin restore the normoxic heart rate in isolated hearts under anoxia. What remains to be investigated is how widespread sAC regulation of f_H is among vertebrates, including fishes, given that sAC activity has a broad functional distribution in non-cardiac tissues (Tresguerres et al., 2014). Such studies may change how we view the evolution of the control of intrinsic f_H.

4.2. Cardiac Innervation and the Intracardiac Nervous System

4.2.1. Phylogeny of Cardiac Control

Evolutionary trends in the anatomy and function of the fish heart, and its neurohormonal control mechanisms, reflect increasing requirements for more efficient exchange of respiratory gases at the gills and for systemic transport of oxygen and metabolic substrates to tissue vascular beds. An index of the effectiveness of these processes is the capacity for dynamic control of $\dot{Q}$ (i.e., the ability to produce rapid responses to effect changes in tissue perfusion, such as occurs during exercise or hypoxia; see Section 3). Such control is provided by autonomic and hormonal modulation of pacemaker discharge rate and myocardial contractility, both of which affect $\dot{Q}$. A capacity for short-term control of $\dot{Q}$ is characteristic of all vertebrates, but cardiac control among fishes displays three phylogenetically distinct and graded stages. They range from hormonal control alone in hagfishes, through combined neural and hormonal control in lampreys and elasmobranch fishes, to bidirectional neural control in the teleost heart *via* cranial and spinal autonomic routes (summarized in Table 6, Fig. 16A–C).

In the first stage, the most ancient vertebrates (the hagfishes), have aneural hearts which respond to catecholamines released from chromaffin cells that are within the cardiac walls (paracrine control), or are associated with the walls of vessels carrying venous blood to the heart (endocrine control), with cardioacceleration and increased inotropy (Section 4.1; Table 6; Nilsson, 1983; Farrell, 2007b). These catecholamines activate β-adrenergic receptors on pacemaker cells and myocytes to augment $\dot{Q}$ through chronotropic and inotropic effects, respectively (Axelsson et al., 1990; Wilson et al., 2016). In fact, the continual release of catecholamines *in vivo* establishes a tonic level of cardiac drive that elevates $\dot{Q}$ above the intrinsic level. This tonus can then

Table 6

Summary of autonomic and non-neuronal factors in extrinsic cardiac control

	Efferent autonomic innervation								Endocrine/Paracrine	
	Cranial				Spinal				Adrenergic	
	Nerve	Pre-junct NT	Myocardial Receptor	Effect	Nerve	Prejunct NT	Myocardial Receptor	Effect	Myocardial Receptor	Effect
Group										
Cyclostomes										
Lamprey	X	ACH	Nicotinic	$f_H\uparrow$	—				β	I↑
Hagfish	Aneural heart								β	I↑
Elasmobranchs	X	ACH	Muscarinic	f_H, I↓	—				β	f_H, I↑
Teleosts	X	ACH	Muscarinic	f_H, I↓	SG→X, BCT, ASN	AD, NAD	β	f_H, I↑	β	f_H, I↑

Abbreviations: *β*, β-adrenergic; *AD*, adrenaline; *ACh*, acetylcholine; *ASN*, anterior spinal nerves; *BCT*, branchiocardiac trunk; f_H, heart rate; *I*, inotropic effect; *NAD*, noradrenaline; *pre-junct NT*, pre-junctional neurotransmitter; *SG*, sympathetic ganglia; *X*, vagal cardiac rami. Summary based on reviews by Nilsson (2011), Nilsson and Holmgren (1994), and Laurent et al. (1983).

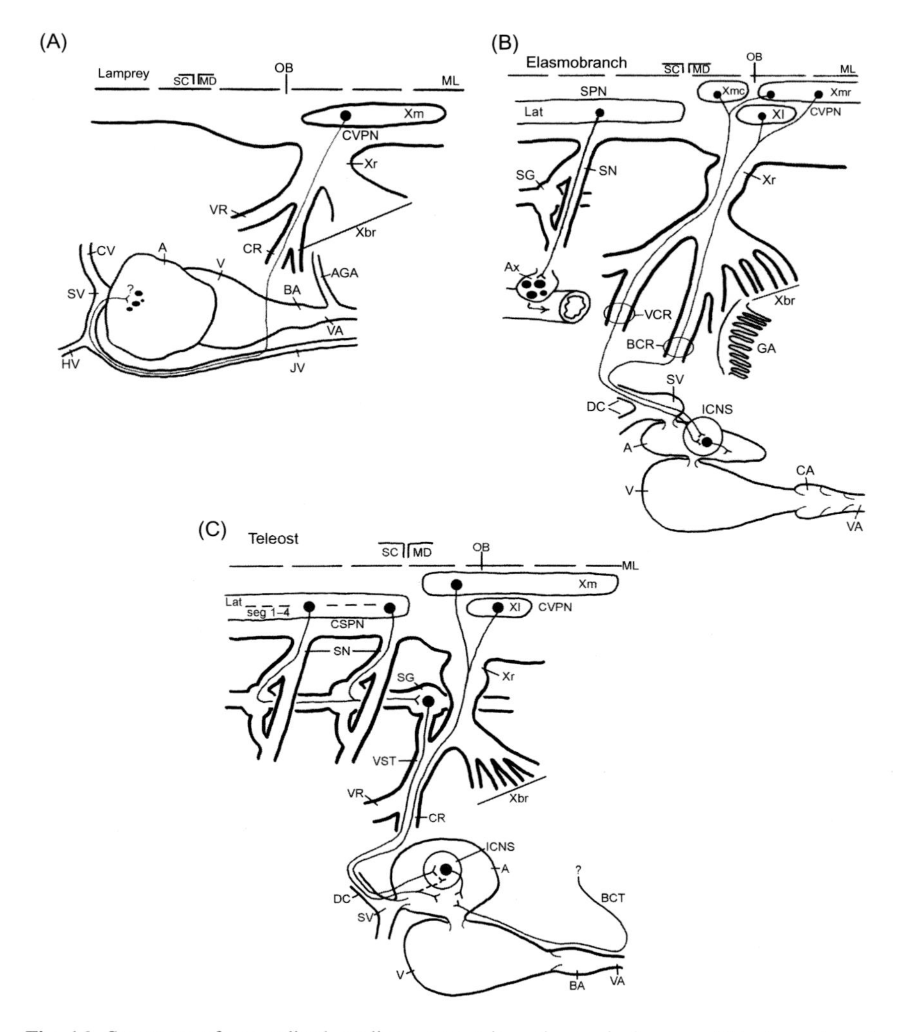

Fig. 16. Summary of generalized cardiac autonomic pathways in lamprey (A), elasmobranch fishes (B), and teleost fishes (C). *Diagrams* show dorsal view of right half of medulla and rostral spinal cord, as well as the vagus nerve (cranial nerve X) and the cranial portion of the paravertebral chain (B and C); rostral is to the right and ventral is toward the bottom in all panels. Abbreviations: *A*, atrium; *AGA*, caudal-most afferent gill artery; *Ax*, axillary body; *BA*, bulbus arteriosus; *BCR*, branchial cardiac ramus of vagus; *BCT*, branchiocardiac trunk; *C*, conus arteriosus; *CR*, vagal cardiac ramus; *CSPN*, cardiac spinal pre-ganglionic neurons; *CV*, cardinal vein; *CVPN*, cardiac vagal pre-ganglionic neurons; *DC*, ducti Cuvier; *GA*, last gill arch; *HV*, hepatic vein; *ICNS*, intracardiac nervous system and neurons (see text and Fig. 17 for details); *JV*, jugular vein; *Lat*, lateral spinal pre-ganglionic motor column (in panel B, "seg 1–4" indicates that CSPN may be located in spinal cord segments 1–4); *MD*, medulla; *ML*, midline of central nervous system; *OB*, short vertical line marks location of obex on midline; *SC*, spinal cord; *SG*, paravertebral (sympathetic) ganglion; *SN*, spinal nerve; *SPN*, spinal pre-ganglionic motor neuron pool innervating axillary body; *SV*, sinus venosus; *V*, ventricle; *VA*, ventral aorta; *VCR*, visceral cardiac ramus of vagus; *VR*, vagal visceral ramus; *VST*, vagosympathetic trunk; *Xbr*, vagal rami innervating gill arches; *Xl*, lateral vagal motor neuron group; *Xm*, medial vagal motor neuron group; *Xmc*, *Xmr*, medial vagal motor neuron group (c, caudal division; r, rostral division, respectively); *Xr*, vagal root. *Filled ovals* (panels A and B) represent chromaffin cells.

be altered by modulating catecholamine release, and adjust cardiac rate and contractility in response to changes in metabolic demand (Farrell, 2007a, b, c). This phylogenetic stage is modified in the heart of their cyclostome cousins, the lampreys.

The second stage is seen in the lamprey and the elasmobranchs. The lamprey heart is innervated by cholinergic axons coursing in the cardiac vagi, but adrenergic innervation is lacking (Table 6; Fig. 16A; Gibbins, 1994; Morris and Nilsson, 1994; Nilsson, 1983). The cardiac response to vagal activation is an increase in heart rate, a situation contrary to that in all other vertebrate groups (Table 6). ACH, released from vagal axon terminals in the lamprey heart, binds to nicotinic postjunctional receptors, but the cells possessing these receptors have not been identified. ACH has no direct actions on the myocardium, while β-adrenergic agents stimulate it (Haverinen et al., 2014). These observations are consistent with the hypothesis that vagal cardioaugmentation in lampreys results from a nicotinically mediated release of catecholamines from the numerous catecholamine-containing cells within the heart (Augustinsson et al., 1956) (Fig. 16A). Neural adjustment of $\dot{Q}$ in lampreys, may thus work by changing vagal tone at the level of the CNS. In the elasmobranch heart, axons in the vagal cardiac rami innervate the venous pole of the heart, causing cardioinhibition (negative inotropy and potent negative chronotropy) when activated (Table 6, Fig. 16B; Morris and Nilsson, 1994; Nilsson, 2011). This prototypical pattern of vagal cardiac innervation has been conserved throughout vertebrate evolution. Adrenergic cardiac innervation is either lacking, or at the best is very sparse, in elasmobranch fishes (Nilsson, 1983). Instead, adrenergic control of cardiac function is provided by catecholamines released from collections of intracardiac chromaffin cells, and those associated with the walls of the central veins including the *ducti Cuvier* (Fig. 16B), as in cyclostomes. This arrangement affords rapid transport of catecholamines to intracardiac effectors (pacemaker cells and myocardium) that determine $\dot{Q}$. Chromaffin cells in the heart and around veins are innervated by axon terminals, and some of these show ultrastructural characteristics of presumptive cholinergic terminals (Laurent et al., 1983; Nilsson, 1983); possibly representing projections of spinal preganglionic autonomic neurons (Fig. 16B). These collections of chromaffin cells, along with their innervation, constitute a neuroendocrine equivalent to the two-neuron spinal autonomic pathway mediating adrenergic control of the heart in more recently evolved vertebrate lineages (Section 4.2.3).

The third phylogenetic step is the development of a basic, two-neuron, architecture for autonomic efferent innervation of the viscera (see Fig. 16C and Section 4.2.3). This was first discovered in mammals more than a century ago (summarized by Cannon, 1929; Gaskell, 1916; Langley, 1921). This architecture is present in all basal actinopterygian and teleost fishes studied to date, which represent the earliest phylogenetic stage to display dual, and functionally antagonistic, autonomic cardiac innervation. This system sets resting or basal heart rate by a combination of maintained sympathetic and parasympathetic tone (summarized in Sandblom and Axelsson, 2011; Table 5), both of which can be modulated differentially to alter $\dot{Q}$ and appropriately adjust tissue perfusion (i.e., rates of O_2 and metabolic substrate delivery).

4.2.2. Methods for Investigating Cardiac Neuroanatomy

Classical studies of cardiac neuroanatomy used fine dissection techniques in combination with standard microscopy. Such historical studies have provided the broad outlines of the pattern of extrinsic and intrinsic cardiac innervation (summarized by Laurent et al., 1983; Nilsson, 1983), but fine detail or phenotypes of axons and neurons in the heart could not be resolved. This level of detail became clearer with the advent of reliable histochemical techniques for detecting either specific neurotransmitters or enzymes in the synthetic pathways for these neurotransmitters. For instance, histochemical reactions with catecholamines (Falck–Hillarp technique, Dahlstrom and Fuxe, 1964) and acetylcholinesterase (e.g., Koelle, 1962) were used to outline the sympathetic and parasympathetic innervation of fish hearts, respectively. Recently, explorations of the architecture of cardiac neuroanatomy have benefited from rapid advances in immunohistochemistry. These include the development of a large number of primary antibodies against neuronal intracellular structures and neurotransmitters, and their corresponding brightly fluorescing secondary antibodies. This technique allows for the visualization of axons, synaptic terminals and neuronal somata, and clarifies their regional cardiac distribution and organization (phenotypes of subpopulations of cardiac neurons). Furthermore, the combination of these neuronal labels with fluorescent markers (i.e., phalloidin, a label for F-actin, Small et al., 1999) for other cardiac elements such as the contractile proteins in cardiac myocytes, allows for detailed analyses of the relationship between cardiac innervation, and its targets from the level of single cells to the whole heart (for examples, see Figs. 17 and 18).

The application of antibodies directed against neuronal and cardiac targets in the fish heart is, however, not without problems. Ideally, to maximize the specificity of antibody–antigen reactions, immunohistochemical studies in fish should use antibodies raised against piscine antigens if not antigens in the actual species of fish under study. An example of this is a commercially available antibody raised in zebrafish against a component of neuronal membranes (zn-12; Metcalfe et al., 1990) that appears to reliably detect axons in a variety of peripheral tissues, as well as in the CNS of many fish species. However, there are still relatively few commercially available antibodies produced in fish, and so, many immunohistochemical studies are of necessity done using antibodies raised in other vertebrate species, most commonly in mammals.

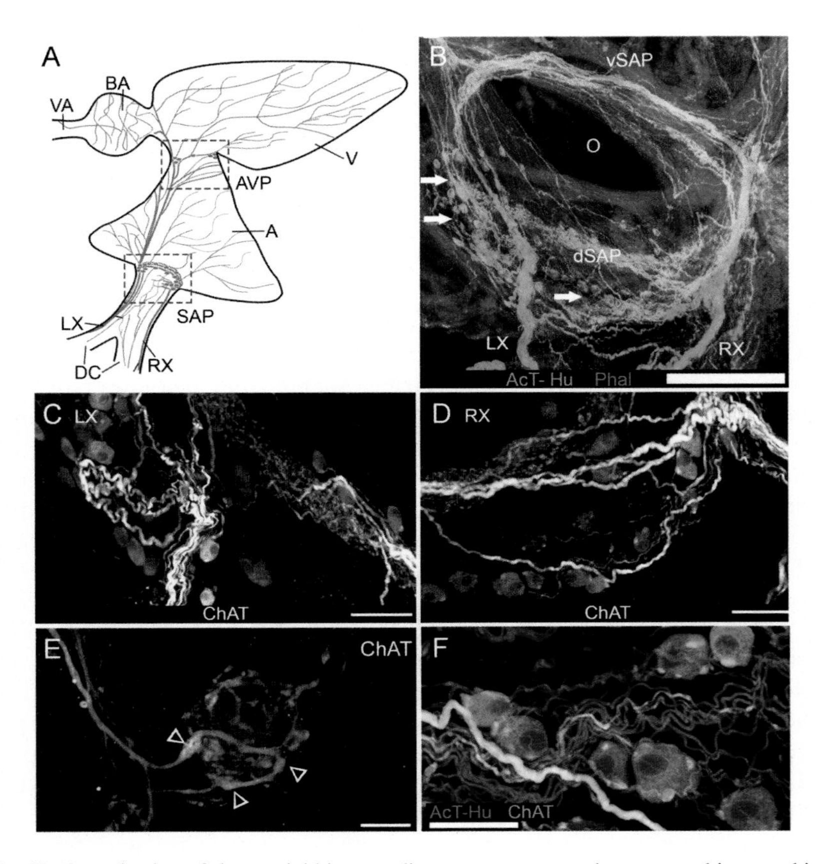

Fig. 17. Organization of the cyprinid intracardiac nervous system, demonstrated immunohistochemically with the aid of confocal microscopy. (A) Schematic overview of the intracardiac innervation pattern in goldfish and zebrafish, showing the heart with major nerve plexi at the sinoatrial (SAP) and atrioventricular (AVP) junctions (*boxes*), as well as regional nerve distribution (A, atrium; BA, bulbus arteriosus; DC, ducti Cuvier; LX, RX, left and right cardiac vagosympathetic rami; V, ventricle; VA, ventral aorta). (B) Whole-mount of zebrafish sinoatrial junction region showing dorsal (dSAP) and ventral (vSAP) regions of SAP surrounding the base of the sinoatrial valve leaflets and projections of LX and RX into the plexus. Somata of intracardiac neurons are indicated by arrows (O, ostium of sinoatrial valve). AcT—Hu, *red*: acetylated tubulin (pan-neuronal marker labelling axons) in combination with human neuronal protein C (labels somata). Phal, magenta: phalloidin (labels F-actin in cardiac myocytes). Scale bar = 100 μm. (C and D) Antibodies against choline acetyltransferase (ChAT) show putative cholinergic axons and intracardiac neuronal somata where LX (C) and RX (D) enter the zebrafish SAP. Scale bars = 100 μm. (E) Detailed view of the soma of a single intracardiac neuron (*center of panel*) surrounded by a basket of cholinergic axons and putative axon-somatic terminals (*arrowheads*) demonstrated with ChAT immunohistochemistry in zebrafish SAP. Scale bar = 10 μm. (F) Combined AcT—Hu (*magenta*) and ChAT (*green*) antibody detection shows part of an intracardiac ganglion in the AVP. Somata are cholinergic, axons are cholinergic and non-cholinergic. Scale bar = 50 μm. Credits: Panels A–D, F, Stoyek, M.R., Croll, R.P., Smith, F.M., 2015. Intrinsic and extrinsic innervation of the heart in zebrafish (*Danio rerio*). J. Comp. Neurol. 523, 1683–1700; panel E, Newton, C.M., Stoyek, M.R., Croll, R.P., Smith, F.M., 2014. Regional innervation of the heart in the goldfish, Carassius auratus: a confocal microscopy study. J. Comp. Neurol. 522, 456–478.

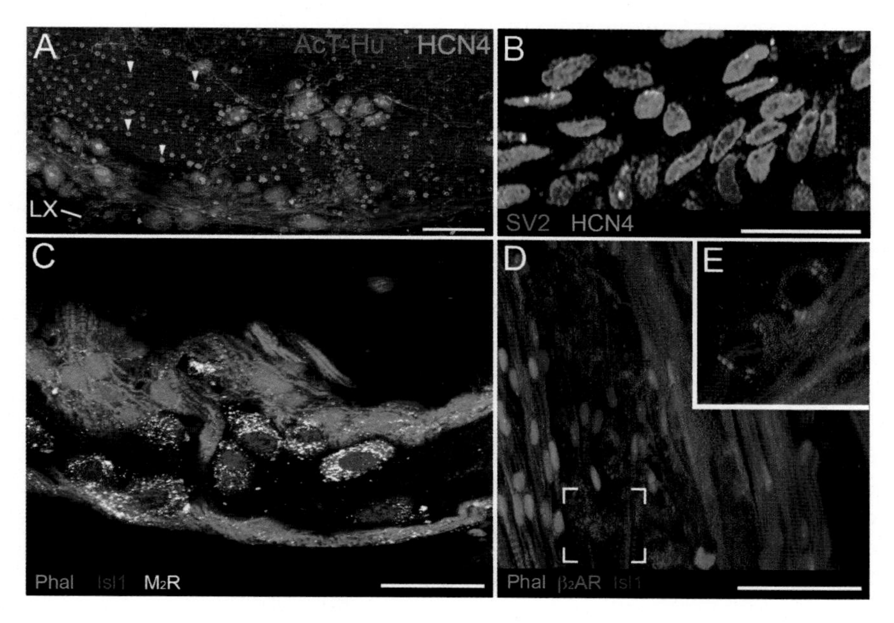

Fig. 18. Autonomic innervation of putative pacemaker cells in the cyprinid heart. Antibodies against hyperpolarization-activated, cyclic nucleotide-gated ion channel 4 (HCN4), and the transcription factor Islet-1 (Isl1) demonstrate the distribution and appearance of putative pacemaker cells. (A) View of a portion of the basal region of one sinoatrial valve and surrounding tissue near the entry of left cardiac vagosympathetic rami (LX) into the sinoatrial plexus in goldfish. This image shows the close proximity of a band of HCN4-positive cells (*green*) to the neuropil and somata of the SAP (AcT—Hu, *red*). Scale bar = 40 μm. (B) Immunoreactivity for the synaptic vesicle marker SV2 (*magenta*) showing that axon terminals are closely associated with HCN4-labelled putative pacemaker cells (*green*) in zebrafish. Scale bar = 50 μm. (C) Antibodies against muscarinic type 2 receptors (M_2R, *white*) indicate that there is a dense localization of receptors on putative pacemaker cells (Isl1 label, *blue*) at the base of zebrafish sinoatrial valve leaflets. Isl1 cells are embedded among cardiac myocytes (Phal, *green*) that express fewer M_2R. Scale bar = 20 μm. (D and E) Putative pacemaker cells in the zebrafish sinoatrial region (Isl1, *blue*) express immunoreactivity for β_2-adrenoreceptors (β_2AR, *red*). Panel E shows a high-magnification image from a single-slice confocal scan (2 μm thick) of the region indicated in the *box* in panel D. Scale bar = 40 μm in (D), 20 μm in (E). Credits: Panel A, Newton, C.M., Stoyek, M.R., Croll, R.P., Smith, F.M., 2014. Regional innervation of the heart in the goldfish, Carassius auratus: a confocal microscopy study. J. Comp. Neurol. 522, 456–478; panel B, Stoyek, M.R., Croll, R.P., Smith, F.M., 2015. Intrinsic and extrinsic innervation of the heart in zebrafish (*Danio rerio*). J. Comp. Neurol. 523, 1683–1700; panels C and D, Stoyek, M.R., Quinn, T.A., Croll, R.P., Smith, F.M., 2016. Zebrafish heart as a model to study the integrative autonomic control of pacemaker function. Am. J. Physiol. Heart Circ. Physiol. 311, H676–H688.

The genomes of mammals (including man) and zebrafish (treated as a "representative" teleost) have considerable homology (Howe et al., 2013). However, because of the gene duplication that occurred during the evolution of teleosts, there may be multiple proteins expressed as a result of the transcription of these duplicated genes. Not all of these proteins are functional, but those that are may have varying functions in fish compared with the protein expressed by the homologous gene in mammals (Kabashi et al., 2011). Therefore, antibodies raised against a mammalian protein may cross-react with multiple epitopes in fish, increasing the likelihood of a false signal. Consequently, appropriate control experiments are essential when using commercially available antibodies raised in members of other vertebrate groups for immunohistochemical studies in fish. General guidelines for controls that are useful in this type of study are given by Saper and Sawchenko (2003) and Saper (2005). The best control for antibody specificity is the preparation of a knockout animal in which the gene that gives rise to the target antigen has been removed; there should be no antibody binding in tissues from these animals. This type of study has been carried out effectively in the zebrafish, which is amenable to genetic modification, but this is more difficult in other fish species. A useful alternative technique is to preadsorb the antibody with the expected antigen, if this is available. If the antigen is unavailable, however, comparison of the pattern of antibody staining with the distribution of mRNA expression for the antigen, obtained by *in situ* hybridization histochemistry, can in some cases verify the specificity of the primary antibody.

In addition to immunohistochemical detection of intracardiac neuronal elements, investigators have used a wide variety of neurotracing techniques to determine the sources and patterns of connectivity of cardiac nerves within the fish heart. All of these techniques involve the application of tracing agents to neural tissues in living animals. Agents are applied to the tissues targeted by peripheral axons, to the axons themselves in peripheral nerves, or to sites in the CNS suspected of originating these axons. Then, after a suitable time is allowed for transport of these agents along axons, they are visualized in the tissues of interest. Among the earliest agents used for such studies was horseradish peroxidase, an enzyme that is actively transported by axons and is detected by its mediation of a peroxide breakdown reaction leaving a visible product in the labeled tissue. This technique was used to determine the location of the neurons in the CNS innervating the heart (see Section 4.2.3). More recently various actively transported neurotracers that fluoresce in tissue under appropriate illumination (e.g., fluorogold, "True Blue," and the carbocyanin dye DiI) have been used to explore the origins and patterns of cardiac innervation (e.g., Goehler and Finger, 1996; Taylor et al., 2009). These agents provide visualization of extra- and intracardiac neural elements with a clarity equivalent to that of fluorescent antibody labeling.

4.2.3. Extrinsic Cardiac Innervation

We summarize the pathways of extrinsic cardiac post-ganglionic innervation and the locations of pre-granglionic neurons in these pathways in

Fig. 16. In lampreys, there is no spinal cardiac innervation. The somata of cardiac vagal pre-ganglionic neurons (CVPN) are located within the vagal motor column, a poorly defined nucleus in the mediodorsal medulla extending rostrally from near the obex to the narrowing of the fourth ventricle (Fig. 16A; Ariens Kappers et al., 1967; Nicol, 1952; Rovainen, 1979). While there have been no definitive studies of cardiotopic organization of the lamprey CVPN within this nucleus, Taylor (1992) and Withington-Wray et al. (1987) suggest that the somata of CVPN are concentrated in the caudal-most portion of the dorsal vagal motor column close to the surface of the fourth ventricle. This hypothesis is supported by functional data presented by Augustinsson et al. (1956). These authors evoked increased f_H in anesthetized lamprey by focal electrical stimulation of the floor of the fourth ventricle at, and just rostral to, the obex close to the proposed anatomical location of CVPN somata. Axons of the CVPN exit the medulla in the caudal-most vagal root, coursing in the epibranchial ramus running along the caudal edge of the last gill pouch. This nerve divides into a visceral and a cardiac ramus, the latter carrying CVPN axons to the jugular vein. These axons then course caudally in the wall of this vein, entering the sinus venosus, and terminating in a plexus in the atrial wall close to collections of intracardiac chromaffin cells (Fig. 16A; Augustinsson et al., 1956; Fange, 1972; Gibbins, 1994). Vagal terminals in the heart release ACH, which purportedly evokes a nicotinically mediated release of catecholamines from adjacent chromaffin cells that increases f_H (Augustinsson et al., 1956; Haverinen et al., 2014). The intracardiac mechanisms associated with this unique vagal augmentation of f_H in lampreys require further investigation.

The detailed organization of the central aspects of cardiac vagal innervation in jawed (Gnathostome) fishes is best understood in elasmobranchs, due largely to an extensive series of studies by Taylor and coworkers. This work has been reviewed previously (Taylor, 2011; Taylor and Wang, 2009; Taylor et al., 1999, 2014), and therefore, it is only briefly described here (Fig. 16B). Studies applying retrograde neurotracers to the branchial and visceral cardiac nerves in elasmobranch fishes show two major groups of CVPN within the vagal motor columns in the medulla. The medial vagal motor neuron group (Xm), which contains the majority of the CVPN, is divided around the obex into rostral and caudal components (Xmr and Xmc, respectively, Fig. 16B). Neurons in the more rostral region of Xmr innervate the heart *via* the branchial cardiac ramus, while neurons in the more caudal portion of this nucleus send cardiofugal axons *via* the visceral cardiac ramus. Neurons located in the Xmc project axons to the heart exclusively *via* the visceral cardiac ramus. A minor part of the CVPN, located in the lateral vagal motor neuron group (Xl in Fig. 16B) and ventrolateral to the Xm, project axons to the heart *via* the branchial cardiac ramus. Electrical recordings of the spontaneous activity of CVPN in these cell groups have shown that neurons in the Xm tend to fire rhythmically, while those in Xl appear to fire non-rhythmically. Complex, but as yet poorly understood, interactions among neurons in these groups are modulated by afferent activity from chemoreceptors and mechanoreceptors associated with breathing and other inputs (Taylor et al., 1999). Considerable speculation exists (without resolution) about the functional roles of the different groups of CVPN; particularly in generating and maintaining cardiorespiratory synchrony during exercise and when exposed to environmental factors such as hypoxia (Taylor et al., 2014). The existence in this relatively primitive vertebrate group of two cardiac vagal centers in the brainstem, that show not only complex intergroup interactions but also interactions with nearby neurons controlling respiration, indicates a sophisticated level of integrative control. This suggests that the elasmobranch CVPN will provide a fertile model for studying the neural substrates underlying coordination between cardiac output and blood oxygenation.

Vagal pre-ganglionic axons in the cardiac nerves of elasmobranch fishes converge in the wall of the sinus venosus and enter the sinoatrial junction region where they form a nerve plexus that is a major component of the intracardiac nervous system (ICNS, Section 4.2.4). However, clear examples of direct spinal innervation of the heart in elasmobranch fishes are lacking (Section 4.2.1 and Fig. 16B). Instead, a group of neurons (spinal pre-ganglionic neurons, SPN, Fig. 16B) located in the lateral cell column of the rostral spinal cord project axons in the most rostral spinal nerve (or nerves, depending on species) to the axillary body, a collection of chromaffin cells associated with the posterior cardinal sinuses (Fig. 16B; Laurent et al., 1983; Morris and Nilsson, 1994; Nicol, 1952; Nilsson, 1983). This arrangement of spinal neurons connecting synaptically with chromaffin cells in the vascular wall is believed to act as a *de facto* "sympathetic cardiac control element." In this situation, action potentials reaching the axon terminals cause the release of ACH, which then acts *via* nicotinic postjunctional receptors on the chromaffin cells to evoke catecholamine secretion into the venous blood returning to the heart. This, in turn, results in β-adrenergic augmentation of cardiac function. Consequently, a form of sympathetic tone is possible in elasmobranch fishes without direct sympathetic cardiac innervation, and its level can be adjusted by altering the activity of the spinal neuron pool targeting the axillary body.

A complete dual (spinal and vagal) autonomic innervation of the heart is present in the majority of teleost fishes. Neurotracer studies in a variety of species reveal that CVPN are distributed in the medial and lateral vagal motor neuron groups (Xm and Xl, respectively, in Fig. 16C), forming a relatively loose cardiotopical organization in the medulla (Szabo and Libouban, 1979; Lazar et al., 1992 [*Gnathonemus*]; Morita and Finger, 1987 [goldfish]; Withington-Wray et al., 1987 [Atlantic cod, rainbow trout]; Kanwal and

Caprio, 1987 and Goehler and Finger, 1996 [catfish], Taylor et al., 2009; Leite et al., 2009 [pacu]). As discussed above for elasmobranch fishes, these cardiopetal neuronal populations may serve different functions that control $\dot{Q}$. For example, they may promote cardiorespiratory interactions partly through differential convergence of visceral and branchial afferent inputs on the two CVPN pools, and partly through interactions of subsets of CVPN with nearby branchial motor neurons (Taylor et al., 2014).

Peripherally, axons from CVPN project to the heart in the (usually bilateral) vagal cardiac rami (Fig. 16C) and release ACH at synapses with postganglionic neurons that are embedded in the walls of the heart as part of the ICNS (Gibbins, 1994; Laurent et al., 1983; Nilsson, 1983). ACH is excitatory on the postsynaptic membranes of these neurons, and acts through nicotinic receptors.

Cardiac spinal pre-ganglionic neurons (CSPN, Fig. 16C) are located in the lateral cell column of the teleost spinal cord, with a highly variable rostrocaudal distribution that apparently depends on species (Funakoshi et al., 1997). CSPN, may thus, be located in spinal cord segments 1–2 (Morris and Nilsson, 1994), 3–4 (Funakoshi et al., 1997; Morris and Nilsson, 1994), or spread throughout the first four segments (Funakoshi and Nakano, 2007). Axons of the CSPN exit the spinal cord in one or more of the most rostral spinal nerves, coursing *via* rami communicantes into the paravertebral ganglion chain where they synapse on the somata of post-ganglionic neurons targeting the heart (Fig. 16C). ACH, released from the pre-ganglionic terminals, acts on nicotinic receptors to excite the post-ganglionic somata. With no clear consensus on the location of these somata, these cells in teleost fishes appear to be located primarily in the chain ganglion that lies proximal to the caudal-most vagal roots (Funakoshi and Nakano, 2007; Morris and Nilsson, 1994; Nilsson, 1983). A ramus connects this ganglion with the main trunk of the vagus nerve, and the axons of the postganglionic neurons then project, *via* that ramus, into the trunk; this nerve is termed the "vagosympathetic trunk."

The majority of cardiac spinal post-ganglionic axons terminate within the heart on myocardial effectors such as pacemaker cells, myocytes or blood vessels. These effectors respond to adrenergic neurotransmitters released from axon terminals. Some cyprinid species have post-ganglionic adrenergic terminals within the sinoatrial plexus that are immediately adjacent to somata of intracardiac neurons (ICN, Fig. 16C; Newton et al., 2014; Stoyek et al., 2015). This raises the possibility that cardiac spinal outflow could affect cholinergic neurotransmission within intracardiac ganglia, providing a potential means of adrenergic modulation of vagal cardioinhibitory signaling at the level of ICN in the teleost heart. Such interactions between the autonomic limbs within the ICNS could aid in fine-tuning the neural control of $\dot{Q}$, but this possibility requires further clarification.

In some teleost species, Gibbins (1994) and Morris and Nilsson (1994) proposed a secondary route for spinal autonomic outflow to the heart; e.g., *via* a subpopulation of post-ganglionic axons projecting to the venous pole of the heart in the anterior spinal nerves. Data on this putative pathway are lacking, and confirmation is required. In addition, a tertiary route for spinal innervation of the heart *via* nerves that course caudally along the cardiac OFT to the arterial pole has been postulated (Nilsson, 1983; Stoyek et al., 2015). Recent functional evidence has provided support of this idea. Stoyek et al. (2016) showed in the isolated zebrafish heart that stimulation of the branchiocardiac nerve trunk (BCT, Fig. 16C) evokes an increase in f_H that is mediated by β-adrenergic receptors. The origin of this pathway and its functional significance clearly deserve further investigation.

4.2.4. Intracardiac Nervous System

Autonomic elements within the heart itself form the final common pathway for neural control of $\dot{Q}$ in all extant fishes, except hagfish. The ICNS consists of one or more ganglionated plexi containing the somata of ICN and axons of both extracardiac and intrinsic origin. The organization of the ICNS remains largely unexplored in most fish species, unlike the anatomy of the extrinsic cardiac autonomic innervation, which is well described for a range of teleost (and some elasmobranch) fishes, and a few species of lamprey (Gibbins, 1994; Nilsson, 1983; Section 4.2.3). One factor hindering studies of the ICNS is that it is embedded in the cardiac walls and is not readily visible. Using new methods of visualizing this system in whole-mount hearts (see Section 4.2.2), initial neuroanatomical studies show that: (1) the ICNS in fishes is neurochemically complex; (2) ICN somata can be broadly distributed in the sinus venosus, the sinoatrial junction, the atrial walls, and the atrioventricular junction depending on species (Table 7); and (3) all parts of the heart and the OFT, as well as the coronary vasculature, are innervated by cholinergic or adrenergic axons; or by both in some species. We review these aspects in more detail below.

Fluorescent immunohistochemical and neurotracing studies in two cyprinid species (goldfish, Newton et al., 2014; zebrafish, Stoyek et al., 2015), silurid and mugilid species (Zaccone et al., 2010), and ray-finned fishes (Zaccone et al., 2009, 2011) all show that the distribution of neuronal somata and the organization of extrinsic and intrinsic neural pathways in the ICNS are complex. The intracardiac plexi consist of: (1) extrinsic cholinergic axons that represent cranial autonomic pre-ganglionic inputs which project *via* the vagosympathetic trunks and synapse on a subpopulation of ICN somata (these are presumptive post-ganglionic inhibitory neurons; Newton et al., 2014; Stoyek et al., 2015); (2) extrinsic adrenergic axons from the vagosympathetic

Table 7
Summary of regional location and transmitter phenotype of intracardiac neurons in fishes

	Soma location and phenotype				
	DC/SV	SAR	A	AVR	V
Group					
Cyclostomes					
Lamprey	ACH[a], ?[b]	?[b]	?[b]		?[b]
Hagfish	Aneural heart[c,d]				
Elasmobranchs	?(ACH)[e,f,g,m]	ACH[f]	ACH[f]		
Teleosts	ACH[h,i,j,m] AD[j],NANC[j]	ACH[m], AD, NANC[h,i]	ACH, AD[h,i]	ACh, AD[h,i]	—
Basal Actinopterigii	ACH[k], AD[k], NANC[k,l]	ACH[k], AD[k], NANC[k,l]	ACH[k]	AD, NANC[k]	AD, NANC[k]

Abbreviations: *AD*, adrenergic/noradrenergic; *ACH*, cholinergic; *A*, atrium; *AVR*, atrioventricular region; *DC/SV*, *ducti of Cuvier*/sinus venosus; *NANC*, non-adrenergic, non-cholinergic; *pheno*, neurotransmitter phenotype expressed in ICN somata; *SAR*, sinoatrial region; *V*, ventricle; ?, ICN present but phenotype not determined.

Sources: Basal Actinopterigii: [k]Zaccone et al. (2009, bichir), [l]Zaccone et al. (2011, 2012, gar).

Teleosts: [h]Newton et al. (2014, goldfish), [i]Stoyek et al. (2015, zebrafish), [j]Zaccone et al. (2010, mullet, Nile catfish).

Elasmobranchs: [m]Yamauchi (1980), [e]Saetersdal et al. (1975), [f]Gallego et al. (1997), [g]Ramos (2004).

Lamprey: [a]Morris and Nilsson (1994, summary of older literature), [b]Augustinsson et al. (1956).

hagfish: [c]Axelsson et al. (1990), [d]Farrell (2007a, b, c).

trunk that target effector cells in the myocardium, and represent spinal autonomic post-ganglionic excitatory inputs (some extrinsic adrenergic axons also terminate on ICN somata; Newton et al., 2014; Stoyek et al., 2015); (3) intrinsic axons that arise from ICN that either terminate on effector cells (representing the projections of efferent neurons to the myocardium) or synapse on the somata of other ICN (representing projections between ICN within the heart); and (4) axons arising from ICN that project into the vagosympathetic trunks, and thus, cranially toward the brainstem (Stoyek et al., 2015). These latter axons represent projections of potential sensory neurons that carry afferent information from the heart to the CNS. Taken together, these results reveal a neural network within the heart that is closely associated with all effector cell types that determine $\dot{Q}$. Admittedly, detailed studies are limited to a few species, but there is no reason to presuppose that the complexity of this system is not typical of teleost fishes in general. Detailing the ICNS neuroanatomy in a variety of fish species will be an essential first step in determining the functional roles of this local neural network in fast, global, and regional beat-to-beat cardiac control.

Descriptions of the regional distribution and major phenotypes of ICN somata within the fish heart (Table 7) are restricted to a few teleost fishes; elasmobranch fishes and lampreys are almost entirely unrepresented. In teleost fishes, the majority of ICN somata are located at the venous pole, largely in a plexus located at the sinoatrial junction (see Fig. 17A and B), or in the wall of the sinus venosus proximal to this region (Laurent et al., 1983; Table 7). ICN in the goldfish and zebrafish heart are clustered in the SAP, with multiple ganglia lying close to the entry of the left and right vagosympathetic trunks into the plexus region (Fig. 17C and D). These neurons typically have spherical or oval somata with multiple axon-somatic terminals in a basket-like formation (Fig. 17E). Half of the neurons in the goldfish heart, and more than 90% of those in the zebrafish heart, are cholinergic (i.e., they express choline acetyltransferase, an enzyme in the synthesis pathway for ACH; Fig. 17C–E; Newton et al., 2014; Stoyek et al., 2015). Therefore, it is likely that many of these neurons function in efferent parasympathetic pathways targeting cardiac effectors.

Outside of the sinoatrial region, there are a few ICN scattered throughout the atrial wall in teleost fishes (Laurent et al., 1983; Stoyek et al., 2015), and in the ventricular wall (usually associated with the subepicardium) in some ray-finned fishes (Zaccone et al., 2009). Interestingly, in teleost fishes, including the zebrafish, there exists a small collection of ganglia concentrated near the atrioventricular valves (Fig. 17F; Laurent et al., 1983; Stoyek et al., 2015). These neurons are largely cholinergic (Stoyek et al., 2015), and the function(s) of this neuronal population are not yet established.

The function-specific roles of subpopulations of ICN in the control of the myocardium have also not been determined. One potential functional implication of the proximity of neuronal ganglia to effector cells, such as the pacemakers in the sinoatrial and atrioventricular pacemaker zones, is that these neurons are involved in the control of f_H. There exists both anatomical and functional evidence to support this possibility. Pacemaker activity originates in tissues proximal to the sinoatrial valves in the hearts of fishes in some major teleost groups (Chapter 3, Volume 36A: Vornanen, 2017; Vornanen et al., 2010 [salmonids]; Saito, 1973; Saito and Tenma, 1976; Sedmera et al., 2003; Tessadori et al., 2012; Stoyek et al., 2016 [cyprinids]). Furthermore, the expression of one or more members of the HCN ion channel family, as well as their co-localization in cells near the sinoatrial valve that express the transcription factor Isl1, are purported anatomical markers for putative pacemakers in cyprinid hearts (Newton et al., 2014; Stoyek et al., 2015; Tessadori et al., 2012). Such cells, being situated in the basal portions of the sinoatrial valve leaflets and at the valve commissures, are closely associated with the SAP in the goldfish and zebrafish hearts (Fig. 18A; Newton et al., 2014; Stoyek et al., 2015). Indeed, these cells are densely innervated by axon terminals, as shown by the proximity of SV2, which is a synaptic vesicle marker (Fig. 18B). Some of these terminals likely represent the intracardiac projections of local inhibitory post-ganglionic neurons, which release ACH that acts on post-junctional muscarinic receptors to depress f_H. In support of this idea, the presence of a high density of M_2R on the membranes of putative pacemaker cells has recently been demonstrated (Fig. 18C; Stoyek et al., 2016). These post-ganglionic ICN would be activated by inputs from cranial pre-ganglionic neurons whose axons are in the vagosympathetic trunks to the heart. The ability of this pathway within the ICNS to control f_H has been demonstrated by the application of hexamethonium (a nicotinic antagonist acting at post-ganglionic receptors within autonomic ganglia). This drug prevents the bradycardia associated with electrical stimulation of the vagosympathetic trunks in the isolated zebrafish heart (Stoyek et al., 2016).

Stimulation of β_2-AR on pacemaker cells, by catecholamines released from local terminals of adrenergic post-ganglionic neurons, also increases f_H. Immunohistochemical staining of the zebrafish heart shows expression of β_2-AR in the membranes of sinoatrial pacemaker cells (Fig. 18D and E; Stoyek et al., 2016). Therefore, pacemakers in the teleost heart are innervated by both limbs of the autonomic nervous system through the ICNS, and respond appropriately to inputs from these limbs.

Given the details on neuroanatomical organization of the fish heart that are emerging, it is tempting to ascribe primary roles in the control of local effectors to subpopulations of ICN that occur in specific regions, as we describe above for ICN that are located proximally to pacemaker cells. However, assumptions about local neuronal function based only on proximity to effectors are premature. Physiological evidence for a function-specific role for any identifiable population of ICN in the fish heart remains an unresolved problem and is a major impediment to our understanding the principles operating in neural control of the heart. Future studies must include analyses of the distribution, electrophysiological characteristics and connectivity of neurons with pacemaker and other effector cells in experimental models such as the zebrafish and other cyprinids in which the basic anatomical organization of the ICNS is known. Likewise, we argue that it is imperative to know whether, and to what degree, the findings from the few species in which these details are known can be generalized across the groups of extant fishes.

5. CARDIAC STROKE VOLUME AND ITS CONTROL

In vivo and *in vitro* studies show that fish hearts have the capacity to increase V_s by a large amount, up to three-fold in some species (Figs. 7 and 9).

Routine V_s is typically $<0.5\,mL\,kg^{-1}$ and rarely increases to $>1\,mL\,kg^{-1}$ in most fishes (Tables 2 and 4). By comparison, routine V_s in a human is about $1\,mL\,kg^{-1}$ and rarely increases by >30% with exercise. While it is clear that the capacity to increase V_s in most fishes is much larger than in mammals, how different fish species use this capacity varies considerably. For example, prolonged exercise increases V_s considerably more than f_H in some species: 200% vs 50% for rainbow trout (Kiceniuk and Jones, 1977), 120% vs 33% for sockeye salmon (Eliason et al., 2013b); 55%–63% vs 7%–15% for cat shark (*S. stellaris*) (Piiper et al., 1977); 33% vs 7% for leopard shark (Lai et al., 1989a); 26% vs 18% (Axelsson and Nilsson, 1986); and 67% vs 39% (Petersen and Gamperl, 2010) for Atlantic cod. Yet, in other species, such as the sea raven (Axelsson et al., 1989) and northern squawfish (Kolok and Farrell, 1994), V_s and f_H increase by similar amounts. Warm acclimation of the large-scale sucker (either from 10 to 16°C or from 4 to 10°C) results in a switch from a primary increase in V_s during swimming to a primary increase in f_H (Kolok et al., 1993), and a similar switch is seen in winter flounder (Joaquim et al., 2004). Nonetheless, some fish increase f_H almost exclusively when swimming [e.g., borch (*Pagothenia borchgrevinki*), Axelsson et al., 1992; tunas, Brill and Bushnell, 1991; sea bass (*Diacentrachus labrax*), Sandblom et al., 2005].

In polar fishes that live in water near 0°C, V_s is unusually large and f_H is slow (typically <30 bpm) (Axelsson, 2005; Axelsson et al., 1992). A maximum V_s of $12\,mL\,kg^{-1}$ for the perfused heart of the hemoglobin-free icefish *Chionodraco hamatus* (Tota et al., 1991; Table 3) is the highest known in any vertebrate. Their ventricle is, however, very thin-walled, despite having a large RVM, and cannot generate a P_{VA} as high as that seen in red-blooded teleost fishes. Hearts of hemoglobin-free icefishes that have no myoglobin, such as *Chaenocephalus aceratus*, have an even lower pressure generating ability and a lower intrinsic f_H (A.P. Farrell, unpublished personal observation).

All vertebrate hearts can normally eject all the blood they receive during diastole. The explanation for this is that the ventricle contracts more forcefully when stretched to a greater degree during filling (Chapter 2, Volume 36A: Shiels, 2017). As a result, a relationship exists between cardiac filling and V_s (Figs. 7 and 9), which is the well-known Frank–Starling curve. The unusually large operational range for V_s in fishes, as compared with other vertebrates, is the result of many cellular factors (see Chapter 2, Volume 36A: Shiels, 2017). Regardless, V_s in all vertebrate hearts, from hagfishes through to mammals, is primarily dependent on venous blood pressure. This is termed a *vis-a-tergo* (i.e., a force from behind) mechanism. Thus, any factor that alters P_{CV}, such as an α-adrenergic venoconstriction (Chapter 7, Volume 36A:

Sandblom and Gräns, 2017), should alter V_s. For example, the increase in P_{CV} that accompanies aerobic exercise (Table 2; Kiceniuk and Jones, 1977; Sandblom et al., 2005) helps to either increase V_s, or to maintain V_s at a higher f_H. However, this response in not universal. P_{CV} is unchanged during acute warming of Chinook salmon, except when the temperature is supra-optimal (25°C) and individuals begin releasing lactate into the blood (Clark et al., 2008).

Atrial contraction amplifies the pressure intrinsic to the venous blood, and factors that increase cardiac contractility (e.g., catecholamines) make the heart more responsive to filling pressure and decrease end-systolic volume. This effect explains why inotropic agents can shift the position of the Frank–Starling curve relative to cardiac filling pressure (Fig. 7). In addition, V_s is largely independent of the output resistance of the vasculature, which means that up to a defined P_{VA}, maximum V_s is maintained, and CPO increases with increasing P_{VA} (this response termed homeometric regulation). Ultimately, cardiac contractility can become insufficient if P_{VA} increases too much. Under such circumstances end-systolic volume increases, and maximum V_s falls (Fig. 9B). Initially, a tradeoff is possible between flow and pressure development at a constant CPO, but eventually CPO declines. The range for homeometric regulation varies considerably among fishes, being lowest for Atlantic hagfish and highest for tunas (Farrell, 1991).

P_{CV} in some fishes, especially elasmobranch fishes, is routinely subambient (see Chapter 7, Volume 36A: Sandblom and Gräns, 2017). This means that an aspiration mechanism (*vis-a-fronte* = force from in front) is needed to fill the ventricle and provide for routine V_s. *Vis-a-fronte* aspiratory filling is possible because venous return is assisted by ventricular contraction. In such fishes, the pericardial cavity is relatively stiff compared with the cardiac chambers, and ventricular contraction generates a subambient pericardial pressure. This subambient pressure acts on, and expands, the wall of the easily distensible atrium (and sinus venosus), drawing in venous blood. While the earliest reports of the importance of this suction mechanism were likely overstated (i.e., stress and anesthesia substantially lower the intrapericardial pressure of elasmobranch fishes), carefully conducted studies performed subsequently show that intrapericardial pressure clearly oscillates around ambient pressure in the leopard shark and albacore tuna (*Thunnus alalunga*). These results confirm that suction (aspiratory filling) is possible, but that it is not as extreme as previously thought (Abel et al., 1987; Brill and Lai, 2016; Lai et al., 1989a, b, 1998). Indeed, the perfused rainbow trout heart can support routine $\dot{Q}$ by aspirating blood at a subambient filling pressure when the pericardium is intact (Farrell et al., 1988a, b), but not when its integrity is compromised. Pericardiectomy also requires a 40% higher filling pressure to generate maximum

$\dot{Q}$, and this results in a 18% lower maximum CPO (Table 3). However, in the short-finned eel, where filling pressure for routine cardiac performance is usually above ambient, pericardiectomy decreases maximum sustainable pressure development without changing maximum CPO (Franklin and Davie, 1991). This is why we argue that cardiac function should only be studied when the pericardium is intact (or it can be replaced *in vitro* by an artificial one, e.g., Tota et al., 1991). Elastic recoil of the ventricle, can likewise, contribute to *vis-a-fronte* filling of this chamber.

The idea that atrial contraction is the sole or primary contributor to ventricular filling (Farrell and Jones, 1992) is no longer accepted because of imaging studies that have visualized the dynamics of cardiac filling. These studies include elasmobranch and teleost species: rainbow trout (Franklin and Davie, 1992); white sturgeon (Gregory et al., 2004); spotted sea bass (*Paralabrax maculatofasciatus*), kelp bass (*Paralabrax clathratus*), barred sand bass (*Paralabrax nebulifer*), snakehead (*Channa micropeltes*), and swamp eel (*Monopterus albus*) (Lai et al., 1998); leopard shark (Lai et al., 1996); smooth-hound shark (*Mustelus henlei*), the horn shark (*Heterodontus francisci*), the swell shark (*Cephaloscyllium ventriosum*), the shortfin mako (*Isurus oxyrinchus*), and the blue shark (Lai et al., 2004); zebrafish (Ho et al., 2002); and monkfish (*Lophius piscatorius*), the Lusitanian toadfish (*Halobatrachus didadtylus*) (Coucelo et al., 1996) and rainbow trout (Cotter et al., 2008). In each study, Doppler imaging shows a clear biphasic filling of the ventricle: the E-wave during atrial diastole, followed by the A-wave associated with atrial contraction. Nevertheless, the relative contribution of the A-wave to ventricular filling varies considerably among studies (as determined by comparing the peak velocity of the two waves and assuming that the atrioventricular canal diameter is unchanged). In some fishes (e.g., toadfish and monkfish), the E:A ratio is >1, which means that atrial contraction is a secondary contributor to ventricular filling. In other species, the E:A ratio is ~ 1 (e.g., shortfin mako), or <1 (e.g., zebrafish, white sturgeon, smooth-hound shark, horn shark, swell shark, and blue shark), both of which suggest a greater role for atrial contraction in ventricular filling. Even though some of these studies make the case that anesthesia or physical restraint did not alter cardiac function, cardiac filling in fish is very sensitive to small changes in P_{CV}, and these results must be treated cautiously. Furthermore, rather than using a comparison of peak velocities, we contend that integrating the velocity–time profile would provide a better comparator for ventricular filling; even though the peak velocities for both waves show good agreement among these studies (ranging from 0.12 to 0.45 m s^{-1}). For example, the peak velocity for the A-wave (0.15 m s^{-1}) of the zebrafish is similar to the peak velocity of blood leaving ventricle (0.12 m s^{-1}), and much greater than the E-wave (0.06 m s^{-1}). Two other concerns exist regarding the shape of the E-wave. First, an explanation is needed for the faster peak blood velocity during atrial diastole than during atrial contraction; it should be the opposite if atrial contraction adds potential energy to the venous blood. Second, an explanation is needed for how steady values of P_{CV} and intraatrial pressure apparently generate the sharp rise in the E-wave profile rather than maintaining a steady blood velocity (Lai et al., 1996, 1998). An obvious explanation for the former is that the peak in the E-wave is a result of aspiration as the ventricle recoils. Indeed, the peak in the E-wave occurs immediately following the T-wave (ventricular relaxation), and the intraventricular pressure of both the leopard shark and sand bass becomes subambient immediately after ventricular contraction. This increases the pressure gradient between the atrium and ventricle without any change in P_{CV} or intraatrial pressure. Consequently, atrial contraction and ventricular recoil may be the two greatest contributors to ventricular filling. Thus, any impairment or improvement in either atrial or ventricular contractility will likely impact cardiac filling, and hence $\dot{Q}$. For example, both E- and A-wave peak velocities decrease substantially when the pericardial fluid volume in leopard sharks and sand bass is increased (Lai et al., 1996, 1998), or when zebrafish are exposed to a suboptimal temperature (Ho et al., 2002).

Higher cardiac filling pressures are needed to increase V_s (Fig. 7), which demonstrates the importance of the *vis-a-tergo* filling mechanisms. It is difficult to say with confidence, however, that *vis-a-tergo* filling alone results in the end-diastolic ventricular volume needed to generate maximum V_s in fishes. This is because ventricular contractility also increases. This, in turn, may lower the minimum intraventricular pressure and increase the peak velocity of the E-wave. In conclusion, atrial contraction, ventricular contraction, ventricular relaxation, and the energy contained in venous blood all contribute to ventricular filling in fishes.

A large operational range for V_s means that changes in diastolic ventricular volume must be accommodated within a relatively rigid pericardial cavity. For teleost fishes, this is possible only if the atrium, or sinus venosus, or both operate with a variable volume (i.e., their diastolic volume decreases whenever V_s increases, Forster and Farrell, 1994). Elasmobranch fishes, however, have a pericardial–peritoneal canal through which pericardial fluid can be displaced into the peritoneal cavity (Abel et al., 1987; Lai et al., 1989a, 1996; reviewed by Brill and Lai, 2016). When sharks are highly stressed, fluid exits the pericardial cavity and, as a result, venous pressure (P_{ven}) and intrapericardial pressures both become considerably more subambient than normal, and *vis-a-fronte* filling becomes all the more important. Catecholamines also increase the sensitivity of the heart to filling pressure, and atrial and ventricular contractility. These effects combine to move the Starling curve to the right and upward, like the effect of temperature (Fig. 7).

6. CORONARY BLOOD FLOW AND ITS CONTROL

The coronary artery supplies oxygenated blood to the heart, which is a more secure source of O_2 than venous blood; venous PO_2 generally decreases with activity whereas arterial PO_2 does not. The proportion of ventricular compact myocardium, and whether or not the cardiac trabeculae are vascularized, determine the dependence of the heart on a coronary circulation. Thus, coronary artery physiology is relevant to the entire myocardium (including the conal myocardium if present) of Type III (and IV) hearts, but only the ventricular compact layer of Type II hearts. While fishes with Type II hearts may have up to 45% of the ventricular myocardium (Table 1) supplied with blood from the coronary artery, sustained CBF is surprisingly not obligatory for life. For example, salmonids survive, and can even swim after the main coronary artery is ligated (Farrell and Steffensen, 1987; Steffensen and Farrell, 1998). All the same, cardiac contractility is compromised by coronary ligation because P_{VA} decreases. This may explain why mature salmon can still function with high levels of arteriosclerotic lesions in their main coronary artery (Farrell and Jones, 1992). Coronary ligation has not been attempted in elasmobranchs, but they may fare worse than salmonids because the entire heart receives CBF.

Despite its importance, CBF is extremely difficult to measure in fishes due to limited vessel access and the small size of the vessels. Also, CBF values derived from *in vitro* coronary perfusion studies with saline are difficult to evaluate, because the dynamic viscosity of blood relative to the perfusate is not known (Farrell, 1986). Nonetheless, these perfusion studies do provide valuable information on the control of CBF and the importance of CBF to cardiac performance (Agnisola et al., 2003; Farrell et al., 1992).

CBF is $0.20\,mL\,min^{-1}\,g^{-1}$ ventricle (1.1% of $\dot{Q}$) in resting coho salmon (*Oncorhynchus kitsuch*) (Axelsson and Farrell, 1993), whereas in resting rainbow trout, CBF is $0.14\,mL\,min^{-1}\,g^{-1}$ ventricle (0.84% of $\dot{Q}$) (Gamperl et al., 1994a). Thus, relative to body mass, the fish heart is overperfused by a factor of about 10 (RVM in salmonids is about 0.1%). Spontaneous activity in coho salmon increases CBF by up to 2.5-fold (Axelsson and Farrell, 1993). Rainbow trout, likewise, more than double CBF when swimming at 1 body length s^{-1} (Gamperl et al., 1995). Hypoxic exposure (PO_2 12 kPa) of rainbow trout halves arterial PO_2, but increases CBF by only 36%, and so, the percentage of $\dot{Q}$ directed to the heart also increases (Gamperl et al., 1994a).

Rainbow trout also increase CBF to a greater extent at a given swimming speed if they are made hypoxic (Gamperl et al., 1995). Thus, salmonids have the capacity to increase CBF almost three-fold and changes in CBF are associated with increases in CPO and hypoxemia.

Measuring CBF in elasmobranch fishes is more of a challenge because only a few elasmobranch species appear to have a single common coronary artery that conveniently accommodates a cuff-type flow probe. The shark heart is also overperfused relative to its body mass, receives a greater proportion of $\dot{Q}$ than does the teleost heart, and has a large CBF reserve. In the anesthetized sandbar shark (*Carcharhinus plumbeus*), CBF averaged 0.78 mL min^{-1} g^{-1} ventricle (3.0% of $\dot{Q}$) with the heart generating 2.2 mW g^{-1} ventricle (Cox et al., 2016b). Similarly, CBF is 0.64 mL min^{-1} g^{-1} ventricle in the anesthetized school shark and increases twofold with acute warming of leopard sharks (Cox, 2015). The fact that CBF goes only to the compact myocardium of the salmonid heart, which is typically <50% of ventricular mass, may explain the higher proportion of $\dot{Q}$ going to CBF in sharks. Although CBF normalized per g ventricular mass is appropriate for Type III (and IV) hearts, except for the fact that the trabeculae with the smallest diameter (<60 μm) in elasmobranch fishes have no coronary vessels (Cox, 2015), comparing Type II and III hearts using this dimension is potentially misleading. This is because at least half of a Type II ventricle receives no CBF. Normalizing CBF per g compact myocardium (Farrell, 1987) could circumvent this difficulty when comparing CBF values for Type II and Type III hearts.

In coho salmon, rainbow trout, sandbar shark, and school shark (*Galeorhinus australis*), as in mammals, the majority of CBF occurs during diastole (Fig. 19) (Axelsson and Farrell, 1993; Cox et al., 2016b; Davie and Franklin, 1993; Gamperl et al., 1995). This is presumably because cardiac contraction compresses the coronary vessels and increases coronary vascular resistance (R_{cor}). Nevertheless, CBF increases when ventricular contraction is stronger and the diastolic period is shorter. A 3.5-fold variation in CBF among individual, anesthetized, sandbar sharks was linearly related with f_H and R_{cor} (Cox, 2015). Therefore, changes in CBF must occur through a change in either P_{DA} or R_{cor}, or some combination thereof. Such changes overcome the compressive forces that are clearly evident in the phasic nature of recordings of CBF.

The origin of all coronary vessels is downstream of the lamellar vessels of the gill circulation, which means that the arterial blood pressure driving CBF is essentially P_{DA}. The direct effect of perfusion pressure on CBF is readily evident from *in vitro* coronary perfusion studies (e.g., Cox et al., 2016b; Davie et al., 1992; Farrell, 1987; Farrell et al., 1992). Nevertheless, changes in P_{DA} are rarely sufficient to effect two-fold to three-fold changes in CBF.

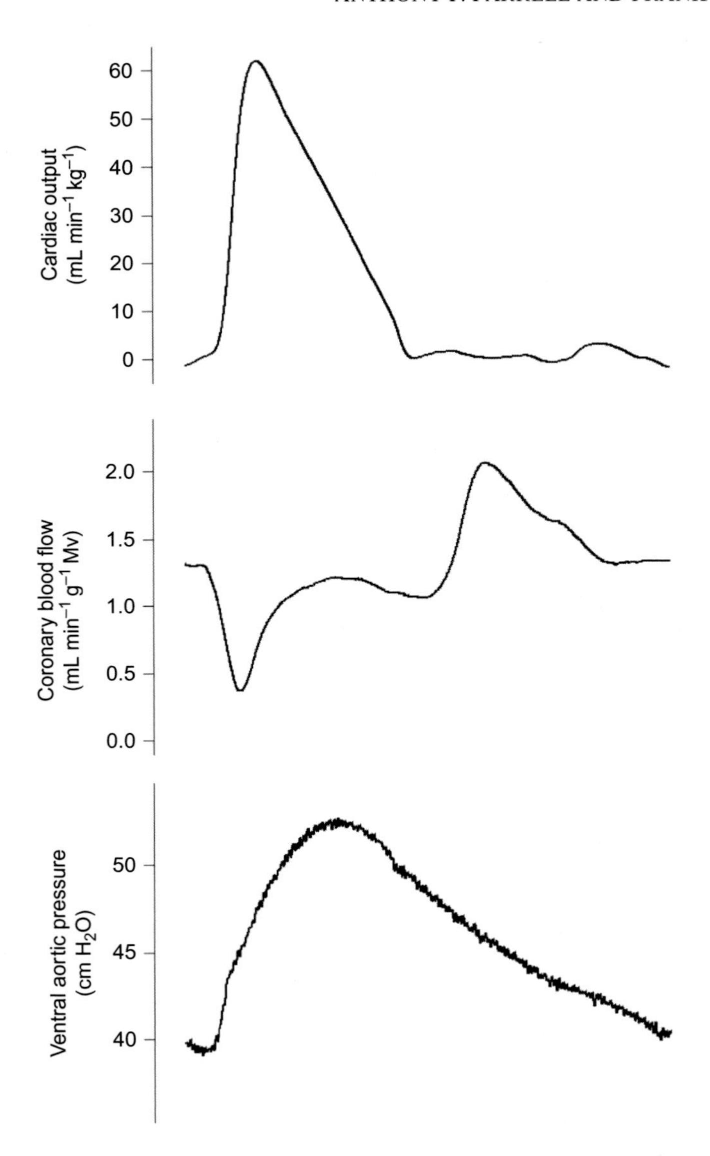

Fig. 19. Pulsatile recording of coronary blood flow, cardiac output and ventral aortic pressure in a shark. Credit: Cox, 2015. The functional significance and evolution of the coronary circulation in sharks. Dissertation, University of British Columbia.

For example, during spontaneous activity in coho salmon, P_{DA} increases by no more than 90%, whereas CBF increases by up to 2.5-times (Gamperl et al., 1995). Consequently, most of the large reserve capacity for CBF is a vasodilatory reserve. Surprisingly, *in vivo* drug injections have not revealed a major coronary vasodilatory mechanism (Axelsson and Farrell, 1993; Cox et al., 2016b; Gamperl et al., 1994b), although isoproterenol (β-adrenoceptor agonist) does reduce R_{cor} by 30% (Axelsson and Farrell, 1993). Injection of adrenaline in rainbow trout increases CBF by 50% despite by an increase in R_{cor} because the associated increase in R_{sys} is superseded a large increase in P_{DA} (Gamperl et al., 1994b). For a subgroup of rainbow trout, however, $\dot{Q}$ initially decreases due to bradycardia (presumably due to a baroreflex that attenuates the increase in P_{DA}) and the associated increase in CBF is delayed for 2–4 min until routine $\dot{Q}$ is restored.

The potential mechanisms responsible for coronary vasoactivity are numerous. The coronary vasculature is innervated (Laurent et al., 1983; Newton et al., 2014; Stoyek et al., 2015) and vascular rings from the main coronary artery respond to a variety of vasoactive agents (Acierno et al., 1990; Farrell and Graham, 1986; Farrell and Davie, 1991; Small et al., 1990; Small and Farrell, 1990; summarized by Farrell, 1991), as do coronary arterioles (Costa et al., 2015). Cox (2015) suggests that CBF in fish can be locally regulated by factors that are influenced by f_H and CPO, but how this regulation is achieved, and how it relates to the known vasoactive mechanisms, remains undescribed, and warrants further study.

Fine regulation of CBF in fish clearly occurs *via* a complex system of neural and hormonal controls. Known vasodilators include purinergic agents (adenosine, adenosine diphosphate [ADP], and ATP), catecholamines (noradrenaline, adrenaline, isoproterenol), NO-related agents (ACH, L-arginine), and other paracrine agents (serotonin, prostacyclin, prostaglandin E_2, thromboxane B_2) (Agnisola et al., 1996; Costa et al., 2015; Davie and Daxboeck, 1984; Farrell and Davie, 1991; Farrell and Graham, 1986; Farrell and Johansen, 1995; Mustafa and Agnisola, 1994, 1998; Mustafa et al., 1997; Small et al., 1990). Because some agents have biphasic responses depending on their concentration, known vasoconstrictors include many of the vasodilators: purinergic agents (adenosine, ADP, and ATP), catecholamines (noradrenaline, adrenaline; phenylephrine, α-adrenergic blockers), NO (nitric oxide)-related agents (ACH), and other paracrine agents (serotonin, indomethacin, cortisol, prostaglandin F_{2a}, carbocyclic thromboxane A_2, pinane, thromboxane A_2) (Agnisola et al., 1996; Costa et al., 2015; Davie and Daxboeck, 1984; Farrell, 1987; Farrell and Davie, 1991; Farrell and Graham, 1986; Farrell and Johansen, 1995; Mustafa et al., 1992; Mustafa and Agnisola, 1994, 1998; Mustafa et al., 1997; Small et al., 1990). The most comprehensive summary and analysis of the involvement of NO-mediated control of coronary vascular resistance for any fish is that for rainbow trout by Agnisola (2005), who summarizes coronary vasoactivity in arterial rings and in working and non-working heart preparations. He concludes that NO mediates the dilations associated with ACH, adenosine, and serotonin, actions that are inhibited by L-NNA (an inhibitor of nitric oxide synthase). Also, while NO release is inversely related to coronary resistance, both vascular stretch and hypoxia trigger NO release, with the NO release under hypoxia being blocked by theophylline, a purinergic antagonist. Thus, NO plays a central role in the control of coronary resistance in rainbow trout, possibly as an amplifier of the adenosine-mediated vasodilation under hypoxia. Correspondingly, data suggest that the hypoxia-induced increase in CBF and decrease in coronary resistance in coho salmon (*Oncorhynchus kisutch*) does not involve either cholinergic or adrenergic mechanisms (Axelsson and Farrell, 1993).

Acclimation temperature will undoubtedly play an important role in deciphering the complex control of CBF. In steelhead trout, the effects of adrenaline, adenosine, and nitric oxide donors are all attenuated by cold acclimation, whereas the constrictor effects of ACH and endothelin are enhanced (Costa et al., 2015).

7. SUMMARY

While we have made enormous progress in understanding the functioning of the fish heart and its control, many fundamental questions remain unanswered. We hope that others will explore our suggestions of areas that are fertile for future research, as noted in each preceding section.

REFERENCES

Abel, D.C., Lowell, W.R., Graham, J.B., and Shabetai, R. 1987. Elasmobranch pericardial function II. The influence of pericardial pressure on cardiac stroke volume in horn sharks and blue sharks. Fish Physiol. Biochem. 4:5–15.

Acierno, R., Agnisola, C., Venzi, R., Tota, B., 1990. Performance of the isolated and perfused working heart of the teleost *Conger*: study of the inotropic effect of prostacyclin. J. Comp. Physiol. B 160, 365–371.

Agnisola, C., 2005. Role of nitric oxide in the control of coronary resistance in teleosts. Comp. Biochem. Physiol. A 142, 178–187.

Agnisola, C., Mustafa, T., Hansen, J.K., 1996. Autoregulatory index, adrenergic responses, and interaction between adrenoreceptors and prostacyclin in the coronary system of rainbow trout. J. Exp. Zool. 275, 239–248.

Agnisola, C., Petersen, L., Mustafa, T., 2003. Effect of coronary perfusion on the basal performance, volume loading and oxygen consumption in the isolated resistance-headed heart of the trout (*Oncorhynchus mykiss*). J. Exp. Biol. 206, 4003–4010.

Alderman, S.L., Harter, T.S., Wilson, J.M., Supuran, C.T., Farrell, A.P., Brauner, C.J., 2016. Evidence for a plasma-accessible carbonic anhydrase in the lumen of salmon heart that may enhance oxygen delivery to the myocardium. J. Exp. Biol. 219, 719–724.

Altimiras, J., Aissaoui, A., Tort, L., Axelsson, M., 1997. Cholinergic and adrenergic tones in the control of heart rate in teleosts. How should they be calculated? Comp. Biochem. Physiol. A 118, 131–139.

Altimiras, J., Axelsson, M., Claireaux, G., Lefrancois, C., Mercier, C., Farrell, A.P., 2002. Cardiorespiratory status of triploid brown trout during swimming at two acclimation temperatures. J. Fish Biol. 60, 102–116.

Altimiras, J., Claireaux, G., Sandblom, E., Farrell, A.P., McKenzie, D.J., Axelsson, M., 2008. Gastrointestinal blood flow and postprandial metabolism in swimming sea bass *Dicentrarchus labrax*. Physiol. Biochem. Zool. 81, 663–672.

Anttila, K., Couturier, C.S., Overli, O., Johnsen, A., Marthinsen, G., Nilsson, G.E., Farrell, A.P., 2014a. Atlantic salmon show capability for cardiac acclimation to warm temperatures. Nat. Commun. 5, 4252.

Anttila, K., Jørgensen, S.M., Casselman, M.T., Timmerhaus, G., Farrell, A.P., Takle, H., 2014b. Association between swimming performance, cardiorespiratory morphometry, and thermal tolerance in Atlantic salmon (*Salmo salar* L.). Front. Mar. Sci. 1 (76). http://dx.doi.org/10.3389/fmars.2014.00076.

Anttila, K., Lewis, M., Prokkola, J.M., Kanerva, M., Seppänen, E., Kolari, I., Nikinmaa, M., 2015. Warm acclimation and oxygen depletion induce species-specific responses in salmonids. J. Exp. Biol. 218, 1471–1477.

Ariens Kappers, C.U., Huber, G.C., Crosby, E.C., 1967. The Comparative Anatomy of the Nervous System of Vertebrates, Including Man. Hafner, New York.

Arthur, P.G., Keen, J.E., Hochachka, P.W., Farrell, A.P., 1992. Metabolic state of the *in situ* perfused trout heart during severe hypoxia. Am. J. Physiol. 263, R798–R804.

Arthur, P.G., Franklin, C.E., Cousins, K.L., Thorarensen, H., Hochachka, P.W., Farrell, A.P., 1997. Energy turnover in the normoxic and anoxic turtle heart. Comp. Biochem. Physiol. A 117, 121–126. Axelsson 1980 cited in Table 2.

Augustinsson, K.-B., Fänge, R., Johnels, A., Östlund, E., 1956. Histological, physiological and biochemical studies on the heart of two cyclostomes, hagfish (*Myxine*) and lamprey (*Lampetra*). J. Physiol. 131, 257–276.

Axelsson, M., 1988. The importance of nervous and humoral mechanisms in the control of cardiac performance in the Atlantic cod, *Gadus morhua*, at rest and during non-exhaustive exercise. J. Exp. Biol. 137, 287–303.

Axelsson, M., 2005. In: Farrell, A.P., Steffensen, J.F. (Eds.), Polar Fishes. Academic Press, New York, pp. 239–280.

Axelsson, M., Farrell, A.P., 1993. Coronary blood flow *in vivo* in the coho salmon (*Oncorhynchus kisutch*). Am. J. Physiol. 264, R963–R971.

Axelsson, M., Fritsche, R., 1991. Effects of exercise, hypoxia and feeding on the gastrointestinal blood flow in the Atlantic cod Gadus morhua. J. Exp. Biol. 158, 181–198.

Axelsson, M., Nilsson, S., 1986. Blood pressure control during exercise in the Atlantic cod, *Gadus morhua*. J. Exp. Biol. 126, 225–236.

Axelsson, M., Ehrenstrom, F., Nilsson, S., 1987. Cholinergic and adrenergic influence on the teleost heart *in vivo*. Exp. Biol. 46, 179–186.

Axelsson, M., Driedzic, W.R., Farrell, A.P., Nilsson, S., 1989. Regulation of cardiac output and gut blood flow in the sea raven, *Hemitripterus americanus*. Fish Physiol. Biochem. 6, 315–326.

Axelsson, M., Farrell, A.P., Nilsson, S., 1990. Effects of hypoxia and drugs on the cardiovascular dynamics of the Atlantic hagfish *Myxine glutinosa*. J. Exp. Biol. 151, 297–316.

Axelsson, M., Davison, W., Forster, M.E., Farrell, A.P., 1992. Cardiovascular responses of the red-blooded Antarctic fishes, *Pagothenia bernacchii* and *P. borchgrevinki*. J. Exp. Biol. 167, 179–201.

Axelsson, M., Davison, W., Franklin, C.E., 2000a. Cholinergic and adrenergic tone on the heart of the Antarctic dragonfish, *Gymnodraco auticeps*, living at sub-zero temperature. Exp. Biol. Online 5, 1–12.

Axelsson, M., Thorarensen, H., Nilsson, S., Farrell, A.P., 2000b. Gastrointestinal blood flow in the red Irish lord, *Hemilepidotus hemilepidotus*: long-term effects of feeding and adrenergic control. J. Comp. Physiol. B 170, 145–152.

Axelsson, M., Altimiras, J., Claireaux, G., 2002. Postprandial blood flow to the gastrointestinal tract is not compromised during hypoxia in the sea bass *Dicentrarchus labrax*. J. Exp. Biol. 205, 2891–2896.

Badr, A., El-Sayed, M.F., Vornanen, M., 2016. Effects of seasonal acclimatization on temperature-dependence of cardiac excitability in the roach, *Rutilus rutilus*. J. Exp. Biol. 219, 1495–1504. http://dx.doi.org/10.1242/jeb.138347.

Baker, D.W., Hanson, L.M., Farrell, A.P., Brauner, C.J., 2011. Exceptional CO_2 tolerance in white sturgeon (*Acipenser transmontanus*) is associated with protection of maximum cardiac performance during hypercapnia *in situ*. Physiol. Biochem. Zool. 84, 239–248.

Bass, A., Ostadal, B., Pelouch, V., Vitek, V., 1973. Differences in weight parameters, myosin ATPase activity and the enzyme pattern of energy supplying metabolism between the compact and spongious cardiac musculature of carp and turtle. Pflugers Arch. 343, 65–77.

Braun, M.H., Brill, R.W., Gosline, J.M., Jones, D.R., 2003a. Form and function of the bulbus arteriosus in yellowfin tuna (*Thunnus albacares*): dynamic properties. J. Exp. Biol. 206, 3311–3326.

Braun, M.H., Brill, R.W., Gosline, J.M., Jones, D.R., 2003b. Form and function of the bulbus arteriosus in yellowfin tuna (*Thunnus albacares*) and blue marlin (*Makaira nigricans*): static properties. J. Exp. Biol. 206, 3327–3335.

Brauner, Harter, 2017. The teleost O_2 and CO_2 transport system and specialized mechanisms to enhance blood O_2 delivery. In: Gamperl, A.K., Gillis, T.E., Farrell, A.P., Brauner, C.J. (Eds.), Fish Physiology. In: The Cardiovascular System: Development, Plasticity and Physiological Responses, vol. 36B. Academic Press, San Diego (In press).

Brill, R.W., Bushnell, P.G., 1989. CARDIO: a Lotus 1-2-3 based computer program for rapid calculation of cardiac output from dye or thermal dilution curves. Comput. Biol. Med. 19 (5), 361–366.

Brill, R.W., Bushnell, P.C., 1991. Metabolic scope of high-energy demand teleosts—the tunas. Can. J. Zool. 69, 2002–2009.

Brill, R.W., Lai, N.C., 2016. Elasmobranch cardiovascular system. In: Shadwick, R.E.F.A.P., Brauner, C.J. (Eds.), Physiology of Elasmobranch Fishes: Internal Processes. Academic Press, New York, pp. 1–82.

Burggren, W.W., Christoffels, V.M., Crossley, D.A., Enok, S., Farrell, A.P., Hedrick, M.S., Hicks, J.W., Jensen, B., Moorman, A.F.M., Mueller, C.A., Skovgaard, N., Taylor, E.W., Wang, T., 2014. Comparative cardiovascular physiology: future trends, opportunities and challenges. Acta Physiol. 210, 257–276.

Burggren, W.W., Dubansky, B., Bautista, N.M., 2017. Cardiovascular development in embryonic and larval fishes. In: Gamperl, A.K., Gillis, T.E., Farrell, A.P., Brauner, C.J. (Eds.), Fish Physiology. In: The Cardiovascular System: Development, Plasticity and Physiological Responses, vol. 36B. Academic Press, San Diego (In press).

Bushnell, P.G., 1988. Cardiovascular and respiratory responses to hypoxia in three species of obligate ram ventilating fishes, skipjack tuna, *Katsuwonus pelamis,* yellowfin tuna, *Thunnus albacares,* and bigeye tuna, *Thunnus obesus*. Ph.D. thesis, University of Hawaii, Honolulu, p. 276.

Bushnell, P.G., Brill, R.W., 1991. Responses of swimming skipjack *(Katsuwonus pelamis)* and yellowfin *(Thunnus albacares)* tunas to acute hypoxia, and a model of their cardiorespiratoiy system. Physiol. Zool. 64, 787–811.

Bushnell, P.G., Brill, R.W., Bourke, R.W., 1990. Cardiorespiratory responses of skipjack tuna *(Katsuwonus pelamis),* yellowfin tuna *(Thunnu salbacares),* and big-eye tuna (*T. obesus)* to acute reductions in ambient oxygen. Can. J. Zool. 68, 1857–1863.

Bushnell, P.G., Jones, D.R., Farrell, A.P., 1992. The arterial system. In: Hoar, W.S., Randall, D.J., Farrell, A.P. (Eds.), Fish Physiology. In: vol. XIIA. Academic Press, San Diego, pp. 89–139.

Butler, P.J., 1986. Exercise. In: Nilsson, S., Holmgren, S. (Eds.), Fish Physiology: Recent Advances. Croom Helm, London, pp. 102–118.

Butler, P.J., Metcalfe, J.D., 1988. Cardiovascular and respiratory systems. In: Shuttleworth, T.V. (Ed.), Physiology of Elasmobranch Fishes. Springer-Verlag, New York, pp. 1–47.

Cameron, J.S., 1979. Autonomic nervous tone and regulation of heart rate in the goldfish, *Carassius auratus*. Comp. Biochem. Physiol. 63C, 341–349.

Cannon, W.B., 1929. Organization for physiological homeostasis. Physiol. Rev. 9, 399–431.

Casselman, M.T., Anttila, K., Farrell, A.P., 2012. Using maximum heart rate as a rapid screening tool to determine optimum temperature for aerobic scope in salmon *Oncorhynchus* spp. J. Fish Biol. 80, 358–377.

Cech, J.J., Bridges, R.W., Rowell, D.M., Baker, P.J., 1976. Cardiovascular responses of winter flounder, *Pseudopleuronectes americanus,* to acute temperature increase. Can. J. Zool. 54, 1383–1388.

Chan, D.K.O., 1986. Cardiovascular, respiratory, and blood adjustments to hypoxia in the Japanese eel, *Anguilla japonica*. Fish Physiol. Biochem. 2, 179–193.

Chen, Z., Snow, M., Lawrence, C., Church, A., Narum, S., Devlin, R., Farrell, A.P., 2015. Selection for upper thermal tolerance in rainbow trout (*Oncorhynchus mykiss* Walbaum). J. Exp. Biol. 218, 803–812.

Claireaux, G., McKenzie, D.J., Genge, A.G., Chatelier, A., Aubin, J., Farrell, A.P., 2005. Linking swimming performance, cardiac pumping ability and cardiac anatomy in rainbow trout. J. Exp. Biol. 208, 1775–1784.

Clark, T.D., Farrell, A.P., 2011. Effects of body mass on physiological and anatomical parameters of mature salmon: evidence against a universal heart rate scaling exponent. J. Exp. Biol. 214, 887–893.

Clark, T.D., Seymour, R.S., Christian, K., Wells, R.M.G., Baldwin, J., Farrell, A.P., 2007. Changes in cardiac output during swimming and aquatic hypoxia in the air-breathing Pacific tarpon. Comp. Biochem. Physiol. A 148, 562–571.

Clark, T.D., Sandblom, E., Cox, G.K., Hinch, S.G., Farrell, A.P., 2008. Circulatory limits to oxygen supply during an acute temperature increase in the Chinook salmon (*Oncorhynchus tshawytscha*). Am. J. Physiol. 295, R1631–1639.

Clark, T.D., Jeffries, K.M., Hinch, S.G., Farrell, A.P., 2011. Exceptional aerobic scope and cardiovascular performance of pink salmon (*Oncorhynchus gorbuscha*) may underlie resilience in a warming climate. J. Exp. Biol. 214, 3074–3081.

Costa, I.A.S.F., Hein, T.W., Gamperl, A.K., 2015. Cold-acclimation leads to differential regulation of the steelhead trout (*Oncorhynchus mykiss*) coronary microcirculation. Am. J. Physiol. 308, R743–R754.

Cotter, P.A., Han, A.J., Everson, J.J., Rodnick, K.J., 2008. Cardiac hemodynamics of the rainbow trout (*Oncorhynchus mykiss*) using simultaneous Doppler echocardiography and electrocardiography. J. Exp. Zool. 309A, 243–254.

Coucelo, J., Coucelo, J., Azevedo, J., 1996. Ultrasonography characterization of heart morphology and blood flow of lower vertebrates. J. Exp. Zool. 275, 73–82.

Cousins, K.L., Farrell, A.P., 1996. Stretch-induced release of atrial natriuretic factor from the heart of rainbow trout (*Oncorhynchus mykiss*). Can. J. Zool. 74, 380–387.

Cousins, K.L., Farrell, A.P., Sweeting, R.W., Vesely, D.L., Keen, J.E., 1997. Release of atrial natiuretic factor (ANF) prohormone peptides 1-30, 31-67 and 99-126 from freshwater- and seawater-acclimated perfused trout (*Oncorhynchus mykiss*) hearts. J. Exp. Biol. 200, 1351–1362.

Cox, 2015. The functional significance and evolution of the coronary circulation in sharks. Dissertation, University of British Columbia.

Cox, G.K., Sandblom, E., Farrell, A.P., 2010. Cardiac responses to anoxia in the Pacific hagfish, *Eptatretus stoutii*. J. Exp. Biol. 213, 3692–3698.
Cox, G.K., Sandblom, E., Richards, J.G., Farrell, A.P., 2011. Anoxic survival of the Pacific hagfish, *Eptatretus stoutii*. J. Comp. Physiol. 181, 361–371.
Cox, G.K., Brill, R.W., Bonaro, K.A., Farrell, A.P., 2017. Determinants of coronary blood flow in the sandbar shark *Carcharhinus plumbeus*. J. Comp. Physiol. B 187, 315–327.
Cox, G.K., Kennedy, G.E., Farrell, A.P., 2016b. Morphological arrangement of the coronary vasculature in a shark (*Squalus suckleі*) and a teleost (*Oncorhynchus mykiss*). J. Morphol. 277, 896–905.
Crocker, C.E., Farrell, A.P., Gamperl, A.K., Cech Jr., J.J., 2000. Cardiorespiratory responses of white sturgeon to environmental hypercapnia. Am. J. Physiol. 279, R617–R628.
Dahlstrom, A., Fuxe, K., 1964. Evidence for the existence of monoamine-containing neurons in the central nervous system: I. Demonstration of monoamines in the cell bodies of brain stem neurons. Acta Physiol. Scand. (Suppl. 232), 1–55.
Davie, P.S., 1987. Coronary supply to the myocardium of the hearts of very active fishes. Proc. Physiol. Soc. N.Z. 6, 36.
Davie, P.S., Daxboeck, C., 1984. Anatomy and adrenergic pharmacology of the coronary vascular bed of Pacific blue marlin (*Makaira nigricans*). Can. J. Zool. 62, 1886–1888.
Davie, P.S., Farrell, A.P., 1991a. The coronary and luminal circulations of the myocardium of fishes. Can. J. Zool. 69, 1993–2001.
Davie, P.S., Farrell, A.P., 1991b. Cardiac performance of an isolated heart preparation from the dogfish (*Squalus acanthias*): the effects of hypoxia and coronary artery perfusion. Can. J. Zool. 69, 1822–1828.
Davie, P.S., Forster, ME., 1980. Cardiovascular responses to swimming in eels. Comp. Biochem. Physiol. 67A, 367–373.
Davie, P.S., Franklin, C.E., 1992. Myocardial oxygen consumption and mechanical efficiency of a perfused dogfish heart preparation. J. Comp. Physiol. B 162, 256–262.
Davie, P.S., Franklin, C.E., 1993. Preliminary observations on blood flow in the coronary arteries of two school sharks (*Galeorhinus australis*). Can. J. Zool. 71, 1238–1241.
Davie, P.S., Forster, M.E., Davison, B., Satchell, G.H., 1987. Cardiac function in the New Zealand hagfish, *Eptatretus cirrhatus*. Physiol. Zool. 60, 233–240.
Davie, P.S., Farrell, A.P., Franklin, C.E., 1992. Cardiac performance of an isolated eel heart: effects of hypoxia and responses to coronary artery perfusion. J. Exp. Zool. 262, 113–121.
Davison, W., 1989. Training and its effects on teleost fish. Comp. Biochem. Physiol. 94A, 1–10.
DeAndres, A.V., Munoz-Chapuli, R., Sans-Coma, V., Garcia-Garrido, L., 1990. Anatomical studies of the coronary system in elasmobranchs: I. Coronary arteries in lamnoid sharks. Am. J. Anat. 187, 303–310.
Diaz, R.J., Breitburg, D.L., 2009. The hypoxic environment. In: Richards, J.G., Farrell, A.P., Brauner, C.J. (Eds.), Hypoxia. Fish Physiology. In: vol. 27. Academic Press, Burlington, pp. 1–23.
Driedzic, W.R., 1983. The fish heart as a model system for the study of myoglobin. Comp. Biochem. Physiol. 76A, 487–493.
Driedzic, W.R., Gesser, H., 1994. Energy metabolism and contractility in ectothermic vertebrate hearts: hypoxia, acidosis, and low temperature. Physiol. Rev. 74, 221–258.
Drost, H.E., Carmack, E.C., Farrell, A.P., 2014. Upper thermal limits of cardiac function for Arctic cod *Boreogadus saida*, a key food web fish species in the Arctic Ocean. J. Fish Biol. 84, 1781–1792.
Drost, H., Lo, M., Carmack, E., Farrell, A.P., 2016. Acclimation potential of Arctic cod (*Boreogadus saida* Lepechin) from the rapidly warming Arctic Ocean. J. Exp. Biol. 219, 3114–3125.

Dupont-Prinent, A., Claireaux, G., McKenzie, D.J., 2009. Effects of feeding and hypoxia on cardiac performance and gastrointestinal blood flow during critical speed swimming in the sea bass *Dicentrarchus labrax*. Comp. Biochem. Physiol. A 155, 233–240.

Egginton, S., Cordiner, S., 1997. Cold-induced angiogenesis in seasonally acclimatized rainbow trout (*Oncorhynchus mykiss*). J. Exp. Biol. 200, 2263–2268.

Egginton, S., Forster, M., Davison, W., 2001. Control of vascular tone in notothenioid fishes is determined by phylogeny, not environmental temperature. Am. J. Physiol. 280, R1197–R1205.

Ekstrom, A., Brijs, J., Clark, T.D., Grans, A., Jutfelt, F., Sandblom, E., 2016. Cardiac oxygen limitation during an acute lethal thermal challenge in the European perch: effects of chronic environmental warming and experimental hyperoxia. Am. J. Physiol. 311, R440–R449.

Eliason, Antilla, 2017. Temperature and the cardiovascular system. In: Gamperl, A.K., Gillis, T.E., Farrell, A.P., Brauner, C.J. (Eds.), Fish Physiology. In: The Cardiovascular System: Development, Plasticity and Physiological Responses, vol. 36B. Academic Press, San Diego (In press).

Eliason, E.J., Farrell, A.P., 2014. Effect of hypoxia on specific dynamic action and postprandial cardiovascular physiology in rainbow trout (*Oncorhynchus mykiss*). Comp. Biochem. Physiol. A 171, 44–50.

Eliason, E.J., Kiessling, A., Karlsson, A., Djordjevic, B., Farrell, A.P., 2007. Validation of the hepatic portal vein cannulation technique using Atlantic salmon (*Salmo salar* L.). 71, 290–297.

Eliason, E.J., Higgs, D.A., Farrell, A.P., 2008. Postprandial gastrointestinal blood flow, oxygen consumption and heart rate in rainbow trout (*Oncorhynchus mykiss*). Comp. Biochem. Physiol. A 149, 380–388.

Eliason, E.J., Clark, T.D., Hague, M.J., Hanson, L.M., Gallagher, Z.S., Jeffries, K.M., Gale, M.K., Patterson, D.A., Hinch, S.G., Farrell, A.P., 2011. Differences in thermal tolerance among sockeye salmon populations. Science 332, 109–112.

Eliason, E.J., Clark, T.D., Hinch, S.G., Farrell, A.P., 2013a. Cardiorespiratory collapse at high temperature in swimming adult sockeye salmon. Conserv. Physiol. 1, cot008.

Eliason, E.J., Clark, T.D., Hinch, S.G., Farrell, A.P., 2013b. Cardiorespiratory performance and blood chemistry during swimming and recovery in three populations of elite swimmers: adult sockeye salmon. Comp. Biochem. Physiol. 166B, 385–397.

Eliason, E.J., Wilson, S.M., Farrell, A.P., Cooke, S.J., Hinch, S.G., 2013c. Low cardiac and aerobic scope in a coastal population of sockeye salmon, *Oncorhynchus nerka*, with a short upriver migration. J. Fish Biol. 82, 2104–2112. Can. J. Fish. Aquat. Sci. 70: 349–357.

Emery, S.H., Mangano, C., Randazzo, V., 1985. Ventricle morphology in pelagic elasmobranch fishes. Comp. Biochem. Physiol. 82A, 635–643.

Fange, R., 1972. The circulatory system. In: Hardisty, M.W., Potter, I.C. (Eds.), The Biology of Lampreys. In: vol. 2. Academic Press, London, pp. 241–259.

Fänge, R., Östlund, E., 1954. The effects of adrenaline, noradrenaline, tyramine and other drugs on the isolated heart from marine vertebrates and a cephalopod (*Eledone cirrosa*). Acta Zool. (Stockholm), 35, 289–305.

Farrell, A.P., 1978. Cardiovascular events associated with air-breathing in two teleosts, *Hoplerythrinus unitaeniatus* and *Arapaima gigas*. Can. J. Zool. 56, 953–958.

Farrell, A.P., 1982. Cardiovascular changes in the unanesthetized ling cod (*Ophiodon elongates*) during short-term, progressive hypoxia and spontaneous activity. Can. J. Zool. 60, 933–941.

Farrell, A.P., 1984. A review of cardiac performance in the teleost heart: intrinsic and humoral regulation. Can. J. Zool. 62, 523–536.

Farrell, A.P., 1985. A protective effect of adrenaline on the acidotic teleost heart. J. Exp. Biol. 116, 503–508.

Farrell, A.P., 1986. Cardiovascular responses in the sea raven, *Hemitripterus americanus*, elicited by vascular compression. J. Exp. Biol. 122, 65–80.
Farrell, A.P., 1987. Coronary flow in a perfused rainbow trout heart. J. Exp. Biol. 129, 107–123.
Farrell, A.P., 1991. From hagfish to tuna: a perspective on cardiac function in fish. Physiol. Zool. 64, 1137–1164.
Farrell, A.P., 2002. Cardiorespiratory performance in salmonids during exercise at high temperature: insights into cardiovascular design limitations in fishes. Comp. Biochem. Physiol. A 132, 797–810.
Farrell, A.P., 2007a. Tribute to P. L. Lutz: a message from the heart—why hypoxic bradycardia in fishes? J. Exp. Biol. 210, 1715–1725.
Farrell, A.P., 2007b. Cardiovascular systems in primitive fishes. In: McKenzie, D.J., Farrell, A.P., Brauner, C. (Eds.), Primitive Fishes. Academic Press, New York, pp. 53–120.
Farrell, A.P., 2007c. Cardiorespiratory performance during prolonged swimming tests with salmonids: a perspective on temperature effects and potential analytical pitfalls. Philos. Trans. R. Soc. B 362, 2017–2030.
Farrell, A.P., 2009. Environment, antecedents and climate change: lessons from the study of temperature physiology and river migration of salmonids. J. Exp. Biol. 212, 3771–3780.
Farrell, A.P., 2016. Pragmatic perspective on aerobic scope: peaking, plummeting, pejus and apportioning. J. Fish Biol. 88, 322–343.
Farrell, A.P., Clutterham, S.M., 2003. On-line venous oxygen tensions in rainbow trout during graded exercise at two acclimation temperatures. J. Exp. Biol. 206, 487–496.
Farrell, A.P., Davie, P.S., 1991. Coronary artery reactivity in the mako shark, *Isurus oxyrinchus*. Can. J. Zool. 69, 375–379.
Farrell, A.P., Graham, M.S., 1986. Effects of adrenergic drugs on the coronary circulation of Atlantic salmon (*Salmo salar*). Can. J. Zool. 64, 481–484.
Farrell, A.P., Johansen, J.A., 1995. Vasoactivity of the coronary artery of rainbow trout, steelhead trout and dogfish: lack of support for non-prostanoid endothelium-derived relaxation factors. Can. J. Zool. 73, 1899–1911.
Farrell, A.P., Jones, D.R., 1992. The heart. In: Hoar, W.S., Randall, D.J., Farrell, A.P. (Eds.), Fish Physiology. In: vol XII, Pt. A. Academic Press, New York, pp. 1–88.
Farrell, A.P., Milligan, C.L., 1986. Myocardial intracellular pH in a perfused rainbow trout heart during extracellular acidosis in the presence and absence of adrenaline. J. Exp. Biol. 125, 347–359.
Farrell, A.P., Richards, J.G., 2009. Defining hypoxia: an integrative synthesis of the responses of fish to hypoxia. In: Richards, J.G., Farrell, A.P., Brauner, C.J. (Eds.), Hypoxia. Fish Physiology. In: vol. 27. Academic Press, Burlington, pp. 487–503.
Farrell, A.P., Stecyk, J.A.W., 2007. The heart as a working model to explore themes and strategies for anoxic survival in ectothermic vertebrates. Comp. Biochem. Physiol. A 147, 300–312.
Farrell, A.P., Steffensen, J.F., 1987. Coronary ligation reduces maximum sustained swimming speed in chinook salmon, *Oncorhynchus tshawytscha*. Comp. Biochem. Physiol. A 87, 35–37.
Farrell, A.P., MacLeod, K.R., Driedzic, W.R., Wood, S., 1983. Cardiac performance in the in situ perfused fish heart during extracellular acidosis: interactive effects of adrenaline. J. Exp. Biol. 107, 415–429.
Farrell, A.P., Wood, S., Hart, T., Driedzic, W.R., 1985. Myocardial oxygen consumption in the sea raven, *Hemitripterus americanus*: the effects of volume loading, pressure loading and progressive hypoxia. J. Exp. Biol. 117, 237–250.
Farrell, A.P., MacLeod, K.R., Chancey, B., 1986. Intrinsic mechanical properties of the perfused rainbow trout heart and the effects of catecholamines and extracellular calcium under control and acidotic conditions. J. Exp. Biol. 125, 319–345.

Farrell, A.P., Johansen, J.A., Graham, M.S., 1988a. The role of the pericardium in cardiac performance of the trout (*Salmo gairdneri*). Physiol. Zool. 61, 213–221.
Farrell, A.P., MacLeod, K.R., Scott, C., 1988b. Cardiac performance of the trout (*Salmo gairdneri*) heart during acidosis: effects of low bicarbonate, lactate and cortisol. Comp. Biochem. Physiol. A 91, 271–277.
Farrell, A.P., Hammons, A.M., Graham, M.S., Tibbits, G.F., 1988c. Cardiac growth in rainbow trout, *Salmo gairdneri*. Can. J. Zool. 66, 2368–2373.
Farrell, A.P., Small, S., Graham, M.S., 1989. Effect of heart rate and hypoxia on the performance of a perfused trout heart. Can. J. Zool. 67, 274–280.
Farrell, A.P., Johansen, J.A., Steffensen, J.F., Moyes, C.D., West, T.G., Suarez, R.K., 1990. Effects of exercise-training and coronary ablation on swimming performance, heart size and cardiac enzymes in rainbow trout, *Oncorhynchus mykiss*. Can. J. Zool. 68, 1174–1179.
Farrell, A.P., Johansen, J.A., Suarez, R.K., 1991. Effects of exercise-training on cardiac performance and muscle enzymes in rainbow trout, *Oncorhynchus mykiss*. Fish Physiol. Biochem. 9, 303–312.
Farrell, A.P., Davie, P.S., Franklin, C.E., Johansen, J.A., Brill, R.W., 1992. Cardiac physiology in tunas. I. *In vitro* perfused heart preparations from yellowfin and skipjack tunas. Can. J. Zool. 70, 1200–1210.
Farrell, A.P., Franklin, C.E., Arthur, P.G., Thorarensen, H., Cousins, 1994. Mechanical performance of an *in situ* perfused heart from the turtle *Chrysemys scripta* during normoxia and anoxia at 5°C and 15°C. J. Exp. Biol. 191, 207–229.
Farrell, A.P., Gamperl, A.K., Hicks, J.M.T., Shiels, H.A., Jain, K.E., 1996. Maximum cardiac performance of rainbow trout, *Oncorhynchus mykiss*, at temperatures approaching their upper lethal limit. J. Exp. Biol. 199, 663–672.
Farrell, A.P., Gamperl, A.K., Birtwell, I.K., 1998. Prolonged swimming, recovery and repeat swimming performance of mature sockeye salmon *Oncorhynchus nerka* exposed to moderate hypoxia and pentachlorophenol. J. Exp. Biol. 201, 2183–2193.
Farrell, A.P., Thorarensen, H., Axelsson, M., Crocker, C.E., Gamperl, A.K., Cech Jr. J.J., 2001. Gut blood flow in fish during exercise and severe hypercapnia. Comp. Biochem. Physiol. A 128, 551–563.
Farrell, A.P., Axelsson, M., Altimiras, J., Sandblom, E., Claireaux, G., 2007a. Maximum cardiac performance and adrenergic sensitivity of the sea bass *Dicentrarchus labrax* at high temperatures. J. Exp. Biol. 210, 1216–1224.
Farrell, A.P., Simonot, D.L., Seymour, R.S., Clark, T.D., 2007b. A novel technique for estimating the compact myocardium in fishes reveals surprising results for an athletic air-breathing fish, the Pacific tarpon. J. Fish Biol. 71, 389–398.
Farrell, A.P., Eliason, E.J., Sandblom, E., Clark, T.D., 2009. Fish cardiorespiratory physiology in an era of climate change. Can. J. Zool. 87, 835–851.
Farrell, A.P., Farrell, N.D., Jourdan, H., Cox, G.K., 2012. A perspective on the evolution of the coronary circulation in fishes and the transition to terrestrial life. In: Sedmera, D., Wang, T. (Eds.), Ontogeny and Phylogeny of the Vertebrate Heart. Springer, New York, pp. 75–102.
Farrell, A.P., Altimiras, J., Franklin, C.E., Axelsson, M., 2013. Niche expansion of the shorthorn sculpin (*Myoxocephalus scorpius*) to Arctic waters is supported by a thermal independence of cardiac performance at low temperature. Can. J. Zool. 91, 573–580.
Farrell, A.P., Eliason, E.J., Clark, T.D., Steinhausen, M.F., 2014. Oxygen removal from water versus arterial oxygen delivery: calibrating the Fick equation in Pacific salmon. J. Comp. Physiol. B 184, 855–864.
Forster, M.E., 1991. Myocardial oxygen consumption and lactate release by the hypoxic hagfish heart. J. Exp. Biol. 156, 583–590.

Forster, M.E., Farrell, A.P., 1994. The volumes of the chambers of the trout heart. Comp. Biochem. Physiol. A 109, 127–132.

Forster, M.E., Axelsson, M., Farrell, A.P., Nilsson, S., 1991. Cardiac function and circulation in hagfishes. Can. J. Zool. 69, 1985–1992.

Forster, M.E., Davison, W., Axelsson, M., Farrell, A.P., 1992. Cardiovascular responses to hypoxia in the hagfish, *Eptatretus cirrhatus*. Respir. Physiol. 88, 373–386.

Foxon, G.E.H., 1950. A description of the coronary arteries in dipnoan fishes and some remarks on their importance from the evolutionary standpoint. J. Anat. 84, 121–131.

Franklin, C.E., Davie, P.S., 1991. The pericardium facilitates pressure work in the eel heart. J. Fish Biol. 39, 559–564.

Franklin, C.E., Davie, P.S., 1992. Dimensional analysis of the ventricle of an in situ perfused trout heart using echocardiography. J. Exp. Biol. 166, 47–60.

Franklin, C.E., Farrell, A.P., Altimiras, J., Axelsson, M., 2013. Thermal dependence of cardiac function in Arctic fish: implications of a warming world. J. Exp. Biol. 216, 4251–4255.

Fritsche, R., Nilsson, S., 1990. Autonomic control of blood pressure and heart rate during hypoxia in cod, *Gadus morhua*. J. Comp. Physiol. B 160, 287–292.

Funakoshi, K., Nakano, M., 2007. The sympathetic nervous system of anamniotes. Brain Behav. Evol. 69, 105–113.

Funakoshi, K., Abe, T., Rahman, M.S., Kishida, R., 1997. Spinal and vagal projections to the sympathetic trunk of the wrasse, *Halichoeres poecilopterus*. JANS 67, 125–129.

Gallaugher, P.E., Thorarensen, H., Kiessling, A., Farrell, A.P., 2001. Effects of high intensity exercise training on cardiovascular function, oxygen uptake, internal oxygen transfer and osmotic balance in chinook salmon (*Oncorhynchus tshawytscha*) during critical speed swimming. J. Exp. Biol. 204, 2861–2872.

Gallego, A., Duran, A.C., De Andres, A.V., Navarro, P., Munoz-Chapuli, R., 1997. Anatomy and development of the sinoatrial valves in the dogfish (*Scyliorhinus canicula*). Anat. Rec. 248, 224–232.

Gamperl, A.K., Driedzic, W.R., 2009. Cardiovascular function and cardiac metabolism. In: Richards, J.G., Farrell, A.P., Brauner, C.J. (Eds.), Hypoxia. Fish Physiology. vol. 27. Academic Press, Burlington, pp. 301–360.

Gamperl, A.K., Farrell, A.P., 2004. Cardiac plasticity in fishes: environmental influences and intraspecific differences. J. Exp. Biol. 207, 2537–2550.

Gamperl, A.K., Shiels, H.A., 2013. Cardiovascular system. In: Evans, D.H., Claiborne, J.B., Currie, S. (Eds.), The Physiology of Fishes, fourth ed. CRC Press, Boca Raton. 491 pp.

Gamperl, A.K., Pinder, A., Boutilier, R., 1994a. Effect of coronary ablation and adrenergic stimulation on *in vivo* cardiac performance in trout (*Oncorhynchus mykiss*). J. Exp. Biol. 186, 127–143.

Gamperl, A.K., Pinder, A., Grant, R., Boutilier, R., 1994b. Influence of hypoxia and adrenaline administration on coronary blood flow and cardiac performance in seawater rainbow trout (*Oncorhynchus mykiss*). J. Exp. Biol. 193, 209–232.

Gamperl, A.K., Wilkinson, M., Boutilier, R.G., 1994c. Beta-adrenoreceptors in the trout (*Oncorhynchus mykiss*) heart: characterization, quantification, and effects of repeated catecholamine exposure. Gen. Comp. Endocrinol. 95, 259–272.

Gamperl, A.K., Axelsson, M., Farrell, A.P., 1995. Effects of swimming and environmental hypoxia on coronary blood flow in rainbow trout. Am. J. Physiol. 269, R1258–R1266.

Gamperl, A.K., Vijayan, M.M., Pereira, C., Farrell, A.P., 1998. ß-receptor and stress protein 70 expression in the hypoxic myocardium of rainbow trout and chinook salmon. Am. J. Physiol. 43, 428–436.

Gamperl, A.K., Swafford, B.L., Rodnick, K.J., 2011. Elevated temperature, pr se, does not limit the ability of rainbow trout to increase stroke volume. J. Therm. Biol. 36, 7–14.

Gaskell, W.H., 1916. The Involuntary Nervous System. Longmans, London, UK.

Gesser, H., 1985. Effects of hypoxia and acidosis on fish heart performance. In: Gilles, R. (Ed.), Respiration and Metabolism. Springer-Verlag, Berlin, Germany, pp. 402–410.

Gesser, H., Jorgensen, E., 1982. pHi, contractility and Ca-balance under hypercapnic acidosis in the myocardium of different vertebrate species. J. Exp. Biol. 96, 405–412.

Gesser, H., Poupa, O., 1983. Acidosis and cardiac muscle contractility: comparative aspects. Comp. Biochem. Physiol. A 76, 559–566.

Gesser, H., Andresen, P., Brams, P., Sund-Laursen, J., 1982. Inotropic effects of adrenaline on the anoxic or hypercapnic myocardium of rainbow trout and eel. J. Comp. Physiol. 147, 123–128.

Gibbins, I., 1994. Comparative anatomy and evolution of the autonomic nervous system. In: Nilsson, S., Holmgren, S. (Eds.), Comparative Physiology and Evolution of the Autonomic Nervous System. Harwood Academic Publishers, Switzerland, pp. 1–67.

Gillis, Johnson, 2017. Cardiac remodelling, protection and regeneration. In: Gamperl, A.K., Gillis, T.E., Farrell, A.P., Brauner, C.J. (Eds.), Fish Physiology. In: The Cardiovascular System: Development, Plasticity and Physiological Responses, vol. 36B. Academic Press, San Diego (In press).

Gillis, T.E., Regan, M.D., Cox, G.K., Harter, T.S., Brauner, C.J., Richards, J.G., Farrell, A.P., 2015. Characterizing the metabolic capacity of the anoxic hagfish heart. J. Exp. Biol. 218, 3754–3761.

Goehler, L.E., Finger, T.E., 1996. Visceral afferent and efferent columns in the spinal cord of the teleost, *Ictalurus punctatus*. J. Comp. Neurol. 371, 437–447.

Gold, D., Loirat, T., Farrell, A.P., 2015. Cardiorespiratory responses to haemolytic anaemia in rainbow trout *Oncorhynchus mykiss*. J. Fish Biol. 87, 848–859.

Gollock, M.J., Currie, S., Petersen, L.H., Gamperl, A.K., 2006. Cardiovascular and haematological responses of Atlantic cod (*Gadus morhua*) to acute temperature increase. J. Exp. Biol. 209, 2961–2970.

Goolish, E.M., 1987. Cold-acclimation increases the ventricle size of carp, *Cyprinus carpio*. J. Therm. Biol. 12, 203–206.

Graham, M.S., Farrell, A.P., 1985. The seasonal intrinsic cardiac performance of a marine teleost. J. Exp. Biol. 118, 173–183.

Graham, M.S., Farrell, A.P., 1989. The effect of temperature acclimation and adrenaline on the performance of a perfused trout heart. Physiol. Zool. 62, 38–61.

Graham, M.S., Farrell, A.P., 1990. Myocardial oxygen consumption in trout acclimated to 5°C and 15°C. Physiol. Zool. 63, 536–554.

Graham, M.S., Farrell, A.P., 1992. Environmental influences on cardiovascular variables in rainbow trout, *Oncorhynchus mykiss* (Walbaum). J. Fish Biol. 41, 851–858.

Graham, M.S., Wood, C.M., Turner, J.D., 1982. The physiological responses of the rainbow trout to strenous exercise: Interactions of water hardness and environmental acidity. Can. J. Zool. 60, 3153–3164.

Gräns, A., Axelsson, M., Pitsillides, K., Olsson, C., Höjesjö, J., Kaufman, R.C., Cech Jr., J.J., 2009. A fully implantable multi-channel biotelemetry system for measurement of blood flow and temperature: a first evaluation in the green sturgeon. Hydrobiologia 619, 11–25.

Gräns, A., Olsson, C., Pitsillides, K., Nelson, H.E., Cech Jr., J.J., Axelsson, M., 2010. Effects of feeding on thermoregulatory behaviours and gut blood flow in white sturgeon (*Acipenser transmontanus*) using biotelemetry in combination with standard techniques. J. Exp. Biol. 213, 3198–3206.

Grant, R.T., Regnier, M., 1926. The comparative anatomy of the cardiac coronary vessels. Heart 14, 285–317.

Greene, C.W., 1902. Contributions to the physiology of the California hagfish, *Polistotrema stouti*—II. The absence of regulative nerves for the systemic heart. Am. J. Physiol. 6, 318–324.

Greer Walker, M., Santer, R.M., Benjamin, M., Norman, D., 1985. Heart structure of some deep-sea fish (Teleostei-Macrouridae). J. Zool. *(London)* 205, 75–89.

Gregory, J.A., Graham, J.B., Cech Jr., J.J., Dalton, N., Michaels, J., Lai, N.C., 2004. Pericardial and pericardioperitoneal canal relationships to cardiac function in the white sturgeon (*Acipensor transmontanus*). Comp. Biochem. Physiol. A 138, 203–213.

Grimes, A.C., Kirby, M.L., 2009. The outflow tract of the heart in fishes: anatomy, genes and evolution. J. Fish Biol. 74, 983–1036.

Halpern, M.H., May, M.M., 1958. Phylogenetic study of the extracardiac arteries to the heart. Am. J. Anat. 102, 469–480.

Hansen, C.A., Sidell, B.D., 1983. Atlantic hagfish cardiac muscle: metabolic basis of tolerance to anoxia. Am. J. Physiol. 244, R356–R362.

Hanson, L.M., Farrell, A.P., 2007. The hypoxic threshold for maximum cardiac performance in rainbow trout *Oncorhynchus mykiss* (Walbaum) during simulated exercise conditions at 18°C. J. Fish Biol. 71, 926–932.

Hanson, L.M., Obradovich, S., Mouniargi, J., Farrell, A.P., 2006. The role of adrenergic stimulation in maintaining maximum cardiac performance in rainbow trout (*Oncorhynchus mykiss*) during hypoxia, hyperkalemia and acidosis at 10°C. J. Exp. Biol. 209, 2442–2451.

Hanson, L.M., Baker, D.W., Kuchel, L.J., Farrell, A.P., Val, A.L., Brauner, C.J., 2009. Intrinsic mechanical properties of the perfused armoured catfish heart with special reference to the effects of hypercapnic acidosis on maximum cardiac performance. J. Exp. Biol. 212, 1270–1276.

Haverinen, J., Egginton, S., Vornanen, M., 2014. Electrical excitation of the heart in a basal vertebrate, the European river lamprey (*Lampetra fluviatilis*). Physiol. Biochem. Zool. 87, 817–828.

Hemmingsen, E.A., Douglas, E.L., 1977. Respiratory and circulatory adaptations to the absence of hemoglobin in Chaenichthyid fishes. In: Llano, G.A. (Ed.), Adaptations Within Antarctic Ecosystems. Smithsonian Institute, Washington, DC, pp. 479–487.

Hemmingsen, E.A., Douglas, E.L., Johansen, K., Millard, R.W., 1972. Aortic blood flow and cardiac output in the hemoglobin-free fish *Chaenocephalus aceratus*. Comp. Biochem. Physiol. 43A, 1045–1051.

Hipkins, S.F., 1985. Adrenergic responses of the cardiovascular system of the eel, *Anguilla australis, in vivo*. J. Exp. Zool. 235, 7–20.

Hipkins, S.F., Smith, D.G., 1983. Cardiovascular events associated with spontaneous apnea in the Australian short-finned eel, *Anguilla australis*. J. Exp. Zool. 227, 339–348.

Hipkins, S.F., Smith, D.G., Evans, B.K., 1986. Lack of adrenergic control of dorsal aortic blood pressure in the resting eel, *Anguilla australis*. J. Exp. Zool. 238, 155–166.

Ho, Y., Shau, Y., Tsai, H., Lin, L., Huang, P., Hsieh, F., 2002. Assessment of zebrafish cardiac performance using Doppler echocardiographic and power angiography. Ultrasound Med. Biol. 28, 1137–1143.

Holeton, G.F., 1970. Oxygen uptake and circulation by a hemoglobinless Antarctic fish, (*Chaenocephalus aceratus*, Lonnberg) compared with three red-blooded Antarctic fish. Comp. Biochem. Physiol. 34, 457–471.

Holeton, G.F., 1975. Respiration and morphometrics of hemoglobinless Antarctic ice-fish. In: Cech, J.J., Bridges, D.W., Horton, D.D. (Eds.), Respiration of Marine Organisms. Proc Marine Sect 1st Maine Biol Sci Symp Portland Me. Trigum Publs., Portland, pp. 198–211.

Houlihan, D.F., Agnisola, C., Lyndon, A.R., Gray, C., Hamilton, N.M., 1988. Protein synthesis in a fish heart: responses to increased power output. J. Exp. Biol. 137, 565–587.

Howe, K., et al., 2013. The zebrafish reference genome sequence and its relationship to the human genome. Nature 496, 498–503.

Hu, N., Yost, H.J., Clark, E.B., 2001. Cardiac morphology and blood pressure in the adult zebrafish. Anat. Rec. 264, 1–12.

Hughes, G.M., Peyraud, C., Peyraud-Waitzenneger, M., Soulier, P., 1982. Physiological evidence for the occurrence of pathways shunting blood away from the secondary lamellae of eel gills. J. Exp. Biol. 98, 277–288.

Icardo, 2017. Heart morphology and anatomy. In: Gamperl, A.K., Gillis, T.E., Farrell, A.P., Brauner, C.J. (Eds.), Fish Physiology. In: The Cardiovascular System: Morphology, Control and Function, vol. 36A. Academic Press, San Diego, pp. 1–54.

Icardo, J.M., Colvee, E., Lauriano, E.R., Fudge, D.S., Glover, C.N., Zaccone, G., 2016. Morphological analysis of the hagfish heart. I. The ventricle, the arterial connection and the ventral aorta. J. Morph. 277, 326–340.

Imbrogno, Cerra, 2017. Hormonal and autacoid control of cardiac function. In: Gamperl, A.K., Gillis, T.E., Farrell, A.P., Brauner, C.J. (Eds.), Fish Physiology. In: The Cardiovascular System: Morphology, Control and Function, vol. 36A. Academic Press, San Diego, pp. 265–315.

Jain, K.E., Farrell, A.P., 2003. Influence of seasonal temperature on the repeat swimming performance of rainbow trout *Oncorhynchus mykiss*. J. Exp. Biol. 206, 3569–3579.

Jain, K.E., Birtwell, I.K., Farrell, A.P., 1998. Repeat swimming performance of mature sockeye salmon following a brief recovery period: a proposed measure of fish health and water quality. Can. J. Zool. 76, 1488–1496.

Jensen, D., 1961. Cardioregulation in an aneural heart. Comp. Biochem. Physiol. 2, 181–201.

Jensen, D., 1965. The aneural heart of the hagfish. Ann. N. Y. Acad. Sci. 127, 443–458.

Joaquim, N., Wagner, G.N., Gamperl, A.K., 2004. Cardiac function and critical swimming speed of the winter flounder (*Pleuronectes americanus*) at two temperatures. Comp. Biochem. Physiol. A 138, 277–285.

Johansen, K., Burggren, W.W., 1980. Cardiovascular function in the lower vertebrates. In: Bourne, G.H. (Ed.), Heart and Heartlike Organs. Academic Press, New York, pp. 61–117.

Johansen, K., Gesser, H., 1986. Fish cardiology: structural, haemodynamic, electromechanical and metabolic aspects. In: Nilsson, S., Holmgren, S. (Eds.), Fish Physiology: Recent Advances. Croom Helm, London, pp. 71–85.

Johansen, K., Lenfant, C., Hanson, D., 1968. Cardiovascular dynamics in lungfishes. Z. vergl. Physiol. 59, 157–186.

Johnsson, M., Axelsson, M., 1996. Control of the systemic heart and the portal heart of *Myxine glutinosa*. J. Exp. Biol. 199, 1429–1434.

Johnston, I.A., Fitch, N., Zummo, G., Wood, R.E., Harrison, P., Tota, B., 1983. Morphometric and ultrastructural features of the ventricular myocardium of the haemoglobinless icefish *Chaenocephalus aceratus*. Comp. Biochem. Physiol. 76, 475–480.

Jones, D.R., Randall, D.J., 1978. The respiratory and circulatory systems during exercise. In: Hoar, W.S., Randall, D.J. (Eds.), Fish Physiology. Academic Press, New York, pp. 425–501.

Jones, D.R., Langille, B.W., Randall, D.J., Shelton, G., 1974. Blood flow in dorsal and ventral aortae of the cod, *Gadus morhua*. Am. J. Physiol. 226, 90–95.

Jones, D.R., Brill, R.W., Bushnell, P.G., 1993. Ventricular and arterial dynamics of anesthetized and swimming tuna. J. Exp. Biol. 182, 97–112.

Jones, D.R., Perbhoo, K., Braun, M.H., 2005. Necrophysiological determination of blood pressure in fishes. Naturwissenschaften 92, 582–585.

Joyce, W., Gesser, H., Bayley, M., Wang, T., 2015. Anoxia and acidosis tolerance of the heart in an air-breathing fish (*Pangasianodon hypophthalmus*). Physiol. Biochem. Zool. 88, 648–659.

Kabashi, E., Brustein, E., Champagne, M., Drapeau, P., 2011. Zebrafish models for the functional genomics of neurogenic disorders. Biochim. Biophys. Acta 1812, 335–345.

Kanwal, J.S., Caprio, J., 1987. Central projections of the glossopharyngeal and vagal nerves in the channel catfish, *Ictalurus punctatus*: clues to differential processing of visceral inputs. J. Comp. Neurol. 264, 216–230.

Karlsson, A., Eliason, E.J., Mydland, L.T., Farrell, A.P., Kiessling, A., 2006. Postprandial changes in plasma free amino acid levels obtained simultaneously from the hepatic portal vein and the dorsal aorta in rainbow trout (*Oncorhynchus mykiss*). J. Exp. Biol. 209, 4885–4894.

Keen, J.E., Farrell, A.P., 1994. Maximum prolonged swimming speed and maximum cardiac performance of rainbow trout, *Oncorhynchus mykiss*, acclimated to two different water temperatures. Comp. Biochem. Physiol. A 108, 287–295.

Keen, A.N., Gamperl, A.K., 2012. Blood oxygenation and cardiorespiratory function in steelhead trout (*Oncorhynchus mykiss*) challenges with an acute temperature increase and zatebradine-induced bradycardia. J. Therm. Biol 37, 201–210.

Keen, J.E., Vianzon, D.-M., Farrell, A.P., Tibbits, G.F., 1993. Thermal acclimation alters both adrenergic sensitivity and adrenoceptor density in cardiac tissue of rainbow trout. J. Exp. Biol. 181, 27–47.

Keen, J.K., Aota, S., Brill, R.W., Farrell, A.P., Randall, D.J., 1995. Cholinergic and adrenergic regulation of resting heart rate and ventral aortic pressure in two species of tropical tunas, *Katsuwonus pelamis* and *Thunnus albacares*. Can. J. Zool. 73, 1681–1688.

Keen, A.N., Fenna, A.J., McConnell, J.C., Sherratt, M.J., Gardner, P., Shiels, H.A., 2015. The dynamic nature of hypertrophic and fibrotic remodeling of the fish ventricle. Front. Physiol. 6, 427.

Kiceniuk, J.W., Jones, D.R., 1977. The oxygen transport system in trout (*Salmo gairdneri*) during sustained exercise. J. Exp. Biol. 69, 247–260.

Klaiman, J.M., Fenna, A.J., Shiels, H.A., Macri, J., Gillis, T.E., 2011. Cardiac remodeling in fish: strategies to maintain heart function during temperature change. PLoS One 6, e24464.

Koelle, G.B., 1962. The use of histochemistry in pharmacological studies. Biochem. Pharmacol. 9, 5–14.

Kolok, A.S., Farrell, A.P., 1994. Individual variation in the swimming performance and cardiac performance of northern squawfish, *Ptychocheilus oregonensis*. Physiol. Zool. 67, 706–722.

Kolok, A.S., Spooner, R.M., Farrell, A.P., 1993. The effect of exercise on the cardiac output and blood flow distribution of the largescale sucker *Catostomus macrocheilus*. J. Exp. Biol. 183, 301–321.

Korajoki, H., Vornanen, M., 2012. Expression of SERCA and phospholamban in rainbow trout (*Oncorhynchus mykiss*) heart: comparison of atrial and ventricular tissue and effects of thermal acclimation. J. Exp. Biol. 215, 1162–1169.

Lague, S.L., Speers-Roesch, B., Richards, J.G., Farrell, A.P., 2012. Exceptional cardiac anoxia tolerance in tilapia (*Oreochromis hybrid*). J. Exp. Biol. 215, 1354–1365.

Lai, N.C., Graham, J.B., Lowell, W.R., Laurs, R.M., 1987. Pericardial and vascular pressures and blood flow in the albacore tuna, *Thunnus alalunga*. Exp. Biol. 46, 187–192.

Lai, N.C., Graham, J.B., Bhargava, V., Lowell, W.R., Shabetai, R., 1989a. Branchial blood flow distribution in the blue shark (*Prionace glauca*) and the leopard shark (*Triakis semifasciata*). Exp. Biol. 48, 273–278.

Lai, N.C., Graham, J.B., Lowell, W.R., Shabetai, R., 1989b. Elevated pericardial pressure and cardiac output in the leopard shark, *Triakis semifasciata*, during exercise: the role of the pericardioperitoneal canal. J. Exp. Biol. 23, 155–162.

Lai, N.C., Graham, J.B., Bhargava, V., Shabetai, R., 1996. Mechanisms of venous return and ventricular filling in elasmobranch fishes. Am. J. Physiol. 270, H1766–H1771.

Lai, N.C., Korsmeyer, K.E., Katz, S., Holts, D.B., Laughlin, L., Graham, J.B., 1997. Hemodynamics and blood respiratory properties in the shortfin mako shark, *Isurus oxyrinchus*. Copeia 1997, 424–428.
Lai, N.C., Graham, J.B., Dalton, N., Shabetai, R., Bhargava, V., 1998. Echocardiographic and hemodynamic determinations of the ventricular filling pattern in some teleost fishes. Physiol. Zool. 71, 157–167.
Lai, N.C., Dalton, N., Lai, Y.Y., Kwong, C., Rasmussen, R., Holts, D., Graham, J.D., 2004. A comparative echocardiographic assessment of ventricular function in five species of sharks. Comp. Biol. Physiol. A 137, 5005–5521.
Langley, J.N., 1921. The Autonomic Nervous System Part I. W Heffer and Sons Ltd, Cambridge, UK.
Laurent, P., 1962. Contribution etude morphologique et physiologique de l'innervation du coeur des teleosteens. Arch. Anat. Microsc. Morphol. Exp. 51, 337–458.
Laurent, P., Holmgren, S., Nilsson, S., 1983. Nervous and humoral control of the fish heart: structure and function. Comp. Biochem. Physiol. A 76, 525–542.
Lazar, G., Szabo, T., Libouban, S., Ravaille-Veron, M., Toth, P., Brandle, K., 1992. Central projections and motor nuclei of the facial, glossopharyngeal, and vagus nerves in the mormyrid fish *Gnathonemus petersii*. J. Comp. Neurol. 325, 343–358.
Lee, C.G., Farrell, A.P., Lotto, A., Hinch, S.G., Healey, M.C., 2003a. Excess post-exercise oxygen consumption in adult sockeye (*Oncorhynchus nerka*) and coho (*O. kisutch*) salmon following critical speed swimming. J. Exp. Biol. 206, 3253–3260.
Lee, C.G., Farrell, A.P., Lotto, A., MacNutt, M.J., Hinch, S.G., Healey, M.C., 2003b. The effect of temperature on swimming performance and oxygen consumption in adult sockeye (*Oncorhynchus nerka*) and coho salmon (*O. kisutch*) salmon stocks. J. Exp. Biol. 206, 3239–3251.
Leite, C.A., Taylor, E.W., Guerra, C.D., Florindo, L.H., Belao, T., Rantin, F.T., 2009. The role of the vagus nerve in the generation of cardiorespiratory interactions in a neotropical fish, the pacu, *Piaractus mesopotamicus*. J. Comp. Physiol. A 195, 721–731.
Lillywhite, H.B., Zippel, K.C., Farrell, A.P., 1999. Resting and maximal heart rates in ectothermic vertebrates. Comp. Biochem. Physiol. A 124, 369–384.
Maldanis, L., Carvalho, M., Almeida, M.R., Freitas, F.I., Gomes de Andrade, J.A.F., Nunes, R.S., Rochitte, C.E., Poppi, R.J., Freitas, R.O., Rodrigues, F., Siljeström, S., Lima, F.A., Galante, D., Carvalho, I.S., Perez, C.A., Rodrigues de Carvalho, M., Bettini, J., Fernandez, V., Xavier-Neto, J., 2016. Heart fossilization is possible and informs the evolution of cardiac outflow tract in vertebrates. eLife 5, e14698.
Mandic, M., Todgham, A.E., Richards, J.G., 2009. Mechanisms and evolution of hypoxia tolerance in fish. Proc. R. Soc. B 276, 735–744.
McKenzie, D.J., Wong, S., Randall, D.J., Egginton, S., Taylor, E.W., Farrell, A.P., 2004. The effects of sustained exercise and hypoxia upon oxygen tensions in the red muscle of rainbow trout. J. Exp. Biol. 207, 3629–3637.
McKenzie, D.J., Skov, P.V., Taylor, E.W., Wang, T., Steffensen, J.F., 2009. Abolition of reflex bradycardia by cardiac vagotomy has no effect on the regulation of oxygen uptake by Atlantic cod in progressive hypoxia. Comp. Biochem. Physiol. A 153, 332–338.
Mendonça, P.C., Gamperl, A.K., 2010. The effects of acute changes in temperature and oxygen availability on cardiac performance in winter flounder (*Pseudopleuronectes americanus*). Comp. Biochem. Physiol. A 155 (2), 245–252. http://dx.doi.org/10.1016/j.cbpa.2009.11.006. Epub 2009 Nov 12.
Mercier, C., Axelsson, M., Imbert, N., Claireaux, G., Lefrancois, C., Altimiras, J., Farrell, A.P., 2002. *In vitro* cardiac performance in triploid brown trout at two acclimation temperatures. J. Fish Biol. 60, 117–133.

Metcalfe, W.K., Myers, P.Z., Trevarrow, B., Bass, M.B., Kimmel, C.B., 1990. Primary neurons that express the L2/HNK-1 carbohydrate during early development in the zebrafish. Development 110, 491–504.

Milligan, C.L., Farrell, A.P., 1991. Lactate utilization by an *in situ* perfused trout heart: effects of workload and blockers of lactate transport. J. Exp. Biol. 155, 357–373.

Milligan, C.L., Wood, C.M., 1986. Intracellular and extracellular acid-base status and H+ exchange with the environment after exhaustive exercise in the rainbow trout. J. Exp. Biol. 123, 93–121.

Morita, Y., Finger, T.E., 1987. Topographic representation of the sensory and motor roots of the vagus nerve in the medulla of goldfish, *Carassius auratus*. J. Comp. Neurol. 264, 231–249.

Morris, J.L., Nilsson, S., 1994. The circulatory system. In: Nilsson, S., Holmgren, S. (Eds.), Comparative Physiology and Evolution of the Autonomic Nervous System. Harwood Academic Publishers, Switzerland, pp. 193–246.

Mustafa, T., Agnisola, C., 1994. Vasoactivity of prostanoids in the trout (*Oncorhynchus mykiss*) coronary system: modification by noradrenaline. Fish Physiol. Biochem. 13, 249–261.

Mustafa, T., Agnisola, C., 1998. Vasoactivity of adenosine in the trout (*Oncorhynchus mykiss*) coronary system: involvement of nitric oxide and interaction with noradrenaline. J. Exp. Biol. 201, 3075–3083.

Mustafa, T., Agnisola, C., Tota, B., 1992. Myocardial and coronary effects of exogenous arachidonic acid on the isolated and perfused heart preparation and its metabolism in the heart of trout (*Oncorhynchus mykiss*). Comp. Biochem. Physiol. C 103, 163–167.

Mustafa, T., Agnisola, C., Hansen, J.K., 1997. Evidence for NO-dependent vasodilation in the trout (*Oncorhynchus mykiss*) coronary system. J. Comp. Physiol. B 167, 98–104. Neely et al. (1973, 1975) cited in Table 4.

Neely, J.R., Rovetto, M.J., Whitmer, J.T., Morgan, H.E., 1973. Effects of ischemia on function and metabolism of the isolated working rat heart. Am. J. Physiol. 225, 651–658.

Neely, J.R., Whitmer, J.T., Rovetto, M.J., 1975. Effect of coronary blood flow on glycolytic flux and intracellular pH in isolated rat hearts. Circ. Res. 37, 733–741.

Newton, C.M., Stoyek, M.R., Croll, R.P., Smith, F.M., 2014. Regional innervation of the heart in the goldfish, *Carassius auratus*: a confocal microscopy study. J. Comp. Neurol. 522, 456–478.

Nicol, J.A.C., 1952. Autonomic nervous systems in lower chordates. Biol. Rev. Camb. Philos. Soc. 27, 1–49.

Nilsson, S., 1983. Autonomic Nerve Function in the Vertebrates. vol. 13. Springer-Verlag, Berlin.

Nilsson, S., 2011. Comparative anatomy of the autonomic nervous system. Auton. Neurosci. 165, 3–9.

Nilsson, S., Holmgren, S. (Eds.), 1994. Comparative Physiology and Evolution of the Autonomic Nervous System. Harwood Academic Publishers, Chur, Switzerland.

O'Donogue, C.H., Abbot, E., 1928. The blood vascular system of the spiny dogfish *Squalus acanthias* and *Squalus suckleyii*. Trans. R. Soc. Edinb. Earth Sci. 55, 823–890.

Olson, K.R., Farrell, A.P., 2006. The cardiovascular system. In: Evans, D.H., Claiborne, J.B. (Eds.), Physiology of Fishes, third ed. CRC Press, Boca Raton, pp. 119–152.

Overgaard, J., Gesser, H., 2004. Force development, energy state and ATP production of cardiac muscle from turtles and trout during normoxia and severe hypoxia. J. Exp. Biol. 207, 1915–1924.

Overgaard, J., Stecyk, J.A.W., Gesser, H., Wang, T., Farrell, A.P., 2004. Effects of temperature and anoxia upon the performance of *in situ* perfused trout hearts. J. Exp. Biol. 205, 655–665.

Parker, T.J., 1886. On the blood vessels of *Mustelus antarticus*. Phil. Trans. R. Soc. Lond. B 177, 685–732.

Parker, G.H., Davis, F.K., 1899. The blood vessels of the heart in *Carcharias*. Raja and Amia. Proc. Boston Soc. Nat. Hist. 29, 163–178.

Penney, C.M., Nash, G.W., Gamperl, A.K., 2014. Cardiorespiratory responses of seawater acclimated adult Arctic Charr (*Salvelinus alpinus*) and Atlantic Salmon (*Salmo salar*) to an acute temperature increase. Can. J. Fish. Aquat. Sci. 71, 1096–1105.

Petersen, L.H., Gamperl, A.K., 2010. Effect of acute and chronic hypoxia on the swimming performance, metabolic capacity and cardiac function of Atlantic salmon. J. Exp. Biol. 213, 808–819.

Petersson, K., Nilsson, S., 1980. Drug induced changes in cardiovascular parameters in the Atlantic cod, *Gadus morhua*. J. Comp. Physiol. 137B, 131–138.

Peyraud-Waitzenegger, M., Soulier, P., 1989. Ventilatory and circulatory adjustments in the European eel (*Anguilla anguilla* L.) exposed to short-term hypoxia. Exp. Biol. 48, 107–122.

Peyraud-Waitzenegger, M., Barthelemy, L., Peyraud, C., 1980. Cardiovascular and ventilatory effects of catecholamines in unrestrained eels (*Anguilla anguilla*). A study of seasonal changes in reactivity. J. Comp. Physiol. 138B, 367–375.

Pieperhoff, S., Bennett, W., Farrell, A.P., 2009. The intercellular organization of the two muscular systems in the adult salmonid heart, the compact and the spongy myocardium. J. Anat. 215, 536–547.

Piiper, J., Meyer, M., Worth, H., Willmer, H., 1977. Respiration and circulation during swimming activity in the dogfish *Scyliorhinus stellaris*. Respir. Physiol. 30, 221–239.

Poupa, O., Johansen, K., 1975. Adaptive tolerance of fish myocardium to hypercapnic acidosis. Am. J. Physiol. 228, 684–688.

Poupa, O., Lindstrom, L., 1983. Comparative and scaling aspects of heart and body weights with reference to blood supply of cardiac fibres. Comp. Biochem. Physiol. 76A, 413–421.

Poupa, O., Gesser, H., Johansen, K., 1978. Myocardial inotrophy of CO_2 in water- and air-breathing vertebrates. Am. J. Physiol. 234, R155–R157.

Poupa, O., Lindstrom, L., Maresca, A., Tota, B., 1981. Cardiac growth, myoglobin, proteins and DNA in developing tuna (*Thunnus thynnus thynnus*). Comp. Biochem. Physiol. 70A, 217–222.

Powell, M.D., Gamperl, A.K., 2016. Effects of *Loma morhua* (microsporidia) infection on the cardiorespiratory performance of Atlantic cod *Gadus morhua* (L). J. Fish Dis. 39, 189–204.

Ramos, C., 2004. The structure and ultrastructure of the sinus venosus in the mature dogfish (*Scyliorhinus canicula*): the endocardium, the epicardium and the subepicardial space. Tissue Cell 36, 399–407.

Randall, D.J., 1970. The circulatory system in fish physiology. In: Hoar, W.S., Randall, D.J. (Eds.), Fish Physiology. Academic Press, New York, pp. 132–172.

Reeves, R.B., 1963. Energy cost of work in aerobic and anaerobic turtle heart muscle. Am. J. Physiol. 205, 17–22. Note: cited as Reeves (1964a) in Table 4.

Rovainen, C.M., 1979. Neurobiology of lampreys. Physiol. Rev. 59, 1007–1077.

Ruben, J.A., Bennett, A.F., 1981. Intense exercise, bone structure, and blood calcium levels in vertebrates. Nature 291, 411–413.

Rummer, J.L., McKenzie, D.J., Innocenti, A., Supuran, C.T., Brauner, C.J., 2013. Root effect hemoglobin may have evolved to enhance general tissue oxygen delivery. Science 340, 1327–1329.

Saetersdal, T.S., Sorensen, E., Myklebust, R., Helle, K.B., 1975. Granule containing cells and fibres in the sinus venosus of elasmobranchs. Cell Tissue Res. 163, 471–490.

Saito, T., 1973. Effects of vagal stimulation on the pacemaker action potentials of carp heart. Comp. Biochem. Physiol. 44A, 191–199.

Saito, T., Tenma, K., 1976. Effects of left and right vagal stimulation on excitation and conduction of the carp heart (*Cyprinus carpio*). J. Comp. Physiol. 111, 39–53.

Sanchez-Quintana, D., Hurle, J.M., 1987. Ventricular myocardial architecture in marine fishes. Anat. Rec. 217, 263–273.

Sandblom, E., Axelsson, M., 2006. Venous hemodynamic responses to acute temperature increase in the rainbow trout. Am. J. Physiol. 292, R2292–R2298.

Sandblom, E., Axelsson, M., 2011. Autonomic control of circulation in fish: a comparative view. Auton. Neurosci. 165, 127–139.

Sandblom, Gräns, 2017. Form, function, and control of the vasculature. In: Gamperl, A.K., Gillis, T.E., Farrell, A.P., Brauner, C.J. (Eds.), Fish Physiology. In: The Cardiovascular System: Morphology, Control and Function, vol. 36A. Academic Press, San Diego, pp. 369–433.

Sandblom, E., Farrell, A.P., Altimiras, J., Axelsson, M., Claireaux, G., 2005. Cardiac preload and venous return in swimming sea bass (*Dicentrarchus labrax* L.). J. Exp. Biol. 208, 1927–1935.

Sandblom, E., Clark, T.D., Hinch, S.G., Farrell, A.P., 2009a. Sex-specific differences in cardiac control and haematology of sockeye salmon (*Oncorhynchus nerka*) approaching their spawning grounds. Am. J. Physiol. 297, R1136–R1143. Not in text.

Sandblom, E., Cox, G.K., Perry, S.F., Farrell, A.P., 2009b. The role of venous capacitance, circulating catecholamines and heart rate in the hemodynamic response to increased temperature and hypoxia in the dogfish. Am. J. Physiol. 296, R1547–R1556.

Santer, R.M., 1985. Morphology and innervation of the fish heart. Adv. Anat. Embryol. Cell Biol. 89, 1–102.

Santer, R.M., Greer Walker, M., 1980. Morphological studies on the ventricle of teleost and elasmobranch hearts. J. Zool. Lond. 190, 259–272.

Santer, R.M., Greer Walker, M.G., Emerson, L., Witthames, P.R., 1983. On the morphology of the heart ventricle in marine teleost fish (Teleostei). Comp. Biochem. Physiol. 76A, 453–457.

Saper, C.B., 2005. Editorial: an open letter to our readers on the use of antibodies. J. Comp. Neurol. 493, 477–478.

Saper, C.B., Sawchenko, P.E., 2003. Magic peptides, magic antibodies: guidelines for appropriate controls for immunohistochemistry. J. Comp. Neurol. 465, 161–163.

Satchell, G.H., 1971. Circulation in Fishes. Cambridge Monographs in Experimental Biology. Cambridge University Press, Cambridge, England.

Satchell, G.H., 1991. Physiology and Form of Fish Circulation. Cambridge University Press, UK.

Satchell, G.H., Hanson, D., Johansen, K., 1970. Differential blood flow through the afferent branchial arteries of the skate, *Raja rhina*. J. Exp. Biol. 52, 721–726.

Scarabello, M., Heigenhauser, G.J., Wood, C.M., 1991. The oxygen debt hypothesis in juvenile rainbow trout after exhaustive exercise. Respir. Physiol. 84, 245–259.

Sedmera, D., Reckova, M., deAlmeida, A., Sedmerova, M., Biermann, M., Volejnik, J., Sarre, A., Raddatz, E., McCarthy, R.A., Gourdie, R.G., et al., 2003. Functional and morphological evidence for a ventricular conduction system in zebrafish and Xenopus hearts. Am. J. Physiol. 284, H1152–H1160.

Seth, H., Axelsson, M., 2010. Sympathetic, parasympathetic and enteric regulation of the gastrointestinal vasculature in rainbow trout (*Oncorhynchus mykiss*) under normal and postprandial conditions. J. Exp. Biol. 213, 3118–3126.

Seth, H., Sandblom, E., Holmgren, S., Axelsson, M., 2008. Effects of gastric distension on the cardiovascular system in rainbow trout (*Oncorhynchus mykiss*). Am. J. Physiol. Regul. Integr. Comp. Physiol. 294, R1648–R1656.

Seth, H., Axelsson, M., Farrell, A.P., 2011. The circulation and metabolism of the gastrointestinal tract. In: Grosell, M., Farrell, A.P., Brauner, C.J. (Eds.), The Multifunctional Gut of Fish. In: Fish Physiology Series, vol. 30. Elsevier, San Diego, pp. 351–393.

Shiels, 2017. Cardiomyocyte morphology and physiology. In: Gamperl, A.K., Gillis, T.E., Farrell, A.P., Brauner, C.J. (Eds.), Fish Physiology. In: The Cardiovascular System: Morphology, Control and Function, vol. 36A. Academic Press, San Diego, pp. 55–98.

Shiels, H.A., Vornanen, M., Farrell, A.P., 2002. The force–frequency relationship in fish hearts—a review. Comp. Biochem. Physiol. A 132, 811–826.

Short, S., Butler, P.J., Taylor, E.W., 1977. The relative importance of nervous, humoral, and intrinsic mechanisms in the regulation of heart rate and stroke volume in the dogfish (*Scyliorhinus caniculii* L). J. Exp. Biol. 70, 77–92.

Sidell, B.D., O'Brien, K.M., 2006. When bad things happen to good fish: the loss of hemoglobin and myoglobin expression in Antarctic icefishes. J. Exp. Biol. 209, 1791–1802.

Simonot, D.L., Farrell, A.P., 2007. Cardiac remodelling in rainbow trout *Oncorhynchus mykiss* Walbaum in response to phenylhydrazine-induced anaemia. J. Exp. Biol. 210, 2574–2584.

Simonot, D.L., Farrell, A.P., 2009. Coronary vascular volume remodelling in rainbow trout *Oncorhynchus mykiss*. J. Fish Biol. 75, 1762–1772.

Small, S.A., Farrell, A.P., 1990. Vascular reactivity of the coronary artery in steelhead trout (*Oncorhynchus mykiss*). Comp. Biochem. Physiol. C 97, 59–63.

Small, S.A., MacDonald, C., Farrell, A.P., 1990. Vascular reactivity of the coronary artery in rainbow trout (*Oncorhynchus mykiss*). Am. J. Physiol. 258, R1402–R1410.

Small, J., Rottner, K., Hahne, P., Anderson, K.I., 1999. Visualising the actin cytoskeleton. Microsc. Res. Tech. 47, 3–17.

Speers-Roesch, B., Sandblom, E., Lau, G.Y., Farrell, A.P., Richards, J.G., 2010. Effects of environmental hypoxia on cardiac energy metabolism and performance in tilapia. Am. J. Physiol. 298, R104–R119.

Speers-Roesch, B., Brauner, C.J., Farrell, A.P., Hickey, A.J.R., Renshaw, G.M.C., Wang, Y.S., Richards, J.G., 2012a. Hypoxia tolerance in elasmobranchs. II. Cardiovascular function and tissue metabolic responses during progressive and relative hypoxia exposures. J. Exp. Biol. 215, 103–114.

Speers-Roesch, B., Richards, J.G., Brauner, C.J., Farrell, A.P., Hickey, A.J.R., Wang, Y.S., Renshaw, G.M.C., 2012b. Hypoxia tolerance in elasmobranchs. I. Critical oxygen tension as a measure of blood oxygen transport during hypoxia exposure. J. Exp. Biol. 215, 93–102.

Stecyk, 2017. Cardiovascular function under limiting oxygen conditions. In: Gamperl, A.K., Gillis, T.E., Farrell, A.P., Brauner, C.J. (Eds.), Fish Physiology. In: The Cardiovascular System: Development, Plasticity and Physiological Responses, vol. 36B. Academic Press, San Diego (In press).

Stecyk, J.A.W., Farrell, A.P., 2002. Cardiorespiratory responses of the common carp (*Cyprinus carpio*) to severe hypoxia at three acclimation temperatures. J. Exp. Biol. 205, 759–768.

Stecyk, J.A.W., Farrell, A.P., 2006. Regulation of the cardiorespiratory system of common carp (*Cyprinus carpio*) during severe hypoxia at three seasonal acclimation temperatures. Physiol. Biochem. Zool. 79, 614–627.

Stecyk, J.A.W., Stensløkken, K.-O., Farrell, A.P., Nilsson, G.E., 2004. Maintained cardiac pumping in anoxic crucian carp. Science 306, 77.

Steffensen, J.F., Farrell, A.P., 1998. Swimming performance, venous oxygen tension and cardiac performance of coronary-ligated rainbow trout, *Oncorhynchus mykiss*, exposed to progressive hypoxia. Comp. Biochem. Physiol. A 119, 585–592.

Steinhausen, M.F., Sandblom, E., Eliason, E.J., Verhille, C., Farrell, A.P., 2008. The effect of acute temperature increases on the cardiorespiratory performance of resting and swimming sockeye salmon (*Oncorhynchus nerka*). J. Exp. Biol. 211, 3915–3926.

Stensløkken, K.-O., Sundin, L., Renshaw, G.M.C., Nilsson, G.E., 2004. Adenosinergic and cholinergic control mechanisms during hypoxia in the epaulette shark (*Hemiscyllium ocellatum*), with emphasis on branchial circulation. J. Exp. Biol. 207, 4451–4461.

Stevens, E.D., Bennion, G.R., Randall, D.J., Shelton, G., 1972. Factors affectingarterial blood pressures and blood flow from the heart in intact, unrestrained ling cod, *Ophiodon elongatus*. Comp. Biochem. Physiol. 43, 681–695.

Stoyek, M.R., Croll, R.P., Smith, F.M., 2015. Intrinsic and extrinsic innervation of the heart in zebrafish (*Danio rerio*). J. Comp. Neurol. 523, 1683–1700.

Stoyek, M.R., Quinn, T.A., Croll, R.P., Smith, F.M., 2016. Zebrafish heart as a model to study the integrative autonomic control of pacemaker function. Am. J. Physiol. Heart Circ. Physiol. 311, H676–H688.

Szabo, T., Libouban, S., 1979. On the course and origin of cranial nerves in the teleost fish *Gnathonemus* determined by ortho- and retrograde horseradish peroxidase axonal transport. Neurosci. Lett. 11, 265–270.

Takei, Y., McCormick, S.D., 2013. Hormonal control of fish euryhalinity. In: McCormick, S.D., Farrell, A.P., Brauner, C. (Eds.), Euryhaline Fishes. Academic Press, New York, pp. 70–123.

Taylor, E.W., 1992. Nervous control of the heart and cardiorespiratory interactions. In: Hoar, W.S., Randall, D.J., Farrell, A.P. (Eds.), Fish Physiology. In: The Cardiovascular System, vol. XII, Pt. B. Academic Press, Toronto, pp. 343–387.

Taylor, E.W., 2011. Central control of cardiorespiratory interactions in fish. In: Farrell, A.P. (Ed.), Encyclopedia of Fish Physiology: from Genome to Environment: The Senses, Supporting Tissues, Reproduction, and Behavior. In: vol. 2. Academic Press, London, pp. 1178–1189.

Taylor, E.W., Wang, T., 2009. Control of the heart and of cardiorespiratory interactions in ectothermic vertebrates. In: Glass, M.L., Wood, S.C. (Eds.), Cardio-Respiratory Control in Vertebrates. Springer-Verlag, Berlin, pp. 285–315.

Taylor, E.W., Jordan, D., Coote, J.H., 1999. Central control of the cardiovascular and respiratory systems and their interactions in vertebrates. Physiol. Rev. 79, 855–916.

Taylor, E.W., Leite, C.A., Florindo, L.H., Belao, T., Rantin, F.T., 2009. The basis of vagal efferent control of heart rate in a neotropical fish, the pacu, *Piaractus mesopotamicus*. J. Exp. Biol. 212, 906–913.

Taylor, E.W., Leite, C.A., Sartori, M.R., Wang, T., Abe, A.S., Crossley 2nd, D.A., 2014. The phylogeny and ontogeny of autonomic control of the heart and cardiorespiratory interactions in vertebrates. J. Exp. Biol. 217, 690–703.

Tessadori, F., van Weerd, J.H., Burkhard, S.B., Verkerk, A.O., de Pater, E., Boukens, B.J., Vink, A., Christoffels, V.M., Bakkers, J., 2012. Identification and functional characterization of cardiac pacemaker cells in zebrafish. PLoS One 7, e47644.

Thorarensen, H., Farrell, A.P., 2006. Postprandial intestinal blood flow, metabolic rates, and exercise in chinook salmon (*Oncorhynchus tshawytscha*). Physiol. Biochem. Zool. 79, 688–694.

Thorarensen, H., McLean, E., Donaldson, E.M., Farrell, A.P., 1991. The blood vasculature of the gastrointestinal tract in chinook, *Oncorhynchus tshawytscha* (Walbaum), and coho, *O. kisutch* (Walbaum), salmon. J. Fish Biol. 38, 525–531.

Thorarensen, H., Gallaugher, P.E., Kiessling, A.K., Farrell, A.P., 1993. Intestinal blood flow in swimming chinook salmon *Oncorhynchus tshawytscha* and the effects of haematocrit on blood flow distribution. J. Biol. 179, 115–129.

Thorarensen, H., Gallaugher, P.E., Farrell, A.P., 1996a. Cardiac output in swimming rainbow trout, *Oncorhynchus mykiss*, acclimated to seawater. Physiol. Zool. 69, 139–153.

Thorarensen, H., Gallaugher, P.E., Farrell, A.P., 1996b. The limitations of heart rate as a predictor of metabolic rate in fish. J. Fish Biol. 49, 226–236.

Tota, B., 1978. Functional cardiac morphology and biochemistry in Atlantic bluefin tuna. In: Sharp, G.D., Dizon, A.E. (Eds.), The Physiological Ecology of Tunas. Academic Press, New York, pp. 89–112.

Tota, B., 1983. Vascular and metabolic zonation in the ventricular myocardium of mammals and fishes. Comp. Biochem. Physiol. A 76, 423–437.

Tota, B., 1989. Myoarchitecture and vascularization of the elasmobranch heart ventricle. J. Exp. Zool. 252, 122–135.

Tota, B., Cimini, V., Salvatore, G., Zummo, G., 1983. Comparative study of the arterial and lacunary systems of the ventricular myocardium of elasmobranch and teleost fishes. Am. J. Anat. 167, 15–32.

Tota, B., Acierno, R., Agnisola, C., 1991. Mechanical performance of the isolated and perfused heart of the haemoglobinless antarctic icefish *Chionodraco hamatus* (Liinnberg): effects of loading conditions and temperature. Phil. Trans. R. Soc. Lond. B 332, 191–198.

Tresguerres, M., Barott, K.L., Barron, M.E., Roa, J.N., 2014. Established and potential physiological roles of bicarbonate-sensing soluble adenylyl cyclase (sAC) in aquatic animals. J. Exp. Biol. 217, 663–672.

Tsukuda, H., Liu, B., Fujii, K.I., 1985. Pulsation rate and oxygen consumption of isolated hearts of the goldfish, *Carassius auratus,* acclimated to different temperatures. Comp. Biochem. Physiol. 82A, 281–283.

van Dijk, P.L.M., Wood, C.M., 1988. The effect of β-adrenergic blockade on the recovery process after strenuous exercise in the rainbow trout, *Salmo gairdneri* Richardson. J. Fish Biol. 32, 557–570.

Verhille, C., Anttila, K., Farrell, A.P., 2013. A heart to heart on temperature. Impaired temperature tolerance of triploid rainbow trout (*Oncorhynchus mykiss*) due to early cardiac collapse. Comp. Biochem. Physiol. A 64, 653–657.

Verhille, C.E., English, K.K., Cocherell, D.E., Farrell, A.P., Fangue, NA., 2016. High thermal tolerance of a rainbow trout population near its southern range limit suggests local thermal adjustment. Conserv. Physiol. 4, cow057–cow057.

Vornanen, M., 2016. The temperature dependence of electrical excitability in fish hearts. J. Exp. Biol. 219, 1941–1952.

Vornanen, M., 2017. Electrical excitability of the fish heart and its autonomic regulation. In: Gamperl, A.K., Gillis, T.E., Farrell, A.P., Brauner, C.J. (Eds.), Fish Physiology. In: The Cardiovascular System: Morphology, Control and Function, vol. 36A. Academic Press, San Diego, pp. 99–153.

Vornanen, M., Halinen, M., Haverinen, J., 2010. Sinoatrial tissue of crucian carp heart has only negative contractile responses to autonomic agonists. BMC Physiol. 10, 10.

Wagner, G.N., Hinch, S.G., Kuchel, L.J., Lotto, A., Jones, S.R.M., Patterson, D.A., Macdonald, J.S., Van der Kraak, G., Shrimpton, M., English, K.K., Larsson, S., Cooke, S.J., Healey, M.C., Farrell, A.P., 2005. Metabolic rates and swimming performance of adult Fraser River sockeye salmon (*Oncorhynchus nerka*) after a controlled infection with *Parvicapsula minibicornis*. Can. J. Fish. Aquat. Sci. 62, 2124–2133.

Wagner, G.N., Kuchel, L.J., Lotto, A., Patterson, D.A., Shrimpton, J.M., Hinch, S.G., Farrell, A.P., 2006. Routine and active metabolic rates of migrating adult wild sockeye salmon (*Oncorhynchus nerka* Walbaum) in seawater and freshwater. Physiol. Biochem. Zool. 79, 100–108.

Watters, K.W., Smith, L.S., 1973. Respiratory dynamics of the starry flounder, *Platichthys stellatus,* in response to low oxygen and high temperature. Mar. Biol. 19, 133–148.

Webber, D.M., Boutilier, R.G., Kerr, SR., 1998. Cardiac output as a predictor of metabolic rate in cod gadus morhua. J. Exp. Biol. 201, 2779–2789.

Wilson, C.M., Farrell, A.P., 2013. Pharmacological characterization of the heartbeat in an extant vertebrate ancestor, the Pacific hagfish, *Eptatretus stoutii.* Comp. Biochem. Physiol. A 164, 258–263.

Wilson, C.M., Stecyk, J.A.W., Couturier, C.S., Nilsson, G.E., Farrell, A.P., 2013. Phylogeny and effects of anoxia on hyperpolarization-activated cyclic nucleotide-gated channel gene expression in the heart of a primitive chordate, the Pacific hagfish (*Eptatretus stoutii*). J. Exp. Biol. 216, 4462–4472.

Wilson, C.M., Roa, J.N., Cox, G.K., Tresguerres, M., Farrell, A.P., 2016. Introducing a novel mechanism to control heart rate in ancestral hagfish. J. Exp. Biol. 219, 3227–3236.

Withington-Wray, D.J., Taylor, E.W., Metcalfe, J.D., 1987. The location and distribution of vagal preganglionic neurones in the hindbrain of lower vertebrates. In: Taylor, E.W. (Ed.), The Neurobiology of the Cardiorespiratory System. Manchester University Press, Manchester, UK, pp. 304–321.

Wood, C.M., 1974. A critical examination of the physical and adrenergic factors affecting blood flow through the gills of the rainbow trout. J. Exp. Biol. 60, 241–265.

Wood, C.M., Milligan, C.L., 1987. Adrenergic analysis of extracellular and intracellular lactate and H^+ dynamics after strenuous exercise in the starry founder, *Platichthys stellatus*. Physiol. Zool. 60, 69–81.

Wood, C.M., Perry, S.F., 1985. Respiratory, circulatory, and metabolic adjustments to exercise in fish. In: Gilles, R. (Ed.), Circulation, Respiration, and Metabolism. Springer-Verlag, Berlin, Germany, pp. 2–22.

Wood, C.M., Shelton, G., 1980. Cardiovascular dynamics and adrenergic responses of the rainbow trout *in vivo*. J. Exp. Biol. 87, 247–270.

Wood, C.M., McMahon, B.R., McDonald, D.G., 1977. An analysis of changes in blood pH following exhausting activity in the starry flounder (*Platichthys stellatus*). J. Exp. Biol. 69, 173–185.

Wood, C.M., Pieprzak, P., Trott, J.N., 1979. The influence of temperature and anaemia on the adrenergic and cholinergic mechanisms controlling heart rate in the rainbow trout. Can. J. Zool. 57, 2440–2447.

Yamauchi, A., 1980. Fine structure of the fish heart. In: Bourne, C.H. (Ed.), Hearts and Heartlike Organs. Academic Press, New York, pp. 119–148.

Zaccone, G., Mauceri, A., Maisano, M., Giannetto, A., Parrino, V., Fasulo, S., 2009. Distribution and neurotransmitter localization in the heart of the ray-finned fish, bichir (*Polypterus bichir bichir* Geoffroy St. Hilaire, 1802). Acta Histochem. 111, 93–103.

Zaccone, G., Mauceri, A., Maisano, M., Giannetto, A., Parrino, V., Fasulo, S., 2010. Postganglionic nerve cell bodies and neurotransmitter localization in the teleost heart. Acta Histochem. 112, 328–336.

Zaccone, G., Marino, F., Zaccone, D., 2011. Intracardiac neurons and neurotransmitters in fish. In: Farrell, A.P. (Ed.), Encyclopedia of Fish Physiology: From Genome to Environment. Academic Press, San Diego, USA, pp. 1067–1072.

Zaccone, D., Grimes, A.C., Farrell, A.P., Dabrowski, K., Marino, F., 2012. Morphology, innervation and its phylogenetic step in the heart of the longnose gar *Lepisosteus osseus*. Acta Zool. 93, 381–389.

5

HORMONAL AND AUTACOID CONTROL OF CARDIAC FUNCTION

SANDRA IMBROGNO[1]
MARIA C. CERRA

University of Calabria, Cosenza, Italy
[1]Corresponding author: sandra.imbrogno@unical.it

This chapter provides an overview of our current knowledge of humoral regulation of fish heart function. The influence exerted by classic cardiac hormones (i.e., catecholamines, angiotensin II, and natriuretic peptides), and the more recently discovered chromogranin A-derived fragments, are discussed. The role of local, i.e., autacoid modulators (the gasotransmitters carbon

The Cardiovascular System: Morphology, Control and Function, Volume 36A
FISH PHYSIOLOGY

DOI: http://dx.doi.org/10.1016/bs.fp.2017.05.001

monoxide, hydrogen sulfide, and nitric oxide), is also summarized. Endocrine and autacoid signaling represent extrinsic and intrinsic regulatory networks, respectively, that are fundamental for preserving cardiac function under developmental and environmental challenges, the latter often extreme. Thus, the evolution of these two humoral systems has allowed for significant variation, and flexibility, in the cardiac function of fishes.

By presenting the most recent data, we will highlight the complexity of some mechanisms involved in the functional modulation of the fish heart under both normal and stress-induced conditions. Although there are many gaps in our knowledge due to the variable information available from different fish groups, our aim is to discuss both classic works and recent advances made by our group and by others. In providing this information, we hope to describe, in a spatiotemporal dimension, the mechanisms involved in the integrated control of the fish heart. The chapter will also emphasize the power of the fish heart as an experimental model for identifying humoral-dependent protective mechanisms that are difficult to explore using more conventional mammalian models and experimental approaches.

1. INTRODUCTION

The design of the heart in adult fish demonstrates remarkable morphofunctional nonuniformity (or heterogeneity) that encompasses all levels of organization from gene transcription and translation to gross morphology (for information on cardiac anatomy, see Chapter 1, Volume 36A: Icardo, 2017). This heterogeneity results from a modular morphogenesis driven by distinct transcriptional regulatory programs that control the regional specification of structures of each anatomical region (see Tota and Garofalo, 2012 for references). Compartmentalized selector genes and signaling systems orchestrate the topological organization of the processes that activate and orient subsequent time-dependent developmental events, thus helping to define the phylotypic cardiac plan (see Tota and Garofalo, 2012 for references). In fish, examples of such cardiac modular morphogenesis include chamber formation, trabeculation pattern, localized pacemaker activity, and organization of the atrial–ventricular region and outflow tract (Icardo, 2006; Icardo and Colvee, 2001).

The three major cardiac tissues, the endocardium, myocardium, and epicardium, are a universal trait of cardiac anatomy. They not only play a key morphogenetic role, like in chamber formation and compartimentalization, but also control the morphofunctional integrity of the adult cardiac pump through their close morphological and functional integration. For example, as documented in zebrafish (references in Poss et al., 2002), the cross-talk between the endocardium, the myocardium and the epicardium provides obligatory growth-modulating signals that direct, with different mechanisms, either the growth of the developing heart or the regeneration of the injured adult ventricle.

To support the varying physiological requirements of the animal, the fish heart adjusts it's metabolic and hemodynamic functions in response to both extrinsic and intrinsic signaling pathways. The former represented by circulating and intracardiac hormones, the latter consisting of humoral autacoids (i.e., locally generated signaling substances) (Ahrens, 1996). Both endocrine and autacoid signalings are responsible for beat-to-beat (e.g., Starling's law of the heart), short-term (E–C coupling, myocardial contractility), and long-term (modified gene expression) cardiac modulation under normal and stress-induced conditions.

Most of the knowledge on the morphofunctional heterogeneity and flexibility of the fish heart is based on studies of teleost fish. Accordingly, this group of fishes will be used as a point of reference to illustrate how endocrine and autacoid circuits contribute to organ integration. Rather than presenting a comprehensive picture, we will focus on classic [i.e., catecholamines (CAT), angiotensin II (AngII), and natriuretic peptides (NPs)] and novel cardiac modulators [i.e., chromogranin-A (CgA) and its derived peptides] that were more recently identified as components of the teleost heart's visceral integration system. The role of locally released autacoids [i.e., carbon monoxide (CO), nitric oxide (NO), and hydrogen sulfide (H_2S)], which evolved as regulatory molecules before the establishment of hormonal networks typical of complex animal body plans, will also be illustrated. The nitric oxide synthase (NOS)/NO system will be highlighted as a major integrative mechanism that contributes to fish heart modulation by coordinating both humoral signals and the cross-talk between the endocardial endothelium (EE) and the myocardium. Where available, data from lampetroids, elasmobranchs, and lungfish will be discussed to highlight how the intrinsic and extrinsic endocrine circuits have connected and been integrated during the evolutionary history of the fish heart.

2. CATECHOLAMINES: BASAL CONTROL, STRESS, AND CARDIOTOXICITY

In general, the function of the vertebrate heart is significantly impacted by CAT, typically adrenaline and noradrenaline that arrive *via* the circulation or are released from sympathetic nerve terminals, and bind to cardiac α- and β-adrenoreceptors (ARs). However, in several poikilotherms, such as cyclostomes, elasmobranchs and, to a lesser extent, teleosts, the sympathetic nervous system (SNS) differs from that of mammals. In many cases, peripheral nerves are replaced by aggregates of chromaffin cells (Burnstock, 1969) that are located in hemodynamically strategic regions of the cardiovascular system, such as the walls of large arteries (e.g., cyclostomes, lungfish, teleosts) and axillary bodies (e.g., elasmobranchs), or are embedded within the cardiac muscle (e.g., elasmobranchs, lungfish, some teleosts). These chromaffin cells often associate with sympathetic nerves and/or receive cholinergic stimulation, and this provides a local site of CAT production that contributes to humoral cardiovascular control (for references, see Abrahamsson et al., 1979; Gannon and Burnstock, 1969; Tota, 1999).

CAT-dependent cardiac modulation moves from a relatively simple design found in lampetroids (e.g., *Lampetra fluvialis*: Augustinsson et al., 1956) to a more complex system as described in teleosts. This evolutionary trend is characterized by important inter- and intraspecific differences that correlate with myoarchitecture, blood supply, and innervation (for an extensive review and references, see Tota et al., 2010). In lampetroids only cholinergic nerves innervate the heart, and their stimulation increases heart rate (f_H), although sometimes a slight bradycardia has been observed either during or after stimulation (see Nilsson, 1983 for references). However, the heart itself possesses chromaffin tissue formed by atrial and ventricular CAT-containing granules (Otsuka et al., 1977), which releases adrenaline in response to intracavity stimuli (Dashow and Epple, 1985). Adrenaline, in turn, stimulates the release of noradrenaline, and probably also of dopamine from other cardiovascular chromaffin cells (Dashow and Epple, 1985). Noradrenaline binds to myocardial β-ARs and increases the rate and strength of cardiac contraction, but this response is less intense than that induced by acetylcholine (ACH) (Nilsson, 1983; Otorii, 1953).

Starting with the elasmobranchs, the complexity of the adrenergic control of the vertebrate heart increases. Cartilaginous fish also possess intracardiac CAT stores located in the sinus venosus, atrium, and ventricle, and their stimulation by cholinergic preganglionic sympathetic fibers induces CAT release (Abrahamsson, 1979; Nilsson et al., 1976; Opdyke et al., 1983; Randall and Perry, 1992). In addition, as shown in most studies on dogfish species (for an exception, see Perry and Gilmour, 1996), physical disturbance, exercise, and hypoxia enhance plasma CAT levels (Butler et al., 1986; Metcalfe and Butler, 1989; Opdyke et al., 1982a; Taylor, 1992). In elasmobranchs, CAT affect heart performance, both indirectly and directly. Indirectly, plasma CAT regulate cardiac filling pressure through an α-adrenergic-dependent vasoconstriction which subsequently causes an increase in venous pressure (*Squalus acanthias*: Sandblom et al., 2006), and by modulating vagal inhibition (*S. acanthias*: Agnisola et al., 2003). Directly, CAT released by both intracardiac stores, and suprarenal and axillary bodies, induce positive chronotropism and inotropism *via* cardiac β-ARs (see Capra and Satchell,

1977; Nilsson et al., 1976; Randall and Perry, 1992; Saetersdal et al., 1975) that resemble the mammalian β_2-type (Ask, 1983). Cardiostimulation occurs by the modulation of myocardial Ca^{2+} transients. After CAT stimulation, there is an initial increase in Ca^{2+} entry into the cell, but this ion is rapidly removed from the cytoplasm, and this accelerates CAT-dependent relaxation (Woo and Morad, 2001). In many cases, an initial bradycardia is observed after CAT administration (see Nilsson, 1983 for references). Since this inhibitory effect was induced by noradrenaline, but not by adrenaline or isoprenaline, an α-ARs-dependent mechanism was postulated (Capra and Satchell, 1977). Cardiac adrenergic effects are also accompanied by vasomodulation of the well-developed coronary system, and this allows for appropriate blood supply to the myocardium during stressful situations (Axelsson, 1995).

An ideal model to study the effects of CAT on cardiac function is the air-breathing lungfish (Dipnoa). This is because this group of fishes routinely experiences high levels of physiological stress caused by prolonged periods of dehydration. This occurs during the dry tropical season, when the animals aestivate within subterranean mud cocoons (Janssens and Cohen, 1968). When water returns, the lungfish rapidly awakens, and is immediately able to swim (for references, see Ip and Chew, 2010). Aestivation requires deep cardiorespiratory and metabolic changes including a decrease in f_H, a drop in blood pressure, the complete reliance on air breathing, and decreased oxygen consumption (for references, see Amelio et al., 2013a). In parallel, the gill secondary lamellae collapse and filaments of the posterior gill arches fuse together. The gills are then bypassed as blood flow is redirected to the lungs, so that the single circulation (typical of the piscine heart) switches to an adult amphibian-like circulation (for references, see Amelio et al., 2013a). These rearrangements may also involve CAT, but few data are available on this aspect. However, it is known in the lungfish heart that CAT (mainly dopamine) are present within chromaffin cell aggregates that are located in the *sinus venosus* and auricle, as described in the African species *Protopterus aethiopicus* and *Protopterus annectens* (see references in Abrahamsson et al., 1979; Larsen et al., 1994). These cells often populate the subendocardium, and this suggests that CAT are released into the atrial lumen, and/or that they paracrinally modulate the myocardium (Fritsche et al., 1993; Larsen et al., 1994). As shown in *Protopterus dolloi*, CAT are mobilized by aerial hypoxia (Perry et al., 2005). In contrast, during terrestrial aestivation, plasma CAT concentrations do not change (Perry et al., 2008).

Teleosts are the first vertebrate group to have true cardiac sympathetic innervation which occurs *via* "vagosympathetic" trunks (Laurent et al., 1983; Newton et al., 2014; Nilsson, 1983; Taylor, 1992), but also have low plasma CAT levels under resting conditions that exert a basal excitatory tone (for references, see Tota et al., 2010). A considerable number of studies in teleosts shows that in stress-intolerant species, physical and environmental challenges (e.g., exhaustive exercise, anemia, hypoxia, hypercapnia) are accompanied by a sudden release of CAT from the chromaffin cells, such as those located in the walls of the posterior cardinal vein of the head kidney (Nandi, 1961). These circulating CAT, plus those released by nerve terminals and cardiac chromaffin cells (Farrell and Jones, 1992; Nilsson and Holmgren, 1992), target the heart. The combined cardioexcitatory, vascular and respiratory responses help alleviate the detrimental effects of the stressor (Farrell et al., 1986). However, stress-tolerant species such as the members of the genus *Anguilla*, do not show increased plasma CAT in response to stressful stimuli (see Imbrogno, 2013 for references). This suggests, that in teleosts, there are different stress-coping strategies involving CAT. Notably, two different styles characterize the teleost response to stress: proactive and reactive (Schjolden et al., 2005 and references therein). The first category describes fish that do not survive the experiment and show strenuous avoidance behavior, whereas reactive individuals show reduced signs of stress and survive. Strikingly, plasma CAT levels are four- to fivefold higher in the former compared to their reactive counterparts (van Raaij et al., 1996).

In homeotherms, the heart is injured by excessive CAT which induce myocardial lesions in the so-called "sympathetic storm" (Samuels, 2007). However, some fish such as the tench, *Tinca tinca*, show a low susceptibility to isoproterenol (ISO)-induced lesions (Ostádal and Rychterová, 1971), suggesting a high resistance to the cardiotoxic action of CAT. Apart from species-specific differences, this may relate to the comparatively low physiological temperatures, f_H's, contraction velocity and oxidative metabolism of the fish heart. This may translate into lower oxygen requirements compared to the hearts of homeotherms. As a result, the negative effects of intracardiac and/or circulating CAT may be diminished.

As observed in several species (e.g., *Anguilla anguilla*: Forster, 1976; Pennec and Le Bras, 1988; *Anguilla dieffenbachii*: Forster, 1981), spontaneously released CAT exert a basal adrenergic tone on the teleost heart. This excitatory tone is elicited through α- and β-ARs located in pacemaker and contractile cells (Axelsson et al., 1987; Gamperl et al., 1994; Pennec and Le Bras, 1984). Stimulation of cardiac β_2-ARs results in positive chronotropism and inotropism under both basal and loading conditions (Farrell et al., 1986; Graham and Farrell, 1989). Interestingly, in teleosts, the effects induced by the stimulation of cardiac ARs are influenced by temperature. As observed during winter in the European eel, *A. anguilla*, adrenaline, and noradrenaline induce an α-ARs-dependent bradycardia, while in summer a tachycardia is observed as a consequence of β-ARs stimulation (Peyraud-Waitzenegger et al., 1980). In trout, the activation of β_2-AR by adrenaline stimulates the contractile activity of the atrium over a wide range of temperature, i.e., from 2 to

17°C. However, this effect is more pronounced at 2°C, indicating that adrenergic stimulation can be particularly relevant in adjusting cardiac performance at lower temperatures (Ask et al., 1981). This is at least partially because the density (B_{max}) and affinity (K_d) of cardiac β-ARs are very temperature dependent, with the former parameter much lower in fishes acclimated at high temperatures (Keen et al., 1993).

As shown in the eel *A. anguilla*, adrenergic control of the teleost heart also occurs through the activation of β_3-ARs (Imbrogno et al., 2006, 2015). These receptors, whose presence in teleosts was first demonstrated in the rainbow trout *Oncorhynchus mykiss* (Nickerson et al., 2003), are highly expressed in the heart. Similar to their homologous mammalian counterpart (for references, see Angelone et al., 2008a), the activation of teleost β_3-ARs decreases cardiac mechanical performance by reducing stroke volume (V_S) and stroke work (W_S) (Fig. 1) (Imbrogno et al., 2006). This effect involves a pertussis toxin (PTx)-sensitive G_i protein mechanism and requires the NO-cyclic guanosine monophosphate (cGMP)-activated protein kinase G (PKG) cascade (Imbrogno et al., 2006). Of note, in *A. anguilla*, while the majority of the *in vitro* cardiac preparations respond to ISO stimulation by increasing

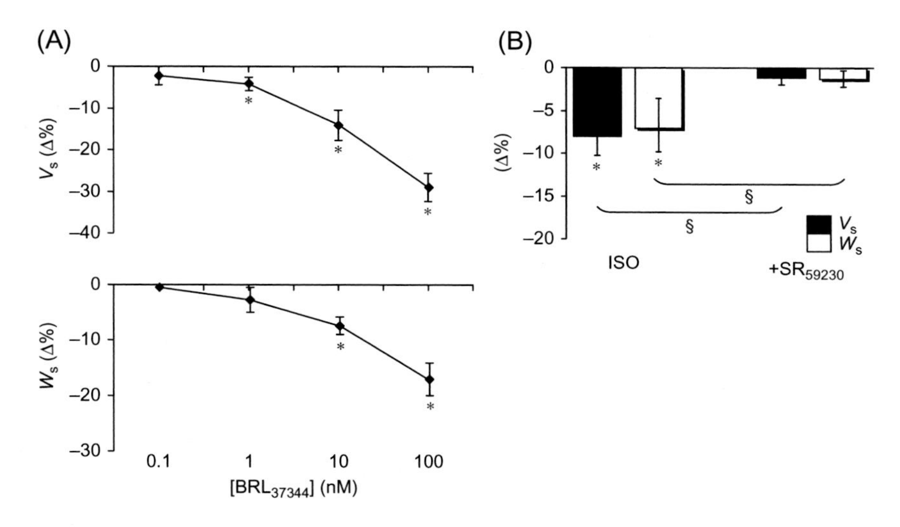

Fig. 1. (A) Cumulative dose–response curve for BRL_{37344} (β_3-adrenoreceptor agonist) on stroke volume (V_S) and stroke work (W_S) in isolated and perfused eel (*A. anguilla*) hearts. (B) Effects of isoproterenol (ISO; 100 nmol L^{-1}) before and after treatment with SR_{59230} (β_3-adrenoreceptor antagonist; 10 nmol L^{-1}) on V_S and W_S in isolated and perfused eel hearts. Modified from Imbrogno, S., Angelone, T., Adamo, C., Pulerà, E., Tota, B., Cerra, M.C., 2006. Beta3-adrenoceptor in the eel (*Anguilla anguilla*) heart: negative inotropy and NO-cGMP-dependent mechanism. J. Exp. Biol. 209, 4966–4973.

contractility, in 30% of the perfused heart preparations contractility is decreased (Imbrogno et al., 2006). This ISO-dependent cardioinhibition is mediated by β_3-AR, as it is abolished by the β_3-AR-specific antagonist SR_{59230} (Fig. 1) (Imbrogno et al., 2006). The presence of β_3-ARs in the teleost heart and the involvement of these receptors in counteracting ISO stimulation suggest that in fishes CAT regulate cardiac performance in a "ying-yang fashion" that involves different types of AR (i.e., β_2 *vs* β_3-ARs), with the stimulation of β_3-ARs providing cardioprotection against overstimulation. In addition, we hypothesize that species-specific differences in the B_{max}/activity of the cardioinhibitory β_3-ARs and/or the presence of "antiadrenergic" counter-regulatory cascades (see below in this chapter) may contribute to reducing the susceptibility of the fish heart to CAT injuries.

3. ANGIOTENSIN II

Angiotensin II (AngII), the principal bioactive component of the renin–angiotensin system (RAS), is a pluripotent hormone whose biological actions are extensively studied in mammals. The RAS cascade starts with the renin-mediated cleavage of the decapeptide angiotensin I (AngI) from angiotensinogen. Then, a C-terminal dipeptide is removed from AngI, generating the octapeptide AngII. Further cleavage of the AngII-N-terminus by aminopeptidases generates the vasoactive angiotensin III (AngIII; Ang 2–8), angiotensin IV (AngIV; Ang 3–8), and angiotensin 1–7 (Ang 1–7) (Nishimura, 2001 for references). AngI/AngII conversion is catalyzed by a rather non-specific dipeptidyl-carboxypeptidase, angiotensin-converting enzyme (ACE: Ehlers and Riordan, 1990), a glycosylated integral membrane protein synthesized in a large variety of cells and tissues. In mammals, the heart itself expresses all RAS components in both myocytes and fibroblasts (Dostal, 2000 and references therein). In addition, in cardiac cells, the enzyme chymase contributes to AngI/II conversion (Varagic and Frohlich, 2002 and references therein). Native AngII is a stable molecule, which shows few structural differences between species, and this suggests that it is highly conserved. Nonetheless, variation in the AngII sequence is present at positions 1 (Asp/Asx/Asn), 3 (Val/Ile/Pro), and 5 (Ile/Val), and AngI shows substitutions at position 9 (His/Ser/Ala/Asn/Tyr/Gly/Thr) (see Kobayashi and Takei, 1996; Nishimura, 2001).

Non-mammalian studies have shown that the RAS has a high phylogenetic continuity. Based on the failure to detect renin activity and to elicit AngII-dependent vasopressor effects in cyclostomes and elasmobranchs, it was initially believed that RAS first appeared in teleosts (Nishimura et al., 1973). However, application of highly sensitive molecular biology techniques allowed AngI to be identified in cyclostomes (Rankin et al., 2004; Takei, 2000) and in *Triakis scyllia* (Takei et al., 1993).

Like mammals, fish possess two RASs: one is localized in the plasma and regulates immediate cardiovascular responses; the second is expressed in the heart and the arteriolar wall of rectal, and interrenal gland, conus arteriosus and gills, and largely contributes to long-term modifications of ion-water and hemodynamic homeostasis (Tota et al., 2010 and references therein; Lacy et al., 2016). RAS is present in the heart of elasmobranchs, sarcopterigians, lungfish, and teleosts (Nishimura et al., 1970, 1973; Takei, 2000; Takei et al., 2004). ACE-like activity has also been detected in the heart of the lamprey *Lampetra fluviatilis* (Cobb et al., 2002), in the branchial heart of the hagfish *Myxine glutinosa* (Cobb et al., 2004), and in the heart of the elasmobranchs *Raja erinacea*, *S. acanthias*, and *Scyliorhinus canicula* (Lipke and Olson, 1988; Uva et al., 1992).

A functional RAS, homologous to that found in mammals, is present in teleosts (see Kobayashi and Takei, 1996) and has been studied due to its role in ion and osmotic regulation, particularly during seawater acclimation (Nishimura, 1985; Olson, 1992; Wong et al., 2006). The principal bioactive RAS component, AngII, has been identified and sequenced in various teleost species, including the chum salmon *Oncorhynchus keta* (Takemoto et al., 1983), the Japanese goosefish *Lophius litulon* (Hayashi et al., 1978), and the American eel *Anguilla rostrata* (Khosla et al., 1985). Notably, other angiotensin peptides are present in teleosts. For example, [Asn^1,Val^5] AngII and [Asp^1, Val^5] AngII circulate in trout and eel plasma (Conlon et al., 1996; Wong and Takei, 2012). In trout, the cleavage of angiotensinogen produces [Asn^1]-AngII that is converted to [Asp^1]-AngII in the plasma by asparaginase (Conlon et al., 1996). In contrast, in eel, plasma asparaginase activity is low and the conversion occurs in liver and kidney with angiotensinogen, not AngII, as a substrate (Wong and Takei, 2012).

The teleost heart possesses an intracardiac RAS. ACE activity is present in the ventricle of a variety of species (e.g., *Heteropneustes fossilis*, *Clarias batrachus*, *Channa gachua*, *Anabas testudineus*, *Notopterus chitala*, *Monopterus cuchia*: Olson et al., 1987), immunoreactive AngII-like material was described in the heart of the Antarctic teleost *Champsocephalus gunnari* (Masini et al., 1997), and cardiac AngII binding sites were reported in the trout (*O. mykiss*; Cobb and Brown, 1992) and in the eel (*A. anguilla*; Imbrogno et al., 2013). This suggests the presence of an AngII-dependent circuit of regulation in the teleost heart.

In mammals, the effects of AngII on the heart are mediated by plasma membrane AT_1 and AT_2 receptors, classified according to the affinity for various antagonists (for references, see Cerra et al., 2001; de Gasparo, 2002). AT_1 receptors are losartan sensitive, while AT_2 receptors selectively bind other antagonists such as CGP_{42112} (see references in de Gasparo, 2002). AT_1 receptors are mainly responsible for AngII-dependent positive myocardial inotropy and myocyte growth. This stimulation is mediated *via* G-protein phospholipase C, Ca^{2+} mobilization, inositol triphosphate, diacylglycerol, and protein kinase C (see Dostal, 2000 for references). Stimulation of these receptors also activates the JAK–STAT pathway, resulting in cardiac growth (Booz et al., 2002). In contrast, cardiac AT_2 receptors, whose expression is reduced after birth, antagonize AT_1 growth promoting effects *via* a number of phosphatases. This receptor is also coupled with the NO/cGMP signaling pathway, either directly or indirectly through enhanced bradykinin or endothelial NOS (eNOS) expression (references in Dostal, 2000).

Cloning and ligand affinity studies indicate that, in nonmammalian species, the pool of ATs is mostly represented by AT_1-like receptors. It is currently believed that a prototypic AT receptor (putatively AT_1) evolved in primitive vertebrates, but then diverged during phylogeny into more than one type to give mammalian ATs (Nishimura, 2001). Based on functional and pharmacological analyses, ATs have been identified in elasmobranch and teleost tissues and organs involved in cardiovascular and hydromineral homeostasis (see Kobayashi and Takei, 1996; Nishimura, 2001 for extensive references).

In elasmobranchs, ATs were identified in the rectal gland and in the interrenal glands (Hazon et al., 1997; Tierney et al., 1997). In *S. acanthias*, ATs are also located at the adrenergic nerve endings and/or in the adrenal medulla (Bernier et al., 1999; Carroll and Opdyke, 1982; Opdyke and Holcombe, 1976). This suggests a functional link with the adrenergic system. In *S. canicula*, the atrium, the ventricular myocardium, and the outer conal layer are able to bind the homologous dogfish Pro3AngII with high-affinity (Cerra et al., 2001). In this species, competition experiments performed with either homologous AngII, or with specific ligands for mammalian AT_1 and AT_2 receptors (CV_{11974} and CGP_{42112}, respectively) indicate a prevalence of CGP_{42112}-selective AngII binding sites in both the inner and outer conal layers (Cerra et al., 2001). Although caution is needed when interpreting the binding data obtained with mammalian selective ATs ligands, these results suggest the presence of an unknown dogfish AT_2-like receptor, thus contradicting the view that non-mammalian ATs are mostly AT_1-like.

There have been several studies, to date, that have attempted to define the molecular and biochemical characteristics of AngII receptors in teleost fish (see, for example, Marsigliante et al., 1996; Tran van Chuoi et al., 1998). These studies were mainly performed using commercial drugs and analogues that often have negative or inconsistent results, most likely due to a lack of specificity between the drugs and receptors (see Russell et al., 2001 for references). Two AngII receptor types were detected in the eel. One, characterized in the European eel (*A. anguilla*), shows a cDNA sequence [GenBank accession

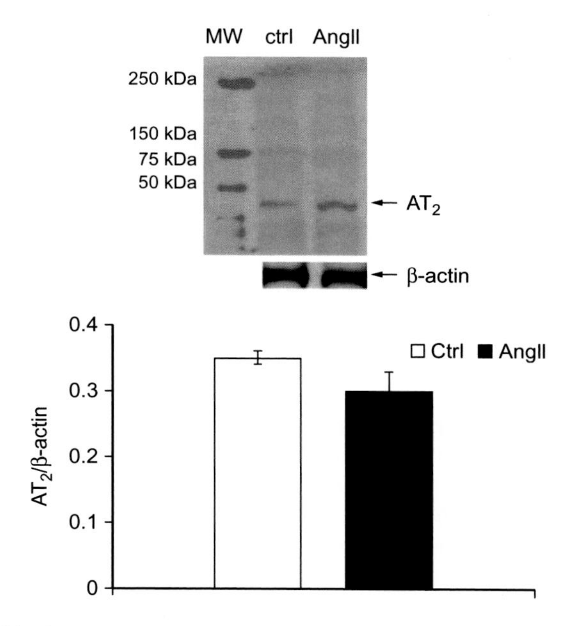

Fig. 2. Western blotting and densitometric analyses of the AT_2 receptor in cardiac homogenates of control and AngII (1.2 nmol g body weight^{-1}) treated eels (*A. anguilla*). From Imbrogno, S., Garofalo, F., Amelio, D., Capria, C., Cerra, M.C., 2013. Humoral control of cardiac remodeling in fish: role of angiotensin II. Gen. Comp. Endocrinol. 194, 189–197. doi:10.1016/j.ygcen.2013.09.009.

number AJ05132 (Tran van Chuoi et al., 1998)] with 60% homology to the mammalian AT_1 receptor (Russell et al., 2001); the other, characterized in the Japanese eel *Anguilla japonica* (Wong and Takei, 2013), shares homology with mammalian AT_2, as demonstrated by molecular phylogenetic and synteny analysis (Wong and Takei, 2013). To the best of our knowledge, only ATs able to bind mammalian anti-AT_2 antibody have been detected by Western blotting in the cardiac tissue of *A. anguilla* (Fig. 2) and were associated with the long-term effects of the peptide (Imbrogno et al., 2013).

3.1. Short-Term Modulation

In all vertebrates, AngII induces cardiovascular actions by directly targeting vascular smooth muscle and cardiac tissues, and by involving the SNS (Carroll and Opdyke, 1982). The identification of both cardiac ACE-like activity (e.g., in *R. erinacea*, *S. acanthias*, and *S. canicula*; Lipke and Olson, 1988; Uva et al., 1992) and AngII binding (*S. canicula*; Cerra et al., 2001) indicates that a local RAS is functioning in the elasmobranch heart. Nevertheless, no corresponding functional information is available, except for one study that reports unchanged heart contractility after AngII administration in *S. acanthias* (Opdyke et al., 1982b). The only available data on Dipnoan cardiovascular function demonstrates that, in the Australian lungfish (*Neoceratodus forsteri*), application of AngII of either the fish-type [Asn^1, Val^5] (Sawyer et al., 1976), the tetrapod-type [Asp^1, Val^5], or the native type [Asn^1, Val^5] (Joss et al., 1999) induces a dose-dependent increase in arterial pressure with no effects on f_H and negligible renal effects. In teleosts, the heart is either directly or indirectly stimulated by homologus [Asn^1, Val^5]-AngII. Indirect effects are mediated either by CAT (Bernier and Perry, 1999; Oudit and Butler, 1995) or by the modulation of cardiac nervous tone (Reid, 1992). In fact, the dose-dependent increments in cardiac output (Q) and V_S observed in *O. mykiss* after the exposure to teleost AngII are reversed by α-adrenergic blockage (Bernier and Perry, 1999). In contrast to these cardiostimulatory effects, AngII administered to the isolated and *in vitro* perfused working heart of the eel (*A. anguilla*) reduced both contractility and f_H. This direct cardioinhibition involves AT_1-like receptors (Fig. 3), $G_{i/o}$ proteins, the cholinergic system, and the EE–NO–cGMP–PKG cascade (Imbrogno et al., 2003). Although the reason of this divergence is unclear, it is reasonable to hypothesize that it is related to species' differences and/or the organizational level under study (intact cardiovascular system *vs in situ* heart *vs* isolated and denervated working cardiac preparations). For example, contrary to *in vitro* experiments, *in vivo* analyses can detect synergism between the adrenergic system and the RAS, both activated under stress and emergency situations (Hazon et al., 1995). Of note, in the European eel, contractility is unaffected by either adrenergic inhibition (i.e., *via* phentolamine, sotalol, or propanolol) or stimulation (i.e., *via* phenylephrine and ISO), and this argues against an intracardiac adrenergic involvement and indicates that CAs are not involved in the direct cardiac effects of AngII (Imbrogno et al., 2003).

3.2. Long-Term Readjustments

In mammals, the adult heart is believed to grow and remodel only through hypertrophy, although this paradigm has been questioned by evidence showing that, during normal cardiac growth, adult myocardiocytes proliferate and die (Nadal-Ginard et al., 2003). In the presence of either cardiac injury, or increased workload, or a number of bioactive molecules, this proliferation contributes significantly to pathologic organ remodeling (Leri et al., 2002 and references therein). In this regard, AngII represents an important endocrine mediator of mammalian cardiac growth (Chintalgattu et al., 2003; Huckle and Earp, 1994; Li et al., 1999; Rademaker et al., 2004). It has been demonstrated that, by directly and indirectly cross-talking with growth-

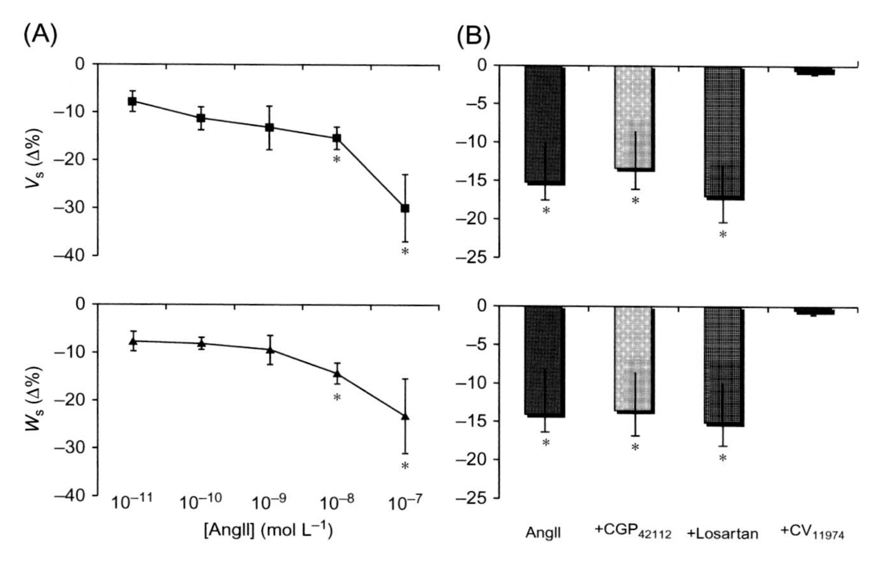

Fig. 3. (A) Cumulative dose–responses for AngII (from 10^{-11} to 10^{-7} mol L^{-1}) on stroke volume (V_S) and stroke work (W_S) in isolated and perfused eel (*A. anguilla*) hearts. (B) Effects of AngII (10^{-8} mol L^{-1}) before and after treatment with the AngII receptor antagonists CGP_{42112} (10^{-6} mol L^{-1}), losartan (10^{-6} mol L^{-1}), and CV_{11974} (10^{-7} mol L^{-1}) on V_S and W_S in isolated and perfused eel hearts. Modified from Imbrogno, S., Cerra, M.C., Tota, B., 2003. Angiotensin II-induced inotropism requires an endocardial endothelium-nitric oxide mechanism in the *in vitro* heart of *Anguilla anguilla*. J. Exp. Biol. 206, 2675–2684.

promoting and growth-inhibiting factors (i.e., NO, IGF-1, ET-1, ANP, and BNP), and with angiogenic molecules (i.e., VEGF, FGF-1), AngII induces cardiac growth and remodeling, particularly under pathologic conditions (Chintalgattu et al., 2003; Huckle and Earp, 1994; Li et al., 1999; Rademaker et al., 2004).

Unlike the mammalian heart, the fish heart can undergo significant structural and functional remodeling to match changes in hemodynamic requirements caused by growth and changes in activity level and environmental conditions. One example of this flexibility is the morphodynamic modifications reported for the eel ventricle during ontogenesis, where ventricular growth is accompanied by an enhancement of hemodynamic performance under both basal and load-stimulated conditions (Cerra et al., 2004). This improvement in hemodymanic performance is characterized by increased basal f_H and V_S, and by an enhanced capacity of the heart from large eels to maintain work against higher output pressures. In fact, in large eels, the ventricle becomes more "muscularized," by increasing both the thickness of the *compacta* and the diameter of the trabeculae in the *spongiosa*. Although few mechanistic data are available, it is expected that the growth-related modifications of the fish heart require precise humoral regulation. In this context, in addition to cortisol (Johansen et al., 2011), a role for AngII can be envisaged. Data obtained on the eel suggests that the teleost heart is a target for AngII-induced long-term effects. In fact, chronic (4 weeks) administration of AngII in *A. anguilla* elicits cardiac morphofunctional readjustments. AngII-treated animals show an enhanced cardiac hemodynamic performance in response to afterload increases. This correlates with an AngII-dependent modification of the expression and localization of molecules that regulate cell growth [such as c-kit, heat shock protein 90 (HSP-90), (eNOS)-like] and apoptosis [i.e., apoptosis repressor with CARD domain (ARC)] (Imbrogno et al., 2013). Interestingly, inhibition of AT_2 by the application of the selective antagonist CGP_{42112}, which cross-reacts with fish cardiac AngII receptors (Cerra et al., 2001; Imbrogno et al., 2003), demonstrates that this receptor subtype is involved in the abovementioned growth-related effects (Imbrogno et al., 2013). This is of relevance, since in mammals, AT_2 generally offsets or opposes AT_1-induced actions on cell growth, blood pressure, and fluid intake, and mediates anti-growth and apoptotic actions (Gallinat et al., 2000). Moreover, AngII-treated eel hearts show an increased expression of a NOS isoform recognized by mammalian anti-eNOS antibody (eNOS-like). Since in mammals, reduced eNOS-derived NO generation is responsible for hypertrophy and reactive oxygen species (ROS) formation (Wenzel et al., 2007), it is possible that in teleosts an AngII-dependent increase in NO bioavailability counter-balances the growth promoting signaling and coordinates cardioprotective programs (Imbrogno et al., 2013). The significance of the growth-promoting actions of AngII on the eel heart *vs* that of whole animal performance is unknown. However, this long-term effect must be considered in the context of the proliferative myocardial response of the teleost heart (see also Chapter 3, Volume 36B: Gillis and Johnson, 2017), which is notably different to that encountered in mammals. In fish, prolonged stimuli, e.g., those elicited by cold acclimation or chronic exercise, trigger physiological hyperplastic and/or hypertrophic cardiac growth (Vornanen et al., 2005; see Chapter 3, Volume 36B, and references therein: Gillis and Johnson, 2017). Thus, in contrast to mammals in which cardiomyocyte hypertrophy results in cardiac dysfunction, in fish, proliferative signals are not detrimental to the heart. In this context, a dichotomy is expected between fish and mammals concerning the long-term effects of neurohumoral agents, including AngII. In fact, while these agents have the potential to accelerate the progression of heart failure in mammals, they may be fundamental components of the stress defense of the cardiocirculatory system of cold-blooded vertebrates (Imbrogno et al., 2013). Accordingly, it can be suggested that during evolutionary diversification this hormone was recruited for different functions.

4. NATRIURECTIC PEPTIDES: INTERFACE BETWEEN MYOCARDIAL PERFORMANCE AND ION/FLUID BALANCE

Natriurectic peptides (NPs) are a group of hormones first identified by de Bold et al. (1981) in rat atrial secretory granules. They are present not only in vertebrates, but also in invertebrates, and in single-celled organisms. NPs consist of at least five members, i.e., atrial NP (ANP), brain NP (BNP), ventricular NP (VNP), C-type NP (CNP), and *Dendroaspis* NP (DNP) (for references, see Tota et al., 2010). Released as a consequence of cardiac stretch induced by hypervolemia and characterized by potent diuretic, natriuretic, and vasorelaxant properties, NPs are currently recognized as a ubiquitous hormonal system in mammalian and non-mammalian vertebrates. Through their multitarget effects on the heart, vasculature, kidney, adrenals, and central and autonomic nervous systems, NPs participate in the homeostatic circuits that link blood volume and myocardial stretch. They are also involved in cardiac morphogenesis and adult heart remodeling (O'Tierney et al., 2010; Tamura et al., 2000). Notably, NPs interact with other endocrine-regulated pathways, including NO, prostaglandins, and other vasodilatory and volume-regulating peptides (i.e., the RAS: Takei et al., 2014), and this amplifies their efficiency in the control of ion/fluid balance and of cardiovascular function. This control is particularly relevant in fish. In fact, most fish are continuously exposed to an osmotic gradient between their internal fluids and the surrounding water. They also show a large range of adaptability to different salinities, i.e., ranging from euryhaline to stenohaline, and this requires a remarkable effort to maintain volume and cardiovascular equilibrium. This feature makes them well suited to decipher the homeostatic significance of NPs as either volume or cardiovascular regulators, or both. Of note, more than in any other vertebrate, in fishes, volume loading and filling pressure are major regulators of cardiac output *via* the Frank–Starling mechanism (see Chapter 4, Volume 36A: Farrell and Smith, 2017). Accordingly, the role of NPs at the interface between myocardial performance and ion/fluid equilibrium may be even more emphasized.

4.1. Structural Traits and Evolutionary Hints of the NP/NPR System

The basic NP structure consists of a highly conserved 17-amino acid intramolecular ring generated by a disulfide bridge between two Cys, and N- and C-terminal sequences of different lengths (Takei and Hirose, 2002). The C-terminus is very long in VNP and DNP, but it is absent in CNP. ANP and CNP are the best-conserved NPs, while the most variable NP is BNP; this latter NP shows a sequence identity of only 35% with human and mouse (see Takei and Hirose, 2002).

Phylogenetic and comparative genomic analyses suggest that the first NP, which appeared in the ancestral vertebrate lineage, was a CNP_4-like peptide (Takei et al., 2011). It is thought that CNP_1, CNP_2, and CNP_3 genes were generated from the CNP_4-like NP gene by chromosome or gene duplications before the divergence of chondrichthyans and Osteichthyes, and that the cardiac NPs were generated by a CNP_3 duplication. Seven NP members existed at the divergence of ray-finned fishes and lobe-finned fishes (tetrapods), but some of the NP genes disappeared during evolution (see Takei et al., 2011 for references).

In parallel with the peptides, a complex set of NP receptors (NPRs) differentiated during evolution. Four types of NPRs, namely NPR-A, B, C, and D, are present from cyclostomes to mammals (see Takei et al., 2011 for references). NPR-A and NPR-B are recognized as the membrane form of guanylate cyclase (GC) (particulate GC: pGC) whose activation increases the intracellular pool of cGMP. They have different affinities for ANP, BNP, and CNP. NPR-A is activated by ANP and BNP; NPR-B is the natural receptor for CNP, and also shows a high affinity for teleost VNP. NPR-C binds all NPs with almost equal affinity, and mainly acts as a clearance receptor for NPs. It does not possess GC activity. However, like NPR-A and NPR-B, its activation contributes to the increase in intracellular cGMP. This occurs through a G_i-dependent activation of the eNOS/NO/soluble GC (sGC) cascade. Apart from several structural similarities with NPR-A and NPR-B, little information on NPR-D is available (for references, see Takei and Hirose, 2002).

Among vertebrates, fish possess the most complex NPs/NPRs system, which is composed of four types of NPs and four types of NPRs. A single peptide is present in the hagfish and in elasmobranchs: EbuNP and CNP, respectively (Kawakoshi et al., 2003; Suzuki et al., 1994; Takei, 2000). By contrast, in the sturgeon *Acipenser transmontanum* ANP-, BNP-, CNP-, and VNP-encoding genes are all expressed (Kawakoshi et al., 2004). In teleosts, apart from a few examples (e.g., medaka which lack ANP), a complete set of cardiac NPs is also present (see references in Tota et al., 2010). Examples of NP sequences in teleosts are provided in Fig. 4. In fish, the gene transcripts for NPR-A have been identified in eel, trout, and medaka (Garg and Pandey, 2005). Also NPR-B and NPR-C are present in teleosts (i.e., in the eel), as well as in hagfish and spiny dogfish (Takei and Hirose, 2002; Takei et al., 2007); NPR-C is present in both elasmobranchs and bony fish (Toop and Donald, 2004), while NPR-D was only found in the eel (Kashiwagi et al., 1995).

The reason for the complexity of the NPs/NPRs system in fish is unknown. From an evolutionary standpoint, it can be hypothesized that the diuretic, natriuretic, and cardiovascular properties of NPs contributed significantly to body fluid homeostasis when fish colonized different habitats. This may have allowed the diversification of such a crucial cardiovascular and water/fluid regulatory system.

Peptide	Sequence
Eel ANP	SKSSSPCFGGKLDRIGSYSGLGCN–SRK
Trout ANP	SKAVSGCFGARMDRIGTSSGLGCSPKRRS
Icefish ANP	KRSSSCFGARMDRIGNASSLGCNNGR
Eel BNP	YSGCFGRKMDRIGSMSSLGCKTVSKRN
Trout BNP	YSGCFGRRMDRIGSMSSLGCTTVGKYNAKTR
Icefish BNP	SSSCFGRRMDRIGSMSSLGCNTVGKYNRK
Eel CNP	GWNRGCFGLKLDRIGSLSGLGC
Trout CNP	GWNRGCFGLKLDRIGSMSGLGC
Icefish CNP	–
Eel VNP	KSFNSCFGTRMDRIGSWSGLGCNSLKNGTKKKIFGN
Trout VNP	KSFNSCFGNRIERIGSWSGLGCNNVKTGNKKRIFNGN
Icefish VNP	–

Cardiac region	IC_{50}
Atrium	eCNP=eANP=eVNP
V. Myocardium	eCNP=eANP=eVNP
V. Endothelial endocardium	eVNP<<eANP<<<eCNP
Bulbus arteriosus (inner layer)	eCNP<<eVNP<<eANP
Bulbus arteriosus (outer layer)	eCNP≤eVNP≤eANP

Fig. 4. Structure of several teleost ANP, BNP, CNP, and VNPs, and half-maximal inhibitory concentrations (IC_{50}; i.e., potency) of eel NPs for [125I]-r-ANP(1–28) binding sites in different cardiac regions of *A. anguilla*. Modified from Takei, Y., Hirose, S., 2002. The natriuretic peptide system in eels: a key endocrine system for euryhalinity? Am. J. Physiol. Regul. Integr. Comp. Physiol. 282, 940–951; Cerra, M.C., Canonaco, M., Takei, Y., Tota, B., 1996. Characterization of natriuretic peptide binding sites in the heart of the eel *Anguilla anguilla*. J. Exp. Zool. 275, 27–35; Inoue, K., Sakamoto, T., Yuge, S., Iwatani, H., Yamagami, S., Tsutsumi, M., Hori, H., Cerra, M.C., Tota, B., Suzuki, N., Okamoto, N., Takei, Y., 2005. Structural and functional evolution of three cardiac natriuretic peptides. Mol. Biol. Evol. 22, 2428–2434.

4.2. Fish Cardiac NP/NPR System

In fish, the heart is a major site of NP production. The peptides are differentially expressed in the various cardiac regions, thus representing an endocrine hallmark of morphofunctional heterogeneity. As shown in lungfish (Larsen et al., 1994; Masini et al., 1996) and teleosts (Donald et al., 1992; Loretz et al., 1997; Reinecke et al., 1985; Uemura et al., 1991), NPs (i.e., ANP) are present in atrial and ventricular myocytes. In general, atrial myocardial NPs production exceeds that of the ventricle, indicating that most of the secretory activity of the heart is concentrated in the atria.

A peptide structurally and functionally similar to mammalian ANP, namely salmon cardiac peptide (SCP), is present in the atrium and ventricle of Atlantic salmon (Arjamaa et al., 2000; Tervonen et al., 1998; Vierimaa et al., 2006). VNP is also present in the atrium and ventricle of salmonids (rainbow trout), as well as in eels and sturgeons (Inoue et al., 2005; Takei, 2000). In *Salmo salar*, both SCP and VNP are located in pacemaker ganglionic cells and in cardiomyocytes, and this suggests a neuromodulatory and/or neurotransmitter role for both peptides (Yousaf et al., 2012). CNP, regardless of its major endothelial production, is also produced by the heart in fish, as shown in the elasmobranchs *S. canicula* and *T. scyllia*, and in the eel and rainbow trout (see Inoue et al., 2003; Loretz and Pollina, 2000).

Like NPs, NPRs are differently expressed in the various regions of the teleost heart. As in mammals, in teleosts the cardiac regional localization of NPs correlates with the sensory ability and the hemodynamic characteristics of the different heart chambers. As shown by the application of homologous peptides for receptor studies, major NP binding sites are located in the atrium, the ventricle, and the outflow tract (*bulbus arteriosus*) (Cerra et al., 1992, 1996). In *A. anguilla*, NPR-C-like receptors are present in atrial and ventricular myocardium, while NPR-A-type receptors are located in the ventricular EE (Cerra et al., 1996). A receptor with high affinity for CNP, presumably NPR-B, is also present in the endothelial and epicardial layers of the *bulbus arteriosus* of the eel (Cerra et al., 1996), in which it may mediate a CNP-dependent modulation of bulbar hemodynamics (e.g., the *Windkessel* function).

4.3. Osmoregulatory *vs* Cardioprotective Functions

In mammals, changes in blood pressure and other physiopathological processes dynamically regulate NPs in the heart, as well as in other tissues (Ruskoaho, 1992). Once released, NPs act on proximal and distal targets to generate complex responses. They coordinate direct chronotropic, inotropic, and vasorelaxant responses, and elicit indirect actions that reduce cardiac hemodynamic loads. This results in a tight control of heart–vessel and blood volume–ion homeostasis.

Like other vertebrates, fish maintain the ionic and osmotic homeostasis of extracellular fluids through sophisticated mechanisms. However, in contrast to terrestrial vertebrates, piscine blood volume has to be maintained in the face of fluctuating ionic/osmotic gradients between the aquatic environment and their body fluids. In teleosts, the mechanisms that preserve ion concentration and osmoregulation, as well as body fluid volume, are highly diversified because of the evolutionary adaptation to various osmotic habitats (Takei et al., 2014). In this context, the teleost NPs represents a crucial regulator of osmotic and cardiovascular homeostasis. However, data from studies on different teleost species do not clarify whether the role of these peptides is mainly osmoregulatory, or cardiovascular, or both. In trout, injection of isotonic saline increases blood volume and ANP secretion (Cousins and Farrell, 1996). In addition, when exogenous ANP is injected into the trout circulation, the consequence is a potent diuresis and natriuresis (Duff and Olson, 1986). In contrast, in *A. japonica*, a major stimulus for ANP secretion is increased plasma osmolality, and not the high blood volume-induced atrial stretch, since blood volume transiently decreases after seawater transfer (Kaiya and Takei, 1996). In the eel, although volume challenges are also stimulatory for NPs release, hyperosmolality represents a stimulus which is more potent than hypervolemia in potentiating ANP release (Cousins and Farrell, 1996; Kaiya and Takei, 1996). Accordingly, these data suggest a key role of NPs in the adaptation to environmental salinity.

In parallel with their effects on osmoregulation (Cousins and Farrell, 1996; Kaiya and Takei, 1996), both direct and indirect evidence in teleosts suggests that NPs are cardioprotective factors that, by reducing blood volume, protect the heart from excessive preload and afterload (Olson et al., 1997). This effect is particularly relevant in fish, where the Frank–Starling mechanism has a significant influence on heart performance. This heterometric regulation allows the myocardium to respond to increased venous return (preload) with a more vigorous contraction of its lengthened fibers, thus increasing V_S and consequently Q (the product of f_H and V_s) (Shiels and White, 2008; also see Chapter 4, Volume 36A: Farrell and Smith, 2017). In rainbow trout, an increase in filling pressure stimulates stretch-dependent cardiac NPs secretion (Cousins and Farrell, 1996; Olson et al., 1997). Once released, NPs relax both arterial and venous branchial vessels, thus reducing gill perfusion pressures and resistance (see Farrell and Olson, 2000). This downstream effect prevents excessive load from being imposed on the heart by elevated afterloads occurring beyond the *Windkessel* control exerted by the *bulbus arteriosus* (see Johansen and Burggren, 1980, for *Windkessel* effect). Given that this regulation is additional and/or alternative to the Frank–Starling response, it contributes significantly to teleost heart performance. Thus, in fish, the stretch-dependent activation of NPs release provides an effective control system for regulating mean circulatory filling pressure, and therefore, venous return (Olson et al., 1997). According to the above data, the increased compliance and the decreased vascular tone of small veins (both reducing venous return) and the branchial vasculature are considered the two prominent cardiovascular actions of NPs. Interestingly, in freshwater fish which experience only volume but not salt load, NPs are constitutively released (trout: Olson and Duff, 1992). These data suggest that teleost NPs represent a fundamental system that protects the heart from overload (Johnson and Olson, 2009).

Very recently, a role for cardiac NPs other than osmoregulation and cardioprotection was proposed based on studies in a teleost model. Knockdown and transgenic analyses in zebrafish revealed the involvement of NPs in the mechanisms that regulate embryonic cardiomyocyte proliferation. It was observed that low concentrations of NPs, by recruiting different receptor types, enhance the proliferation of embryonic cardiomyocytes *via* cAMP signaling; by contrast, high NP concentrations elicit the opposite effect through a PKG-mediated pathway (Becker et al., 2014). Consistent with a regulatory role in myocardial growth, knocking-down BNP and CNP3 genes in medaka (*Oryzias latipes*) revealed the importance of these NPs for normal heart development and performance, a role usually taken by ANP in other vertebrates (Miyanishi et al., 2013). Thus, in fish, NPs may also be considered a cardiac growth factor. This function is retained during evolution, as exemplified by BNP that in the mammalian and human heart drives morphogenesis and differentiation, as well as the physiopathological remodeling accompanied by "fetal reprogramming" (see Tota et al., 2010 for references).

5. CHROMOGRANIN A-DERIVED PEPTIDES AS CARDIAC STABILIZERS

CgA is a member of the granin family, a group of acidic soluble proteins found in secretory granules of endocrine, neuroendocrine, and neuronal cells, and is co-stored and co-released with hormones, neurotransmitters, and/or amines in response to specific stimuli (see Helle et al., 2007 for references). CgA, the first member of the family to be isolated and characterized, is a 48 kDa protein identified almost 5 decades ago (Banks and Helle, 1965; Winkler and Fischer-Colbrie, 1992). Immunological and sequencing studies have demonstrated that CgA has a broad distribution across phylogenetic groups, from invertebrates to fish to mammals, and that there is significant conservation of protein sequence (for references, see Tota et al., 2010). In fish, the protein has been identified in teleosts, such as the coho salmon *Oncorhynchus kisutch* (Deftos et al., 1987) and the zebrafish (Xie et al., 2008). It shows a very high interspecific homology, indicative of a long evolutionary history. It has also been recently demonstrated that CgA-like proteins are present in coelenterates, the nematode parasite *Ascaris suum*, and the protozoan *Paramecium tetraurelia* (see Imbrogno et al., 2016 for references).

Apart from other important functions, e.g., biogenesis and exocytosis of chromaffin granules as well as being a diagnostic marker for neuroendocrine tumors, CgA acts as a prohormone (Eiden, 1987) and precursor of a number of regulatory peptides (Fig. 5). These peptides are generated by cell-, tissue-, and species-specific proteolytic processes operated by proteases and prohormone convertases (see Helle et al., 2007 and references therein). Full-length CgA,

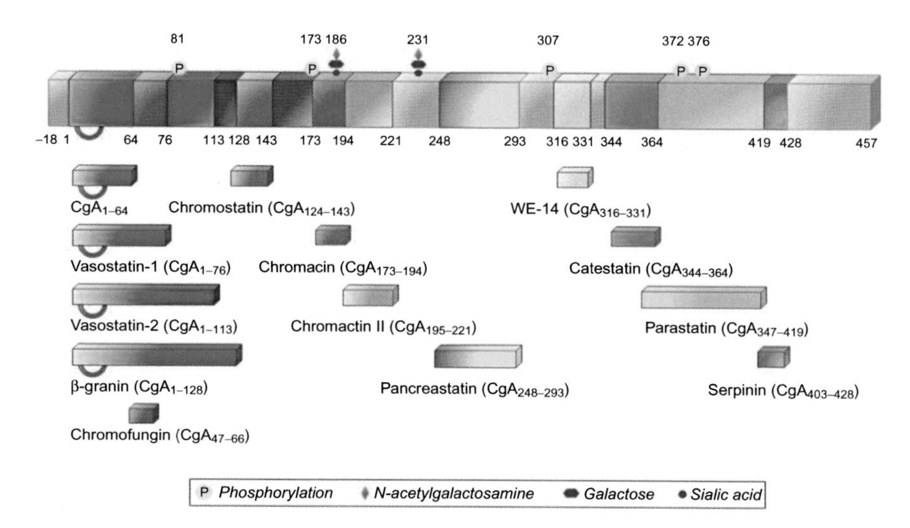

Fig. 5. Chromogranin A (CgA) and its derived peptides. The N-terminal disulfide bridge and the posttranslational modifications of the bovine sequence are indicated. Figure shows: the dysglycemic hormone pancreastatin; the vasodilators, and cardioinhibitors, Vasostatin 1 (VS-1) and Vasostatin 2 (VS-2); the antimicrobial agent chromacin, the catecholamine release inhibitor and cardioprotective peptide catestatin (CST); the antifungal fragment with cardiac function chromofungin and its well-conserved domains WE-14, parastatin and GE-25 (whose role is still to be identified); and the recently identified C-terminal serpinin peptides (Serp and pGlu-Serp) with important roles in neuroendocrine cell granule biogenesis, cell death, and cardiac stimulation. For an extensive review, see Tota et al. (2014). Modified from Tota, B., Cerra, M.C., Gattuso, A., 2010. Catecholamines, cardiac natriuretic peptides and chromogranin A: evolution and physiopathology of a 'whip-brake' system of the endocrine heart. J. Exp. Biol. 213, 3081–3103.

as well as several derived peptides including the N-terminal fragments vasostatin 1 (VS-1) and vasostatin 2 (VS-2) (CgA1–76 and 1–113, respectively), and the middle domain catestatin (CST; CgA344–364) have been intensively analyzed in mammals for their role as cardiovascular stabilizers. As shown in the rat, VS-1 and CST modulate basal cardiac performance and exert anti-adrenergic cardiosuppressive actions (Angelone et al., 2008b; Cerra et al., 2006; Pasqua et al., 2013; Tota et al., 2003). They also protect the myocardium against the damage induced by ischemia/reperfusion (I/R) (Bassino et al., 2015; Cappello et al., 2007; Penna et al., 2010, 2012, 2014). Accordingly, in the mammalian heart, CgA and its derived peptides are considered important humoral regulatory factors under normal conditions and in the presence of stress (Mazza et al., 2010; Tota et al., 2014). As illustrated here, the fish heart is not an exception.

5.1. Vasostatins and Catestatin Cardiac Influences

In the isolated *ex vivo* perfused working heart of the eel (*A. anguilla*), VSs and CST induce a dose-dependent reduction of V_S and W_S (Fig. 6). They also counteract the effects of β-adrenergic (ISO) stimulation, revealing an anti-adrenergic effect (Imbrogno et al., 2004, 2010; Tota et al., 2004) (Fig. 6). The cardiac actions of VSs and CTs on the eel heart resemble those reported for the isolated working heart of the frog (*Rana esculenta*) and on isolated and Langendorff perfused rat cardiac preparations, and thus, they represent a conserved functional trait of these peptides (see Tota et al., 2004 for references). Interestingly, analysis of the cardiac signal transduction mechanisms evoked by VSs and CST in teleosts shows aspects of uniformity, and diversity, with

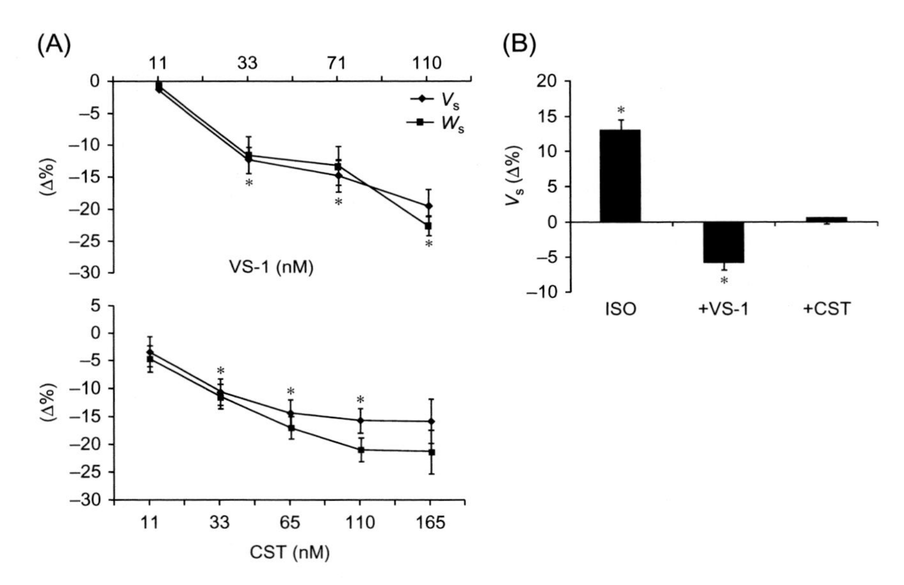

Fig. 6. (A) Cumulative concentration–response curve for VS-1 (from 11 to 110 nmol L^{-1}) and CST (from 11 to 165 nmol L^{-1}) on stroke volume (V_S) and stroke work (W_S) in isolated and perfused eel (*A. anguilla*) hearts. (B) Effects of isoproterenol (ISO) alone and in presence of VS-1 (33 nmol L^{-1}) or CST (110 nmol L^{-1}) on V_S in isolated and perfused eel hearts. Modified from Imbrogno, S., Angelone, T., Corti, A., Adamo, C., Helle, K.B., Tota, B., 2004. Influence of vasostatins, the chromogranin A-derived peptides, on the working heart of the eel (*Anguilla anguilla*): negative inotropy and mechanism of action. Gen. Comp. Endocrinol. 139, 20–28; Imbrogno, S., Garofalo, F., Cerra, M.C., Mahata, S.K., Tota, B., 2010. The catecholamine release-inhibitory peptide catestatin (chromogranin A344–363) modulates myocardial function in fish. J. Exp. Biol. 213, 3636–3643. doi:10.1242/jeb.045567; Tota, B., Angelone, T., Cerra, M.C., 2014. The surging role of chromogranin A in cardiovascular homeostasis. Front. Chem. 2, 64. doi:10.3389/fchem.2014.00064).

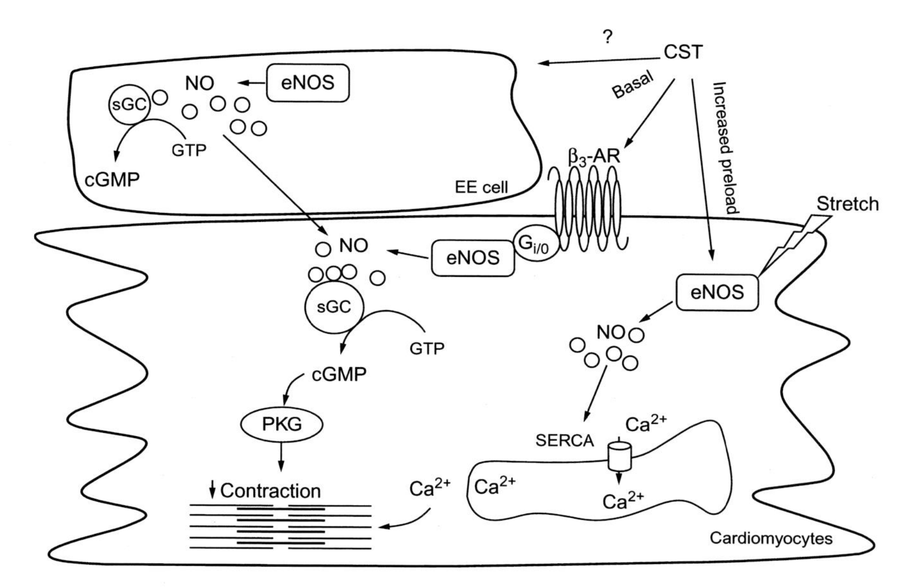

Fig. 7. Schematic diagram of putative CST signaling in myocardial cells and its modulatory action on the mechanical performance of the eel (*A. anguilla*) heart. Under basal conditions, the cardioinhibitory action of CST involves β_3-ARs, $G_{i/o}$ proteins, and the NO-cGMP pathway. Under increased preload conditions, CST enhances the heart's sensitivity to the Frank–Starling response through a NO-dependent regulation of SR Ca^{2+} re-uptake. β_3-AR, β_3-adrenergic receptor; EE, endocardial endothelium, eNOS, endothelial NOS; PKG, protein kinase G; SERCA, sarco(endo)plasmic reticulum Ca^{2+}-ATPases; sGC, soluble guanylate cyclase; cGMP, cyclic guanosine monophosphate; GTP, guanosine triphosphate; NO, nitric oxide; $G_{i/o}$, $G_{i/o}$ proteins. Modified from Imbrogno, S., Garofalo, F., Cerra, M.C., Mahata, S.K., Tota, B., 2010. The catecholamine release-inhibitory peptide catestatin (chromogranin A344–363) modulates myocardial function in fish. J. Exp. Biol. 213, 3636–3643. doi:10.1242/jeb.045567.

respect to amphibians and mammals. In fact, similar to the rat heart, the reduction of contractility elicited by VS-1 and CST in the eel involves β-ARs, $G_{i/o}$ proteins, and the NO/cGMP/PKG pathway (see Fig. 7 for CST). This differs from the mechanism described in the frog heart in which the CgA7–57-dependent cardioinhibition is independent of ARs, $G_{i/o}$ proteins, and NO (Corti et al., 2002, 2004).

The cardiodepressive action exerted by VSs and CST on the unstimulated and adrenergically stimulated teleost heart is of relevance, since it adds to the possibilities for cardioprotection in the presence of excessive excitatory challenges. Notably, the two CgA fragments use apparently redundant molecular strategies for cardiac control, whose biological significance remains unclear. As in mammals (see Mazza et al., 2015; Tota et al., 2014 for references), in teleosts VS-1 and CST may induce intracardiac actions following a spatiotemporal and tissue-specific scheme. By acting on either overlapping or different sites, the two peptides may elicit either summation and synergism or potentiation of target responses. A similar case is illustrated in mammals by the proANF precursor, whose cleavage gives rise to the major form of circulating ANP (ANP1–28), plus several biologically active peptides (proANF1–30, a long-acting sodium stimulator; proANF31–67, a vessel dilator; and proANF79–98, a kaliuretic stimulator) (Vesely et al., 1994). At least two of them, i.e., the vessel dilator and ANP, are vasodilatory, diuretic, and natriuretic, and thus, show almost overlapping properties (Vesely, 2006). So far, no conclusive information is available in teleosts showing that, like in mammals, CgA acts as a prohormone for bioactive fragments. However, it is reasonable to assume that, also in fish, the processing of more than one cardioactive peptide from the same prohormone is advantageous in maintaining homeostasis.

6. GASOTRANSMITTERS AS CARDIAC MODULATORS

Gasotransmitters are a group of membrane permeable gases, namely carbon monoxide (CO), hydrogen sulfide (H_2S), and NO, that are endogenously produced *in vivo* during the catabolic processing of common biological molecules. Initially, they were considered toxicants common to extreme environments. For organisms living in such environments, they represent important physiochemical stressors, which generate conditions often incompatible with life. However, some animal species are able to withstand such stressful conditions. One example is the cyprinodontiform, *Poecilia mexicana*, which is adapted to cope with the high H_2S levels present in freshwater springs of at least three Mexican river drainages (Palacios et al., 2013). Other examples are the Crucian carp *Carassius carassius* and the goldfish *Carassius auratus*, which are exceptional among vertebrates for their ability to tolerate long periods of anoxia (the complete absence of O_2); this lack of oxygen is often associated with dramatic increases in the nitrite concentration of several tissues (see Fago et al., 2012 for references).

Gasotransmitters play a fundamental biological role in fish due to their influence on the activities of cells, tissues, and organs and on organism homeostasis. Studies on this topic started with the seminal discovery of the biological properties of NO (see Tota et al., 2010 for references), and now are extended to H_2S and CO. All of them are under investigation as neuroactive and vasoactive molecules, mainly in relation to cardiorespiratory function and, as in the case of CO and NO, in O_2 sensing (see Perry and Tzaneva, 2016, and references contained therein).

Fish live in an environment that is becoming increasingly polluted *via* anthropogenic sources, and as a result, are being exposed to abnormally high levels of gasotransmitters. As specified below, this represents a severe challenge for animal fitness since gasotransmitters affect cell activity, and also alter critical cell signaling cascades. In fish, many studies have demonstrated that NO is highly involved in cardiovascular control under basal and stressful conditions (for review, see Tota et al., 2010). In contrast, there have been limited studies on the influence of CO and H_2S on the fish heart (but see Tzaneva and Perry, 2016; Wu et al., 2016).

6.1. Nitric Oxide as an Autocrine/Paracrine/Endocrine Mediator and Cardioprotective Factor

NO was the first gasotransmitter to be characterized as a biological mediator. Since Furchgott and Zawadzki's discovery concerning the importance of the endothelium in mediating ACH-induced vasorelaxation (Furchgott and Zawadzki, 1980), and the identification of NO as the endothelium-derived relaxation factor (Palmer et al., 1987), an impressive amount of data has unequivocally described this gas as one of the most important biological molecules in vertebrates. In almost every animal tissue, a group of NOS isoenzymes, namely the constitutive endothelial and neuronal NOS (eNOS and nNOS, respectively), and the inducible NOS (iNOS), use L-Arg as a substrate in combination with O_2 to generate NO. In the steady state, nanomolar concentrations are produced. However, micromolar cytotoxic amounts may also be generated, as in the case for iNOS activation following inflammatory and immunologic stimuli (Vallance et al., 2000). Thanks to their spatial intracellular compartmentalization, NOS isoenzymes produce NO close to their molecular targets, and this allows for interaction with many extrinsic and intrinsic humoral and neuroendocrine pathways (for references, see Tota et al., 2010). To elicit its effects, NO interacts with its major target, sGC, to generate cGMP. Alternatively, NO can by-pass sGC and directly interact with heme proteins, non-heme iron, or free thiol residues on signaling proteins and on ion channels (Balligand and Cannon, 1997; Hare, 2003).

As has been largely shown in mammals, the heart is a major NO producer, and a target of its actions. NO is generated both basally, and following chemical and physical stimulation. This is achieved thanks to the activity of NOS isoforms expressed in myocardial and non-myocardial tissues (i.e., vascular and endocardial endothelial cells, interstitial cells, coronary vessels, and myocardial neurons). NOS activation induces the release of NO, which exerts a wide repertoire of actions that result in the modulation of the three main mechanisms of cardiac adaptation: i.e., beat-to-beat (regulated through the Frank–Starling mechanism); short-term (i.e., through effects on excitation–contraction coupling, cardiac contractility, and relaxation); and long-term-adaptation as a result of altered gene expression (Balligand, 2000;

Hare, 2003; Khan et al., 2003 and references therein). As described in the paragraph below, NOS/NO-induced cardiac regulation follows a spatiotemporal pattern that takes advantage of the peculiar spatial and temporal compartmentalization of NOS isoforms.

6.1.1. Teleost NOS/NO System

In fish, the role of NO as a major organizer of basal cardiac physiology and as a key coordinator of physically and chemically activated intracellular signaling has been established (for references, see Imbrogno et al., 2011; Tota et al., 2010). However, some controversy still exists on aspects of the teleost NOS/NO system. For example, a canonical eNOS appears to be absent in teleosts as well as in agnathans and chondrichthyans (see Andreakis et al., 2011). In fact, while it is accepted that nNOS and iNOS derive from a large-scale duplication event before the teleost-tetrapod split, the eNOS isoform is thought to appear during tetrapod evolution (Andreakis et al., 2011). Nevertheless, studies mainly performed by our research group using physiopharmacological approaches, as well as NADPH-diaphorase and immunolocalization with heterologous mammalian antibodies, have revealed the presence of an "eNOS-like" activity in the heart of several teleost species (Amelio et al., 2006, 2008, 2013b; Garofalo et al., 2009; Imbrogno et al., 2011, 2014; Tota et al., 2005). Hopefully, ongoing studies will clarify this aspect.

Tota et al. (2005) and Amelio et al. (2006, 2008) largely analyzed the cardiac expression and localization of NOS in fish species with distinct phylogeny and ecophysiological habitats. These include extreme stenotherms, such as teleosts endemic in Antarctic waters [(i.e., the red-blooded *Trematomus bernacchii*, and species with the naturally occurring knockout for the oxygen binding chromoproteins, *Chionodraco hamatus* (hemoglobin-free) and *Chaenocephalus aceratus* (hemoglobin-free/myoglobin-free)], the African aestivating lungfish (*P. dolloi*), and the temperate eurytherms *Thunnus thynnus thynnus* and *A. anguilla* (Fig. 8). In all species, the enzyme localizes in the EE which lines the ventricular trabeculae and, to a lesser extent, in the cardiomyocytes of both the atrium and the ventricle where it mainly associates with the plasmalemma and is less evident within the cytoplasm. An eNOS-like isoform was detected by immunofluorescence on the visceral pericardium of all species examined (Amelio et al., 2006, 2008, 2013b), further supporting autocrine–paracrine activity in teleosts, as shown for other humoral cardiac regulators (e.g., ANPs in *T. bernacchii* and *C. hamatus*: Cerra et al., 1997). This zonal NOS expression is consistent with a tissue-selective intracardiac production of NO. In the teleost heart, in which the coronary system is only present in approximately one-third of fish species (see Chapter 4, Volume 36A: Farrell and Smith, 2017), this intra-cavitary NO generation contributes to the contractile modulation of myocardial performance. However, in fish showing

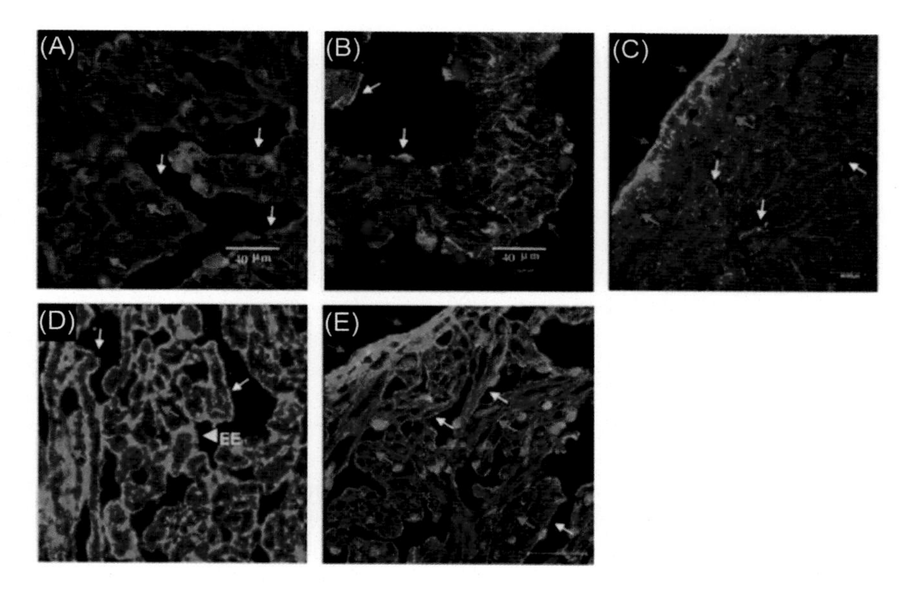

Fig. 8. Representative images of eNOS-like immunolabeling in ventricular sections of *C. aceratus* (A), *C. hamatus* (B), *T. bernacchii* (C), *A. anguilla* (D), and *P. dolloi* (E). In (A, B, and E), the nuclei (*red*) are counterstained with propidium iodide. *Yellow arrow*: EE; *pink arrow*: cardiomyocytes; *blue arrow*: epicardium. Modified from Imbrogno, S., Tota, B., Gattuso, A., 2011. The evolutionary functions of cardiac NOS/NO in vertebrates tracked by fish and amphibian paradigms. Nitric Oxide 25, 1–10. doi:10.1016/j.niox.2011.05.001.

a well-developed coronary circulation (e.g., rainbow trout), NO also profoundly affects microvessel tone/resistance (Costa et al., 2015). This contributes to the local adjustment of blood flow, and thus, of myocardial function.

6.1.1.1. Autocrine and Paracrine NO. NO, generated in one cell, is able to act on the cell itself, eliciting autocrine effects, or on adjacent cells, inducing paracrine modulation. As compared to the vascularized homeotherm heart, the fish heart is characterized by a much higher EE surface-to-myocardial volume ratio (see Imbrogno et al., 2011 for references). This offers an excellent opportunity for investigating the role of NO in paracrine signaling. In fact, the comparatively greater amount of EE covering the internal cavity of the teleost heart represents an important source of NO. This gas, by acting as a freely diffusible messenger, modulates the performance of the subjacent myocardium.

In fish, the EE is a paracrine source of NO. This has been revealed in *A. anguilla* by exposing the luminal surface of the ventricle to Triton X-100. This detergent, used at low concentrations, impairs EE function, but does not alter cell structure and ultrastructure, or pacemaker activity (for references, see Tota et al., 2005). In the presence of a functionally altered EE, the eel heart increases its contractility, probably due to an interruption of the depressive basal NO tone, consequent to the deactivation of EE-NOS signaling (Tota et al., 2005).

Of note, in addition to its role as a NO generator, the EE of the teleost heart is necessary for the coupling of intra-cavitary stimuli to the NOS/NO system. This is illustrated by the positive inotropic response induced by nanomolar concentrations of ACH, which is abolished by EE damage induced by Triton X-100 (*A. anguilla*: Imbrogno et al., 2001). Similarly, the integrity of the EE is a prerequisite for transducing the reduction in myocardial contractility that in *A. anguilla* is induced by luminal exposure to endocrines such as VS-1 (Imbrogno et al., 2004) and AngII (Imbrogno et al., 2003). This supports the idea that the EE is involved in transducing chemical stimuli to the teleost myocardium through NO–cGMP-dependent mechanisms.

The spatial confinement of NOSs within cardiomyocytes ensures that NO is able to finely modulate intracellular colocalized effectors, resulting in autocrine modulation. For example, nNOS located on the SR (Xu et al., 1999) promotes left ventricular relaxation by mainly regulating the rate of SR Ca^{2+} reuptake by SR Ca^{2+}-ATPase (SERCA2a) (Zhang et al., 2008). In contrast, myocardial eNOS is typically located in specific intracellular membrane domains, including the Golgi apparatus and plasma membrane caveolae, and primarily mediates the inotropic response to sustained stretch through a mechanism which involves NO-dependent *S*-nitrosylation of thiol residues in the ryanodine receptor's Ca^{2+} release channels (Massion et al., 2005).

In mammals, the autocrine role of NO has mainly been described in relation to the involvement of eNOS and nNOS in the stretch-induced increase in contractility described by the Frank–Starling response (Seddon et al., 2007). During the two phases of this myocardial response, both constitutive NOSs are activated. In the early phase, autocrine NO generated by nNOS located on the SR regulates intracellular Ca^{2+} by modulating the phospholamban (PLB)–SERCA2a complex. This occurs *via* a cGMP-independent action. During the subsequent slow phase, when the myocardial fibers are stretched, activated eNOS sustains the late increase in the Ca^{2+} transient, and force generation *via* the NO/sGC/cGMP pathway. This allows spatial and temporal modulation of autocrine NO production, which supports the different moments of the stretch response (for references, see Gattuso et al., 2016).

A NO-dependent autocrine pathway has also emerged in fish. For example, Garofalo et al. (2009) reported that the beat-to-beat regulation of the *in vitro* working heart of *A. anguilla* involves myocardial autocrine NO that modulates relaxation through the regulation of PLB *S*-nitrosylation-dependent Ca^{2+} reuptake by SERCA2a pumps. This resembles the situation found in mammals and supports the notion that autocrine NO-dependent myocardial regulation is tightly linked to the compartmentation of specific NOS isoforms, their different stimulation, and the recruitment of distinct downstream pathways that represent a crucial and early event during vertebrate evolution.

6.1.1.2. Nitrite as a Bioactive NO Source and Endocrine Mediator. Nitrite (NO_2^-) represents a key intrinsic signaling molecule in many biological processes, including hypoxic vasodilation, the inhibition of mitochondrial respiration, cytoprotection following I/R, and the regulation of gene and protein expression (for references, see Angelone et al., 2012).

In various cells and tissues, NO_2^- represents an important physiological reservoir of NO. It can be reduced, independent of NOS, to NO through both non-enzymatic and enzymatic pathways such as acidic disproportionation and conversion *via* xanthine-oxidoreductase, mitochondrial enzymes, deoxyhemoglobin, and deoxymyoglobin (for references, see Angelone et al., 2012). Moreover, circulating NO_2^- may represent a way by which the NO signal is delivered through the organism to mediate actions distal from its production site, thus providing the gas with endocrine properties (Elrod et al., 2008). For example, in mice overexpressing the transcript for human eNOS in cardiomyocytes, NO produced in the heart, and transported in the blood as NO_2^- and/or *S*-nitrosothiol, elicits a cytoprotective response against ischemic injury in the liver (Elrod et al., 2008).

When compared to terrestrial animals, water-breathing organisms are more likely to be exposed to environmental NO_2^-, and as a result, have an additional mechanism for uptake of exogenous NO_2^- from the respiratory surfaces (Jensen, 2009). It has been reported in zebrafish that exposure to water NO_2^- is accompanied by significant NO production and increased Hb–NO levels, vasodilation, and decreased blood pressure (Jensen, 2007). Accordingly, acting in combination with endogenous NO, NO_2^- significantly influences cardiovascular function. This effect may be amplified in the presence of chronic environmental hypoxia and acidosis that elicit NO_2^- reduction to NO (see Imbrogno et al., 2014 for references). Therefore, fish need to balance the advantage of an ambient pool of NO_2^- for internal NO production, with the potentially dangerous consequences of an imbalance of the NO/NO_2^- equilibrium, particularly in NO_2^--polluted habitats (see Jensen and Hansen, 2011).

The importance of NO_2^- in the NO-dependent control of cardiac function has been reported by Cerra et al. (2009) in *A. anguilla*, and in the icefish (*C. hamatus*). These species are particularly suited to studying NO/NO_2^- signaling. The eel experiences considerable fluctuations in environmental O_2 and is a champion of prolonged acidosis and hypoxia tolerance. In this teleost, under normoxic conditions, NO_2^- induces a reduction in contractility of the isolated perfused heart, as shown by a dose-dependent decrease of V_S

and W_S which resembles that elicited by NO. These effects depend on NOS activity, are mediated by the cGMP–PKG pathway, and are accompanied by increased protein *S*-nitrosylation (Cerra et al., 2009). In contrast, the icefish is an extreme stenotherm that is endemic to the stable, icy, and richly oxygenated water of the Antarctic. It also lacks Hb, and thus, is deprived of a key protein in NO homeostasis which is not only able to scavenge NO, but also to generate this gasotransmitter from NO_2^- (for references, see Cerra et al., 2009). Interestingly, and contrary to the eel, but similar to the effect induced in icefish by NO, NO_2^- induces a NOS-dependent increase in contractility. It is assumed that, in the absence of Hb, the reduction of NO_2^- to NO occurs through cardiac myoglobin that in icefish may represent the predominant form of NO_2^- reductase (for references, see Cerra et al., 2009).

A synopsis, based on both mammalian and nonmammalian data of NO as an autocrine/paracrine/endocrine mediator is provided in Fig. 9.

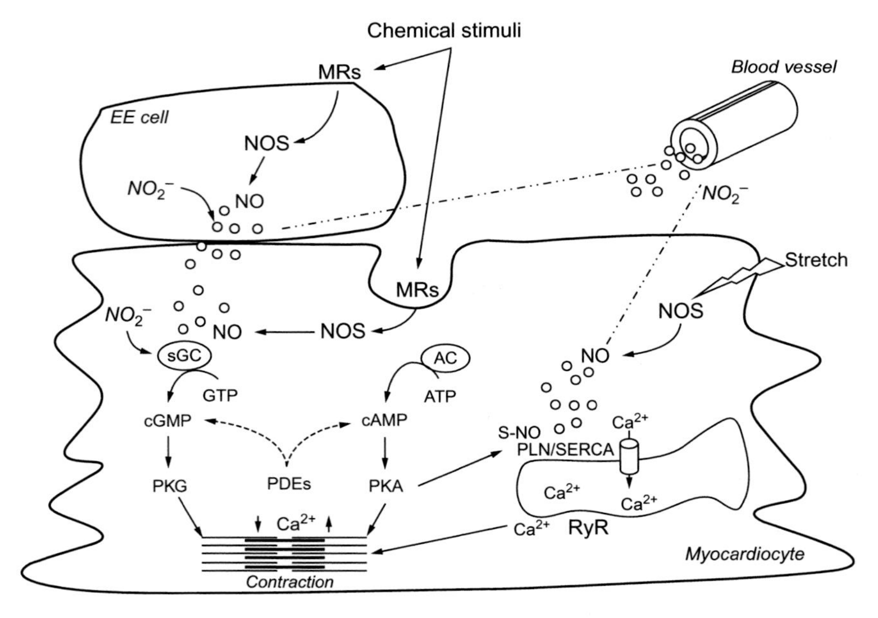

Fig. 9. Schematic diagram showing the role of NO as a spatial paracrine/autocrine/endocrine cardiac integrator. Chemical stimuli converge on the NOS–NO signaling located in the EE and/or in myocardial cell caveolae. Stretch conditions activate the release of autocrine NO which directly modulates SR Ca^{2+} reuptake through PLB *S*-nitrosylation (eel: Garofalo et al., 2009). Moreover, NO produced in the heart can be transported in the blood as nitrite, thus mediating endocrine actions distal from its site of production. Modified from Imbrogno, S., Tota, B., Gattuso, A., 2011. The evolutionary functions of cardiac NOS/NO in vertebrates tracked by fish and amphibian paradigms. Nitric Oxide 25, 1–10. doi:10.1016/j.niox.2011.05.001.

6.1.1.3. NO as a Cardioprotective Factor. Recent experimental evidence suggests that, like in mammals, the teleost NOS/NO system not only modulates heart performance under basal conditions, and in the presence of chemical and physical stimulation, but preserves cardiac function under different environmental challenges; e.g., such as those imposed by changes in temperature and oxygen availability. By using the isolated working heart of *A. anguilla*, Amelio et al. (2013b) explored the influence of temperature on NO-dependent modulation of the Frank–Starling response. In long-term temperature acclimated fish (i.e., spring animals perfused at 20°C and winter animals perfused at 10°C) inhibition of NO production significantly reduced the Frank–Starling response, whereas under thermal shock (spring animals perfused at 10 or 15°C and winter animals perfused at 15 or 20°C), NOS inhibition by L-N5 (1-iminoethyl)-ornithine (L-NIO) was without effect on the preload-induced increase of V_S. Moreover, eels exposed to thermal shock showed decreased NOS expression, which was associated with a decrease in pAkt and an increase in HSP90 expression. These data strongly suggest that NOS/NO-dependent modulation of the Frank–Starling mechanism in fish is sensitive to thermal stress.

Data on the goldfish (*C. auratus*) and the Crucian carp (*C. carassius*) exposed to hypoxia/anoxia further support a protective role for the intracardiac NOS/NO system (Hansen and Jensen, 2010; Sandvik et al., 2012). It is known that reduced oxygen availability compromises NOS-mediated NO generation. Under these conditions, an increased cardiac NOS expression or, alternatively, a reduction of NO_2^- to NO conversion may contribute to the stabilization of myocardial NO levels (Hansen and Jensen, 2010). In line with these observations, Imbrogno et al. (2014) showed that the isolated and perfused working goldfish heart exposed to both normoxia and hypoxia is under a tonic NO–cGMP modulation that also enhances the Frank–Starling response. In the presence of reduced oxygen availability, cardiac NOS expression is increased, and this suggests an elevation in NO production that may contribute to adjusting heart performance during hypoxic challenges.

6.2. Carbon Monoxide

Studies on mammals have demonstrated that CO is generated during heme catabolism by heme oxygenase (HO) enzymes, and that this is accomplished *via* oxidation at the alpha-methene bridge which also generates biliverdin and free Fe^{2+}. Three HO isoforms have been identified in mammals, namely HO-1 (inducible), and HO-2 and HO-3 (both constitutive), which are the products of individual genes (for references, see Wu and Wang, 2005). Only HO-1 and HO-2 are present in teleost fish (*C. auratus*: Wang et al., 2008; *Dicentrarchus labrax:* Prevot-D'Alvise et al., 2013; *Cyprinus carpio*: Jancsó and Hermesz, 2015; *Danio rerio*: Tzaneva and Perry, 2016).

In mammals, CO elicits multiple effects, including the regulation of neurotransmitter and neuropeptide release, learning and memory, and odor response adaptation. It is also active in the immune, gastrointestinal, kidney, reproductive, and liver systems (for references, see Wu and Wang, 2005), and plays an important role in respiration by controlling carotid body activity (Prabhakar et al., 1995). In addition to these effects, CO shows vasoactive and cardioprotective properties. It is generally considered to be a vasodilator (Dombkowski et al., 2009) acting through sGC and Ca^{2+}-activated K^+ (K_{Ca}) channels (Dombkowski et al., 2009). However, its effects are heterogeneous, even within similar vascular districts. For example, piglet cerebral arterioles are dilated by CO, whereas cerebral vessels of rabbits and dogs are unaffected by this gas (Dombkowski et al., 2009).

In mammals, CO reduces the contractility of papillary muscles (Liu et al., 2001), and protects myocardial cells and isolated rat and mouse hearts against ischemia/reperfusion (I/R) damage (Guo et al., 2004). Several investigations have demonstrated that CO has similar regulatory roles in teleosts as it does in mammals. For example, in goldfish and zebrafish, the gas was found to induce a temperature- and species-specific inhibition of ventilation in response to changes in water O_2 (references in Perry and Tzaneva, 2016). At the cardiovascular level, it elicits vasorelaxation, as observed on efferent branchial and celiacomesenteric arteries of the rainbow trout, and of the dorsal aorta of the sea lamprey exposed to the CO donor CORM-3 (Dombkowski et al., 2009). This effect is presumably mediated by HO-generated gas which is also responsible for the tonic control of vascular motility. As in mammals, CO-induced relaxation partially depends on activation of GC, and of either K^+ or Cl^- channels (Dombkowski et al., 2009).

Although data supporting a direct effect of CO on the teleost heart are very limited, recent investigations on zebrafish larvae suggest that this gas influences *in vivo* teleost cardiac activity. In fact, under normoxia, in zebrafish unable to produce CO because of HO-1 knockdown, f_H is significantly higher when compared with sham-treated fish. This CO-dependent effect on f_H is supported by the identification of HO-1 in cardiac pacemaker cells, and by the return to normal f_H in the presence of CO. In addition, HO-1 knockdown zebrafish larvae have larger ventricles and generate higher CO and V_S when exposed to hypoxia than control animals (Tzaneva and Perry, 2016).

6.3. Hydrogen Sulfide

H_2S is a potent membrane permeable poison, even more toxic than cyanide. This gases toxicity is due to its interference with metalloproteins, such as cytochrome *c* oxidase (COX), and with cysteine-rich proteins. The latter effect acutely alters ion channels with critical implications for cardiac and brain functions, and this easily and rapidly results in severe functional damage and death (Haouzi et al., 2016). Natural environments usually have low or transient H_2S concentrations. However, these concentrations may increase, thus establishing extreme conditions, lethal for most life (Riesch et al., 2015; Tobler and Plath, 2011).

H_2S is naturally produced by anaerobic digestion of organic substrates, as often occurs in sewers and swamps, and through inorganic reactions in volcanic and natural gas and well waters (for references, see Mancardi et al., 2009). Accordingly, it represents an environmental hazard. A few animals have colonized H_2S-rich habitats by developing multiple adaptive strategies which include modifications of: (i) the integumentary system and respiratory surfaces directly exposed to the environment to generate a barrier against H_2S; (ii) mechanisms involved in the active elimination of H_2S; and (iii) H_2S toxicity targets, making them less sensitive to damage induced by high H_2S amount. An intriguing example of H_2S-adapted species is those of the genus *Poecilia*. These species (i.e., *P. mexicana*, *Poecilia sulphuraria*, and *P. thermalis*) independently colonized at least three Mexican springs with toxic concentrations of the gas, giving rise to species with peculiar phenotypic traits. These sulfide-adapted fishes have enlarged heads and an increased gill surface area, which facilitates oxygen acquisition. They also show physiological and biochemical adaptations that reduce the impacts of sulfide toxicity, including a decreased H_2S susceptibility of COX (for references, see Kelley et al., 2016; Palacios et al., 2013; Pfenninger et al., 2014).

Only in the last few years, thanks to biomedically oriented studies, has attention focused on the function of H_2S as an endogenous gasotransmitter. Since 1930, it has been suggested that H_2S is basally produced by animal tissues (Sluiter, 1930), and it is now clear that the gas is generated as a side product of cysteine metabolic enzymes, namely cystathionine beta-synthase (CBS), and cystathionine gamma-lyase (CSE) (Erickson et al., 1990; Stipanuk and Beck, 1982; Swaroop et al., 1992). Studies in mammals demonstrated that, once produced, H_2S influences the cardiovascular system by inducing a hypometabolic state associated with antiinflammatory, anti-apoptotic, and antioxidant effects (Elrod et al., 2007; Snijder et al., 2013; Wang et al., 2011). It also inhibits the β-AR system, and elicits protection against I/R injury through a modulation of K_{ATP} channels opening, an inhibition of mitochondrial respiration, and the preservation of mitochondrial membrane integrity (for references, see Mancardi et al., 2009).

Apart from the above-cited studies aimed at describing the mechanisms of adaptation to high environmental H_2S, the cardiovascular activity of H_2S in non-mammalian vertebrates has scarcely been investigated. However, data on amphibians suggest a role as a cardiac modulator. As observed in the frog (*R. esculenta*), H_2S generation by NaHS (H_2S donor) dose-dependently (10^{-12}–10^{-7} M) decreases the contractile performance of the isolated and perfused working heart, with a decrement of about 40% in V_S and W_S at the highest concentration tested. Since the frog heart is avascular, this effect is suggested to be the result of a direct action on the myocardium (Mazza et al., 2013). It was also found that the gas reduces the contractile force of isolated frog ventricle myocardial strips. In both cases, K_{ATP} channels were involved (see Mazza et al., 2013 for comments and references).

In teleosts, several investigations have proposed a role for H_2S in ion transfer regulation (Kumai et al., 2015; Kwong and Perry, 2015) and O_2 sensing (Olson et al., 2006). However, data on lamprey and rainbow trout show that it is enzymatically produced at the level of the vascular smooth muscle, and this suggests a cardiovascular function (see Olson et al., 2006 and references therein). In addition, teleost vessels are sensitive to H_2S as they show either monophasic contraction and relaxation, or multiphasic effects, when exposed to the gas (Olson and Donald, 2009). These effects are dependent on K_{ATP} channels and/or prostanoids (trout efferent branchial arteries relaxation), and/or are endothelium dependent (lamprey and hagfish aorta contraction) (Olson and Donald, 2009).

In teleosts, the heart is also a site of H_2S generation. In trout, under hypoxic conditions, application of a polarographic sensor revealed local myocardial production of the gas in the presence of cysteine. Since the gas was not detected in the plasma, it was suggested that it does not serve as a circulating signaling molecule (Whitfield et al., 2008). It is known that the trout myocardium is extremely sensitive to hypoxia, but is protected by a preconditioning-like mechanism (Gamperl et al., 2001). Accordingly, it was hypothesized that in fish, like in mammals (Mancardi et al., 2009), H_2S paracrinally protects the myocardium in the presence of ischemia (Whitfield et al., 2008). Very recently, Wu et al. (2016) revealed that H_2S also influences the development of cardiac function. By exposing zebrafish embryos to different concentrations of H_2S derived from exogenous Na_2S administration, they observed that low sulfide levels (0.02, 0.05, and 0.08 mM) increase f_H and development velocity. This was attributed to the cardioprotective properties of this gas. In contrast, exposure to higher H_2S concentrations (0.10 and 0.20 mM) is accompanied by a decrease in f_H, when compared with control embryos transferred to reconstituted water.

7. INTEGRATED CARDIAC HUMORAL SIGNALING: THE "KNOT" OF THE NOS–NO SYSTEM

An intriguing aspect of the circuits that in teleosts—as in mammals—sustain cardiac homeostasis and function in response to internal and environmental requirements is the convergence on common molecular circuits. According to the system biology perspective (see Tota et al., 2010 for references), this requires that information is continuously transferred among the different parts of the whole system, and that few points of convergence are recruited to elaborate inputs and to generate a variety of responses. By including redundancies as safety factors, this increases the flexibility and the robustness of biological systems. For the continuously beating heart this is a crucial homeostatic task, that in the teleost is challenged not only by immediate (i.e., beat-to-beat) requirements, but also by many stressors such as variable hemodynamic loads, changes in water temperature, pH, osmolarity, O_2 and CO_2 availability, etc. As illustrated by the examples described in this chapter, the teleost heart acts as a sensor/transducer able to integrate and coordinate signals that allow homeostatic plasticity. In this context, the intracardiac NOS/NO/cGMP system, located downstream and at the crossroads of a large number of regulatory neuroendocrine pathways, represents a "knot" for convergent molecular signaling. This occurs also in fish, in which this cascade contributes importantly to the effects observed in the presence of β_3-ARs stimulation (Imbrogno et al., 2006), exposure to AngII (Imbrogno et al., 2003), and CgA-derived peptides (Imbrogno et al., 2004, 2010) (Fig. 10). Once synthesized by the teleost heart, NO binds to sGC and elicits a several hundredfold increase in cGMP content, and an activation of cGMP-dependent pathways. It also directly modulates protein function by generating *S*-nitrosothiol (Garofalo et al., 2009). The result is the modulation of intracellular targets, such as membrane ion channels, transporters, and contractile proteins, and a fine control of myocardial contractility, relaxation, and stretch-induced modulation.

Unfortunately, there are no data on fish with respect to if and how the NOS/NO system cross-talks with the other gasotransmitters. However, in the rat, CO vasodilatory actions are elicited by stimulating the same intracellular target of NO, namely sGC, and induce the same subsequent decrease in intracellular Ca^{2+} and cell hyperpolarization (Wang et al., 1997). In addition, CO modulates sGC activated by NO, with both gases interacting with the heme group of the cyclase (Kharitonov et al., 1995). Interaction also occurs between the NOS–NO system and H_2S. NO enhances H_2S release from vascular tissues by increasing CSE expression. In turn, H_2S may either enhance or reduce NO-dependent relaxation. H_2S also induces vasoconstriction at low concentrations, an effect, which involves NO (see for references, Mazza et al., 2013). Several additional/alternative mechanisms are proposed for this effect. One is an H_2S-dependent suppression of tonic NO-dependent vasodilation; another is the reaction of the gas with NO to form nitrosothiol; the last is direct NOS inhibition (see for references, Mazza et al., 2013). Interestingly, H_2S is thought to stimulate eNOS activity through an Akt-dependent

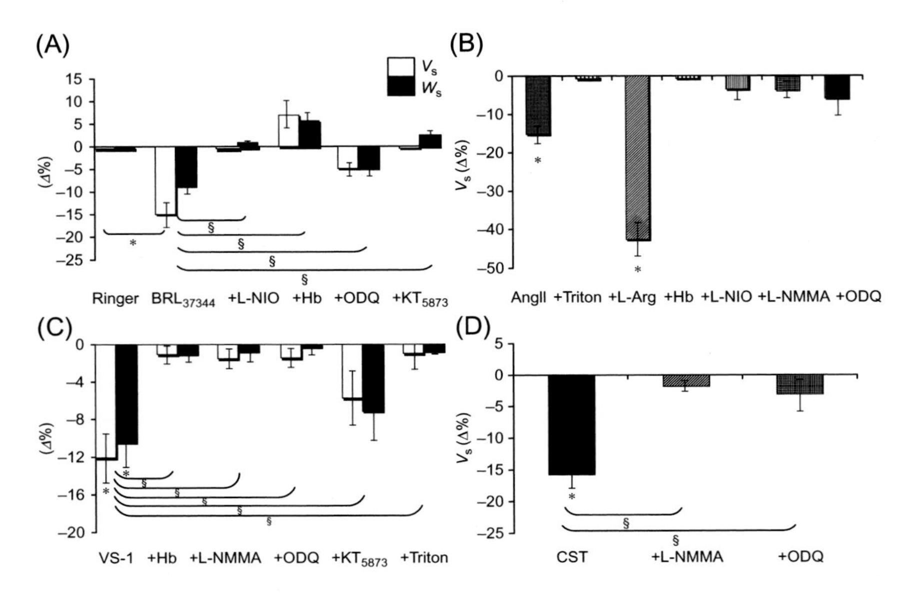

Fig. 10. (A) Effects of BRL_{37344} (10 nmol L^{-1}) before and after treatment with L-N5-1-iminoethyl) ornitine (L-NIO; 10 μmol L^{-1}), hemoglobin (Hb; 1 μmol L^{-1}), 1H-(1,2,4)oxadiazolo-(4,3-a) quinoxalin-1-one (ODQ; 10 μmol L^{-1}), and KT_{5823} (100 nmol L^{-1}) on stroke volume (V_S) and stroke work (W_S) in isolated and perfused eel (*A. anguilla*) hearts. (B) Effects of AngII (10^{-8} mol L^{-1}) before and after injection with Triton X-100 (0.05%), or perfusion with L-arginine (L-Arg; 10^{-6} mol L^{-1}), Hb (10^{-6} mol L^{-1}), L-NIO (10^{-5} mol L^{-1}), NG-monomethyl-larginine (L-NMMA; 10^{-5} mol L^{-1}), and ODQ (10^{-5} mol L^{-1}) on V_S in isolated and perfused *A. anguilla* hearts. (C) Effects of VS-1 (33 nM) before and after treatment with Hb (1 μM), L-NMMA (10 μM), ODQ (10 μM), KT_{5823} (100 nM), and Triton X-100 (0.05%) on V_S and W_S of the isolated and perfused *A. anguilla* heart. (D) Effects of CST (110 nmol L^{-1}) on V_S before and after treatment with L-NMMA (10 μmol L^{-1}) or ODQ (10 μmol L^{-1}). From Imbrogno, S., Cerra, M.C., Tota, B., 2003. Angiotensin II-induced inotropism requires an endocardial endothelium-nitric oxide mechanism in the in-vitro heart of *Anguilla anguilla*. J. Exp. Biol. 206, 2675–2684; Imbrogno, S., Angelone, T., Corti, A., Adamo, C., Helle, K.B., Tota, B., 2004. Influence of vasostatins, the chromogranin A-derived peptides, on the working heart of the eel (*Anguilla anguilla*): negative inotropy and mechanism of action. Gen. Comp. Endocrinol. 139, 20–28; Imbrogno, S., Angelone, T., Adamo, C., Pulerà, E., Tota, B., Cerra, M.C., 2006. Beta3-adrenoceptor in the eel (*Anguilla anguilla*) heart: negative inotropy and NO-cGMP-dependent mechanism. J. Exp. Biol. 209, 4966–4973; Imbrogno, S., Garofalo, F., Cerra, M.C., Mahata, S.K., Tota, B., 2010. The catecholamine release-inhibitory peptide catestatin (chromogranin A344–363) modulates myocardial function in fish. J. Exp. Biol. 213, 3636–3643. doi:10.1242/jeb.045567.

mechanism, which results in increased NO generation. This stimulatory effect on NO production also characterizes the H_2S-dependent inotropic response of the non-mammalian heart. In fact, in the avascular heart of the frog (*R. esculenta*), H_2S elicits negative inotropism by activating an Akt-dependent

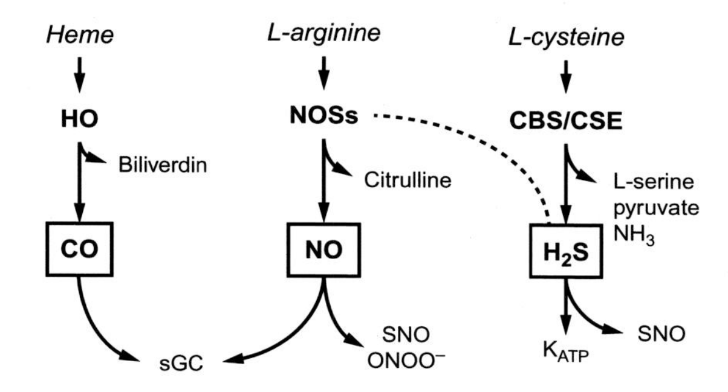

Fig. 11. Simplified scheme, based on data in mammals, showing the biosynthetic pathways for CO, H_2S, and NO generation, and the intracellular targets of the three gasotransmitters. The convergence of CO and NO on sGC, and the inhibitory relationship between H_2S and the NOSs system, is also reported. For details and references, see the text.

eNOS, the cGMP/PKG pathway, and an intact EE (Mazza et al., 2013). A simplified scheme of the interactions among the three gasotransmitters is provided in Fig. 11.

8. CONCLUSIONS

Over the past few years, cardiovascular research has provided novel and crucial pieces of information on the hormonal and autacoid circuits that sustain fish heart homeostasis. A large variety of extrinsic and intrinsic signals allow for fine beat-to-beat, medium-term, and long-term adaptation of fish heart function to internal and external stimuli. These signals converge on common molecular interfaces that operate for connecting and integrating information. This is epitomized by the NOS/NO system, which is located downstream and at the crossroads of complex neuroendocrine networks, and acts as a major organizer of cardiac responses as those activated by β_3-ARs stimulation, and AngII and CgA-derived peptides. The examples provided here are far from exhaustive, but are indicative of a dynamic balance between endocrine substances, receptors, and autocrine/paracrine pathways that results in fine control of fish myocardial contractility, relaxation, and stretch-induced modulation, and an appropriate response to environmental challenges. Further research in fish will better decipher the spatiotemporally integrated molecular strategies involved in cardiac regulation. From an evolutionary and comparative perspective, it is also important to understand how in vertebrates, including mammals, the heterogeneities of cardiac form and function support the uniformity of whole heart function (i.e., pumping).

REFERENCES

Abrahamsson, T., 1979. Phenylethanolamine-N-methyl transferase (PNMT) activity and catecholamine storage and release from chromaffin tissue of the spiny dogfish. Squalus acanthias. Comp. Biochem. Physiol. C Comp. Pharmacol. 64 (1), 169–172.

Abrahamsson, T., Holmgren, S., Nilsson, S., Pettersson, K., 1979. On the chromaffin system of the African lungfish, *Protopterus aethiopicus*. Acta Physiol. Scand. 107, 135–139.

Agnisola, C., Randall, D.J., Taylor, E.W., 2003. The modulatory effects of noradrenaline on vagal control of heart rate in the dogfish, *Squalus acanthias*. Physiol. Biochem. Zool. 76, 310–320.

Ahrens, F.A., 1996. Pharmacology. Wiley-Blackwell, Philadelphia, ISBN: 978-0-683-00085-6, p. 45.

Amelio, D., Garofalo, F., Pellegrino, D., Giordano, F., Tota, B., Cerra, M.C., 2006. Cardiac expression and distribution of nitric oxide synthases in the ventricle of the cold-adapted Antarctic teleosts, the hemoglobinless *Chionodraco hamatus* and the red-blooded *Trematomus bernacchii*. Nitric Oxide 15, 190–198.

Amelio, D., Garofalo, F., Brunelli, E., Loong, A.M., Wong, W.P., Ip, Y.K., Tota, B., Cerra, M.C., 2008. Differential NOS expression in freshwater and aestivating *Protopterus dolloi* (lungfish): heart vs kidney readjustments. Nitric Oxide 18, 1–10.

Amelio, D., Garofalo, F., Wong, W.P., Chew, S.F., Ip, Y.K., Cerra, M.C., Tota, B., 2013a. Nitric oxide synthase-dependent "on/off" switch and apoptosis in freshwater and aestivating lungfish, *Protopterus annectens*: skeletal muscle versus cardiac muscle. Nitric Oxide 32, 1–12. http://dx.doi.org/10.1016/j.niox.2013.03.005.

Amelio, D., Garofalo, F., Capria, C., Tota, B., Imbrogno, S., 2013b. Effects of temperature on the nitric oxide-dependent modulation of the Frank-Starling mechanism: the fish heart as a case study. Comp. Biochem. Physiol. 164A, 356–362. http://dx.doi.org/10.1016/j.cbpa.2012.10.037.

Andreakis, N., D'Aniello, S., Albalat, R., Patti, F.P., Garcia-Fernàndez, J., Procaccini, G., Sordino, P., Palumbo, A., 2011. Evolution of the nitric oxide synthase family in metazoans. Mol. Biol. Evol. 28, 163–179.

Angelone, T., Filice, E., Quintieri, A.M., Imbrogno, S., Recchia, A., Pulerà, E., Mannarino, C., Pellegrino, D., Cerra, M.C., 2008a. Beta (3)-Adrenoceptors modulate left ventricular relaxation in the rat heart *via* the NO-cGMP-PKG pathway. Acta Physiol. 193, 229–239.

Angelone, T., Quintieri, A.M., Brar, B.K., Limchiyawat, P.T., Tota, B., Mahata, S.K., Cerra, M.C., 2008b. The antihypertensive chromogranin a peptide catestatin acts as a novel endocrine/paracrine modulator of cardiac inotropism and lusitropism. Endocrinology 149, 4780–4793.

Angelone, T., Gattuso, A., Imbrogno, S., Mazza, R., Tota, B., 2012. Nitrite is a positive modulator of the Frank-Starling response in the vertebrate heart. Am. J. Physiol. 302, R1271–R1281.

Arjamaa, O., Sormunen, R., Lehto, V.P., Vuolteenaho, O., 2000. Localization of salmon cardiac peptide (sCP) in the heart of salmon (*Salmo salar* L.). Gen. Comp. Endocrinol. 120, 276–282.

Ask, J.A., 1983. Comparative aspects of adrenergic receptors in the hearts of lower vertebrates. Comp. Biochem. Physiol. A Physiol. 76, 543–552.

Ask, J.A., Stene-Larsen, G., Helle, K.B., 1981. Temperature effects on the b2-adrenoceptors of the trout atrium. J. Comp. Physiol. 143, 161–168.

Augustinsson, K.B., Fänge, R., Johnels, A., Östlund, E., 1956. Histological, physiological and biochemical studies on the heart of two cyclostomes, hagfish (Myxine) and lamprey (Lampetra). J. Physiol. 131 (2), 257–276.

Axelsson, M., 1995. The coronary circulation: a fish perspective. Braz. J. Med. Biol. Res. 28, 1167–1177.

Axelsson, M., Ehrenström, F., Nilsson, S., 1987. Cholinergic and adrenergic influence on the teleost heart *in vivo*. Exp. Biol. 46, 179–186.

Balligand, J.L., 2000. Regulation of cardiac function by nitric oxide. In: Mayer, B. (Ed.), Nitric Oxide. Handbook of Experimental Pharmacology, vol. 143. Springer-Verlag, Berlin, Heidelberg, New York, pp. 206–234.

Balligand, J.L., Cannon, P.J., 1997. Nitric oxide synthases and cardiac muscle. Autocrine and paracrine influences. Arterioscler. Thromb. Vasc. Biol. 17, 1846–1858.

Banks, P., Helle, K., 1965. The release of protein from the stimulated adrenal medulla. Biochem. J. 97, 40C–41C.

Bassino, E., Fornero, S., Gallo, M.P., Gallina, C., Femminò, S., Levi, R., Tota, B., Alloatti, G., 2015. Catestatin exerts direct protective effects on rat cardiomyocytes undergoing ischemia/reperfusion by stimulating PI3K-Akt-GSK3β pathway and preserving mitochondrial membrane potential. PLoS One 10 (3), e0119790. http://dx.doi.org/10.1371/journal.pone.0119790.

Becker, J.R., Chatterjee, S., Robinson, T.Y., Bennett, J.S., Panáková, D., Galindo, C.L., Zhong, L., Shin, J.T., Coy, S.M., Kelly, A.E., Roden, D.M., Lim, C.C., MacRae, C.A., 2014. Differential activation of natriuretic peptide receptors modulates cardiomyocyte proliferation during development. Development 141, 335–345. http://dx.doi.org/10.1242/dev.100370.

Bernier, N.J., Perry, S.F., 1999. Cardiovascular effects of angiotensin-II-mediated adrenaline release in rainbow trout *Oncorhynchus mykiss*. J. Exp. Biol. 202, 55–66.

Bernier, N.J., Mckendry, J.E., Perry, S.F., 1999. Blood pressure regulation during hypotension in two teleost species: differential involvement of the renin–angiotensin and adrenergic systems. J. Exp. Biol. 202, 1677–1690.

Booz, G.W., Day, J.N., Baker, K.M., 2002. Interplay between the cardiac renin angiotensin system and JAK-STAT signalling: role in cardiac hypertrophy, ischemia/reperfusion dysfunction, and heart failure. J. Mol. Cell. Cardiol. 34, 1443–1453. Review.

Burnstock, G., 1969. Evolution of the autonomic innervation of visceral and cardiovascular systems in vertebrates. Pharmacol. Rev. 21, 247–324.

Butler, P.J., Metcalfe, J.D., Ginley, S.A., 1986. Plasma catecholamines in the lesser spotted dogfish and rainbow trout at rest and during different levels of exercise. J. Exp. Biol. 123, 409–421.

Cappello, S., Angelone, T., Tota, B., Pagliaro, P., Penna, C., Rastaldo, R., Corti, A., Losano, G., Cerra, M.C., 2007. Human recombinant chromogranin A-derived vasostatin-1 mimics preconditioning via an adenosine/nitric oxide signaling mechanism. Am. J. Physiol. Heart Circ. Physiol. 293, 719–727.

Capra, M.F., Satchell, G.H., 1977. Adrenergic and cholinergic responses of the isolated saline-perfused heart of the elasmobranch fish, *Squalus acanthias*. Gen. Pharmacol. 8, 59–65.

Carroll, R.G., Opdyke, D.F., 1982. Evolution of angiotensin II-induced catecholamine release. Am. J. Physiol. 243, R54–R69.

Cerra, M.C., Canonaco, M., Tota, B., 1992. A quantitative autoradiographic study of 125I atrial natriuretic factor in the heart of a teleost fish (*Conger conger*). J. Exp. Zool. 263, 215–219.

Cerra, M.C., Canonaco, M., Takei, Y., Tota, B., 1996. Characterization of natriuretic peptide binding sites in the heart of the eel *Anguilla anguilla*. J. Exp. Zool. 275, 27–35.

Cerra, M.C., Canonaco, M., Acierno, R., Tota, B., 1997. Different binding activity of A- and B-type natriuretic hormones in the heart of two Antartic teleosts, the red blooded *Trematomus bernacchii* and the haemoglobinless *Chionodraco hamatus*. Comp. Biochem. Physiol. 118, 993–999.

Cerra, M.C., Tierney, M.L., Takei, Y., Hazon, N., Tota, B., 2001. Angiotensin II binding sites in the heart of *Scyliorhinus canicula*: an autoradiographic study. Gen. Comp. Endocrinol. 121, 126–134.

Cerra, M.C., Imbrogno, S., Amelio, D., Garofalo, F., Colvee, E., Tota, B., Icardo, J.M., 2004. Cardiac morpho-dynamic remodelling in the growing eel. J. Exp. Biol. 207, 2867–2875.

Cerra, M.C., De Iuri, L., Angelone, T., Corti, A., Tota, B., 2006. Recombinant N-terminal fragments of chromogranin-A modulate cardiac function of the Langendorff perfused rat heart. Basic Res. Cardiol. 101, 43–52.

Cerra, M.C., Angelone, T., Parisella, M.L., Pellegrino, D., Tota, B., 2009. Nitrite modulates contractility of teleost (*Anguilla anguilla* and *Chionodraco hamatus*, i.e. the Antarctic hemoglobinless icefish) and frog (*Rana esculenta*) hearts. Biochim. Biophys. Acta Theriol. 1787, 849–855.

Chintalgattu, V., Nair, D.M., Katwa, L.C., 2003. Cardiac myofibroblasts: a novel source of vascular endothelial growth factor (VEGF) and its receptors Flt-1 and KDR. J. Mol. Cell. Cardiol. 35, 277–286.

Cobb, C.S., Brown, J.A., 1992. Angiotensin II binding to tissues of the rainbow trout, *Oncorhynchus mykiss*, studied by autoradiography. J. Comp. Physiol. B 162, 197–202.

Cobb, C.S., Frankling, S.C., Rankin, J.C., Brown, J.A., 2002. Angiotensin converting enzyme-like activity in tissues from the river lamprey or lampern, *Lampetra fluviatilis*, acclimated to freshwater and seawater. Gen. Comp. Endocrinol. 127 (1), 8–15.

Cobb, C.S., Frankling, S.C., Thorndyke, M.C., Jensen, F.B., Rankin, J.C., Brown, J.A., 2004. Angiotensin I-converting enzyme-like activity in tissues from the Atlantic hagfish (*Myxine glutinosa*) and detection of immunoreactive plasma angiotensins. Comp. Biochem. Physiol. B Biochem. Mol. Biol. 138 (4), 357–364.

Conlon, J.M., Yano, K., Olson, K.R., 1996. Production of [Asn1, Val5] angiotensin II and [Asp1, Val5] angiotensin II in kallikrein-treated trout plasma (T60K). Peptides 17, 527–530.

Corti, A., Mannarino, C., Mazza, R., Colombo, B., Longhi, R., Tota, B., 2002. Vasostatins exert negative inotropism in the working heart of the frog. Ann. N.Y. Acad. Sci. 971, 362–365.

Corti, A., Mannarino, C., Mazza, R., Angelone, T., Longhi, R., Tota, B., 2004. Chromogranin A N-terminal fragments vasostatin-1 and the synthetic CGA7–57 peptide act as cardiostatins on the isolated working frog heart. Gen. Comp. Endocrinol. 136, 217–224.

Costa, I.A., Hein, T.W., Secombes, C.J., Gamperl, A.K., 2015. Recombinant interleukin-1β dilates steelhead trout coronary microvessels: effect of temperature and role of the endothelium, nitric oxide and prostaglandins. J. Exp. Biol. 218, 2269–2278. http://dx.doi.org/10.1242/jeb.119255.

Cousins, K.L., Farrell, A.P., 1996. Stretch-induced release of atrial natriuretic factor from the heart of rainbow trout (*Oncorhynchus mykiss*). Can. J. Zool. 74, 380–387.

Dashow, L., Epple, A., 1985. Plasma catecholamines in the lamprey: intrinsic cardiovascular messengers? Comp. Biochem. Physiol. C Pharmacol. Toxicol. Endocrinol. 82, 119–122.

de Bold, A.J., Borenstein, H.B., Veress, A.T., Sonnenberg, H., 1981. A rapid and potent natriuretic response to intravenous injection of atrial myocardial extract in rats. Life Sci. 28, 89–94.

de Gasparo, M., 2002. Angiotensin II and nitric oxide interaction. Heart Fail. Rev. 7, 347–358.

Deftos, L.J., Björnsson, B.T., Burton, D.W., O'Connor, D.T., Copp, D.H., 1987. Chromogranin A is present in and released by fish endocrine tissue. Life Sci. 40, 2133–2136.

Dombkowski, R.A., Whitfield, N.L., Motterlini, R., Gao, Y., Olson, K.R., 2009. Effects of carbon monoxide on trout and lamprey vessels. Am. J. Physiol. Regul. Integr. Comp. Physiol. 296, 141–149. http://dx.doi.org/10.1152/ajpregu.90507.2008.

Donald, J.A., Vomachka, A.J., Evans, D.H., 1992. Immunohistochemical localisation of natriuretic peptides in the brains and hearts of the spiny dogfish *Squalus acanthias* and the Atlantic hagfish *Myxine glutinosa*. Cell Tissue Res. 270, 535–545.

Dostal, D.E., 2000. The cardiac renin–angiotensin system: novel signalling mechanisms related to cardiac growth and function. Regul. Pept. 91, 1–11.

Duff, D.W., Olson, K.R., 1986. Trout vascular and renal responses to atrial natriuretic factor and heart extract. Am. J. Physiol. Regul. Integr. Comp. Physiol. 251, R639–R642.

Ehlers, M.R.W., Riordan, J.F., 1990. Angiotensin converting enzyme: biochemistry and molecular biology. In: Laragh, J.H., Brenner, B.M. (Eds.), Hypertension: Pathophysiology, Diagnosis, and Management. Raven, New York, pp. 1217–1231.
Eiden, L.E., 1987. Is chromogranin a prohormone? Nature 325 (6102), 301. No abstract available.
Elrod, J.W., Calvert, J.W., Morrison, J., Doeller, J.E., Kraus, D.W., Tao, L., Jiao, X., Scalia, R., Kiss, L., Szabo, C., Kimura, H., Chow, C.W., Lefer, D.J., 2007. Hydrogen sulfide attenuates myocardial ischemia reperfusion injury by preservation of mitochondrial function. Proc. Natl. Acad. Sci. U.S.A. 104, 15560–15565.
Elrod, J.W., Calvert, J.W., Gundewar, S., Bryan, N.S., Lefer, D.J., 2008. Nitric oxide promotes distant organ protection: evidence for an endocrine role of nitric oxide. Proc. Natl. Acad. Sci. U.S.A. 105, 11430–11435.
Erickson, P.F., Maxwell, I.H., Su, L.J., Baumann, M., Glode, L.M., 1990. Sequence of cDNA for rat cystathionine gamma-lyase and comparison of deduced amino acid sequence with related *Escherichia coli* enzymes. Biochem. J. 269, 335–340.
Fago, A., Jensen, F.B., Tota, B., Feelisch, M., Olson, K.R., Helbo, S., Lefevre, S., Mancardi, D., Palumbo, A., Sandvik, G.K., Skovgaard, N., 2012. Integrating nitric oxide, nitrite and hydrogen sulfide signalling in the physiological adaptations to hypoxia: a comparative approach. Comp. Biochem. Physiol. A Mol. Integr. Physiol. 162, 1–6. http://dx.doi.org/10.1016/j.cbpa.2012.01.011.
Farrell, A.P., Jones, D.R., 1992. The Heart. Fish Physiology, vol. 12A. Academic Press, San Diego, pp. 1–88.
Farrell, A.P., Olson, K.R., 2000. Cardiac natriuretic peptides: a physiological lineage of cardioprotective hormones? Physiol. Biochem. Zool. 73, 1–11.
Farrell, A.P., Smith, F., 2017. Cardiac form, function and physiology. In: Gamperl, A.K., Gillis, T.E., Farrell, A.P., Brauner, C.J. (Eds.), Fish Physiology. In: The Cardiovascular System: Morphology, Control and Function, vol. 36A. Academic Press, San Diego, pp. 155–264.
Farrell, A.P., MacLeod, K.R., Chancey, B., 1986. Intrinsic mechanical properties of the perfused rainbow trout heart and the effects of catecholamines and extracellular calcium under control and acidotic conditions. J. Exp. Biol. 125, 319–345.
Forster, M.E., 1976. Effects of catecholamines on the heart and on branchial and peripheral resistance of the eel, *Anguilla anguilla* (L.). Comp. Biochem. Physiol. 55, 27–32.
Forster, M.E., 1981. Effects of catecholamines on the hearts and ventral aortas of the eel, *Anguilla australis schmidtii* and *Anguilla dieffenbachii*. Comp. Biochem. Physiol. 70C, 85–90.
Fritsche, R., Axelsson, M., Franklin, C.E., Grigg, G.G., Holmgren, S., Nilsson, S., 1993. Respiratory and cardiovascular responses to hypoxia in the Australian lungfish. Respir. Physiol. 94, 173–187.
Furchgott, R.F., Zawadzki, J.V., 1980. The obligatory role of endothelial cells in the relaxation of arterial smooth muscle by acetylcholine. Nature 288, 373–376.
Gallinat, S., Busche, S., Raizada, M.K., Sumners, C., 2000. The angiotensin II type 2 receptor: an enigma with multiple variations. Am. J. Physiol. Endocrinol. Metab. 278, E357–E374.
Gamperl, A.K., Wilkinson, M., Boutilier, R.G., 1994. Beta adrenoreceptors in the trout (*Oncorhynchus mykiss*) heart: characterization, quantification, and effects of repeated catecholamine exposure. Gen. Comp. Endocrinol. 95, 259–272.
Gamperl, A.K., Todgham, A.E., Parkhouse, W.S., Dill, R., Farrell, A.P., 2001. Recovery of trout myocardial function following anoxia: preconditioning in a non-mammalian model. Am. J. Physiol. Regul. Integr. Comp. Physiol. 281, R1755–R1763.
Gannon, J.B., Burnstock, G., 1969. Exitatory adrenergic innervation of the fish heart. Comp. Biochem. Physiol. 29, 765–773.
Garg, R., Pandey, K.N., 2005. Regulation of guanylyl cyclase/natriuretic peptide receptor-A gene expression. Peptides 26, 1009–1023.

Garofalo, F., Parisella, M.L., Amelio, D., Tota, B., Imbrogno, S., 2009. Phospholamban S-nitrosylation modulates Starling response in fish heart. Proc. R. Soc. B 276, 4043–4052.

Gattuso, A., Angelone, T., Cerra, M.C., 2016. Methodological challenges in the *ex vivo* hemodynamic evaluation of the myocardial stretch response: the case of catestatin-induced modulation of cardiac contractility. Nitric Oxide 53, 4–5. http://dx.doi.org/10.1016/j.niox.2015.12.008.

Gillis, T.E., Johnson, E.F., 2017. Cardiac remodelling, protection and regeneration. In: Gamperl, A.K., Gillis, T.E., Farrell, A.P., Brauner, C.J. (Eds.), Fish Physiology. In: The Cardiovascular System: Development, Plasticity and Physiological Responses, vol. 36B. Academic Press, San Diego, (In press).

Graham, M.S., Farrell, A.P., 1989. Effect of temperature acclimation and adrenaline on the performance of a perfused trout heart. Physiol. Zool. 62, 38–61.

Guo, Y., Stein, A.B., Wu, W.J., Tan, W., Zhu, X., Li, Q.H., Dawn, B., Motterlini, R., Bolli, R., 2004. Administration of a CO-releasing molecule at the time of reperfusion reduces infarct size *in vivo*. Am. J. Physiol. Heart Circ. Physiol. 286, H1649–H1653.

Hansen, M.N., Jensen, F.B., 2010. Nitric oxide metabolites in goldfish under normoxic and hypoxic conditions. J. Exp. Biol. 213, 3593–3602. http://dx.doi.org/10.1242/jeb.048140.

Haouzi, P., Sonobe, T., Judenherc-Haouzi, A., 2016. Developing effective countermeasures against acute hydrogen sulfide intoxication: challenges and limitations. Ann. N. Y. Acad. Sci. http://dx.doi.org/10.1111/nyas.13015.

Hare, J.M., 2003. Nitric oxide and excitation–contraction coupling. J. Mol. Cell. Cardiol. 35, 719–729.

Hayashi, T., Nakayama, T., Nakajima, T., Sokabe, H., 1978. Comparative studies on angiotensins V. Structure of angiotensin formed by the kidney of Japanese goosefish and its identification by dansyl method. Chem. Pharm. Bull. (Tokyo) 26, 215–229.

Hazon, N., Tierney, M.L., Hamano, K., Ashida, K., Takei, Y., 1995. Endogenous angiotensins, angiotensin II competitive binding inhibitors and converting enzyme inhibitor in elasmobranch fish. Neth. J. Zool. 45, 117–120.

Hazon, N., Cerra, M.C., Tierney, M.L., Tota, B., Takei, Y., 1997. Elasmobranch renin angiotensin system and the angiotensin receptor. In: Kawashima, S., Kikuyama, S. (Eds.), Advances in Comparative Endocrinology. Proceedings of the XIII International Congress of Comparative Endocrinology. Monduzzi Editore, Bologna, Italy, pp. 1307–1312.

Helle, K.B., Corti, A., Metz-Boutigue, M.H., Tota, B., 2007. The endocrine role for chromogranin A: a prohormone for peptides with regulatory properties. Cell. Mol. Life Sci. 64, 2863–2886.

Huckle, W.R., Earp, H.S., 1994. Regulation of cell proliferation and growth by angiotensin II. Prog. Growth Factor Res. 5, 177–194.

Icardo, J.M., 2006. Conus arteriosus of the teleost heart: dismissed, but not missed. Anat. Rec. A. Discov. Mol. Cell. Evol. Biol 288 (8), 900–908.

Icardo, J.M., 2017. Heart morphology and anatomy. In: Gamperl, A.K., Gillis, T.E., Farrell, A.P., Brauner, C.J. (Eds.), Fish Physiology. In: The Cardiovascular System: Morphology, Control and Function, vol. 36A. Academic Press, San Diego, pp. 1–54.

Icardo, J.M., Colvee, E., 2001. Origin and course of the coronary arteries in normal mice and in iv/iv mice. J. Anat. 199 (Pt. 4), 473–482.

Imbrogno, S., 2013. The eel heart: multilevel insights into functional organ plasticity. J. Exp. Biol. 216, 3575–3586. http://dx.doi.org/10.1242/jeb.089292.

Imbrogno, S., De Iuri, L., Mazza, R., Tota, B., 2001. Nitric oxide modulates cardiac performance in the heart of *Anguilla anguilla*. J. Exp. Biol. 204, 1719–1727.

Imbrogno, S., Cerra, M.C., Tota, B., 2003. Angiotensin II-induced inotropism requires an endocardial endothelium-nitric oxide mechanism in the in-vitro heart of *Anguilla anguilla*. J. Exp. Biol. 206, 2675–2684.

Imbrogno, S., Angelone, T., Corti, A., Adamo, C., Helle, K.B., Tota, B., 2004. Influence of vasostatins, the chromogranin A-derived peptides, on the working heart of the eel (*Anguilla anguilla*): negative inotropy and mechanism of action. Gen. Comp. Endocrinol. 139, 20–28.

Imbrogno, S., Angelone, T., Adamo, C., Pulerà, E., Tota, B., Cerra, M.C., 2006. Beta3-adrenoceptor in the eel (*Anguilla anguilla*) heart: negative inotropy and NO-cGMP-dependent mechanism. J. Exp. Biol. 209, 4966–4973.

Imbrogno, S., Garofalo, F., Cerra, M.C., Mahata, S.K., Tota, B., 2010. The catecholamine release-inhibitory peptide catestatin (chromogranin A344-363) modulates myocardial function in fish. J. Exp. Biol. 213, 3636–3643. http://dx.doi.org/10.1242/jeb.045567.

Imbrogno, S., Tota, B., Gattuso, A., 2011. The evolutionary functions of cardiac NOS/NO in vertebrates tracked by fish and amphibian paradigms. Nitric Oxide 25, 1–10. http://dx.doi.org/10.1016/j.niox.2011.05.001.

Imbrogno, S., Garofalo, F., Amelio, D., Capria, C., Cerra, M.C., 2013. Humoral control of cardiac remodeling in fish: role of Angiotensin II. Gen. Comp. Endocrinol. 194, 189–197. http://dx.doi.org/10.1016/j.ygcen.2013.09.009.

Imbrogno, S., Capria, C., Tota, B., Jensen, F.B., 2014. Nitric oxide improves the hemodynamic performance of the hypoxic goldfish (*Carassius auratus*) heart. Nitric Oxide 42, 24–31. http://dx.doi.org/10.1016/j.niox.2014.08.012.

Imbrogno, S., Gattuso, A., Mazza, R., Angelone, T., Cerra, M.C., 2015. β3-AR and the vertebrate heart: a comparative view. Acta Physiol (Oxf.) 214, 158–175.

Imbrogno, S., Mazza, R., Pugliese, C., Filice, M., Angelone, T., Loh, Y.P., Tota, B., Cerra, M.C., 2016. The Chromogranin A-derived sympathomimetic serpinin depresses myocardial performance in teleost and amphibian hearts. Gen. Comp. Endocrinol. 240, 1–9.

Inoue, K., Russell, M.J., Olson, K.R., Takei, Y., 2003. C-type natriuretic peptide of rainbow trout (*Oncorhynchus mykiss*): primary structure and vasorelaxant activities. Gen. Comp. Endocrinol. 130, 185–192.

Inoue, K., Sakamoto, T., Yuge, S., Iwatani, H., Yamagami, S., Tsutsumi, M., Hori, H., Cerra, M.C., Tota, B., Suzuki, N., Okamoto, N., Takei, Y., 2005. Structural and functional evolution of three cardiac natriuretic peptides. Mol. Biol. Evol. 22, 2428–2434.

Ip, Y.K., Chew, S.F., 2010. Nitrogen metabolism and excretion during aestivation. Prog. Mol. Subcell. Biol. 49, 63–94.

Jancsó, Z., Hermesz, E., 2015. Impact of acute arsenic and cadmium exposure on the expression of two haeme oxygenase genes and other antioxidant markers in common carp (*Cyprinus carpio*). J. Appl. Toxicol. 35, 310–318. http://dx.doi.org/10.1002/jat.3000.

Janssens, P.A., Cohen, P.P., 1968. Biosynthesis of urea in the estivating African lungfish and in *Xenopus laevis* under conditions of water shortage. Comp. Biochem. Physiol. 24, 887–898.

Jensen, F.B., 2007. Nitric oxide formation from nitrite in zebrafish. J. Exp. Biol. 210, 3387–3394.

Jensen, F.B., 2009. The role of nitrite in nitric oxide homeostasis: a comparative perspective. Biochim. Biophys. Acta 1787, 841–848.

Jensen, F.B., Hansen, M.N., 2011. Differential uptake and metabolism of nitrite in normoxic and hypoxic goldfish. Aquat. Toxicol. 101, 318–325.

Johansen, K., Burggren, W., 1980. In: Bourne, G.H. (Ed.), Cardiovascular Function in Lower Vertebrates. Hearts and Heart-Like Organs, vol. 1. Academic Press, London, pp. 61–118.

Johansen, I.B., Lunde, I.G., Røsjø, H., Christensen, G., Nilsson, G.E., Bakken, M., Overli, O., 2011. Cortisol response to stress is associated with myocardial remodeling in salmonid fishes. J. Exp. Biol. 214 (Pt. 8), 1313–1321. http://dx.doi.org/10.1242/jeb.053058.

Johnson, K.R., Olson, K.R., 2009. The response of non-traditional natriuretic peptide production sites to salt and water manipulations in the rainbow trout. J. Exp. Biol. 212, 2991–2997.

Joss, J.M., Itahara, Y., Watanabe, T.X., Nakajima, K., Takei, Y., 1999. Teleost-type angiotensin is present in Australian lungfish, *Neoceratodus forsteri*. Gen. Comp. Endocrinol. 114, 206–212.

Kaiya, H., Takei, Y., 1996. Osmotic and volaemic regulation of atrial and ventricular natriuretic peptide secretion in conscious eels. J. Endocrinol. 149, 441–447.

Kashiwagi, M., Katafuchi, T., Kato, A., Inuyama, H., Ito, T., Hagiwara, H., Takei, Y., Hirose, S., 1995. Cloning and properties of a novel natriuretic peptide receptor, NPR-D. Eur. J. Biochem. 233, 102–109.

Kawakoshi, A., Hyodo, S., Yasudal, A., Takei, Y., 2003. A single and novel natriuretic peptide is expressed in the heart and brain of the most primitive vertebrate, the hagfish (*Eptatretus burgeri*). J. Mol. Endocrinol. 31, 209–222.

Kawakoshi, A., Hyodo, S., Inoue, K., Kobayashi, Y., Takei, Y., 2004. Four natriuretic peptides (ANP, BNP, VNP and CNP) coexist in the sturgeon: identification of BNP in fish lineage. J. Mol. Endocrinol. 32, 547–555.

Keen, J.E., Vianzon, D.-M., Farrell, A.P., Tibbits, G.F., 1993. Thermal acclimation alters both adrenergic sensitivity and adrenoreceptor density in cardiac tissue of rainbow trout. J. Exp. Biol. 181, 27–47.

Kelley, J.L., Arias-Rodriguez, L., Patacsil Martin, D., Yee, M.C., Bustamante, C.D., Tobler, M., 2016. Mechanisms underlying adaptation to life in hydrogen sulfide-rich environments. Mol. Biol. Evol. 33, 1419–1434. http://dx.doi.org/10.1093/molbev/msw020.

Khan, S.A., Skaf, M.W., Harrison, R.W., Lee, K., Minhas, K.M., Kumar, A., Fradley, M., Shoukas, A.A., Berkowitz, D.E., Hare, J.M., 2003. Nitric oxide regulation of myocardial contractility and calcium cycling. Independent impact of neuronal and endothelial nitric oxide synthases. Circ. Res. 92, 1322–1329.

Kharitonov, V.G., Sharma, V.S., Pilz, R.B., Magde, D., Koesling, D., 1995. Basis of guanylate cyclase activation by carbon monoxide. Proc. Natl. Acad. Sci. U.S.A. 92, 2568–2571.

Khosla, M.C., Nishimura, H., Hasegawa, Y., Bumpus, F.M., 1985. Identification and synthesis of [1-asparagine, 5-valine, 9-glycine] angiotensin I produced from plasma of American eel *Anguilla rostrata*. Gen. Comp. Endocrinol. 57, 223–233.

Kobayashi, H., Takei, Y., 1996. Biological actions of ANGII. In: Bradshaw, S.D., Burggren, W., Heller, H.C., Ishii, S., Langer, H., Neuweiler, G., Randall, D.J. (Eds.), The Renin–Angiotensin System. Comparative Aspects. Zoophysiology, vol. 35. Springer Verlag, Berlin, Heildleberg, pp. 113–171.

Kumai, Y., Porteus, C.S., Kwong, R.W., Perry, S.F., 2015. Hydrogen sulfide inhibits Na^+ uptake in larval zebrafish, *Danio rerio*. Pflugers Arch. 467, 651–664. http://dx.doi.org/10.1007/s00424-014-1550-y.

Kwong, R.W., Perry, S.F., 2015. Hydrogen sulfide promotes calcium uptake in larval zebrafish. Am. J. Physiol. Cell Physiol. 309, C60–C69. http://dx.doi.org/10.1152/ajpcell.00053.2015.

Lacy, E.R., Reale, E., Luciano, L., 2016. Immunohistochemical localization of renin-containing cells in two elasmobranch species. Fish Physiol. Biochem. 42 (3), 995–1004. http://dx.doi.org/10.1007/s10695-015-0191-1.

Larsen, T.H., Helle, K.B., Saetersdal, T., 1994. Immunoreactive atrial natriuretic peptide and dopamine beta-hydroxylase in myocytes and chromaffin cells of the heart of the African lungfish, *Protopterus aethiopicus*. Gen. Comp. Endocrinol. 95, 1–12.

Laurent, P., Holmgren, S., Nilsson, S., 1983. Nervous and humoral control ofthe fish heart: structure and function. Comp. Biochem. Physiol. 76, 525–542.

Leri, A., Kajstura, J., Anversa, P., 2002. Myocyte proliferation and ventricular remodelling. J. Card. Fail. 8 (Suppl. 6), S518–S525.

Li, J.Y., Avallet, O., Berthelon, M.C., Langlois, D., Saez, J.M., 1999. Transcriptional and translational regulation of angiotensin II type 2 receptor by angiotensin II and growth factors. Endocrinology 140, 4988–4994.

Lipke, D.W., Olson, K.R., 1988. Distribution of angiotensin-converting enzyme-like activity in vertebrate tissues. Physiol. Zool. 61, 420–428.

Liu, H., Song, D., Lee, S.S., 2001. Role of heme oxygenase-carbon monoxide pathway in pathogenesis of cirrhotic cardiomyopathy in the rat. Am. J. Physiol. Gastrointest. Liver Physiol. 280, G68–G74.

Loretz, C.A., Pollina, C., 2000. Natriuretic peptides in fish physiology. Comp. Biochem. Physiol. A Physiol. 125, 169–187.

Loretz, C.A., Pollina, C., Kaiya, H., Sakaguchi, H., Takei, Y., 1997. Local synthesis of natriuretic peptides in the eel intestine. Biochem. Biophys. Res. Commun. 238, 817–822.

Mancardi, D., Penna, C., Merlino, A., Del Soldato, P., Wink, D.A., Pagliaro, P., 2009. Physiological and pharmacological features of the novel gasotransmitter: hydrogen sulfide. Biochim. Biophys. Acta 1787, 864–872. http://dx.doi.org/10.1016/j.bbabio.2009.03.005.

Marsigliante, S., Muscella, A., Vilella, S., Nicolardi, G., Ingrosso, L., Ciardo, V., Zonno, V., Vinson, G.P., Ho, M.M., Storelli, C., 1996. A monoclonal antibody to mammalian angiotensin II AT1 receptor recognizes one of the angiotensin II receptor isoforms expressed by the eel (*Anguilla anguilla*). J. Mol. Endocrinol. 16, 45–56.

Masini, M.A., Sturla, M., Napoli, L., Uva, B.M., 1996. Immunoreactive localization of vasoactive hormones (atrial natriuretic peptide and endothelin) in the heart of *Protopterus annectens*, an African lungfish. Cell Tissue Res. 284 (3), 501–507.

Masini, M.A., Sturla, M., Uva, B.M., 1997. Vasoactive peptides in the heart of *Champsocephalus gunnari*. Comp. Biochem. Physiol. A Physiol. 118, 1083–1086.

Massion, P.B., Pelat, M., Belge, C., Balligand, J.L., 2005. Regulation of the mammalian heart function by nitric oxide. Comp. Biochem. Physiol. 142A, 144–150.

Mazza, R., Imbrogno, S., Tota, B., 2010. The interplay between chromogranin A-derived peptides and cardiac natriuretic peptides in cardioprotection against catecholamine-evoked stress. Regul. Pept. 165, 86–94. http://dx.doi.org/10.1016/j.regpep.2010.05.005. Review.

Mazza, R., Pasqua, T., Cerra, M.C., Angelone, T., Gattuso, A., 2013. Akt/eNOS signalling and PLN S-sulfhydration are involved in H_2S-dependent cardiac effects in frog and rat. Am. J. Physiol. Regul. Integr. Comp. Physiol. 305, R443–R451. http://dx.doi.org/10.1152/ajpregu.00088.2013.

Mazza, R., Tota, B., Gattuso, A., 2015. Cardio-vascular activity of catestatin: interlocking the puzzle pieces. Curr. Med. Chem. 22, 292–304.

Metcalfe, J.D., Butler, P.J., 1989. The use of alpha-methylp-tyrosine to control circulating catecholamines in the dogfish *Scyliorhinus canicula*: the effects on gas exchange in normoxia and hypoxia. J. Exp. Biol. 141, 21–32.

Miyanishi, H., Okubo, K., Nobata, S., Takei, Y., 2013. Natriuretic peptides in developing medaka embryos: implications in cardiac development by loss-of-function studies. Endocrinology 154 (1), 410–420. http://dx.doi.org/10.1210/en.2012-1730.

Nadal-Ginard, B., Kajstura, J., Leri, A., Anversa, P., 2003. Myocyte death, growth, and regeneration in cardiac hypertrophy and failure. Circ. Res. 92, 139–150.

Nandi, J., 1961. New arrangement of interrenal and chromaffin tissues of teleost fishes. Science 134, 389–390.

Newton, C.M., Stoyek, M.R., Croll, R.P., Smith, F.M., 2014. Regional innervation of the heart in the goldfish, *Carassius auratus*: a confocal microscopy study. J. Comp. Neurol. 522, 456–478. http://dx.doi.org/10.1002/cne.23421.

Nickerson, J.G., Dugan, S.G., Drouin, G., Perry, S.F., Moon, T.W., 2003. Activity of the unique beta-adrenergic Na^+/H^+ exchanger in trout erythrocytes is controlled by a novel beta3-AR subtype. Am. J. Physiol. 285, 526–535.

Nilsson, S., 1983. Autonomic Nerve Function in Vertebrates. Spinger-Verlag, Berlin.

Nilsson, S., Holmgren, S., 1992. Cardiovascular control by purines, 5-hydroxytryptamine, and neuropeptides. In: Hoar, W.S., Randall, D.J., Farrell, A.P. (Eds.), Fish Physiology, vol. 12B. Academic Press, San Diego, pp. 301–341.

Nilsson, S., Abrahamsson, T., Grove, D.J., 1976. Sympathetic nervous control of adrenaline release from the head kidney of the cod, *Gadus morhua*. Comp. Biochem. Physiol. 55, 123–127.

Nishimura, H., 1985. Evolution of the renin angiotensin system and its role in control of cardiovascular function in fishes. In: Foreman, R.E., Gorbman, A., Dodd, J.M., Olson, R. (Eds.), Evolutionary Biology of Primitive Fishes. Plenum, New York, pp. 275–293.

Nishimura, H., 2001. Angiotensin receptors—evolutionary overview and perspectives. Comp. Biochem. Physiol. A Mol. Integr. Physiol. 128 (1), 11–30.

Nishimura, H., Oguri, M., Ogawa, M., Sokabe, H., Imai, M., 1970. Absence of renin in kidneys of elasmobranchs and cyclostomes. Am. J. Physiol. 218 (3), 911–915.

Nishimura, H., Ogawa, M., Sawyer, W.H., 1973. Renin–angiotensin system in primitive bony fishes and a holocephalian. Am. J. Physiol. 224 (4), 950–956.

Olson, K.R., 1992. Blood and extracellular fluid regulation. In: Hoar, W.S., Randall, D.J., Farrell, A.P. (Eds.), Fish Physiology: The Cardiovascular System, vol. 12B. Academic Press, New York, pp. 135–254.

Olson, K.R., Donald, J.A., 2009. Nervous control of circulation—the role of gasotransmitters, NO, CO, and H2S. Acta Histochem. 111, 244–256. http://dx.doi.org/10.1016/j.acthis.2008.11.004.

Olson, K.R., Duff, D.W., 1992. Cardiovascular and renal effects of eel and rat atrial natriuretic peptide in rainbow trout, *Salmo gairdneri*. J. Comp. Physiol. B Biochem. Syst. Environ. Physiol. 162, 408–415.

Olson, K.R., Lipke, D., Datta Munshi, J.S., Moitra, A., Ghosh, T.K., Kunwar, G., Ahmad, M., Roy, P.K., Singh, O.N., Nasar, S.S., et al., 1987. Angiotensin-converting enzyme in organs of air-breathing fish. Gen. Comp. Endocrinol. 68, 486–491.

Olson, K.R., Conklin, D.J., Farrell, A.P., Keen, J.E., Takei, Y., Weaver Jr., L., Smith, M.P., Zhang, Y., 1997. Effects of natriuretic peptides and nitroprusside on venous function in trout. Am. J. Physiol. 273, 527–539.

Olson, K.R., Dombkowski, R.A., Russell, M.J., Doellman, M.M., Head, S.K., Whitfield, N.L., Madden, J.A., 2006. Hydrogen sulfide as an oxygen sensor/transducer in vertebrate hypoxic vasoconstriction and hypoxic vasodilation. J. Exp. Biol. 209, 4011–4023.

Opdyke, D.F., Holcombe, R., 1976. Response to angiotensins I and II and to AI converting-enzyme inhibitor in a shark. Am. J. Physiol. 231, 1750–1753.

Opdyke, D.F., Carroll, R.G., Keller, N.E., 1982a. Catecholamine release and blood pressure changes induced by exercise in dogfish. Am. J. Physiol. 242, 306–310.

Opdyke, D.F., Wilde, D.W., Holcombe, R.F., 1982b. Effect of angiotensin II on vascular resistance in whole-body perfused dogfish. Comp. Biochem. Physiol. C 73 (1), 45–49.

Opdyke, D.F., Bullock, J., Keller, N.E., Holmes, K., 1983. Effect of ganglionic blockade on catecholamine secretion in exercised dogfish. Am. J. Physiol. 245 (6), R915–R919.

Ostádal, B., Rychterová, V., 1971. Effect of necrogenic doses of isoproterenol on the heart of the tench (*Tinca tinca-osteoichthyes*), the frog (*Rana temporariaanura*) and the pigeon (*Columba livia-aves*). Physiol. Bohemoslov. 20, 541–547.

O'Tierney, P.F., Chattergoon, N.N., Louey, S., Giraud, G.D., Thornburg, K.L., 2010. Atrial natriuretic peptide inhibits angiotensin II-stimulated proliferation in fetal cardiomyocytes. J. Physiol. 588, 2879–2889.

Otorii, T., 1953. Pharmacology of the heart of *Entosphenus japonicus*. Acta Med. Biol. (Jap.) 1, 51–59.

Otsuka, N., Chihara, J., Sakurada, H., Kanda, S., 1977. Catecholamine-storing cells in the cyclostome heart. Arch. Histol. Jpn. 40, 241–244.

Oudit, G.Y., Butler, D.G., 1995. Angiotensin II and cardiovascular regulation in a freshwater teleost, *Anguilla rostrata* LeSueur. Am. J. Physiol. 269, R726–R735.

Palacios, M., Arias-Rodriguez, L., Plath, M., Eifert, C., Lerp, H., Lamboj, A., Voelker, G., Tobler, M., 2013. The rediscovery of a long described species reveals additional complexity in speciation patterns of poeciliid fishes in sulfide springs. PLoS One 8 (8), e71069. http://dx.doi.org/10.1371/journal.pone.0071069.

Palmer, R.M., Ferrige, A.G., Moncada, S., 1987. Nitric oxide release accounts for the biological activity of endothelium-derived relaxing factor. Nature 327, 524–526.

Pasqua, T., Corti, A., Gentile, S., Pochini, L., Bianco, M., Metz-Boutigue, M.H., Cerra, M.C., Tota, B., Angelone, T., 2013. Full-length human chromogranin-A cardioactivity: myocardial, coronary, and stimulus-induced processing evidence in normotensive and hypertensive male rat hearts. Endocrinology 154, 3353–3365. http://dx.doi.org/10.1210/en.2012-2210.

Penna, C., Alloatti, G., Gallo, M.P., Cerra, M.C., Levi, R., Tullio, F., Bassino, E., Dolgetta, S., Mahata, S.K., Tota, B., Pagliaro, P., 2010. Catestatin improves post-ischemic left ventricular function and decreases ischemia/reperfusion injury in heart. Cell. Mol. Neurobiol. 30 (8), 1171–1179.

Penna, C., Tullio, F., Perrelli, M.G., Mancardi, D., Pagliaro, P., 2012. Cardioprotection against ischemia/reperfusion injury and chromogranin A-derived peptides. Curr. Med. Chem. 19, 4074–4085. Review.

Penna, C., Pasqua, T., Amelio, D., Perrelli, M.G., Angotti, C., Tullio, F., Mahata, S.K., Tota, B., Pagliaro, P., Cerra, M.C., Angelone, T., 2014. Catestatin increases the expression of anti-apoptotic and pro-angiogenetic factors in the post-ischemic hypertrophied heart of SHR. PLoS One 9 (8), e102536.

Pennec, J.P., Le Bras, Y.M., 1984. Storage and release of catecholamines by nervous endings in the isolated heart of the eel (*Anguilla anguilla* L.). Comp. Biochem. Physiol. 77C, 167–171.

Pennec, J.P., Le Bras, Y.M., 1988. Diel and seasonal rhythms of the heart ratein the common eel (*Anguilla anguilla* L.): role of cardiac innervation. Exp. Biol. 47, 155–160.

Perry, S.F., Gilmour, K.M., 1996. Consequences of catecholamine release on ventilation and blood oxygen transport during hypoxia and hypercapnia in an elasmobranch (*Squalus acanthias*) and a teleost (*Oncorhynchus mykiss*). J. Exp. Biol. 199, 2105–2118.

Perry, S.F., Tzaneva, V., 2016. The sensing of respiratory gases in fish: mechanisms and signalling pathways. Respir. Physiol. Neurobiol. 224, 71–79. http://dx.doi.org/10.1016/j.resp.2015.06.007.

Perry, S.F., Gilmour, K.M., Vulesevic, B., McNeill, B., Chew, S.F., Ip, Y.K., 2005. Circulating catecholamines and cardiorespiratory responses in hypoxic lungfish (*Protopterus dolloi*): a comparison of aquatic and aerial hypoxia. Physiol. Biochem. Zool. 78, 325–334.

Perry, S.F., Euverman, R., Wang, T., Loong, A.M., Chew, S.F., Ip, Y.K., Gilmour, K.M., 2008. Control of breathing in African lungfish (*Protopterus dolloi*): a comparison of aquatic and cocooned (terrestrialized) animals. Respir. Physiol. Neurobiol. 160, 8–17.

Peyraud-Waitzenegger, M., Barthelemy, L., Peyraud, C., 1980. Cardiovascular and ventilatory effects of catecholamines in unrestrained eels (*Anguilla anguilla* L.). J. Comp. Physiol. 138, 367–375.

Pfenninger, M., Lerp, H., Tobler, M., Passow, C., Kelley, J.L., Funke, E., Greshake, B., Erkoc, U.K., Berberich, T., Plath, M., 2014. Parallel evolution of cox genes in H2S-tolerant fish as key adaptation to a toxic environment. Nat. Commun. 5, 3873. http://dx.doi.org/10.1038/ncomms4873.

Poss, K.D., Wilson, L.G., Keating, M.T., 2002. Heart regeneration in zebrafish. Science 298, 2188–2190.

Prabhakar, N.R., Dinerman, J.L., Agani, F.H., Snyder, S.H., 1995. Carbon monoxide: a role in carotid body chemoreception. Proc. Natl. Acad. Sci. U.S.A. 92, 1994–1997.

Prevot-D'Alvise, N., Richard, S., Coupé, S., Bunet, R., Grillasca, J.P., 2013. Acute toxicity of a commercial glyphosate formulation on European sea bass juveniles (*Dicentrarchus labrax* L.): gene expressions of heme oxygenase-1 (ho-1), acetylcholinesterase (AChE) and aromatases (cyp19a and cyp19b). Cell. Mol. Biol. (Noisy-le-grand) 59, 1906–1917.

Rademaker, M.T., Charles, C.J., Espiner, E.A., Frampton, C.M., Nicholls, M.G., Richards, A.M., 2004. Combined inhibition of angiotensin II and endothelin suppresses the brain natriuretic peptide response to developing heart failure. Clin. Sci. (Lond.) 106, 569–576.

Randall, D.J., Perry, S.F., 1992. Catecholamines. In: Hoar, W.S., Randall, D.J., Farrell, A.P. (Eds.), Fish Physiology. Academic Press, San Diego, CA, pp. 255–300.

Rankin, J.C., Watanabe, T.X., Nakajima, K., Broadhead, C., Takei, Y., 2004. Identification of angiotensin I in a cyclostome, *Lampetra fluviatilis*. Zoolog. Sci. 21 (2), 173–179.

Reid, I.A., 1992. Interactions between ANG II, sympathetic nervous system and baroreceptor reflexes in regulation of blood pressure. Am. J. Physiol. 262, E763–E778.

Reinecke, M., Nehls, M., Forssmann, W.G., 1985. Phylogenetic aspects of cardiac hormones as revealed by immunohistochemistry, electronmicroscopy, and bioassay. Peptides 6 (Suppl. 3), 321–331.

Riesch, R., Tobler, M., Plath, M., 2015. Hydrogen sulfide-toxic habitats. In: Riesch, R., Tobler, M., Plath, M. (Eds.), Extremophile Fishes: Ecology, Evolution, and Physiology of Teleosts in Extreme Environments. Springer, Heidelberg, Germany, pp. 137–159.

Ruskoaho, H., 1992. Atrial natriuretic peptide: synthesis, release, and metabolism. Pharmacol. Rev. 44, 479–602.

Russell, M.J., Klemmer, A.M., Olson, K.R., 2001. Angiotensin signalling and receptor types in teleost fish. Comp. Biochem. Physiol. 128A, 41–51.

Saetersdal, T.S., Sorensen, E., Myklebust, R., Helle, K.B., 1975. Granule containing cells and fibres in the sinus venosus of elasmobranchs. Cell Tissue Res. 163, 471–490.

Samuels, M.A., 2007. The brain–heart connection. Circulation 116, 77–84.

Sandblom, E., Axelsson, M., Farrell, A.P., 2006. Central venous pressure and mean circulatory filling pressure in the dogfish *Squalus acanthias*: adrenergic control and role of the pericardium. Am. J. Physiol. Regul. Integr. Comp. Physiol. 291, 1465–1473.

Sandvik, G.K., Nilsson, G.E., Jensen, F.B., 2012. Dramatic increase of nitrite levels in hearts of anoxia-exposed crucian carp supporting a role in cardioprotection. Am. J. Physiol. Regul. Integr. Comp. Physiol. 302, R468–R477. http://dx.doi.org/10.1152/ajpregu.00538.2011.

Sawyer, W.H., Blair-West, J.R., Simpson, P.A., Sawyer, M.K., 1976. Renal responses of Australian lungfis to vasotocin, angiotensin II, and NaCl infusion. Am. J. Physiol. 231, 593–602.

Schjolden, J., Stoskhus, A., Winberg, S., 2005. Does individual variation in stress responses and agonistic behavior reflect divergent stress coping strategies in juvenile rainbow trout? Physiol. Biochem. Zool. 78, 715–723.

Seddon, M., Shah, A.M., Casadei, B., 2007. Cardiomyocytes as effectors of nitric oxide signalling. Cardiovasc. Res. 75, 315–326. http://dx.doi.org/10.1016/j.cardiores.2007.04.031.

Shiels, H.A., White, E., 2008. The Frank–Starling mechanism in vertebrate cardiac myocytes. J. Exp. Biol. 211, 2005–2013.

Sluiter, E., 1930. The production of hydrogen sulphide by animal tissues. Biochem. J. 24, 549–563.

Snijder, P.M., de Boer, R.A., Bos, E.M., van den Born, J.C., Ruifrok, W.P., Vreeswijk-Baudoin, I., van Dijk, M.C., Hillebrands, J.L., Leuvenink, H.G., van Goor, H., 2013. Gaseous hydrogen sulfide protects against myocardial ischemia-reperfusion injury in mice partially independent from hypometabolism. PLoS One 8 (5), e63291. http://dx.doi.org/10.1371/journal.pone.0063291.

Stipanuk, M.H., Beck, P.W., 1982. Characterization of the enzymic capacity for cysteine desulphhydration in liver and kidney of the rat. Biochem. J. 206, 267–277.

Suzuki, R., Togashi, K., Ando, K., Takei, Y., 1994. Distribution and molecular forms of C-type natriuretic peptide in plasma and tissue of a dogfish, *Triakis scyllia*. Gen. Comp. Endocrinol. 96, 378–384.

Swaroop, M., Bradley, K., Ohura, T., Tahara, T., Roper, M.D., Rosenberg, L.E., Kraus, J.P., 1992. Rat cystathionine beta-synthase. Gene organization and alternative splicing. J. Biol. Chem. 267, 11455–11461.

Takei, Y., 2000. Structural and functional evolution of the natriuretic peptide system in vertebrates. Int. Rev. Cytol. 194, 1–66.

Takei, Y., Hirose, S., 2002. The natriuretic peptide system in eels: a key endocrine system for euryhalinity? Am. J. Physiol. Regul. Integr. Comp. Physiol. 282, 940–951.

Takei, Y., Hasegawa, Y., Watanabe, T.X., Nakajima, K., Hazon, N., 1993. A novel angiotensin I isolated from an elasmobranch fish. J. Endocrinol. 139 (2), 281–285.

Takei, Y., Joss, J.M.P., Kloas, W., Rankin, J.C., 2004. Identification of angiotensin I in several vertebrate species: its structural and functional evolution. Gen. Comp. Endocrinol. 135, 286–292.

Takei, Y., Ogoshi, M., Inoue, K., 2007. A 'reverse' phylogenetic approach for identification of novel osmoregulatory and cardiovascular hormones in vertebrates. Front. Neuroendocrinol. 28, 143–160.

Takei, Y., Inoue, K., Trajanovska, S., Donald, J.A., 2011. B-type natriuretic peptide (BNP), not ANP, is the principal cardiac natriuretic peptide in vertebrates as revealed by comparative studies. Gen. Comp. Endocrinol. 171, 258–266. http://dx.doi.org/10.1016/j.ygcen.2011.02.021.

Takei, Y., Hiroi, J., Takahashi, H., Sakamoto, T., 2014. Diverse mechanisms for body fluid regulation in teleost fishes. Am. J. Physiol. Regul. Integr. Comp. Physiol. 307, R778–R792. http://dx.doi.org/10.1152/ajpregu.00104.

Takemoto, Y., Nakajima, T., Hasegawa, Y., Watanabe, T.X., Sokabe, H., Kumagae, S., Sakakibara, S., 1983. Chemical structures of angiotensins formed by incubating plasma with the kidney and the corpuscles of Stannius in the chum salmon, *Oncorhynchus keta*. Gen. Comp. Endocrinol. 51, 219–227.

Tamura, N., Ogawa, Y., Chusho, H., Nakamura, K., Nakao, K., Suda, M., Kasahara, M., Hashimoto, R., Katsuura, G., Mukoyama, M., et al., 2000. Cardiac fibrosis in mice lacking brain natriuretic peptide. Proc. Natl. Acad. Sci. U.S.A. 97, 4239–4244.

Taylor, E.W., 1992. The cardiovascular system. In: Hoar, W.S., Randall, D.J., Farrell, A.P. (Eds.), Fish Physiology, vol. 12. Academic Press, San Diego, pp. 343–387.

Tervonen, V., Arjamaa, O., Kokkonen, K., Ruskoaho, H., Vuolteenaho, O., 1998. A novel cardiac hormone related to A-, B- and C-type natriuretic peptides. Endocrinology 139, 4021–4025.

Tierney, M., Takei, Y., Hazon, N., 1997. The presence of angiotensin II receptors in elasmobranchs. Gen. Comp. Endocrinol. 105, 9–17.

Tobler, M., Plath, M., 2011. Living in extreme habitats. In: Evans, J., Pilastro, A., Schlupp, I. (Eds.), Ecology and Evolution of Poeciliid Fishes. University of Chicago Press, Chicago, IL, pp. 120–127.

Toop, T., Donald, J.A., 2004. Comparative aspects of natriuretic peptide physiology in non-mammalian vertebrates: a review. J. Comp. Physiol. B Biochem. Syst. Environ. Physiol. 174, 189–204.

Tota, B., 1999. Sharks, skates, and rays. The biology of elasmobranch fishes. In: Heart. Johns Hopkins University Press, Baltimore, MD, pp. 238–272 (Chapter 10).

Tota, B., Garofalo, F., 2012. Fish Heart growth and function: from gross morphology to cell signaling and back. In: Sedmera, D., Wang, T. (Eds.), Ontogeny and Phylogeny of the Vertebrate Heart. Springer, New York/Heidelberg/Dordrecht/London, pp. 55–74.

Tota, B., Mazza, R., Angelone, T., Nullans, G., Metz-Boutigue, M.H., Aunis, D., Helle, K.B., 2003. Peptides from the N-terminal domain of chromogranin A (Vasostatins) exert negative inotropic effects in the isolated frog heart. Regul. Pept. 114, 123–130.

Tota, B., Imbrogno, S., Mannarino, C., Mazza, R., 2004. Vasostatins and negative inotropy in vertebrate hearts. Curr. Med. Chem. – Immunol., Endocr. Metab. Agents 4, 195–201.

Tota, B., Amelio, D., Pellegrino, D., Ip, Y.K., Cerra, M.C., 2005. NO modulation of myocardial performance in fish hearts. Comp. Biochem. Physiol. 142, 164–177.

Tota, B., Cerra, M.C., Gattuso, A., 2010. Catecholamines, cardiac natriuretic peptides and chromogranin A: evolution and physiopathology of a 'whip-brake' system of the endocrine heart. J. Exp. Biol. 213, 3081–3103.

Tota, B., Angelone, T., Cerra, M.C., 2014. The surging role of chromogranin A in cardiovascular homeostasis. Front. Chem. 2, 64. http://dx.doi.org/10.3389/fchem.2014.00064.

Tran van Chuoi, M., Dolphin, C.T., Barker, S., Clark, A.J., Vinson, G.P., 1998. Molecular cloning and characterization of the cDNA encoding the angiotensin II receptor of European eel (*Anguilla anguilla*). J. Endocrinol. 156, 227. submitted (April 1998) to the EMBL/GenBank/DDBJ databases.

Tzaneva, V., Perry, S.F., 2016. Evidence for a role of heme oxygenase-1 in the control of cardiac function in zebrafish (*Danio rerio*) larvae exposed to hypoxia. J. Exp. Biol. 219, 1563–1571. http://dx.doi.org/10.1242/jeb.136853.

Uemura, H., Naruse, M., Takei, Y., Nakamura, S., Hirohama, T., Ando, K., Aoto, T., 1991. Immunoreactive and bioactive atrial natriuretic peptide in the carp heart. Zool. Sci. 8, 885–891.

Uva, B., Masini, M.A., Hazon, N., O'Toole, L.B., Henderson, I.W., Ghiani, P., 1992. Renin and angiotensin converting enzyme in elasmobranchs. Gen. Comp. Endocrinol. 86 (3), 407–412.

Vallance, P., Rees, D., Moncada, S., 2000. Therapeutic potential of NOS inhibitors in septic shock. In: Mayer, B. (Ed.), Handbook of Experimental Pharmacology, vol. 143. Springer-Verlag, Berlin, pp. 385–397.

van Raaij, M.T.M., Pit, D.S.S., Balm, P.H.M., Steffens, A.B., van den Thillart, G.E.E.J.M., 1996. Behavioural strategy and the physiological stress response in rainbow trout exposed to severe hypoxia. Horm. Behav. 30, 85–92.

Varagic, J., Frohlich, E.D., 2002. Local cardiac renin–angiotensin system: hypertension and cardiac failure. J. Mol. Cell. Cardiol. 34, 1435–1442.

Vesely, D.L., 2006. Which of the cardiac natriuretic peptides is most effective for the treatment of congestive heart failure, renal failure and cancer? Clin. Exp. Pharmacol. Physiol. 33, 169–176.

Vesely, D.L., Douglass, M.A., Dietz, J.R., Gower Jr., W.R., McCormick, M.T., Rodriguez-Paz, G., Schocken, D.D., 1994. Three peptides from the atrial natriuretic factor prohormone amino terminus lower blood pressure and produce diuresis, natriuresis, and/or kaliuresis in humans. Circulation 90, 1129–1140.

Vierimaa, H., Ronkainen, J., Ruskoaho, H., Vuolteenaho, O., 2006. Synergistic activation of salmon cardiac function by endothelin and beta-adrenergic stimulation. Am. J. Physiol. Heart Circ. Physiol. 291, H1360–H1370.

Vornanen, M., Hassinen, M., Koskinen, H., Krasnov, A., 2005. Steady-state effects of temperature acclimation on the transcriptome of the rainbow trout heart. Am. J. Physiol. Regul. Integr. Comp. Physiol. 289, R1177–R1184.

Wang, R., Wang, Z., Wu, L., 1997. Carbon monoxide-induced vasorelaxation and the underlying mechanisms. Br. J. Pharmacol. 121, 927–934.

Wang, D., Zhong, X.P., Qiao, Z.X., Gui, J.F., 2008. Inductive transcription and protective role of fish heme oxygenase-1 under hypoxic stress. J. Exp. Biol. 211, 2700–2706. http://dx.doi.org/10.1242/jeb.019141.

Wang, X., Wang, Q., Guo, W., Zhu, Y.Z., 2011. Hydrogen sulfide attenuates cardiac dysfunction in a rat model of heart failure: a mechanism through cardiac mitochondrial protection. Biosci. Rep. 31, 87–98. http://dx.doi.org/10.1042/BSR20100003.

Wenzel, S., Rohde, C., Wingerning, S., Roth, J., Kojda, G., Schlüter, K.D., 2007. Lack of endothelial nitric oxide synthase-derived nitric oxide formation favors hypertrophy in adult ventricular cardiomyocytes. Hypertension 49, 193–200.

Whitfield, N.L., Kreimier, E.L., Verdial, F.C., Skovgaard, N., Olson, K.R., 2008. Reappraisal of H2S/sulfide concentration in vertebrate blood and its potential significance in ischemic preconditioning and vascular signalling. Am. J. Physiol. Regul. Integr. Comp. Physiol. 294, R1930–R1937. http://dx.doi.org/10.1152/ajpregu.00025.2008.

Winkler, H., Fischer-Colbrie, R., 1992. The chromogranin A and B: the first 25 years and future perspectives. Neuroscience 49, 479–528.

Wong, M.K., Takei, Y., 2012. Changes in plasma angiotensin subtypes in Japanese eel acclimated to various salinities from deionized water to double-strength seawater. Gen. Comp. Endocrinol. 178, 250–258. http://dx.doi.org/10.1016/j.ygcen.2012.06.007.

Wong, M.K., Takei, Y., 2013. Angiotensin AT2 receptor activates the cyclic-AMP signalling pathway in eel. Mol. Cell. Endocrinol. 365, 292–302. http://dx.doi.org/10.1016/j.mce.2012.11.009.

Wong, M.K., Takei, Y., Woo, N.Y., 2006. Differential status of the renin–angiotensin system of silver sea bream (*Sparus sarba*) in different salinities. Gen. Comp. Endocrinol. 149, 81–89.

Woo, S.H., Morad, M., 2001. Bimodal regulation of Na(+)-Ca(2+) exchanger by beta-adrenergic signalling pathway in shark ventricular myocytes. Proc. Natl. Acad. Sci. U.S.A. 98, 2023–2028.

Wu, L., Wang, R., 2005. Carbon monoxide: endogenous production, physiological functions, and pharmacological applications. Pharmacol. Rev. 57, 585–630.

Wu, L., Shao, Y., Hu, Z., Gao, H., 2016. Effects of soluble sulfide on zebrafish (*Danio rerio*) embryonic development. Environ. Toxicol. Pharmacol. 42, 183–189. http://dx.doi.org/10.1016/j.etap.2016.01.019.

Xie, J., Wang, W.Q., Liu, T.X., Deng, M., Ning, G., 2008. Spatio-temporal expression of chromogranin A during zebrafish embryogenesis. J. Endocrinol. 198, 451–458. http://dx.doi.org/10.1677/JOE-08-0221.

Xu, K.Y., Huso, D.L., Dawson, T.M., Bredt, D.S., Becker, L.C., 1999. Nitric oxide synthase in cardiac sarcoplasmic reticulum. Proc. Natl. Acad. Sci. U.S.A. 96, 657–662. http://dx.doi.org/10.1073/pnas.96.2.657.

Yousaf, M.N., Amin, A.B., Koppang, E.O., Vuolteenaho, O., Powell, M.D., 2012. Localization of natriuretic peptides in the cardiac pacemaker of Atlantic salmon (*Salmo salar* L.). Acta Histochem. 114, 819–826. http://dx.doi.org/10.1016/j.acthis.2012.02.002.

Zhang, Y.H., Zhang, M.H., Sears, C.E., Emanuel, K., Redwood, C., El-Armouche, A., Kranias, E.G., Casadei, B., 2008. Reduced phospholamban phosphorylation is associated with impaired relaxation in left ventricular myocytes from neuronal NO synthase deficient mice. Circ. Res. 102, 242–249. http://dx.doi.org/10.1161/CIRCRESAHA.107.164798.

6

CARDIAC ENERGY METABOLISM

KENNETH J. RODNICK[*,1]
HANS GESSER[†]

[*]Idaho State University, Pocatello, ID, United States
[†]University of Aarhus, Aarhus C, Denmark
[1]Corresponding author: rodnkenn@isu.edu

"The heart is a restless organ, in a constant state of mechanical and metabolic flux."

(Opie, 1991)

The Cardiovascular System: Morphology, Control and Function, Volume 36A
FISH PHYSIOLOGY

DOI: http://dx.doi.org/10.1016/bs.fp.2017.04.003

The heart must provide its own energy requirements to sustain its continuous contractile performance and other physiological functions. This chapter provides an integrated overview of cardiac energy metabolism in fish with a particular emphasis on: maintenance of cardiac energy state; biochemical strategies for energy production; and energetic requirements to maintain contractile function, ion pumping across membranes, and protein synthesis. Major advances in our understanding of fish cardiac metabolism have come from studies of cellular ultrastructure, ion regulation, enzyme activities, select proteins involved in energy metabolism, fuel selection and energy reserves, intracellular metabolites, cardiac adaptability, and physiological performance of cardiac preparations and intact animals. Total energy expenditure of the contracting heart varies between species and includes basal and active components. The importance of myosin-bound ATP phosphatase (ATPase) for contractility, Na^{+}/K^{+}-ATPase for cellular ion homeostasis, and Ca^{2+}-ATPases for myocardial relaxation are highlighted. Fish hearts rely on well-established metabolic pathways to regenerate ATP, however, species differences are apparent. Under normoxic conditions, mitochondria produce most of the ATP used by the fish heart using aerobic metabolism and a variety of energy substrates. However, during O_2-limiting conditions, anaerobic metabolism (glycolysis) becomes the major source of ATP production, despite an inherently limited capacity compared with oxidative phosphorylation. Cold temperature can also compromise several cellular processes related to cardiac energy metabolism, and yet, some fish demonstrate positive compensation of enzyme activities following cold acclimation and acclimatization. Overall, there are numerous, inter-related factors that underlie cardiac energy production and utilization.

1. INTRODUCTION

Physiologists have long appreciated the importance of the heart and cardiovascular system, and the necessity of convective O_2 transport to cells of the vertebrate body for aerobic metabolism. Continuous performance of the fish heart as a pump is necessary for gas exchange at the gills, as well as delivery of a variety of molecules (e.g., ions, nutrients, and hormones) to cells and the removal of waste products. With well over 30,000 fish species (Nelson et al., 2016), and with them occupying a wide variety of aquatic environments and exhibiting a wide spectrum of swimming abilities, it's not surprising that fish possess diverse cardiac anatomy and performance capabilities. Over the last 50+ years, research on fish cardiac energy metabolism has focused on: (1) defining cardiac mechanical performance under different conditions; (2) linking anatomical features, cardiac performance, and biochemical characteristics of the heart; and (3) understanding the needs and provisions of adenosine triphosphate (ATP) to support continuous yet variable energy requirements for cardiac function. Compared with the extensive peer-reviewed literature on mammalian hearts, most of our knowledge of cardiac energy metabolism in fish is derived from a relatively few studies on limited species, and using a small number of animals.

A single circuit, closed, circulatory system first appeared in fish, and provided the physical means to produce higher pressure and increased blood flow to the gills and other tissues compared with most invertebrates. Although quite variable between fish species, the heart itself generally consists of four structures/chambers in series—the sinus venosus, the atrium, the ventricle, and either a bulbus arteriosus (in teleosts) or a conus arteriosus (in elasmobranchs). Not surprisingly, the fish ventricle is highly variable, both in terms of size and structure, and functional capabilities (Farrell and Jones, 1992; Santer, 1985; see Chapter 1, Volume 36A: Icardo, 2017; Chapter 4, Volume 36A: Farrell and Smith, 2017). Identical to mammals and other vertebrates, every cardiac myocyte in fish is activated during each contraction cycle (beat). The heart must maintain its own energy status to sustain its continuous (yet variable) contractile performance, and other functions under different physiological conditions. For the majority of fish hearts, contractile cells are provided with only venous, partially deoxygenated, blood within the inter-trabecular spaces of the ventricle lumen. Select active fish, and all elasmobranchs, also have a ventricle with a compact layer and a coronary circulation that provides oxygenated blood from the gills (see Chapter 1, Volume 36A: Icardo, 2017; Chapter 4, Volume 36A: Farrell and Smith, 2017). Under normoxic conditions, mitochondria and aerobic metabolism produce most of the ATP used by the fish heart. However, during hypoxic/anoxic conditions, anaerobic metabolism (glycolysis) becomes the major source of ATP production, despite an inherently limited capacity compared with oxidative phosphorylation (see Chapter 3, Volume 36B: Gillis and Johnson, 2017). Cold environmental temperatures can also reduce the potential for both cardiac energy production and utilization, and therefore, limit cardiovascular performance. However, it is well documented that some fish compensate for, or even negate, the suppressive effects of cold temperature on cardiac energy metabolism through positive temperature acclimation or acclimatization (e.g., Driedzic, 1992; Driedzic and Gesser, 1994; Kalinin et al., 2009). Overall, cardiac energy metabolism is important because of the direct links between ATP production and mechanical performance at the molecular, cellular and whole heart levels.

In this chapter, we provide an overview of the current understanding of cardiac energy metabolism in fish, with a particular emphasis on biochemical principles and perturbations and the strategies employed by various fishes to maintain metabolic homeostasis and cardiac performance. Known determinants of cardiac energy metabolism in fish are summarized in Table 1.

Table 1

Inter-related factors that underlie cardiac energy consumption and production in fish

Energy (ATP) consumption
Power output (PO) = $(Q \times (P_{out} - P_{in})) \times 0.0167$/mass of ventricle ($M_v$)
A. Cardiac output (Q) = heart rate (f_H) × stroke volume (V_S)
B. Ventral aorta pressure or output pressure (P_{out})
Intrinsic modulators of power output
Adrenergic and cholinergic stimulation
Enzymatic activities, responsible for the breakdown of ATP, especially the phosphatases myosin ATPase, Ca^{2+}-ATPase, and Na^+/K^+-ATPase
Intracellular pH, [Ca^{2+}], and [inorganic phosphate, P_i]
Heart size (M_v)
Fish size
Extrinsic modulators of power output
Environmental temperature and temperature history
O_2 availability (environmental)
Energy (ATP) production
Intrinsic modulators
O_2 availability (tissue oxygenation)
Enzymatic activities required for the production of ATP
Anaerobic
Aerobic
Metabolism fuel availability and preference
Intracellular pH
Fish sex
Extrinsic modulators
Environmental temperature and temperature history
O_2 availability (environmental)

Characteristics of the mammalian heart and other species will be provided for mechanistic insights and comparative perspectives throughout the document. Unfortunately, this chapter does not exhaustively review all of the pertinent literature or provide a complete perspective. Several reviews and monographs have been published about cardiac energetics and metabolism in fish (Driedzic, 1992; Driedzic and Gesser, 1994; Gamperl and Driedzic, 2009; Gesser and Overgaard, 2009; Moyes, 1996), and we recommend that readers pursue these references for greater depth, historical perspective, and additional information.

2. CARDIAC ENERGY STATE AND FUNDAMENTALS OF CELLULAR ENERGY METABOLISM

Cardiac myocytes are the predominant cell type in the vertebrate heart and require ATP to support continuous contractile function, ion pumping across membranes, protein synthesis, and other maintenance functions. Ultimately, there must be an ongoing balance between cellular energy demands and supply by each cardiac myocyte if normal and elevated levels of contractile performance are to be sustained. Performance is dependent, in large part, on cellular energy status. However, under some O_2-limiting conditions, the rate of ATP production does not necessarily limit contractile function, and therefore cardiac performance.

Data on energy metabolism in the mammalian cardiac myocyte suggest that the ΔG_{ATP} available for contractile work and overall energy homeostasis is remarkably constant (Balaban, 2012; Wikman-Coffelt et al., 1983). While the current understanding of cardiac energy state comes from studies of the mammalian heart, the derived equations and related principles should be applicable to the fish heart.

ATP is the fundamental link between muscle function and catabolic metabolic responses in all vertebrates. ATP hydrolysis, beginning far from equilibrium, only proceeds when coupled to reactions catalyzed by selective ATPases and provides the energy needed to support different processes. The products of ATP hydrolysis are adenosine diphosphate (ADP), inorganic phosphate (P_i), and a proton (H^+):

$$\mathrm{ATP} + \mathrm{H_2O} \rightarrow \mathrm{ADP} + \mathrm{P_i} + \alpha H^+ \tag{1}$$

where α is less than 1, Meyer and Foley (1996).

One instructive example of ATP hydrolysis and free energy transduction is the myosin ATPase that is responsible for the sliding filament mechanism and the contraction of cardiac myocytes (see below). It is also noteworthy, and underappreciated, that P_i release from myosin energizes force development and that the energy delivered by ATP hydrolysis is mainly the result of the high solubility of P_i, which loses energy when released in water (Eisenberg and Hill, 1985). Energy-demanding processes ultimately rely on the cellular energy state to supply energy. Energy state, which can be expressed by the phosphorylation potential, is the maximal amount of usable energy or the "Gibbs free energy per mol of hydrolyzed ATP":

$$\Delta G_{\mathrm{ATP}} = \Delta G_{0\mathrm{ATP}} - R \times T \times \ln \frac{\mathrm{ATP}}{\mathrm{ADP} \times \mathrm{P_i} \times \alpha \times \mathrm{H}^+} \tag{2}$$

This equation is based upon cytoplasmic concentrations of "free" ATP and ADP complexed with Mg^{2+} (Alberty and Goldberg, 1992). It is important to note that small changes in ATP concentration may result in proportionately larger changes in [ADP]. Changes in ADP in combination with those in P_i - due to the creatine kinase (CK) catalyzed buffering of ATP - will also amplify the effect of ATP changes on ΔG_{ATP}. However, an increase in

ADP may be counteracted by the conversion to AMP by the reaction catalyzed by adenylate kinase: 2ADP $\leftrightarrow$ ATP + AMP (see below). Under severe conditions, ADP may also be degraded through other pathways (Wyss and Kaddurah-Daouk, 2000). A study of four teleost species indicates that this degradation occurs in some teleost hearts under anoxia (Hartmund and Gesser, 1996). ΔG_{ATP} can be estimated in muscles by the reactants and products of the CK catalyzed reaction that regenerates ATP:

$$H^{+} + \text{creatine phosphate(CP)} + \text{ADP} \leftrightarrow \text{creatine(C)} + \text{ATP} \qquad (3)$$

With an equilibrium constant $\left(K_{eq} = \frac{ATP \times C}{ADP \times CP \times H^{+}}\right)$ of about 100 at a physiological pH of 7.4, the CK reaction favors rephosphorylation of ADP to ATP. It is therefore reasonable to assume that the CK reaction is close to equilibrium (Dobson et al., 1992). This assumption allows ΔG_{ATP} to be estimated from the concentrations of CP, C, and "free" P_i, whereby the latter approximates C, and only CP and C have to be recorded (Meyer, 1988). Equilibrium is also assumed in using the CK reaction to estimate the concentration ADP in ^{31}P NMR measurements of energy state. Alternatively, the energy state has been assessed by the "adenylate energy charge" given by the ratio of the phosphorylated adenylates able to provide energy and total phosphorylated adenylates $\left(\frac{ATP + 0.5\,ADP}{ATP + ADP + AMP}\right)$. ADP multiplied by 0.5 refers to the complementary adenylate kinase reaction: 2ADP $\leftrightarrow$ ATP + AMP (Atkinson, 1968). This ratio is fairly constant in muscle cells as it is buffered by the CK reaction. Here the cellular energy content varies mainly with the CP/total C ratio. Only following large reductions of CP do changes in phosphorylated adenylates have to be taken in account (Connett, 1988). Thus, the CP/total C ratio has been a useful parameter to approximate energy state since both the concentration of ADP and AMP in muscle are very low and difficult to estimate directly. It is also noteworthy that the CK reaction binds protons released by these ATPase reactions (see below).

3. COUPLING BETWEEN CELLULAR PRODUCTION AND CONSUMPTION OF ATP

ATP is the exclusive substrate for myofibrillar ATPase, and is therefore required for cardiac muscle contraction. Because of its continuous mechanical work, the fish ventricle has a high rate of ATP hydrolysis under aerobic conditions [e.g., 9.2 μmol ATP $min^{-1} g^{-1}$ in the perfused rainbow trout (*Oncorhynchus mykiss*) heart (Arthur et al., 1992)]. A similar value (13 μmol

ATP $min^{-1} g^{-1}$) has been recorded in the tilapia (*Oreochromis* hybrid) heart (Lague et al., 2012). However, stored high-energy phosphates (HEPs) in the heart are limited and can be rapidly exhausted without rapid ATP regeneration (rephosphorylation of ATP). Cardiac concentrations of phosphorylated adenylates ATP, ADP, and AMP in tissue strips are lower than ventricular halves from intact tissue. The values obtained from ventricular halves do not vary greatly between fish species (e.g., total adenylates = 4.4 – 5.7 μmol g wet wt^{-1}), with the exception of the hagfish [total adenylates = 3 μmol g wet wt^{-1} (Christensen et al., 1994)]. As a result, ATP and other HEPs would be rapidly consumed (<2 min) at the measured rates of ATP turnover in most fish hearts under aerobic conditions, or even anaerobic conditions [(~10 μmol ATP min^{-1} g $ventricle^{-1}$ in the anoxic tilapia (*Oreochromis* hybrid) heart (Lague et al., 2012) and ~4 μmol ATP min^{-1} g $ventricle^{-1}$ in the anoxic rainbow trout heart (Arthur et al., 1992; Overgaard and Gesser, 2004)]. A more extensive summary of estimated cardiac ATP turnover rates, along with various measures of cardiac performance during normoxia and hypoxia, are provided in Table 2.

Cardiac myocytes are not unique, and rely on a variety of well-established metabolic pathways to regenerate ATP (Fig. 1). In O_2-supplied cardiac myocytes, energy state and ATP production are almost exclusively supported by mitochondrial respiration (Driedzic, 1992; Lague et al., 2012). However, if O_2 availability decreases, a drop in cell respiration can occur, and ATP production is replaced to a varying extent by anaerobic glycolysis. Despite considerable changes in contractile strength and frequency, the mammalian heart shows only minor changes in energy state. For example, under aerobic conditions the concentrations of ATP and CP, as well as the metabolic intermediates ADP, P_i, and C, remain remarkably stable during physiological changes in workload (Balaban, 2012). Although not confirmed, it is reasonable to posit that energy state stability also occurs in fish cardiac muscle. This seems to be the case in the tilapia (*Oreochromis* hybrid) heart as myocardial O_2 consumption ($\dot{V}O_2$) correlates with work (Lague et al., 2012). Such stability reflects dynamic regulation and requires signaling molecules and several complex mechanisms that link contractile state and energy state.

First and foremost, studies of mammalian hearts suggest that Ca^{2+} is a key molecule for maintaining cardiac metabolic homeostasis (Balaban, 2012). This should also be true for fish hearts, where Ca^{2+} has a central role in the regulation of mechanical activity [e.g., zebrafish (Genge et al., 2016) and rainbow trout (Klaiman et al., 2014)]. Studies of the mammalian heart indicate that intracellular free Ca^{2+} $[Ca^{2+}]_i$ determines muscle contractility and facilitates proportional mitochondrial ATP production. $[Ca^{2+}]_i$ is a main determinant of the entry of Ca^{2+} into mitochondria, which enters through a pore in the outer membrane. Within the mitochondrial matrix, Ca^{2+} stimulates three

Table 2

Cardiac performance indices and energy requirements for select fishes and a mammal using different methods at different temperatures

Species	Method	Temperature (°C)	Body mass (g)	Q (mL kg^{-1} min^{-1})	f_H (bpm)	V_S (mL kg^{-1})	Power output (mW g^{-1})	ATP turnover* (μmol min^{-1} g^{-1})	References
Myxine glutinosa	*In vivo*—resting and normoxic	10	55–91	8.7	22	0.4	0.15	0.5	Axelsson et al. (1990)
Eptratretus stouti	*In vivo*—resting and normoxic	10	160	12.3	8.3	1.3	0.42	1.5	Cox et al. (2010)
Squalus acanthias	Perfused heart—normoxic	15	1520	31.6	41	0.78	2.3	8.3	Davie and Franklin (1992)
Hemiscyllium ocellatum	*In vivo*—resting and normoxic	28	1290	45	62	0.74	5.8	20.9	Speers-Roesch et al. (2012)
	In vivo—resting and hypoxic	28	1290	18	20	0.9	1.4	10.9	Speers-Roesch et al. (2012)
Aptychotrema rostrata	*In vivo*—resting and normoxic	28	1540	38	56	0.72	3.9	14.0	Speers-Roesch et al. (2012)
	In vivo—resting and hypoxic	28	1540	20	22	1.0	1.2	9.4	Speers-Roesch et al. (2012)
Hemitripterus americanus	Perfused heart—normoxic	10	1130	10.6	41	0.26	1.1	4.0	Farrell et al. (1985)
	Perfused heart—normoxic	10	1100	24	30	0.66*	1.2	4.3	Bailey and Driedzic (1989)

Oncorhynchus mykiss	*In vivo*—swimming	11	1200	52.6	51	1.03	7.2	25.9	Kiceniuk and Jones (1977)
	Perfused heart—normoxic	15	531	44.6	68	0.82	4.6	16.6	Graham and Farrell (1989)
	Perfused heart—normoxic	15	554	42*	60*	0.7*	4*	14.4	Graham and Farrell (1990)
	Perfused heart—normoxic	16	561	22.2	78	0.28	2.6	9.4	Arthur et al. (1992)
	Perfused heart—hypoxic	16	561	24.3	76	0.32	0.5	3.9	Arthur et al. (1992)
	Perfused heart—normoxic	10	508	43.9	55	0.89	5.1	19.4	Overgaard et al. (2004)
	Perfused heart—hypoxic	10	450	5	40*	0.10	0.5	3.9	Overgaard et al. (2004)
Cyprinus carpio	*In vivo*—resting and normoxic	10	1300	9.5	16.4	0.61	0.71	2.6	Stecyk and Farrell (2006)
	In vivo—resting and hypoxic	10	1300	3	4.5	0.70	0.17	1.3	Stecyk and Farrell (2006)

(*Continued*)

Table 2 (Continued)

Species	Method	Temperature (°C)	Body mass (g)	Q (mL kg^{-1} min^{-1})	f_H (bpm)	V_S (mL kg^{-1})	Power output (mW g^{-1})	ATP turnover* (μmol min^{-1} g^{-1})	References
Orechromis hybrid sp.	*In vivo*—resting and normoxic	22	709	10.5	35	0.3	1.3	4.7	Speers-Roesch et al. (2010)
	In vivo—resting and hypoxic	22	709	4.8	16	0.28	0.5	3.9	Speers-Roesch et al. (2010)
	Perfused heart—normoxic	22	475	12	55	0.22	1.4	5.0	Lague et al. (2012)
	Perfused heart—hypoxic	22	475	10.4	48	0.22	1.4	10.9	Lague et al. (2012)
Rattus norvegicus	Perfused heart—normoxic	37	150–350	156	285	0.66	27.3	98	Neely et al. (1967)

Atlantic hagfish (*Myxine glutinosa*), Pacific hagfish (*Eptratretus stouti*), spiny dogfish (*Squalus acanthias*), epaulet shark (*Hemiscyllium ocellatum*), shovelnose ray (*Aptychotrema rostrata*), sea raven (*Hemitripterus americanus*), rainbow trout (*Oncorhynchus mykiss*), common carp (*Cyprinus carpio*), tilapia (*Orechromis* hybrid sp.), and laboratory rat (*Rattus norvegicus*). Ventricular power output was calculated as $PO = Q \times (P_{out} - P_{in}) \times 0.0167/M_v$, where Q is in mL min^{-1}, P_{in} and P_{out} are in kPa, M_v is ventricle mass, and 0.0167 is the conversion factor to mW.
*Estimated; PO was converted to ATP equivalents using either 60 nmol ATP s^{-1} mW^{-1} (normoxic) or 130 nmol ATP s^{-1} mW^{-1} (hypoxic conditions) (see Arthur et al., 1992; Farrell and Stecyk, 2007).

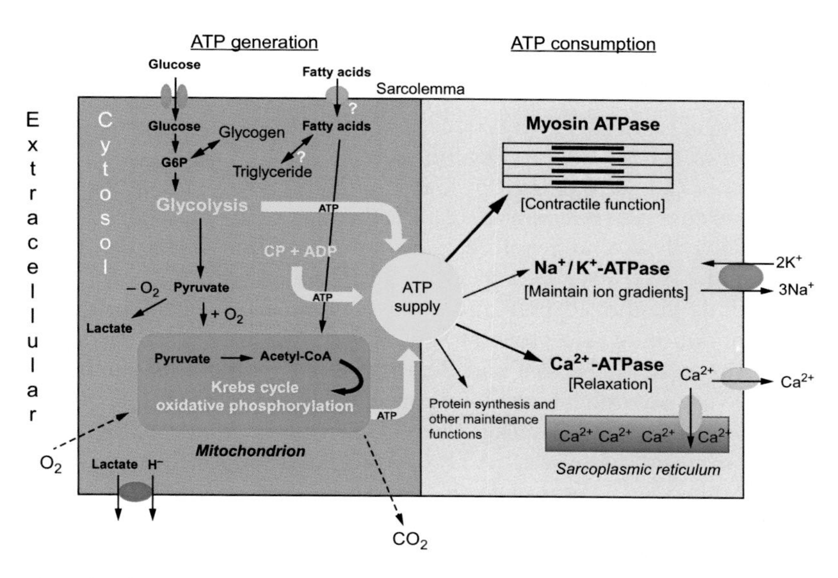

Fig. 1. A simplified summary diagram of intracellular mechanisms for ATP generation and consumption in the teleost cardiomyocyte, plus molecular interactions between the cytosolic and extracellular compartments. Continuous regeneration of cellular ATP supply comes from cytosolic CP, the glycolytic pathway, and mitochondrial oxidative phosphorylation. Depending on the species, major myocardial fuels include extracellular glucose, intracellular glycogen, extracellular fatty acids, and possibly intracellular triglyceride. Extracellular glucose enters the cell *via* facilitated diffusion and glucose transporter proteins, and can be used in glycolysis or the reversible production of glycogen. In the presence of O_2, glucose and glycogen are broken down by glycolysis to pyruvate in the cytosol and then to acetyl CoA in the mitochondria for ATP production. In the absence of O_2, pyruvate is converted to lactate, which can exit the cell *via* the lactate/proton symporter. The mechanism for trans-sarcolemmal fatty acid uptake is unknown, but cytosolic fatty acids can be oxidized in mitochondria. Although not shown, extracellular ketone bodies, but not fatty acids, are important energy substrates for the elasmobranch heart. Also not shown, during high intensity exercise or post-exercise, extracellular lactate may increase significantly, enter the cardiomyocyte, be converted to pyruvate, and provide yet another fuel source for mitochondrial oxidative phosphorylation. Major mechanisms of ATP consumption include the myosin ATPase for contractile function, the sarcolemmal Na^+/K^+-ATPase that maintains trans-sarcolemmal ion gradients for Na^+ and K^+, and Ca^{2+}-ATPases on the sarcolemma and sarcoplasmic reticulum that reduce cytosolic Ca^{2+} and allow for cardiac relaxation. Other ATP demands include protein synthesis and other maintenance functions. Arrow thickness reflects the relative ability of the different ATPases to support ATP-demanding functions. Overall, concurrent and parallel activation of multiple metabolic processes occurs for the production and consumption of ATP. It is also important to note that only about 20% of ATP consumption under aerobic conditions is actually converted into mechanical work, defining mechanical efficiency. The remainder of the energy is liberated as heat. ADP, adenosine diphosphate; ATP, adenosine triphosphate; ATPase, adenosine triphosphatase; CoA, coenzyme A; CP, creatine phosphate; G6P, glucose-6-phosphate.

dehydrogenases of the citric acid cycle (*via* dephosphorylation) and the pyruvate dehydrogenase complex, which commits pyruvate to oxidative metabolism. This, in turn, causes an increase in the concentration of NADH and the net driving force for ATP production; which may be supported by Ca^{2+} stimulation of ATP synthase (Denton, 2009). At steady state, the mitochondrial influx of Ca^{2+} is balanced by Ca^{2+} efflux, mainly by the mitochondrial Na^{+}/Ca^{2+} exchanger (Balaban, 2012). Furthermore, Ca^{2+} activates glycogenolysis and glycolysis, and therefore, alters mitochondrial substrate availability (Dhalla et al., 1977; Friesen et al., 1969).

Second, as mentioned previously, the enzymatic CK reaction favors the rephosphorylation of ADP to ATP and serves as an energy buffer in muscle cells (Meyer et al., 1984; Wu and Beard, 2009). The buffer capacity of CK appears to vary according to the total concentration of creatine (CP + C, which ranges between 5 and 15 mmol kg^{-1} wet weight), while the total concentration of phosphorylated adenylates (ATP + ADP + AMP) is much more constant [about 5 mmol kg^{-1} wet weight in hearts of fish and other vertebrates (Christensen et al., 1994)]. The CK reaction also amplifies variations in P_i, a variation that probably plays an important role in the coupling between energy consumption and mitochondrial ATP production (Wu and Beard, 2009). CK can be located close to different sarcoplasmic ATPase sites such as myofibrillar ATPase in the mammalian heart (Krause and Jacobus, 1992) and the heart of rainbow trout (Haagensen et al., 2008), where it offers an efficient energy supply since a colocalization of enzymes and structures contained in an unstirred layer enhances their coupling and interaction (Goldman and Katchalski, 1971). However, studies of rainbow trout, Atlantic cod (*Gadus morhua*) and Antarctic species suggest that CK is not associated with mitochondrial ATP synthase in fish cardiac myocytes, and apparently, the outer mitochondrial membrane does not restrict the diffusion of ATP and ADP. If this is the case, CK in fish hearts differs from that in adult mammalian hearts, but resembles CK in cardiac muscle of newborn mammals (O'Brien et al., 2014; Sokolova et al., 2009). When free in the cytoplasm, CK may facilitate the transport of ATP and ADP by coupling them to the diffusion of CP and C (Meyer et al., 1984), but this transport function is controversial during full oxygenation (Birkedal et al., 2014; Wu and Beard, 2009). Myokinase or adenylate kinase also act as energy buffers, although quantitatively less so than CK (Connett, 1988).

Third, the glycolytic pathway protects the energy state not only during hypoxia, but also under full oxygenation (see Section 4.2).

Fourth, considerable evidence indicates that cardiac myocytes also maintain energy state by subcellular compartmentation, which provides an efficient transfer mechanism for ATP, ADP and P_i between cytoplasmic ATPases and nearby mitochondria. This mechanism exists in fish cardiac myocytes, despite the fact that they have smaller cross-sections with presumably shorter diffusion distances than mammalian cardiac myocytes (Birkedal et al., 2014; Guzun et al., 2015; Sokolova et al., 2009).

Fifth, myoglobin may serve as an additional energy buffer in muscle cells because it stabilizes cell respiration by buffering cellular O_2 tension. However, based upon different estimates of myoglobin diffusibility, its role in cellular O_2 transport may be limited during full oxygenation (Chung et al., 2006; Wu and Beard, 2009). Myoglobin diffusion appears to support O_2 consumption and contractility in hypoxic fish hearts (Bailey et al., 1990; Helbo et al., 2013; Legate et al., 1998). Furthermore, myoglobin participates in NO metabolism and the removal of reactive O_2 species (Flögel et al., 2013). For unclear reasons, cardiac myoglobin (Mb) content varies considerably among fish species and is even lacking in some species (Macqueen et al., 2014), but this may be related to whole organism activity level (Giovane et al., 1980).

A sixth possibility is that the recently observed flexibility of the organization of the respiratory chain and ATP synthase might be of importance for mitochondrial respiration under different conditions (Cogliati et al., 2016).

4. ENERGY DEMANDS OF CARDIAC PERFORMANCE AND HOMEOSTASIS

4.1. Myosin ATPase

The heart is a biological pump that transforms chemical energy into mechanical energy that is used for the continuous movement of blood. Here, we highlight the importance of ATP-driven free energy transduction and the activity of myosin-bound ATP phosphatase (also called myosin ATPase, myofibrillar ATPase, or actomyosin ATPase) for cardiac myocyte contraction and shortening. A brief discussion of proposed mechanisms of the cyclic interaction of ATP and actin with myosin is also provided. It is clear that ATP consumption by myosin ATPase, which results in cardiac myocyte force production, is the major determinant of energy expenditure in vertebrate cardiac muscle (Suga, 1979). In addition, the molecular interaction of myosin and actin during muscle contraction, and its regulation by Ca^{2+}, appear to be conserved in vertebrates (Genge et al., 2016).

The rhythmical and coordinated contraction of atrial and ventricular chambers is a key determinant of cardiac performance and a major contributor to total contraction-related energy expenditure. Biochemical and histochemical studies in fish and other vertebrates highlight the importance of contractile proteins and associated enzymes for cardiac energetics. The rate and duration of striated muscle contraction and mechanical energy generation are determined, in part, by the attachment/formation and detachment of actinomyosin cross-bridges through sliding filament mechanisms (Eisenberg and Hill, 1985; Gordon et al., 2001). The head region of the myosin protein exerts force on nearby actin and is viewed as a molecular motor (Fig. 2). Movement of the myosin head, and subsequent detachment of myosin from the actin filament, is a cyclical ATP-dependent process that relies upon the myosin ATPase and hydrolysis of ATP to ADP and P_i. Initial ATP binding to myosin is rapid, reversible, and causes the detachment of actin from the actin–myosin–ATP complex (Gordon et al., 2001). The subsequent formation of a strong cross-bridge and cardiac myocyte force development depends upon the concentration of intracellular Ca^{2+}. Specifically, when Ca^{2+} is bound to troponin C, movement of the tropomyosin/troponin complex allows myosin access to binding sites on actin.

The generally accepted paradigm is that this process operates with a fixed stoichiometry of one ATP per cross-bridge cycle (Kammermeier, 1993). However, contrary to initial expectations, ATP hydrolysis by myosin ATPase does not appear to occur during the force-generating step of the contractile cycle, and it is not clear whether the myosin head takes a discrete step for each ATP hydrolyzed. In each ATP hydrolysis cycle, it appears that myosin alternates between a weak-binding conformation and strong-binding conformation (Fig. 2). As a result of this continuous conformation change, the elastic myosin–actin bridge is strained. Furthermore, the overall mechanism of myosin–actin interaction is comparable with the ATPase cycle used to drive the active transport of Ca^{2+} across membranes (Eisenberg and Hill, 1985).

While our understanding of the energy transduction system and cross-bridge cycling mechanism has not been validated in fish striated muscle, myosin ATPase activity is clearly the major contributor to the energy requirements of the contracting fish heart. Mammalian studies suggest that this enzyme accounts for 76% of contraction-related heat production under aerobic conditions (Schramm et al., 1994). Comparative studies also suggest that the ATPase activity of myosin defines the intrinsic speed of muscle contraction (Bárány, 1967) and is proportional to heart rate (Degn and Gesser, 1997). Myosin ATPase, along with the corresponding myosin heavy chain isoforms, define the contractile properties of fish cardiac muscle (Vornanen, 1994). There is evidence from multiple species of freshwater teleosts that biochemical differences exist in myosin ATPase activity and myosin isoforms between atrial and ventricular tissues (Karasiński et al., 2001). On the functional side, physiological studies demonstrate that the rate of isometric contraction is much faster in atrial *vs* ventricular preparations (Aho and Vornanen, 1999; Tiitu and Vornanen, 2001) and that maximal myosin ATPase activity is significantly higher in the atrium (Aho and Vornanen, 1999; Degn and Gesser, 1997). It is also apparent that the catalytic properties of myosin ATPase can vary—in a species and chamber specific manner—in response

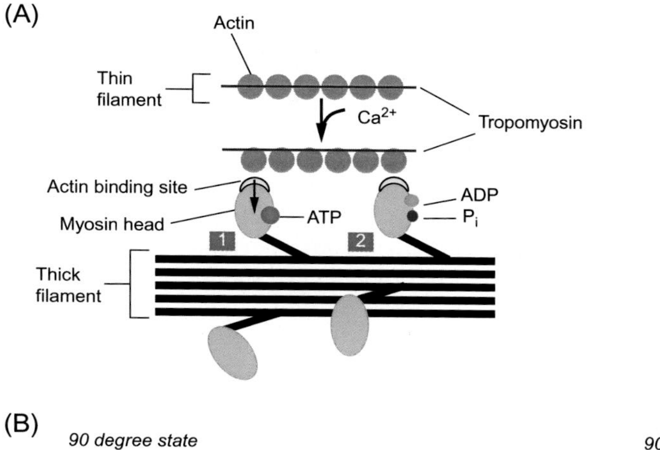

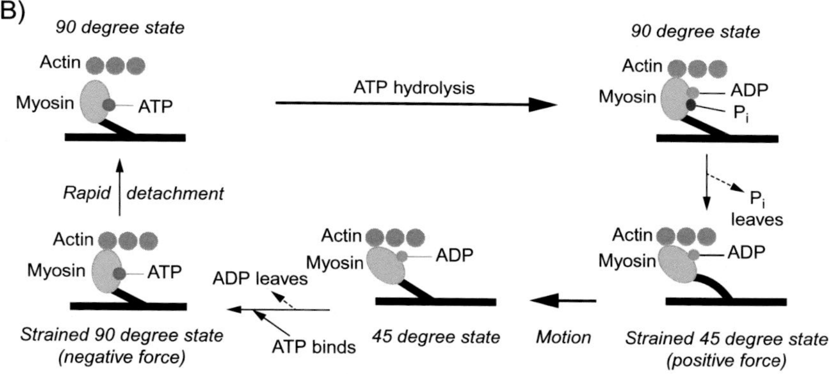

Fig. 2. Energy requirements and conformation states for thin and thick filament interactions during the cross-bridge cycle. (A) In the absence of Ca^{2+}, tropomyosin on the thin filament blocks myosin's access to actin. Ca^{2+} binding to troponin (not shown) allows tropomyosin to move and exposes binding sites for myosin. 1. Prior to each power stroke, ATP binds rapidly and reversibly to the myosin head and causes myosin detachment from actin. 2. ATP hydrolysis to ADP and P_i occurs while bound by way of the myosin ATPase (not shown). (B) A more specific cross-bridge model by Eisenberg and Hill (1985) highlighting the sequential molecular transitions and multiple energy states. Once again, binding of ATP to the cross-bridge induces a weak-binding conformation and the rapid detachment of myosin from actin. Following ATP hydrolysis, P_i leaves the myosin and the cross-bridge undergoes the transition from the 90 degree state to the strained 45 degree state that exerts positive force (a power stroke) while bound. A key component of this model is the two-stage transition from the 90 degree state to the 45 degree state. To complete the cycle, ADP leaves the cross-bridge, another ATP binds, and the cross-bridge enters a strained 90 degree conformation prior to detachment. ADP, adenosine diphosphate; ATP, adenosine triphosphate; P_i, inorganic phosphate. Panel (B) was adapted with permission from Eisenberg, E., Hill, T.L., 1985. Muscle contraction and free energy transduction in biological systems. Science 227, 999–1006.

to changes in water temperature. For example, long-term (>4 weeks) acclimation of the cold-active rainbow trout to cold temperature (4°C) improves atrial and ventricular contractility (Aho and Vornanen, 1999) and increases heart rate (Aho and Vornanen, 2001). Cold acclimation also increases the Ca^{2+} sensitivity of the contractile system and increases ventricular pressure development in the rainbow trout myocardium (Klaiman et al., 2014). Conversely, cold acclimation prolongs myofibrillar contraction kinetics in atrial and ventricular tissue and decreases heart rate of cold-inactive crucian carp (*Carassius carassius*) (Tiitu and Vornanen, 2001; Vornanen, 1994). Inverse thermal compensation in the heart of crucian carp should help reduce cardiac metabolic costs, improve the energy economy of contraction, and ultimately help this species conserve energy during the winter months and anoxic conditions (Tiitu and Vornanen, 2001). Please see Section 7 for a more extensive review of cold adaptations in the hearts of various fishes.

4.2. Cellular Ion Regulation

Intracellular ion homeostasis also appears to be a major requirement for ATP utilization in fish cardiac myocytes. Specifically, the ability of monovalent ions (Na^+, K^+, and Cl^-) and the divalent cation Ca^{2+} to pass selectively across the sarcolemma plays a crucial role in cardiac myocyte function (see Chapter 3, Volume 36A: Vornanen, 2017) and energy expenditure is required to maintain ionic gradients in all cells. Relative to the extracellular fluid, there is an excess of intracellular molecules and molecular aggregates, which are locked inside the cell and together carry a negative charge. This charge attracts positive ions and, if passively distributed, the concentrations of Na^+, K^+, and Ca^{2+} would be higher inside the cell than at the outside. Electrochemical equilibrium would be achieved when the energy liberated as the ion moves down its concentration gradient equals the energy needed to move it against the electrical potential gradient. Here the membrane potential, E_m, equals the equilibrium potential:

$$E_{ion} = \frac{R \times T}{z \times F} \ln \frac{ion_o}{ion_i}; \tag{4}$$

where R is the gas constant, T is temperature (K), z is charge on the ion, F is Faradays constant, ion_i and ion_o are intracellular and extracellular "free" ion concentrations.

The most important role of K^+ in striated muscle function is to determine the E_m, which is the basis for myocyte excitability, and therefore, contractile function. In the mammalian heart, this role requires that $[K^+]_i$ is kept high (100–160 mmol L^{-1}) while $[K^+]_o$ must be much lower (~4 mmol L^{-1}) and cell membrane permeability must be much higher than for other cations (Sejersted and Sjøgaard, 2000). Regarding the first requirement, the authors are unaware of any study that has measured the intracellular and corresponding extracellular concentrations of ions in the fish heart. For the mammalian heart, extracellular concentrations of ionized Na^+ and K^+ are 140 and 4 mmol L^{-1}, respectively. Conversely, intracellular concentrations are about 6 mmol L^{-1} for Na^+ and 80 mmol L^{-1} for K^+ (Opie, 1991). Respective concentrations of Na^+ and K^+ inside frog ventricular cells are about 10 and 136 mmol L^{-1}, and 130 and 2.6 mmol L^{-1}, outside the cells in the plasma (Donohoe et al., 2000). The corresponding intra- and extracellular values in fish are probably similar, although this needs to be confirmed. To the best of our knowledge, the question of membrane permeability to Na^+ and K^+ in the nonexcited cardiac myocyte has not been addressed in fishes. It has been measured in embryonic chickens, and in this bird, the cardiac muscle sarcolemma K^+ permeability (2.4×10^{-7} cm s^{-1}) is about 50 times higher than that for Na^+ (4.6×10^{-9} cm s^{-1}) (Carmeliet et al., 1976). Nonetheless, despite being actively transported by the sarcolemmal Na^+/K^+-ATPase, K^+ is close to its equilibrium potential of about −90 mV and is the main determinant of the resting membrane potential (−80 mV) in rainbow trout cardiac myocytes (Møller-Nielsen and Gesser, 1992). In contrast, the extracellular concentrations (activities) of Na^+ and Ca^{2+} are one to several times higher than intracellular values due to their low permeabilities and the activities of the sarcolemmal Na^+/K^+-ATPase and Ca^{2+}-ATPase.

4.2.1. Na^+/K^+-ATPase

Na^+/K^+-ATPase is located in the plasma membrane of all animal cells and actively pumps three Na^+ out of the cell and two K^+ into the cell at the cost of one ATP (Glitch, 2001). First and foremost, Na^+/K^+-ATPase maintains ion gradients across the plasma membrane. Na^+/K^+-ATPase activity also prohibits cell swelling by maintaining the intracellular osmotic pressure equal to the extracellular one by counteracting an accumulation of positive ions. Furthermore, such an accumulation would change the membrane potential in the positive direction whereby Cl^- would also move into the cell and contribute to the osmotic disequilibrium (MacKnight and Leaf, 1977; Terashima et al., 2006).

Because the stoichiometry of the Na^+/K^+-ATPase is three Na^+ exchanged for two K^+, one net charge is pumped for each enzyme cycle. The pump, therefore, creates a current and will also be sensitive to voltage (Sejersted and Sjøgaard, 2000). The work carried out by Na^+/K^+-ATPase can be estimated by the equation:

$$\text{Watts (Joules)} = 3 \times F \times (E_m - E_{Na^+}) + 2 \times F \times (E_m - E_{K^+}) \qquad (5)$$

where F is Faradays constant, E_m is membrane potential, E_{Na^+} and E_{K^+} are equilibrium potential for Na^+ and K^+, respectively.

Given that K^+ is close to electrochemical equilibrium, whereas Na^+ is about 100 mV above electrochemical equilibrium, the Na^+ transport requires almost all of the energy supplied by ATP hydrolysis (Ingwall and Balschi, 2006). The majority of this energy is maintained in a steep electrochemical Na^+ gradient during cardiac diastole, which enables action potential development and provides energy for different sarcolemmal co- and counter-transporters that regulate intracellular [Ca^{2+}] and pH (Despa and Bers, 2013; Fuller et al., 2013; Skulachev, 1978). Excitable cells possess relatively high Na^+/K^+-ATPase capacities (Clausen, 1998), and this should be accentuated in fish cardiac myocytes with their high surface to volume ratio and predominant reliance on sarcolemmal Ca^{2+} regulation (Vornanen, 1997; Vornanen et al., 2002). In general, however, cells of ectotherms - including cardiac myocytes - display lower Na^+/K^+-ATPase activities than those of endotherms. The lower activities are associated with lower membrane permeabilities for Na^+ and K^+ (Else et al., 1996).

Overall, nearly half of the O_2 consumption coupled to ATP synthesis, and 30% of lactate production *via* anaerobic glycolysis, supports the sarcolemmal Na^+/K^+ ATPase and selective monovalent ion homeostasis in resting rainbow trout atrial myocytes (Mortensen and Gesser, 1999). One apparent advantage to supporting membrane pumps with glycolytically produced ATP includes the proximity of membrane-bound glycolytic enzymes to pump sites (Campbell and Paul, 1992). It is also worth mentioning that the Na^+/K^+-ATPase is stimulated by catecholamines (Sejersted and Sjøgaard, 2000) and inhibited by hypoxia (Wheaton and Chandel, 2011) in mammalian cells. Further studies will be required to confirm whether these changes occur in fishes, and may help to define hormone- and hypoxia-mediated changes in cardiac myocyte ATP utilization.

4.2.2. Na^+/Ca^{2+}-Exchanger

Although not directly responsible for ATP hydrolysis, the sarcolemmal Na^+/Ca^{2+}-exchanger (NCX) is crucial for regulation of the concentration of Ca^{2+} in cardiac myocytes (see Chapter 3, Volume 36B: Gillis and Johnson, 2017) and relies on the activity of the Na^+/K^+-ATPase. Although it is frequency dependent, diastolic intracellular Ca^{2+} concentration is normally $<1\ \mu mol\ L^{-1}$ in rainbow trout atrial myocytes (Shiels et al., 2002a). Thus, extracellular values which are between 1 and 2 mmol L^{-1} result in over a 1000-fold inward Ca^{2+} gradient with an equilibrium potential more positive than the membrane potential can ever attain. NCX catalyzes the electrogenic countertransport of three Na^+ for one Ca^{2+} (Reeves and Hale, 1984). The direction of cation transport depends on the transmembrane concentration gradients for Na^+ and Ca^{2+} and membrane potential, with NCX functioning in either the Ca^{2+}-efflux (forward) or Ca^{2+}-influx (reverse) mode. Hence Ca^{2+} is forced out of the cytoplasm when the membrane potential is below the reversal potential, and into the cell when it is above the reversal potential (Despa and Bers, 2013). While the contribution of the NCX to cardiac contractility in mammals is limited, NCX in reverse mode does play a significant role in defining the activation and degree of contraction in ventricular myocytes from crucian carp (Vornanen, 1999), atrial myocytes of rainbow trout (Hove-Madsen et al., 2000), and zebrafish cardiac myocytes (Genge et al., 2016). Thus, by modifying trans-sarcolemmal Ca^{2+} transport, intracellular Ca^{2+} concentration, and sarcoplasmic reticulum Ca^{2+} content (Hove-Madsen et al., 2000), NCX partially regulates ATP use by the cardiac contractile machinery (myosin ATPase) and cellular ion homeostasis.

4.2.3. Ca^{2+}-ATPases

As mentioned above, repetitive cardiac contraction cycles require significant and rapid changes in cytosolic Ca^{2+}. Myocardial relaxation is the energy-demanding component of excitation–contraction coupling. For myocardial relaxation to occur between contraction events, cytosolic Ca^{2+} concentrations must be reduced to diastolic levels. In addition to the sarcolemmal NCX operating in forward mode, reductions of cytosolic Ca^{2+} concentrations can occur by two mechanisms that require ATP and two different Ca^{2+}-ATPases: (1) actively transporting Ca^{2+} across the sarcolemma and (2) pumping Ca^{2+} into the sarcoplasmic reticulum (SR, a specialized organelle for Ca^{2+} storage and Ca^{2+} release in striated muscle). The sarcolemmal Ca^{2+}-ATPase forces one Ca^{2+} out of the cell per ATP molecule whereas the SR Ca^{2+}-ATPase (SERCA) transports two Ca^{2+} from the cytoplasm into the SR lumen per ATP (Bers, 2014). Together, these two energy-dependent mechanisms create a steep (1000-fold) inward Ca^{2+} concentration gradient, which allows Ca^{2+} to enter the cytoplasm through sarcolemmal Ca^{2+} channels and through channels in the SR opened during excitation (Bers, 2008). It should also be noted that SERCA, and probably other membrane transport ATPases, are very sensitive to decreases in energy state (G_{ATP}) (Kuum et al., 2009). In addition, similar to the Na^+/K^+-ATPase, Ca^{2+}-ATPases in the mammalian heart appear to depend on local ATP production by nearby glycolytic enzymes (Dhar-Chowdhury et al., 2007; Pierce and Philipson, 1985; Sepp et al., 2010, 2014). Evidence suggests that this finding extends to the fish myocardium (Farrar et al., 2006; Gesser, 2002).

While relaxation of the mammalian cardiac myocyte is accomplished by pumping Ca^{2+} back into the SR *via* SERCA or the transportation of Ca^{2+} across the sarcolemma by NCX and the sarcolemmal Ca^{2+}-ATPase, Ca^{2+} transport mechanisms in fish cardiac myocytes vary between species and have not been resolved completely (Shiels and Galli, 2014). Fish cardiac myocytes have a high surface area-to-volume ratio and a small elliptical cross-section

(Clark and Rodnick, 1998; Santer, 1985; Shiels and Galli, 2014), which should promote relatively short diffusion distances for the trans-sarcolemmal Ca^{2+} flux to and from the contractile apparatus. Generally speaking, fish hearts also have reduced SR content compared with mammals, and cardiac relaxation in fish appears to rely predominantly on the NCX and sarcolemmal Ca^{2+} pumps (Tibbits et al., 1991). At physiological temperatures, the rate of SR Ca^{2+} uptake is significantly lower in fish cardiac tissue homogenates than in those of mammals (Aho and Vornanen, 1998). However, active fishes such as bluefin tuna (*Thunnus thynnus*) with elevated cardiac function have a more developed SR and rely much more on SR Ca^{2+} cycling, and therefore, SERCA activity than other less active species (Farrell, 1996). β-Adrenergic stimulation in fishes increases peak intracellular $[Ca^{2+}]$, myocardial twitch force, and augments cardiac energy consumption of the contractile system. β-Adrenergic stimulation also enhances the deactivation of contractility, which together with the augmented peak $[Ca^{2+}]_i$, increases the energy demand of ion transport. For example, results from rainbow trout atrial myocytes suggest that β-adrenergic stimulation increases SERCA activity and SR Ca^{2+} uptake, leading to a faster decay of the Ca^{2+} transient and a faster relaxation rate (Llach et al., 2004).

The contribution of the SR SERCA to relaxation and associated metabolic costs of Ca^{2+} cycling is also affected by environmental temperature and temperature acclimation in select fishes. Studies on the sternothermal burbot (*Lota lota*), which inhabits cold water, suggest that the SR plays an important role in cardiac Ca^{2+} cycling and excitation–contraction coupling (Shiels et al., 2006; Tiitu and Vornanen, 2002). Whether species differences in sarcolemmal Ca^{2+} pumps or cold temperatures affect the relative contribution of NCX, SR Ca^{2+} uptake, and the ATP requirements for relaxation under different environmental conditions are not known. Clearly, additional studies would provide new insights into the functional characteristics and plasticity of Ca^{2+} homeostasis in the fish heart. In summary, during excitation Na^+ and Ca^{2+} enter and K^+ leaves the cytoplasm, both sets of cations moving down their electrochemical gradients. At steady state, the Na^+/K^+-ATPase, NCX, sarcolemmal Ca^{2+}-ATPase, and SERCA maintain intracellular ion homeostasis, and restore resting state by pumping the same amount of respective ions in the opposite direction.

4.3. Basal Energy Requirements and Myocardial O_2 Consumption

Total energy expenditure of the contracting heart includes basal and active components and is related to cardiac performance. Basal metabolism occurs under non-physiological conditions and reflects energy expenditure of the quiescent, non-contracting heart, independent of electrical and mechanical events occurring during each cardiac cycle. This activity is different than cardiac metabolism in a resting animal, and has been estimated by the rate of O_2 consumption or heat production in isolated preparations. Similar to cardiac power output, myothermic measurements of isolated muscle energy metabolism are expressed in Watts (W) per unit weight of cardiac tissue. Unfortunately, very few studies have estimated basal energy requirements of non-contracting hearts, cardiac tissue, or isolated cardiac myocytes from fish. Studies of mammalian hearts indicate that basal cardiac metabolism is variable and depends upon temperature, energy substrate, and O_2 availability (Gibbs and Kotsanas, 1986). Thus, multiple mechanisms likely determine and regulate the "non-mechanical" components of cardiac energy metabolism in fishes.

Resting energy metabolism of the fish heart, much like the mammalian heart, represents a significant fraction of active metabolism. The rate of O_2 consumption for non-contracting atrial myocytes from rainbow trout is approximately one-fourth of the corresponding maximum respiratory capacity between 5 and 20°C (Mortensen and Gesser, 1999). This one-fourth value agrees closely with data for the rate of heat production of isolated rat papillary muscles at 27°C under basal *vs* active conditions (~7 *vs* 30 mW g tissue^{-1}, Gibbs, 1978). Resting ventricular rings from rainbow trout consume O_2 at a rate of 0.09 μmol min^{-1} g^{-1} (predicted ATP turnover rate of ~0.5 μmol min^{-1} g^{-1}, assuming a P:O ratio of three for mitochondrial oxidative phosphorylation) at 16°C. O_2 consumption also increases nearly threefold (to 0.24 μmol O_2 min^{-1}g^{-1}; ~1.4 μmol ATP min^{-1} g^{-1}) in isometric preparations developing force (Kalinin and Gesser, 2002). In this case, resting energy requirements (based upon O_2 consumption) constitute roughly one-third of a working preparation, but not a maximal value. The rate of O_2 consumption recorded in ventricle strips developing isometric force is lower (by ~0.5–0.6 μmol ATP min^{-1} g^{-1}) than the values obtained for contracting perfused hearts from chain pickerel (*Esox niger*) and American eel (*Anguilla rostrata*) at 15°C (Bailey et al., 1991). It should be noted that both resting strips of cardiac tissue and isometric preparations producing force provide values that are lower than corresponding values for perfused hearts (see below).

Less than half of the O_2 consumed by resting atrial myocytes from rainbow trout goes toward ATP synthesis, and when ATP synthase activity is limited, the proton (H^+) gradient across the inner mitochondrial membrane increases to enhance H^+ leak (Mortensen and Gesser, 1999). The presence of proton leak across the inner mitochondrial membrane agrees with previous studies of mammals, and may provide a valuable mechanism for accurately tracking changes in cardiac energy demand with increasing contractile function (Brand, 2005). O_2 consumption data should be interpreted carefully, as variations in proton leak and the resulting coupling ratios render a strict translation of O_2 to ATP production difficult (Brand, 2005). Furthermore, 10%–20% of total ATP production in resting myocytes from rainbow trout is apparently supported by anaerobic glycolysis (Mortensen and Gesser, 1999). Measurements of O_2 consumption could, therefore, underestimate actual ATP requirements.

Basal O_2 consumption of the fish heart has also been estimated by extrapolation of the regression between cardiac power output and O_2 consumption to the ordinate. The value for dogfish (*Squalus acanthias*) hearts at 15°C is 0.34 μmol $O_2 min^{-1} g^{-1}$ (Davie and Franklin, 1992). Estimates for rainbow trout hearts at 15°C range from 0.14 μmol $O_2 min^{-1} g^{-1}$ (Houlihan et al., 1988a,b) to 0.40 μmol $O_2 min^{-1} g^{-1}$ (Graham and Farrell, 1990). These estimates of basal cardiac O_2 consumption, which equate to ATP requirements of 0.8–2.4 μmol $min^{-1} g^{-1}$, are higher than rainbow trout ventricle rings (see above) and yet are comparable to basal O_2 consumption values for mammalian hearts at 27°C (0.33–0.62 μmol $O_2 min^{-1} g^{-1}$, Gibbs and Kotsanas, 1986).

4.4. Myocardial O_2 Consumption and Power Output Under Working Conditions

The active component of cardiac energy metabolism adds the ATP requirements for mechanical contraction (i.e., cross-bridge cycling and myocyte shortening) and ion pumping (for myocyte relaxation and cellular homeostasis) to the basal component. Given the continuous contractile activity and work output of the vertebrate heart, it is not surprising that the energy demands are met predominantly by mitochondrial respiration. ATP must be replenished and regulated despite periods of altered demand and substrate availability (see Section 5). Given this metabolic challenge, even a slight mismatch between the workload and the reaction rate of enzymes in metabolic pathways could severely affect metabolite levels and contractile performance. Fish cardiac myocytes are similar to mammalian cardiac myocytes in that both cells devote considerable cytosolic volume to mitochondria for energy conversion.

Morphological and ultrastructural studies provide insights into the energy requirements and aerobic capacities of fish cardiac myocytes. Fish cardiac myocytes are relatively thin, elongated cells with an elliptical cross-section, and this should promote the trans-sarcolemmal exchange of respiratory gases and cations by diffusion. The contractile elements (myofibrils) are found predominantly near the sarcolemma whereas mitochondria are more centrally located (Clark and Rodnick, 1998; Santer, 1985). The contribution of mitochondria to myocardial cell volume in the rainbow trout ventricle (26%–29%) is slightly less than that in the rat ventricle (32%–37%, Page et al., 1974), whereas myofibrillar volume density is remarkably similar between these two animals (41%–45%, and 44%–49%, respectively; Page et al., 1974). However, it is noteworthy that variability in cellular ultrastructure has been documented in the hearts of Antarctic fishes that differentially express the O_2-binding proteins hemoglobin (Hb) and Mb (O'Brien and Sidell, 2000). Specifically, mitochondrial volume density varies in cardiac myocytes from just 16% in *Gobionotothen gibberifrons* (+Hb/+Mb) to 37% in *Chaenocephalus aceratus* (−Hb/−Mb)], and this is associated with the opposite trend in myofibrillar densities (40.1% and 25.1% for *G. gibberifrons* and *C. aceratus*, respectively). This high mitochondrial density and membranous network may promote intracellular O_2 movement in *C. aceratus* hearts. Interestingly, despite striking differences in cardiac cellular architecture and the presence or absence of O_2-binding proteins, the aerobic metabolic capacity for ATP production (per g of tissue) appears to be conserved among Antarctic species (O'Brien and Sidell, 2000).

Myocardial O_2 consumption under aerobic conditions is an indirect measure of ATP supply and demand during steady-state conditions (Farrell and Stecyk, 2007), as the variable coupling ratio for mitochondria bioenergetics (ATP made per each oxygen atom, or P/O ratio) needs to be appreciated (Brand, 2005). Myocardial O_2 consumption has been measured for working perfused hearts from a variety of fish species and is directly proportional to cardiac power output under aerobic conditions. Power output is an integrated measure of heart performance and is the product of either (1) stroke work and heart rate or (2) cardiac output (heart rate × stroke volume) and ventral aortic pressure (Table 1). The vast majority of power output estimates in fish are derived from measurements of cardiac output and ventral aortic pressure (see Chapter 4, Volume 36A: Farrell and Smith, 2017). Identical to myothermic measurements, power output is expressed in W per unit mass of cardiac tissue.

Experimental evidence suggests that the heart of most fish functions aerobically, with a few notable exceptions. However, fish hearts develop less power and consume less O_2 than hearts from endothermic vertebrates such as birds and mammals (Driedzic et al., 1987, Table 2). Numerous efforts have been devoted to defining the energy requirements of fish hearts under working conditions. At one extreme are the hagfishes. The branchial hearts of Atlantic hagfish (*Myxine glutinosa*) and Pacific hagfish (*Eptatretus stouti*) have very low performance capabilities *in vivo* (0.15–0.4 $mW\,g^{-1}$, Axelsson et al., 1990; Cox et al., 2010) and cardiac energy demands (~1–4 μmol ATP$g^{-1} min^{-1}$) at 10°C. Further, unlike the majority of fish hearts, it has been estimated that anaerobic glycolysis could support a significant portion (up to 52%) of the peak power output recorded in the isolated, fully oxygenated, hagfish heart (Forster, 1989) and that the ATP requirements lie within the glycolytic generating capacity of the cardiac tissue (Forster, 1991; Hansen and Sidell, 1983). A recent study using direct calorimetry on excised hearts from Pacific hagfish confirmed the low metabolic demands of this organ (<0.33 μmol ATP $g^{-1} min^{-1}$) and its exceptional ability to use anaerobic cardiac energy metabolism and stored glycogen to sustain cardiac function during anoxia (Gillis et al., 2015). Perfused hearts from dogfish (*S. acanthias*), sea raven (*Hemitripterus americanus*) and rainbow trout at physiological temperatures (10–16°C) all have much higher power outputs and predicted energy requirements than hagfish hearts, but much lower values than perfused hearts from laboratory rats at 37°C (Table 2). It is noteworthy that the heart rates of unpaced, perfused, fish hearts ranged from just 8–12 beats per min (bpm, hagfish) to 60–78 bpm (rainbow trout), whereas rat hearts were paced at 285 bpm. All of these heart rates represent physiological frequencies. It is also noteworthy that hypoxia reduces *in vivo* cardiac performance in several fishes (Table 2 and see Section 6).

A fundamental difference between the heart of fish and endothermic vertebrates is the lower resting and maximal heart rates in fishes (Lillywhite et al., 1999), and this is likely an important factor in determining species differences in cardiac energy metabolism. In mammals, a higher heart rate at any given work level is accompanied by increased myocardial O_2 consumption (Sarnoff et al., 1958), and the relationship between O_2 consumption and heart rate is essentially linear (Davie and Franklin, 1992; Farrell and Jones, 1992). However, while birds and mammals primarily increase heart rate to increase cardiac output, and therefore power output, teleost fish also have the capacity to increase stroke volume under certain conditions (e.g., exercise; see Chapter 4, Volume 36A: Farrell and Smith, 2017).

Ventricular power output during swimming exercise also varies considerably between fish species, ranging from just 0.27 mW g ventricle^{-1} in hagfish (*M. glutinosa*, Axelsson et al., 1990), to 7.03 mW g ventricle^{-1} in rainbow trout (Kiceniuk and Jones, 1977; Table 2), and 15–18 mW g ventricle^{-1} for various tuna species (Farrell and Jones, 1992; Chapter 4, Volume 36A: Farrell and Smith, 2017). At maximum exercise levels, O_2 uptake by the rainbow trout heart increases by a factor of 9.6 over resting levels, whereas total O_2 consumption of the whole fish increases 7.8 times (Kiceniuk and Jones, 1977).

4.5. Protein Synthesis

To maintain cellular homeostasis, viability, and the mechanical function of cardiac structures, there is a continuous requirement for the synthesis and degradation of numerous proteins. Protein synthesis is an energetically expensive process, and mitochondria production of ATP for the purposes of cellular protein synthesis constitutes a significant component of the energy expenditure of the non-beating mammalian heart (20%–25% of the total, Gibbs and Loiselle, 2001). Although this has not been assessed in the non-beating fish heart, protein synthesis does account for a considerable proportion of the *whole animal* metabolic requirement of fishes (e.g., 24%–42% of O_2 consumption in Atlantic cod, Houlihan et al., 1988a,b). Protein synthesis requires an estimated four ATP equivalents per peptide bond formed, and another ATP equivalent for transport processes (Reeds et al., 1985). The corresponding theoretical minimum cost is ~8.3 μmol O_2 or 50 μmol ATP per mg of protein synthesized (Reeds et al., 1985). However, different fish tissues have different energetic costs for protein synthesis, and the energy costs of protein synthesis may vary inversely with the rate of protein synthesis (reviewed in Houlihan et al., 1995). For the beating fish heart, it is likely that the relatively high energetic cost of contraction and ion regulation relegates protein synthesis to a small, and variable component, of the overall energy demands of the fish heart.

Estimates of the energy demands of protein synthesis and/or the rates of protein synthesis in fish hearts have only been reported by a few investigators, and on a limited number of species. For example, Houlihan et al. (1988b) used an *in vitro* perfused preparation and reported that the energy cost of protein synthesis is just 2.6% of total aerobic energy consumption of the contracting rainbow trout heart. Although this value appears numerically small, the importance of protein synthesis should not be underestimated given the central role of proteins in the maintenance of cardiac myocyte structure and dynamic function. *In vivo* estimates of the fractional rate of protein synthesis are 2%–5% day^{-1} in hearts of resting rainbow trout, and this value declines with increasing body size and fish growth rate.

Protein synthesis also varies with work output and cardiac contractility (Houlihan and Laurent, 1987; Houlihan et al., 1988b). Doubling power output and O_2 consumption in the perfused rainbow trout increase the fractional rate of protein synthesis in the atrium and ventricle by 2.5-fold (Houlihan et al., 1988b). Furthermore, both short-term swimming [40–60 min at 1.25–1.5 body length s^{-1} (BL s^{-1}), Houlihan et al., 1988b] and longer training (swimming continuously at 1 BL s^{-1} for 6 weeks, Houlihan and Laurent, 1987) stimulate the rate of ventricular protein synthesis. These results highlight the stimulating effect of increases in contractility and volume loading on cardiac protein synthesis and suggest that the cost of protein synthesis and the overall energy budget of the fish heart will increase during swimming activity. However, the relationship between cardiac O_2 consumption and rates of protein synthesis is also still a matter of controversy.

It is important to mention that measurements of protein synthesis provide an incomplete perspective on cellular protein turnover and its energetic cost. Very little is known about protein half-life, lability, and degradation rates in the fish heart. Houlihan and Laurent (1987) estimated that the ventricle of spontaneously active rainbow trout retains 18% of the protein synthesized as growth, the remainder being degraded. Indirect measurements under aerobic conditions suggest that ATP-dependent protein degradation can account for 22% of total ATP turnover in hepatocytes from western painted turtle

(*Chrysemys picta bellii*) (Land and Hochachka, 1994). It is highly unlikely that this number (22%) applies to the contracting fish heart, and further investigation is needed to extend our understanding of the energetic costs of protein turnover in fish hearts and how environmental stressors such as low O_2 and temperature variations (low and high) impact protein half-life and proteolytic rate.

5. ENERGY SUBSTRATES AND SYSTEMS USED TO REGENERATE ATP

5.1. Substrate Selectivity

What carbon substrates do mitochondria in fish cardiac myocytes use for ATP regeneration? Potential substrates include carbohydrates, fatty acids, ketone bodies, and amino acids, from either extracellular or intracellular sources. When oxidized, these substrates produce reducing equivalents required for ATP synthesis *via* oxidative phosphorylation and ATP synthase—an enzyme and ion pump embedded in the inner membrane of mitochondria. The question of energy substrate selectivity is important because of differences in potential ATP yield: 129 for each molecule of 16:0 palmitate, 38 for each glucose molecule, and 12 for each molecule of acetyl-CoA (Newsholme and Leech, 1983). Moreover, the yield of ATP per O_2 molecule varies with substrate (3.17 for glucose, 2.83 for palmitate, and 2.50 for acetate, Starnes et al., 1985), and should define the energy economy of the fish heart under different environmental and cellular conditions. When O_2 is limiting, a fuel preference in the heart for glucose and glycogen may be of particular value for maintaining cardiac myocyte energy production.

In the oxygenated mammalian heart, fatty acid catabolism by mitochondrial β-oxidation generally accounts for 60%–90% of total energy production (van der Vusse et al., 1992), while glucose metabolism through glycolysis accounts for 10%–40% of total energy production (Gertz et al., 1988). Moreover, there is a competitive relationship between glucose and fatty acid metabolism in the mammalian heart (Neely and Morgan, 1974). Much less is known about the fish heart, although efforts have been made to identify key substrates for cardiac oxidative metabolism in a limited number of species. To date, the focus on fuel utilization in fish hearts has been on the carbohydrates glucose and glycogen, the carbohydrate metabolite lactate, free long-chain fatty acids, and ketone bodies. However, it is unknown whether amino acid oxidation contributes to cardiac ATP regeneration in fishes. Unlike the mammalian heart, hearts from several fishes can perform quite well *in vitro* under aerobic conditions in the absence of extracellular substrates (Bailey et al., 2000; Becker et al., 2013; Clow et al., 2004; Driedzic and Hart, 1984; Farrell et al., 1988).

This ability is probably due to the much lower energy needs of fish hearts compared with mammals, and the effective use of endogenous glycogen and or triglyceride to support oxidative phosphorylation.

Efforts to define the fuel use and preferences of fish hearts have included several indirect and direct approaches at multiple levels of organization. These studies include biochemical measurements of potential energy substrates and key metabolites in fish cardiac tissue and blood, identification of specific intracellular and extracellular transport proteins, *in vitro* activities of metabolic enzymes in cardiac tissue homogenates, respiration studies on isolated mitochondria, performance and metabolic studies on isolated heart preparations, and tracer studies *in vitro* and *in vivo*.

Cardiac energy metabolism for sustaining contractile activity can be supported by oxidation of exogenous glucose, palmitate and/or lactate in several species of teleost (Bailey and Driedzic, 1993; Driedzic and Hart, 1984; Gesser and Poupa, 1975; Lanctin et al., 1980; Milligan and Farrell, 1991). The polar glucose molecule crosses myocyte membranes by either simple diffusion down a favorable concentration gradient or facilitated diffusion mediated by membrane bound glucose transporter proteins (GLUTs) (Becker et al., 2013; Clow et al., 2004, 2016; Rodnick et al., 1997). The pathways for glucose and fatty acid metabolism differ, and yet, meet at the TCA/Krebs cycle, both providing acetyl–coenzyme A (CoA) for oxidative phosphorylation. In addition, the first phase of the complete oxidation of glucose—the conversion of glucose to pyruvate—is identical to the glycolytic pathway. Pyruvate, with assistance from CoA-SH and nicotinamide adenine dinucleotide (NAD^+), is then converted to acetyl–CoA, plus nicotinamide adenine dinucleotide (reduced, NADH) and CO_2, by the mitochondrial multienzyme complex pyruvate dehydrogenase (PDH, Eq. 6).

$$\text{Pyruvate} + \text{CoA-SH} + \text{NAD}^+ \xrightarrow{\text{Pyruvate dehydrogenase}} \text{Acetyl-CoA} + \text{NADH} + \text{CO}_2 \tag{6}$$

PDH, therefore, marks entry into the TCA/Krebs cycle within the mitochondrial matrix and provides a vital metabolic link between glycolysis and glucose oxidation. However, there is growing evidence from diverse teleosts [sea raven (Sephton et al., 1990); rainbow trout (West et al., 1993); Atlantic cod (Clow et al., 2016); and short-horned sculpin (Clow et al., 2016)] that extracellular glucose is used predominantly to support lactate production and not aerobic metabolism. Studies on rainbow trout hearts with dichloroacetate (a stimulator of PDH activity) also suggest that myocardial oxidation of pyruvate may be limited, and related to relatively high activities of the competing enzyme lactate dehydrogenase (LDH, Battiprolu and Rodnick, 2014). Further, although demonstrated over 40 years ago in pink salmon (*Oncorhynchus gorbuscha*) by Patton et al. (1975), it appears that endogenous fatty acids (*via* triglyceride) may be an underappreciated source of chemical energy in the fish heart under oxygenated conditions (Clow et al., 2016).

A common impression is that lactate is produced predominantly under anaerobic environmental conditions. For example, several studies suggest that lactate production *via* glycolysis accounts for only ~5%, or a minimal contribution, of total ATP production in aerobic fish hearts (Arthur et al., 1992; Driedzic et al., 1983; Lague et al., 2012; Overgaard et al., 2004; Speers-Roesch et al., 2013; West et al., 1993). Nonetheless, plasma lactate may be elevated in fishes after high intensity swimming, and can be taken up from the blood by cardiac myocytes *via* monocarboxylate transporters (MCTs, Omlin and Weber, 2013), converted to pyruvate by the reverse reaction of LDH (Eq. 2), and used in aerobic metabolism (Gesser and Poupa, 1975; Lanctin et al., 1980; Milligan and Farrell, 1991).

$$\text{Pyruvate} + \text{NADH} + \text{H}^+ \xleftrightarrow{\text{Lactate dehydrogenase}} \text{Lactate} + \text{NAD}^+ \qquad (7)$$

The working cardiac muscle of Atlantic hagfish (*M. glutinosa*) prefers exogenous glucose to long-chain fatty acids when both substrates are present (Sidell et al., 1984). This preference for carbohydrate is further demonstrated by the utilization of carbohydrate stores (endogenous glycogen) in the presence of palmitate, but not glucose (Sidell et al., 1984). In addition to glucose, ketone bodies may serve as predominant oxidative fuels in the cardiac muscle of elasmobranch fishes (Driedzic, 1978; Zammit and Newsholme, 1979). The absence of plasma fatty acids, the binding protein albumin, and the key mitochondrial enzyme carnitine palmitoyltransferase (CPT) probably limits the use of exogenous long-chain fatty acids by many species of elasmobranchs (Speers-Roesch and Treberg, 2010). Conversely, fatty acid fuels are oxidized preferentially to carbohydrates in the heart of the Antarctic fish *G. gibberifrons* (Sidell et al., 1995), and evidence from temperate species suggests that fatty acid fuels become more important at low body temperatures (Bailey and Driedzic, 1993; Kleckner and Sidell, 1985; Sephton and Driedzic, 1991), whereas glucose becomes less important due to limitations in membrane permeability (Becker et al., 2013; Hall et al., 2004). Hearts from much warmer Amazonian fishes appear to rely more on anaerobic glucose metabolism than aerobic metabolism for energy production (Driedzic and de Almeida-Val, 1996). Overall, the use of exogenous fuels for aerobic cardiac energy metabolism in fishes is extremely diverse, and likely depends upon the species studied, their thermal history and sex, in addition to experimental conditions such as temperature, O_2 levels, and available substrates. Readers should also appreciate the challenges of studying fish cardiac energy metabolism under physiological conditions, whereby a contracting heart receives a full complement of extracellular factors including all circulating substrates, hormones, and their binding proteins. Furthermore, comparatively little is known about the dynamic utilization and resynthesis of endogenous energy stores, and the mechanisms that regulate overall energy substrate preference in fish hearts.

5.2. Biochemical Assays as Indicators of Cardiac Energy Metabolism

Another approach to investigate cardiac energy metabolism is to measure the activities of key enzymes of energy metabolism. Enzyme capacities measured *in vitro* may be of particular value from a comparative perspective, and may highlight the biochemical basis for interspecies differences in cardiac energy metabolism and whole animal performance, but should be interpreted with caution. While it is likely that maximal ATPase activity—measured *in vitro*—reflects maximal myosin ATPase activities, it is important to note that biochemical assays of ATPase activity cannot accurately assess the dynamic energy needs of contracting myofibrils *in vivo* (Houadjeto et al., 1991).

Enzymes ultimately define the catalytic (kinetic) capacities and physiologically relevant rates of ATP production (e.g., glycolytic and mitochondrial pathways) and consumption (myofibrillar and ion-transporting ATPases) of the myocardium. Estimates of the total or maximal ATP demands of fish cardiac myocytes have been performed using biochemical assays of "total" ATPase in crude homogenates of ventricular tissue under optimized conditions (Driedzic et al., 1987; Sidell et al., 1987). Similar to measurements of ventricular power output, the expression of enzyme activities per gram wet mass of ventricle provides the means to compare cellular energy metabolism across species. Generally speaking, the ventricular ATPase activity of temperate fishes is related to life style and swimming activity. For example, sluggish, bottom-dwelling marine fishes such as the Atlantic hagfish and ocean pout (*Macrozoarces americanus*) have considerably lower ATPase activities at 15°C (5.3 and 11.7 U min^{-1} g $ventricle^{-1}$, respectively) compared with pelagic active species such as the sea bass (*Dicentrarchus labrax*, 31.2 U min^{-1} g $ventricle^{-1}$) and striped bass (*Morone saxatilis*, 35.8 U min^{-1} g $ventricle^{-1}$) (Sidell et al., 1987). While it is clear that there are pronounced biochemical differences in cardiac energy transduction across diverse species of fishes, it appears that maximal ATPase activity increases exponentially with respect to resting cardiac power output across a wide spectrum of vertebrates (Driedzic et al., 1987). For example, based upon measurements of maximum O_2 consumption, calculated rates of ATP supply closely match maximum ATPase activity in both sea raven and rat hearts (Driedzic et al., 1987; Farrell et al., 1985).

Total cardiac ATPase activity in fish hearts also correlates with the maximal activities of two metabolic enzymes: hexokinase (HK) and CPT (Sidell et al., 1987), which are predictors of ATP produced *via* aerobic pathways from carbohydrate and long-chain fatty acids, respectively (Crabtree and Newsholme, 1972a,b). Similar relationships exist between HK, CPT and total ATPase activities in the hearts of Antarctic teleosts (*Notothenia rossii* and *C. aceratus*) measured at 0°C (Sidell et al., 1987). It appears that there is expansion of specific metabolic enzymes in cardiac myocytes to catabolize both carbohydrate and fatty acids as ATP demand and power output increases. This trend is different for mammals, where only the capacity for fatty acid metabolism is expanded as cardiac work capacity increases (Driedzic et al., 1987). Nonetheless, when interpreting enzyme activities measured *in vitro* under non-physiological conditions, the importance of kinetic properties and impacts of many intracellular factors (e.g., substrate levels, cofactors, inhibitors, and pH) should be considered, as well as the lack of a comprehensive understanding of cellular performance *in vivo*. For instance, a study of three tuna species showed no relation between the cardiac enzyme activities measured *in vitro* and the influences of environmental temperature and O_2 on cardiac performance (Swimmer et al., 2004).

6. HYPOXIA

Cardiac hypoxia is the main theme of Chapter 5, Volume 36B: Stecyk (2017) and is also briefly addressed in Chapter 4, Volume 36A: Farrell and Smith (2017). Therefore, the following section only addresses basic aspects of the impact of hypoxia on cardiac energy turnover. Hypoxia is one of the most significant environmental challenges faced by water breathing fishes. Two strategies are employed to meet the challenge of hypoxia and involve the inhibition of aerobic energy metabolism to different degrees (Farrell and Stecyk, 2007). One strategy is to decrease energy consumption. The second strategy is to enhance anaerobic energy production. A problem with the latter (i.e., anaerobic glycolysis) is that the yield of ATP from glucose (2 ATP per molecule) and glycogen (3 ATP per glycosyl unit) is much lower than with mitochondrial oxidative phosphorylation (up to 36 ATP per glucose molecule). This means that anaerobic glycolytic flux must proceed at a rate of about 10 times oxidative phosphorylation to fully compensate for aerobic ATP production.

Glycolysis, beginning with glucose, is cytosolic and involves 10 enzymatic steps that culminate in the production of pyruvate. The two glycolytic reactions that regenerate ATP are the phosphoglycerate kinase and the pyruvate kinase (PK) reactions (Newsholme and Leech, 1983). Under aerobic conditions, glycolysis appears to be essential for normal excitation–contraction coupling in the heart of rainbow trout (Farrar et al., 2006; Gesser, 2002). Under O_2 limiting conditions, mitochondria will not take up pyruvic acid, and most pyruvic acid serves as an electron acceptor and is reduced to lactic acid *via* the LDH reaction (Eq. 7). Hence, the production of lactic acid is important because it results in the oxidation of NADH back to NAD^+, and, at the same time, prevents the buildup of pyruvic acid and product inhibition of glycolysis at the PK step (Gevers, 1977; Williamson, 1965). However, hypoxia also imposes an intracellular acid load, and this is problematic because of the allosteric proton inhibition of the glycolytic regulatory enzyme 6-phosphofructokinase (PFK, Newsholme and Leech, 1983). Proton liberation correlates with the production of lactic acid, but the principal source of elevated proton activity remains controversial and may be directly due to lactate acid accumulation (e.g., Marcinek et al., 2010) and/or ATP dephosphorylation (e.g., Robergs et al., 2004). Although not confirmed, fish cardiac myocytes may counteract intracellular acidification by passive proton buffering, active proton extrusion *via* sarcolemmal Na^+/H^+ exchange (energetically supported by the Na^+/K^+-ATPase) or lactate/H^+ cotransport. Lactate/proton transport has a stoichiometry of 1 proton:1 lactate and is thought to be important for intracellular pH recovery after ischemic events in mammalian cardiac muscle (Juel, 1997). It is noteworthy that environmental hypoxia increases the plasma concentration of glucose in many fishes (MacCormack and Driedzic, 2007; VanRaaij et al., 1996), and this in turn may increase cardiac myocyte glucose uptake *via* facilitated diffusion- and concentration-dependent mechanisms (Becker et al., 2013). Numerous studies have also shown that the concentration of cardiac glycogen decreases, while that of lactate increases when a variety of fish species are exposed to environmental hypoxia (Driedzic and Gesser, 1994).

Moreover, the capacity to maintain cardiac contractility under hypoxic conditions (a measure of cardiac hypoxia tolerance or sensitivity) varies dramatically between species of fishes (Driedzic and Gesser, 1994). In general, hearts of more active species are hypoxia sensitive, and yet, some hearts from hypoxia-sensitive species are capable of surviving severe hypoxia. For many, but not all fishes, the absence of O_2 leads to a rapid reduction in contractile performance *in vitro* and a reflex bradycardia *in vivo*, which decreases the actinomyosin cross-bridge cycling rate, and therefore, the ATP demands of myofibrillar myosin ATPases and ion pumps (Na^+/K^+-ATPases and Ca^{2+}-ATPases).

What impairs the function of fish hearts during hypoxia or anoxia? The primary cause is not typically a lack of ATP (Nielsen and Gesser, 1984; Turner and Driedzic, 1980) or an increase in intracellular proton concentration (Arthur et al., 1992). Instead, despite a well-maintained ATP level and the protection of the energy state (ΔG_{ATP}) by a possible counteraction of increases in ADP (see Section 2), the cellular energy state (ΔG_{ATP}) may decrease due to a net decrease of CP and a corresponding increase in P_i (e.g., Marban and Kusuoka, 1987). Furthermore, increases in P_i depress maximal Ca^{2+}-activated force and Ca^{2+} sensitivity of the myofilaments (Kentish, 1986). During progressive hypoxia, the cardiac energy state (recorded with respect to the influence of ADP and P_i) and cardiac power output are maintained at higher levels in the hypoxia-tolerant epaulet shark (*Hemiscyllium ocellatum*) than in the hypoxia-sensitive shovelnose ray (*Aptychotrema rostrata*). The shark also showed a lower accumulation of cardiac lactate relative to the ray, which may reflect superior O_2 delivery to the shark heart during hypoxia exposure (Speers-Roesch et al., 2012). It should also be noted that the energy demand of contraction might be reduced during hypoxia through a depressed Ca^{2+} transient, and thus, less actin–myosin interaction (Movafagh and Morad, 2010). Interestingly, severe hypoxia does not necessarily evoke maximal anaerobic capacity in the fish heart as both contractility and glycolytic activity may be further enhanced by adrenaline or an increase in extracellular Ca^{2+} (Driedzic and Gesser, 1994).

Biochemical studies suggest that species differences in the ability to counteract the negative effect of hypoxia are related to the cardiac capacity for glycolytic ATP production. For example, a positive relation has been found between the maintenance of anoxic contractility and the ratio of glycolytic to mitochondrial enzyme capacities (Driedzic and Gesser, 1994). Among nine species of fish, this enzyme ratio was the highest in myocardial tissue from Atlantic hagfish and crucian carp (Christensen et al., 1994), both of which have exceptional hypoxic tolerance (Hansen and Sidell, 1983; Vornanen et al., 2009). Notably, though, this trend does not include the hypoxia-tolerant myocardium of the European eel (*Anguilla anguilla*, Christensen et al., 1994), and this suggests the existence of other protective mechanisms. Indirect evidence also indicates that the CK reaction is coupled to hypoxic tolerance. Myocardial activities of CK and a key glycolytic enzyme (pyruvate kinase) increased following acclimation to severe hypoxia (Birkedal and Gesser, 2004). The ratio of CK to aerobic capacity is also particularly high in crucian carp and Atlantic hagfish, but again the European eel myocardium is an exception (Christensen et al., 1994). In addition to energy buffering (Wu and Beard, 2009), the CK reaction counteracts cellular acidification because the rephosphorylation of ADP to ATP binds H^+ ions (Wallimann et al., 1992). However, the results from European eel suggest that enzyme activities measured *in vitro* may not always reflect cardiac metabolic capacities in fish (Christensen et al., 1994).

As mentioned previously, acidosis is known to inhibit glycolysis, with adverse effects on cellular energy state and cardiac contractility in most fish (Driedzic and Gesser, 1994; Hartmund and Gesser, 1995). The effects of acidosis may be mediated directly through competitive binding of hydrogen ions to troponin-C binding Ca^{2+} sites (Orchard and Kentish, 1990). In species such as flounder (*Platichthys flesus*) and armored catfish (*Liposarcus pardalis*), though, increases in cellular Ca^{2+} counteract this decrease (Driedzic and Gesser, 1994), and should also protect cellular energy state by stimulating cardiac energy metabolism (Balaban, 2012; Williams et al., 2015).

If there is access to some O_2 during hypoxia, cardiac muscle can utilize oxidative phosphorylation, and the corresponding P/O ratios (number of ATP produced per oxygen atom by mitochondria) varies considerably (Brand, 2005). Hypoxia may augment this ratio by reducing the mitochondrial proton gradient and in turn proton leak. In addition, stimulation of the CK reaction during hypoxia increases P_i, a substrate for oxidative phosphorylation (Gnaiger, 2001; Wu and Beard, 2009). Together these effects may explain the decreased energy cost of contraction that occurred when snapper (*Pagrus auratus*) were acclimated to hypoxia (Cook et al., 2013). A similar decrease in the energy cost of contraction during hypoxia is observed in turtle (*Trachemys scripta*) cardiac muscle (Overgaard and Gesser, 2004). Although mammalian cardiac mitochondria are damaged and lose their capacity to produce energy following reoxygenation, this is not the case in hypoxia-tolerant ectotherms or perhaps all ectotherms. Among other things, reoxygenation in hypoxia-tolerant fishes does not result in Ca^{2+} overload or the reversal of ATP-synthase into an ATPase, and mitochondrial proton leak is kept low (Galli and Richards, 2014). Furthermore, cardiac mitochondria from the hypoxia-tolerant epaulet shark (*H. ocellatum*) produce less reactive oxidative species, both under normoxia and hypoxia, than mitochondria from the hypoxia-sensitive shovelnose ray (*Aptychotrema rostrata*) (Hickey et al., 2012). Mitochondrial respiration in the epaulet shark, therefore, appears to be more resistant to stresses associated with hypoxia than the shovelnose ray. Recent findings of organizational flexibility of the respiratory chain and ATP synthase (Cogliati et al., 2016) might also be relevant for mitochondrial respiration in fish hearts under different conditions such as hypoxia (see Section 3).

7. COLD TEMPERATURE

Diurnal and seasonal changes in ambient temperature are fundamental characteristics of many aquatic environments. Given that most fish are ideal ectotherms (in continuous thermal equilibrium with their environment), temperature changes can have immediate and direct effects on cardiac energy metabolism and cardiovascular function. Chapter 4, Volume 36B: Eliason and Anttila (2017) focuses on how the cardiovascular system of fish responds to changes in environmental temperature, especially increasing temperature. This section complements their presentation and provides a biochemical perspective of cardiac energy metabolism due to cold or reduced temperatures in temperate zone fishes. A predictable observation has been that physiological rate functions are reduced at cold temperatures. Nevertheless, it is well documented that some fish can compensate for, or even negate, the suppressive effects of cold temperature on cardiac energy metabolism through positive temperature acclimation or acclimatization. However, this is not a universal finding. For example, not all fish maintain their metabolic activity and swimming performance at cold temperatures, with cardiac activity and energy metabolism functioning at reduced levels.

7.1. Impacts of Cold Temperature on Cardiac Energy Metabolism

Cold temperature presents a kinetic constraint for cardiac energy metabolism because of predicted thermal sensitivities of several biochemical processes. This can be demonstrated by the temperature coefficient Q_{10}, which defines the rate of change of a biochemical process due to a 10°C change in temperature. A Q_{10} value of 2 implies that a biochemical process, such as diffusion or an enzymatic reaction, increases two-fold for each increase of 10°C or is reduced by 50% for each 10°C decrease. Conversely, a Q_{10} value of 1 implies that the process is independent of temperature change. A general finding is that acute reductions in temperature decrease the capacity for cardiac ATP production, contractile activity, and power output. Thus, both ATP demand and supply are reduced as temperature is decreased within the physiological range.

Cold temperature increases cytoplasmic viscosity, and therefore, reduces diffusion. Diffusion is a temperature-sensitive process, and diffusion coefficients in fish skeletal muscle predict that temperature impacts the movement of small molecules (e.g., metabolites and respiratory gases) through the aqueous cytoplasm (Sidell and Hazel, 1987). Kinematic viscosity (and therefore diffusive resistance) increases with a Q_{10} of 1.35 between 25 and 5°C. Diffusion coefficients of several metabolites—including ATP ($Q_{10}=1.55$ between 25 and 5°C)—also increase at colder temperature (Sidell and Hazel, 1987; Hubley et al., 1996) and might limit the molecular communication between cellular compartments (e.g., exchange between cytosol and mitochondria) and reduce the rate of ATP production and intracellular transport. Without any adaptive responses, increases in cytoplasmic viscosity could also impede the movement of Ca^{2+} between myofibrils and the sarcolemma or sarcoplasmic reticulum, and limit respiratory gas exchange with mitochondria in fish cardiac myocytes. The diffusion limitations for intracellular Ca^{2+} at cold temperature may also be significant in the case of the sarcolemmal and SR pumping of Ca^{2+} *via* Ca^{2+}-ATPases. For example, energy-dependent SERCA activity is temperature dependent in the hearts of cold (4°C) acclimated rainbow trout ($Q_{10}=1.64$) and crucian carp ($Q_{10}=1.75$), making Ca^{2+} transport into the SR less efficient as temperature decreases (Aho and Vornanen, 1998). However, Da Silva et al. (2011) report a Q_{10} value of just 1.25 for cardiac SERCA2 activity in rainbow trout acclimated at 16–22°C, suggesting that thermal history can influence the participation and energy requirements of SR Ca^{2+} uptake for intracellular Ca^{2+} homeostasis.

Cold temperature also affects cardiac metabolic enzyme activity and fuel utilization. A persistent but unanswered question in temperate zone fish is whether cardiac energy metabolism is more reliant on oxidation of lipid substrates *vs* carbohydrate substrates. A general finding has been higher Q_{10}s for enzymes involved in anaerobic (glucose and glycogen only) *vs* aerobic metabolism involving carbohydrate and lipid substrates (Blier and Guderley, 1988). More specifically, the Q_{10} for cardiac hexokinase (a possible rate-limiting step for glycolysis) was 2.63 ± 0.29 for seven species of teleosts, a value quite similar to the observed data for glycogen phosphorylase at or below the acclimation temperature (Driedzic, 1992). Conversely, with a Q_{10} of 1.60 ± 0.09 for eight species of teleosts, temperature change will have less of an impact on mitochondrial enzymes such as citrate synthase (Driedzic, 1992). Additional evidence for the reduced use of carbohydrate as an energy source at cold body temperature comes from findings that the trans-sarcolemmal uptake of exogenous glucose in rainbow trout cardiac tissue decreases by 45% ($Q_{10}=1.91$) with acute reductions in temperature between 14 and 4°C (Becker et al., 2013).

7.2. Impacts of Cold Temperature Acclimation on Cardiac Energy Metabolism

How do fish compensate for diffusional and catalytic limitations in cardiac muscle at cold temperature? A growing body of evidence suggests that extensive modification of myocyte gene expression occurs and is responsible for temperature-induced changes in the cardiac phenotype, including energy metabolism (Jayasundara et al., 2013, 2015b; Korajoki and Vornanen, 2013; Vornanen et al., 2005). At the morphological level, enlargement of the ventricle appears to be an important factor for the maintenance of cardiac output and performance in some eurythermal fishes that remain active at cold temperatures (Aho and Vornanen, 1999; Driedzic, 1992; Gamperl and Farrell, 2004; Graham and Farrell, 1990; Klaiman et al., 2011; Rodnick and Sidell, 1997; see Chapter 3, Volume 36B: Gillis and Johnson, 2017). In the case of rainbow trout, an increase in ventricular mass occurs during cold acclimation and may be specific for males (Klaiman et al., 2011). Cardiac enlargement may require additional factors related to seasonal acclimitization (i.e., shortened photoperiod) and not just reduced temperature, per se (Gamperl and Farrell, 2004). Ultimately, the larger heart of cold-acclimated rainbow trout has a lower mass-specific rate of O_2 consumption that reduces the effect of the greater ventricle mass on total myocardial O_2 demand (Graham and Farrell, 1990). This energetic advantage is further realized at maximum myocardial power output when mechanical efficiency is improved substantially. However, contrary to expectations, the hearts of Arctic charr (*Salvelinus alpinus*) reared at 15°C are 15%–30% larger, not smaller, than the hearts of fish reared at 5°C (Ruiz and Thorarensen, 2001).

In some species, cold acclimation modifies the *in vitro* activities of enzymes responsible for cardiac contractile function and aerobic metabolism. A general response of fish to extended exposure to cold temperature is an increase in enzyme capacity. However, the biochemical strategies are highly variable between species, and may relate directly to cardiac function. In the cold-active rainbow trout, cold acclimation increases the activity of the myofibrillar ATPase (Aho and Vornanen, 1999; Klaiman et al., 2011). Cold acclimation also expands the cardiac myocyte SR and/or capabilities for Ca^{2+} uptake in eurythermal perch (*Perca fluviatilis*, Bowler and Tirri, 1990), rainbow trout (Aho and Vornanen, 1998; Shiels et al., 2002a,b), and bluefin tuna (Shiels et al., 2011), and increases the expression of SERCA proteins in the heart of cold-active burbot (*L. lota*). Conversely, cold acclimation does not increase the rate of SR Ca^{2+} uptake or SERCA protein expression in the cardiac tissue of the cold-dormant crucian carp (Aho and Vornanen, 1998; Korajoki and Vornanen, 2013). This may be adaptive in crucian carp and reduce cardiac energy consumption during winter dormancy.

Increases in the activity of metabolic enzymes following cold acclimation occurs in the heart of some, but not all species tested (Driedzic, 1992; Driedzic and Gesser, 1994; Rodnick and Sidell, 1997; Tschantz et al., 2002). Increases in aerobic enzyme activities during cold acclimation can occur in the absence of increases in mitochondrial volume density in cardiac myocytes (Rodnick and Sidell, 1997), in the presence of fewer mitochondrial gene transcripts (Vornanen et al., 2005), and be associated with a lower rate of mitochondrial protein synthesis (West and Driedzic, 1999). Cold acclimation has also been shown to increase enzymatic indicators of anaerobic metabolism (Pierce and Crawford, 1997) and gene transcripts of enzymes involved in aerobic glycolysis (Vornanen et al., 2005).

Protein synthesis does not always vary with temperature acclimation and thermal history. Isolated hearts from rainbow trout acclimated to 15°C exhibit two-fold higher rates of protein synthesis at 15°C compared with hearts from fish acclimated to 5°C and tested at 5°C (Sephton and Driedzic, 1995). Interestingly, mitochondria isolated from hearts of sexually immature rainbow trout acclimated to 13°C synthesize protein at the same rate at 25 and 15°C, but the rate of protein synthesis is decreased dramatically at 5°C (West and Driedzic, 1999). In addition, corresponding rates of protein synthesis in isolated, non-contracting, cardiac myocytes from rainbow trout were: (1) reduced dramatically at 5°C *vs* 15°C and 25°C, which were similar and (2) not directly coupled to measurements of O_2 consumption, and therefore, rates of ATP synthesis (West and Driedzic, 1999). Conversely, *in vivo* measurements for the Atlantic cod heart suggest that the fractional rate of protein synthesis is identical in fish acclimated at 5°C or 15°C (Foster et al., 1992). Whether temperature effects—acute or chronic—on cardiac protein synthesis are specific to the preparation employed, directly influenced by contractile activity, or exhibit species-specific thermal compensation will require further study.

Decreasing of cardiac protein synthesis may serve as an important component of metabolic depression and energy conservation in select species. Cunner (*Tautogolabrus adspersus*) decrease cardiac protein synthesis *in vivo* by 50%–60% as ambient water temperatures are reduced seasonally from 8°C, when fish are active and feeding, to 4°C, when fish naturally enter winter dormancy (Lewis and Driedzic, 2007). Acute hypothermia also depresses cardiac protein synthesis in this species by 50% (Lewis and Driedzic, 2010). Conversely, hyperactivation of cardiac protein synthesis (by 175%) occurs when cunner return to 8°C. Similar to cold temperature, exposure to reduced O_2 also reduces cardiac protein synthesis in several fishes: crucian carp (Smith et al., 1996), oscar (*Astronotus ocellatus*) (Lewis et al., 2007), and cunner (Lewis and Driedzic, 2010). Modulation of cardiac protein synthesis by temperature and O_2 concentration may share a conserved cellular pathway in some fishes that ultimately reduces protein synthesis during environmental hypothermia and hypoxia.

8. BODY SIZE AND SEX DIFFERENCES IN CARDIAC ENERGY METABOLISM

Biochemical assays suggest that ventricular cardiac energy metabolism changes in some fishes during normal growth and aging. For example: maximal *in vitro* activities of HK, PK, LDH, HOAD, and CS are higher in the spongy layer of the adult Atlantic salmon (*Salmo salar*) ventricle as compared to that in parr (Ewart and Driedzic, 1987); the Mb content of the spongy myocardium of adult Atlantic salmon hearts was significantly higher than in parr; and the spongy layer of the heart from adult brook trout (*Salvelinus fontinalis*) had higher mitochondrial activities of CPT, CS, and HOAD as compared with immature fish (Ewart and Driedzic, 1987). Activities of CPT, cytochrome oxidase (COX), but not CS, also increased progressively in the ventricular spongy tissue of female rainbow trout as fish grew from ~15 g to 3 kg (Rodnick and Williams, 1999). Accordingly, these investigations suggest that there are size-dependent increases in the enzymatic capacity for energy and aerobic metabolism of both carbohydrates and fatty acids in the ventricular spongy tissue of salmonid fishes, possibly to support increased myocardial power output. Although not tested, it is possible that larger fish experience higher cardiac energetic costs at high swim velocities compared with smaller fish, and that this may be related to size-dependent changes in cardiac enzyme activities and metabolism.

However, the expanded potential for cardiac energy production in larger hearts is not a universal finding in fish. In contrast to active salmonid fish, the sea raven is a sedentary benthic species with a relatively small ventricle without an outer compact layer or a well-developed coronary vasculature. Activities of glycolytic enzymes HK and PFK, the mitochondrial enzymes CS, HOAD and malate dehydrogenase (MDH), and Mb content of the sea raven heart are independent of body mass over a 20-fold range (Ewart et al., 1988). Moreover, the smaller hearts of sea raven utilize more O_2 on a per gram tissue basis than larger hearts at elevated levels of cardiac output, and both PK and COX activities scaled negatively with body mass. Cardiac oxidative potential is, therefore, reduced is this species during growth and energy metabolism in small hearts may be supported predominantly by glucose catabolism. Conversely, cardiac LDH activity increased with body mass and may ultimately facilitate the use of exogenous lactate produced by skeletal muscle in larger sea raven (Ewart et al., 1988). A more recent allometric study showed equivalent scaling exponents for whole animal metabolic rates and whole heart O_2 consumption rates with body mass in five teleost species (*Danio rerio*, *Fundulus heteroclitus*, *Gambusia holbrooki*, *Oryzias latipes*, and *Pimephales promelas*) (Jayasundara et al., 2015a). Additional studies will be required to identify the mechanisms underlying size-dependent plasticity, or the stability of cardiac energy metabolism in fishes.

Finally, while the study of sex differences in cardiovascular energetics has received considerable attention in mammals, very few studies have been conducted on fish. Nevertheless, recent studies on ventricular cardiac tissue from rainbow trout have shown that: (1) male rainbow trout store more glycogen than females (Harmon et al., 2011); (2) males also have higher activities of the mitochondrial enzymes CS and HOAD per gram of ventricular tissue (Battiprolu and Rodnick, 2007); (3) females prefer aerobic glycolysis and exogenous glucose for ATP production and have increased lactate production and LDH activity compared with males (Battiprolu and Rodnick, 2007); and finally (4) female hearts also appear more resistant than males to the negative effects of hypoxia on contractile function (Battiprolu and Rodnick, 2014). It remains to be determined whether these *in vitro* results for rainbow trout extend to other fish species and define meaningful sex differences in terms of cardiac function.

9. REMAINING QUESTIONS, CHALLENGES, AND FUTURE DIRECTIONS

The current understanding of fish cardiac energy metabolism has focused on a diverse set of physiological categories and variables (Fig. 3). Although progress continues, there are still many unanswered questions and challenges ahead. The heart of fishes has been of particular interest to comparative physiologists because of variability in environmental O_2 and temperature, and a direct link between cardiac pumping ability and aerobic swimming performance. However, given the extreme diversity of fish, their habitats, and corresponding physiologies, it is clear that we should not make the simple assumption that cardiac energy metabolism, except for very basic features, is identical between fishes. Also, many fundamental differences are already known to exist between fish and mammals.

Mechanisms that regulate energy substrate preference under different environmental conditions, and varying rates of cardiac energy expenditure, are still poorly understood in fishes. To move forward, there should be a greater appreciation of the simultaneous use of multiple substrates—extracellular and cytosolic—involving several metabolic pathways. A promising target for future research is the AMP-activated protein kinase, a key enzyme and central regulator of cellular ATP regeneration in fish skeletal muscle (Magnoni et al., 2014). With variability in mind, future studies should include considerations of inter- and intraspecific differences in cardiac energy metabolism, fish size, sex, nutrition, stress, domestication, and environmental history. Another continuing challenge is to develop physiologically relevant *in vitro* studies that can transfer meaningful observations to intact hearts and fish. Defining the effects of different experimental conditions such as hypoxia and cold temperature on cardiac mechanical efficiency, mitochondrial coupling ratio, and protein degradation should be of particular value for our understanding of cardiac energy metabolism in fish. Recent applications of molecular techniques, such as targeted gene expression and metabolomic profiling, should also bring new and comprehensive perspectives to fish physiology and our growing appreciation of the metabolic phenotype of fish hearts.

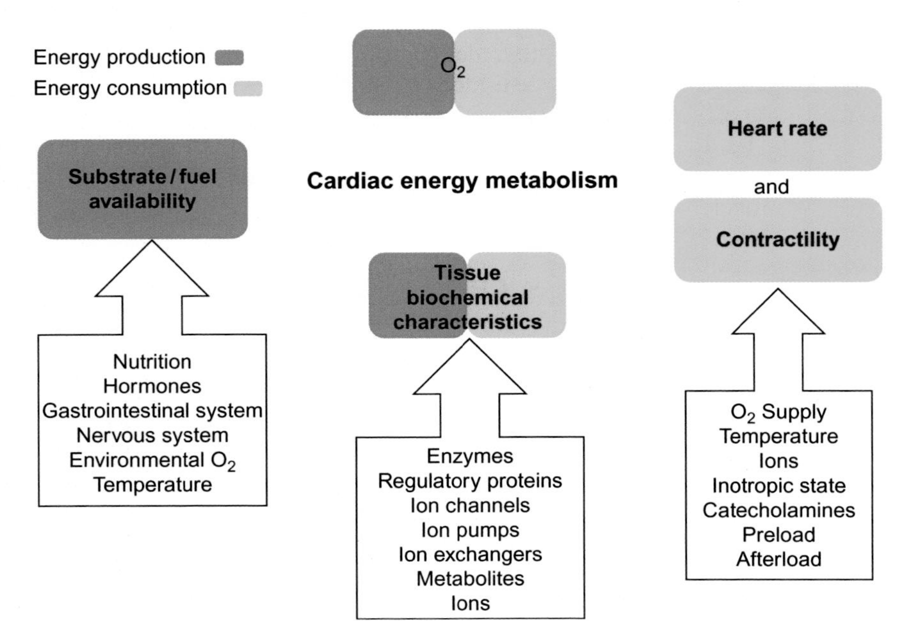

Fig. 3. A summary of the major determinants of cardiac energy metabolism in fish. Categories are shown as energy producing and/or energy consuming with select variables listed below categories. This diagram highlights the diversity of interrelated variables that define cardiac metabolism. Note: some variables (e.g., ions, O_2, and temperature) are involved in multiple categories and represent a large component of our current understanding of myocardial energetics. Although few studies exist, other variables such as ventricular geometry and mechanical efficiency of the heart should also impact fish cardiac energy metabolism.

REFERENCES

Aho, E., Vornanen, M., 1998. Ca^{2+}-ATPase activity and Ca^{2+} uptake by sarcoplasmic reticulum in fish heart: effects of thermal acclimation. J. Exp. Biol. 201, 525–532.

Aho, E., Vornanen, M., 1999. Contractile properties of atrial and ventricular myocardium of the heart of rainbow trout *Oncorhynchus mykiss*: effects of thermal acclimation. J. Exp. Biol. 202, 2663–2677.

Aho, E., Vornanen, M., 2001. Cold acclimation increases basal heart rate but decreases its thermal tolerance in rainbow trout (*Oncorhynchus mykiss*). J. Comp. Physiol. B 171, 173–179.

Alberty, R.A., Goldberg, R.N., 1992. Standard thermodynamic formation properties for the adenosine 5-triphosphate series. Biochemistry 31, 10610–10615.

Arthur, P.G., Keen, J.E., Hochachka, P.W., Farrell, A.P., 1992. The metabolic state of the *in situ* perfused heart during severe hypoxia. Am. J. Physiol. 263, R798–R804.

Atkinson, D.E., 1968. Energy charge of adenylate pool as a regulatory parameter. Interaction with feedback modifiers. Biochemistry 7, 430–434.

Axelsson, M., Farrell, A.P., Nilsson, S., 1990. Effects of hypoxia and drugs on the cardiovascular dynamics of the Atlantic hagfish *Myxine glutinosa*. J. Exp. Biol. 151, 297–316.

Bailey, J.R., Driedzic, W.R., 1989. Effects of acute temperature change on cardiac performance and oxygen consumption of a marine fish, the sea raven (*Hemitripterus americanus*). Physiol. Zool. 62, 1089–1101.

Bailey, J.R., Driedzic, W.R., 1993. Influence of low temperature acclimation on fate of metabolic fuels in rainbow trout (*Oncorhynchus mykiss*) hearts. Can. J. Zool. 71, 2167–2173.

Bailey, J.R., Sephton, D.H., Driedzic, W.R., 1990. Oxygen uptake by isolated perfused fish hearts with differing myoglobin concentrations under hypoxic conditions. J. Mol. Cell. Cardiol. 22, 1125–1134.

Bailey, J., Sephton, D., Driedzic, W.R., 1991. Impact of an acute temperature-change on performance and metabolism of pickerel (*Esox niger*) and eel (*Anguilla rostrata*) hearts. Physiol. Zool. 64, 697–716.

Bailey, J.R., Rodnick, K.J., MacDougall, R., Clowe, S., Driedzic, W.R., 2000. Anoxic performance of the American eel (*Anguilla rostrata* L.) heart requires extracellular glucose. J. Exp. Zool. 286, 699–706.

Balaban, R.S., 2012. Metabolic homeostasis of the heart. J. Gen. Physiol. 139, 407–414.

Bárány, M., 1967. ATPase activity of myosin correlated with speed of muscle shortening. J. Gen. Physiol. 50, 197–218.

Battiprolu, P.K., Rodnick, K.J., 2007. Sex differences in energy metabolism and performance of teleost cardiac tissue. Am. J. Physiol. 292, R827–R836.

Battiprolu, P.K., Rodnick, K.J., 2014. Dichloroacetate selectively improves cardiac function and metabolism in female and male rainbow trout. Am. J. Physiol. 307, H1401–H1411.

Becker, T.A., DellaValle, B., Gesser, H., Rodnick, K.J., 2013. Limited effects of exogenous glucose during severe hypoxia and a lack of hypoxia-stimulated glucose uptake in isolated rainbow trout cardiac muscle. J. Exp. Biol. 216, 3422–3432.

Bers, D.M., 2008. Calcium cycling and signaling in cardiac myocytes. Ann. Rev. Physiol. 70, 23–49.

Bers, D.M., 2014. Calcium sarcoplasmic reticulum calcium leak: basis and roles in cardiac dysfunction. Ann. Rev. Physiol. 76, 107–127.

Birkedal, R., Gesser, H., 2004. Effects of hibernation on mitochondrial regulation and metabolic capacities in myocardium of painted turtle (*Chrysemys picta*). Comp. Biochem. Physiol. A 139, 285–291.

Birkedal, R., Laasmaa, M., Vendelin, M., 2014. The location of energetic compartments affects energetic communication in cardiomyocytes. Front. Physiol. 5, 1–9.

Blier, P., Guderley, H., 1988. Metabolic responses to cold acclimation in the swimming musculature of lake whitefish, *Coregonus clupeaformis*. J. Exp. Zool. 246, 244–252.

Bowler, K., Tirri, R., 1990. Temperature dependence of the heart isolated from the cold or warm acclimated perch (*Perca fluviatilis*). Comp. Biochem. Physiol. 96, 177–180.

Brand, M.D., 2005. The efficiency and plasticity of mitochondrial energy transduction. Biochem. Soc. Trans. 33, 897–904.

Campbell, J.D., Paul, R.J., 1992. The nature of fuel provision for the Na^+,K^+-ATPase in porcine vascular smooth muscle. J. Physiol. (London) 447, 67–82.

Carmeliet, E.E., Horres, C.R., Lieberman, M., Vereecke, J.S., 1976. Developmental aspects of potassium flux and permeability of embryonic chick heart. J. Physiol. (London) 254, 673–692.

Christensen, M., Hartmund, T., Gesser, H., 1994. Creatine kinase, energy-rich phosphates and energy metabolism in heart muscle of different vertebrates. J. Comp. Physiol. B 164, 118–123.

Chung, Y., Huang, S.J., Glabe, A., Jue, T., 2006. Implication of CO inactivation on myoglobin function. Am. J. Physiol. 290, C1616–C1624.

Clark, R.J., Rodnick, K.J., 1998. Morphometric and biochemical characteristics of ventricular hypertrophy in male rainbow trout (*Oncorhynchus mykiss*). J. Exp. Biol. 201, 1541–1552.

Clausen, T., 1998. Clinical and therapeutic significance of the Na^+,K^+ pump. Clin. Sci. 95, 3–17.

Clow, K.A., Rodnick, K.J., MacCormack, T.J., Driedzic, W.R., 2004. The regulation and importance of glucose uptake in the isolated Atlantic cod heart: rate-limiting steps and effects of hypoxia. J. Exp. Biol. 207, 1865–1874.

Clow, K.A., Short, C.E., Driedzic, W.R., 2016. Extracellular glucose supports lactate production but not aerobic metabolism in cardiomyocytes from both normoglycemic Atlantic cod and low glycemic short-horned sculpin. J. Exp. Biol. 219, 1384–1393.

Cogliati, S., Enriquez, J.A., Scorrano, L., 2016. Mitochondrial cristae: where beauty meets functionality. Trends Biochem. Sci. 41, 261–273.

Connett, R.J., 1988. Analysis of metabolic control: new insights using scaled creatine kinase model. Am. J. Physiol. 254, R949–R959.

Cook, D.G., Iftikar, F.I., Baker, D.W., Hickey, A.J.R., Neill, A., Herbert, N.A., 2013. Low O_2 acclimation shifts the hypoxia avoidance behaviour of snapper (*Pagrus auratus*) with only subtle changes in aerobic and anaerobic function. J. Exp. Biol. 216, 369–378.

Cox, G.K., Sandblom, E., Farrell, A.P., 2010. Cardiac responses to anoxia in the Pacific hagfish, *Eptatretus stouti*. J. Exp. Biol. 213, 3692–3698.

Crabtree, B., Newsholme, E.A., 1972a. Activities of lipases and carnitine palmitoyltransferase in muscles from vertebrates and invertebrates. Biochem. J. 130, 697–705.

Crabtree, B., Newsholme, E.A., 1972b. Activities of phosphorylase, hexokinase, phosphofructokinase, lactate dehydrogenase and glycerol 3-phosphate dehydrogenases in muscles from vertebrates and invertebrates. Biochem. J. 126, 49–58.

Da Silva, D., Costa, D.C.F., Alves, C.M., Block, B.A., Landira-Fernandez, A.M., 2011. Temperature dependence of cardiac sarcoplasmic reticulum Ca^{2+}-ATPase from rainbow trout *Oncorhynchus mykiss*. J. Fish Biol. 79, 789–800.

Davie, P.S., Franklin, C.E., 1992. Myocardial oxygen consumption and mechanical efficiency of a perfused dogfish heart preparation. J. Comp. Physiol. B 162, 256–262.

Degn, P., Gesser, H., 1997. Ca^{2+} activated myosin-ATPase in cardiac myofibrils of rainbow trout, freshwater turtle and rat. J. Exp. Zool. 278, 381–390.

Denton, R.M., 2009. Regulation of mitochondrial dehydrogenases by calcium ions. Biochim. Biophys. Acta 1787, 1309–1316.

Despa, S., Bers, D.M., 2013. Na^+ transport in the normal and failing heart—remember the balance. J. Mol. Cell. Cardiol. 61, 2–10.

Dhalla, N.S., Yates, J.C., Proveda, V., 1977. Calcium linked changes in myocardial metabolism in isolated perfused rat heart. Can. J. Physiol. Pharmacol. 55, 925–933.

Dhar-Chowdhury, P., Malester, B., Rajacic, P., Coetzee, W.A., 2007. The regulation of ion channels and transporters by glycolytically derived ATP. Cell. Mol. Life. Sci. 64, 3069–3083.

Dobson, G.P., Veech, R.L., Passonneau, J.V., Kobayashi, K., Inubushi, T., Wehrli, S., Nioka, S., Chance, B., 1992. 31P NMR and enzymatic analysis of cytosolic phosphocreatine, ATP, Pi and intracellular pH in the isolated working perfused rat heart. NMR Biomed. 5, 20–28.

Donohoe, P.H., West, T.G., Boutilier, R.G., 2000. Factors affecting membrane permeability and ionic homeostasis in the cold-submerged frog. J. Exp. Biol. 203, 405–414.

Driedzic, W.R., 1978. Carbohydrate metabolism in perfused dogfish heart. Physiol. Zool. 51, 42–50.

Driedzic, W.R., 1992. Cardiac energy metabolism. In: Hoar, W.S., Randall, D.J., Farrell, A.P. (Eds.), In: Fish Physiology, vol. XIIA. Academic Press, San Diego, pp. 219–266.

Driedzic, W.R., de Almeida-Val, V.M.F., 1996. Enzymes of cardiac energy metabolism in Amazonian teleosts and the fresh-water stingray (*Potamotrygon hystixi*). J. Exp. Zool. 274, 327–333.

Driedzic, W.R., Gesser, H., 1994. Energy metabolism and contractility in ectothermic vertebrate hearts: hypoxia, acidosis and low temperature. Physiol. Rev. 74, 221–258.

Driedzic, W.R., Hart, T., 1984. Relationship between exogenous fuel availability and performance by teleost and elasmobranch hearts. J. Comp. Physiol. B 154, 593–599.

Driedzic, W.R., Scott, D.L., Farrell, A.P., 1983. Aerobic and anaerobic contributions to energy metabolism in perfused sea raven (*Hemitripterus americanus*) hearts. Can. J. Zool. 61, 1880–1883.

Driedzic, W.R., Sidell, B.D., Stowe, D., Branscombe, R., 1987. Matching of vertebrate cardiac energy demand to energy metabolism. Am. J. Physiol. 252, R930–R937.

Eisenberg, E., Hill, T.L., 1985. Muscle contraction and free energy transduction in biological systems. Science 227, 999–1006.

Eliason, E.J., Anttila, K., 2017. Temperature and the cardiovascular system. In: Gamperl, A.K., Gillis, T.E., Farrell, A.P., Brauner, C.J. (Eds.), Fish Physiology. In: The Cardiovascular System: Development, Plasticity and Physiological Responses, vol. 36B. Academic Press, San Diego (In press).

Else, P.L., Windmill, D.J., Markus, V., 1996. Molecular activity of sodium pumps in endotherms and ectotherms. Am. J. Physiol. 271, R1287–R1294.

Ewart, H.S., Driedzic, W.R., 1987. Enzymes of energy metabolism in salmonid hearts: spongy versus cortical myocardia. Can. J. Zool. 65, 623–627.

Ewart, H.S., Canty, A.A., Driedzic, W.R., 1988. Scaling of cardiac oxygen consumption and enzyme activity levels in sea raven (*Hemitripterus americanus*). Physiol. Zool. 61, 50–56.

Farrar, R.S., Battiprolu, P.K., Pierson, N.S., Rodnick, K.J., 2006. Steroid induced cardiac contractility requires exogenous glucose, glycolysis and the sarcoplasmic reticulum in rainbow trout. J. Exp. Biol. 209, 2114–2128.

Farrell, A.P., 1996. Features heightening cardiovascular performance in fishes; with special reference to tunas. Comp. Biochem. Physiol. A 113, 61–67.

Farrell, A.P., Jones, D.R., 1992. The heart. In: Hoar, W.S., Randall, D.J., Farrell, A.P. (Eds.), In: Fish Physiology, vol. XIIA. Academic Press, San Diego, pp. 1–88.

Farrell, A.P., Smith, F., 2017. Cardiac form, function and physiology. In: Gamperl, A.K., Gillis, T.E., Farrell, A.P., Brauner, C.J. (Eds.), Fish Physiology. In: The Cardiovascular System: Morphology, Control and Function, vol. 36A. Academic Press, San Diego, pp. 155–264.

Farrell, A.P., Stecyk, J.A.W., 2007. The heart as a working model to explore themes and strategies for anoxic survival in ectothermic vertebrates. Comp. Biochem. Physiol. A 147, 300–312.

Farrell, A.P., Wood, S., Hart, T., Driedzic, W.R., 1985. Myocardial oxygen consumption in the sea raven *Hemitripterus americanus*: the effects of volume loading, pressure loading and progressive hypoxia. J. Exp. Biol. 117, 237–250.

Farrell, A.P., MacLeod, K.R., Scott, C., 1988. Cardiac performance of the trout (*Salmo gairdneri*) heart during acidosis: effects of low bicarbonate, lactate and cortisol. Comp. Biochem. Physiol. A 91, 271–277.

Flögel, U., Fago, A., Rassaf, T., 2013. Keeping the heart in balance: the functional interactions of myoglobin with nitrogen oxides. J. Exp. Biol. 213, 2726–2733.

Forster, M.E., 1989. Performance of the heart of the hagfish, *Eptatretus cirrhatus*. Fish. Physiol. Biochem. 6, 327–331.

Forster, M.E., 1991. Myocardial oxygen consumption and lactate release by the hypoxic hagfish heart. J. Exp. Biol. 156, 583–590.

Foster, A.R., Houlihan, D.F., Hall, S.J., Burren, L.J., 1992. The effects of temperature acclimation on protein synthesis rates and nucleic acid content of juvenile cod (*Gadus morhua*). Can. J. Zool. 70, 2095–2102.

Friesen, A.J.D., Oliver, N., Allen, G., 1969. Activation of cardiac glycogen phosphorylase by calcium. Am. J. Physiol. 217, 445–450.

Fuller, W., Tulloch, L.B., Shattock, M.J., Calaghan, S.C., Howie, J., Wypijewski, K.J., 2013. Regulation of the cardiac sodium pump. Cell. Mol. Life Sci. 70, 1357–1380.

Galli, G.L.J., Richards, J.G., 2014. Mitochondria from anoxia-tolerant animals reveal common strategies to survive without oxygen. J. Comp. Physiol. B 184, 285–302.

Gamperl, A.K., Driedzic, W.R., 2009. Cardiovascular function and cardiac metabolism. In: Farrell, A.P., Brauner, C.J. (Eds.), In: Fish Physiology, vol. 27. Academic Press, San Diego, pp. 301–360.

Gamperl, A.K., Farrell, A.P., 2004. Cardiac plasticity in fishes; environmental influences and intraspecific differences. J. Exp. Biol. 207, 2539–2550.

Genge, C.E., Lin, E., Lee, L., Sheng, X.Y., Rayani, K., Gunawan, M., Stevens, C.M., Li, A.Y., Talab, S.S., Claydon, T.W., Hove-Madsen, L., Tibbits, G.F. (2016). The Zebrafish heart as a model of mammalian cardiac function. *In*: Reviews of Physiology Biochemistry and Pharmacology (eds. Nilius, B., DeTombe, P., Gudermann, T., Jahn, R., Lill, R., Petersen, O.H.) vol. 171, 99–136, Cham, Switzerland: Springer International.

Gertz, E.W., Wisneski, J.A., Stanley, W.C., Neese, R.A., 1988. Myocardial substrate utilization during exercise in humans. Dual carbon-labeled carbohydrate isotope experiments. J. Clin. Invest. 82, 2017–2025.

Gesser, H., 2002. Mechanical performance and glycolytic requirement in trout ventricular muscle. J. Exp. Zool. 293, 360–367.

Gesser, H., Overgaard, J., 2009. Comparative aspects of hypoxia tolerance of the ectothermic vertebrate heart. In: Glass, M.L., Wood, S.C. (Eds.), Cardio-Respiratory Control in Vertebrates, Springer-Verlag, Berlin, pp. 263–284.

Gesser, H., Poupa, O., 1975. Lactate as substrate for force development in hearts with different isozyme patterns of lactate dehydrogenase. Comp. Biochem. Physiol. B 52, 311–313.

Gevers, W., 1977. Generation of protons by metabolic processes in heart cells. J. Mol. Cell. Cardiol. 9, 867–874.

Gibbs, C.L., 1978. Cardiac energetics. Physiol. Rev. 58, 174–254.

Gibbs, C.L., Kotsanas, G., 1986. Factors regulating basal metabolism of the isolated perfused rabbit heart. Am. J. Physiol. 250, H998–H1007.

Gibbs, C.L., Loiselle, D.S., 2001. Cardiac basal metabolism. Jpn. J. Physiol. 51, 399–426.

Gillis, T.E., Johnson, E.F., 2017. Cardiac remodelling, protection and regeneration. In: Gamperl, A.K., Gillis, T.E., Farrell, A.P., Brauner, C.J. (Eds.), Fish Physiology. In: The Cardiovascular System: Development, Plasticity and Physiological Responses, vol. 36B. Academic Press, San Diego (In press).

Gillis, T.E., Regan, M.D., Cox, G.K., Harter, T.S., Brauner, C.J., Richards, J.G., Farrell, A.P., 2015. Characterizing the metabolic capacity of the anoxic hagfish heart. J. Exp. Biol. 218, 3754–3761.

Giovane, A., Greco, G., Maresca, A., Tota, B., 1980. Myoglobin in the heart ventricle of tuna and other fishes. Experientia 36, 219–220.

Glitch, H.G., 2001. Electrophysiology of the sodium-potassium-ATPase in cardiac cells. Physiol. Rev. 81, 1791–1826.

Gnaiger, E., 2001. Bioenergetics at low oxygen: dependence of respiration and phosphorylation on oxygen and adenosine diphosphate supply. Resp. Physiol. 128, 277–297.

Goldman, R., Katchalski, E., 1971. Kinetic behavior of a two-enzyme membrane carrying out a consecutive set of reactions. J. Theor. Biol. 32, 243–257.

Gordon, A.M., Regnier, M., Homsher, E., 2001. Skeletal and cardiac muscle contractile activation: tropomyosin "rocks and rolls". News Physiol. Sci. 16, 49–55.

Graham, M.S., Farrell, A.P., 1989. The effect of temperature acclimation and adrenaline on the performance of a perfused trout heart. Physiol. Zool. 62, 38–61.

Graham, M.S., Farrell, A.P., 1990. Myocardial oxygen consumption in trout acclimated to 5° and 15°C. Physiol. Zool. 63, 536–554.

Guzun, R., Kaambre, T., Bagur, R., Grichine, A., Usson, Y., Varikmaa, M., Anmann, T., Tepp, K., Timohhina, N., Shevchuk, I., Chekulayev, V., Boucher, F., Dos Santos, P., Schlattner, U., Wallimann, T., Kuznetsov, A.V., Dzeja, P., Aliev, M., Saks, V., 2015. Modular organization of cardiac energy metabolism: energy conversion, transfer and feedback regulation. Acta Physiol. 213, 84–106.

Haagensen, L., Jensen, D.H., Gesser, H., 2008. Dependence of myosin-ATPase on structure bound creatine kinase in cardiac myfibrils from rainbow trout and freshwater turtle. Comp. Biochem. Physiol. A 150, 404–409.

Hall, J.R., MacCormack, T.J., Barry, C.A., Driedzic, W.R., 2004. Sequence and expression of a constitutive facilitated glucose transporter (GLUT1) in Atlantic cod *Gadus morhua*. J. Exp. Biol. 207, 4697–4706.

Hansen, C.A., Sidell, B.D., 1983. Atlantic hagfish cardiac muscle: metabolic basis of tolerance to anoxia. Am. J. Physiol. 244, R356–R362.

Harmon, K.J., Bolinger, M.T., Rodnick, K.J., 2011. Carbohydrate energy reserves and effects of food deprivation in male and female rainbow trout. Comp. Biochem. Physiol. A 158, 423–431.

Hartmund, T., Gesser, H., 1995. Acidosis, glycolysis and energy state in anaerobic heart tissue from rainbow trout. J. Comp. Physiol. B 165, 219–229.

Hartmund, T., Gesser, H., 1996. Cardiac force and high-energy phosphates under metabolic inhibition in four ectothermic vertebrates. Am. J. Physiol. 271, R946–R954.

Helbo, S., Fago, A., Gesser, H., 2013. Myoglobin-dependent O_2 consumption of the hypoxic trout heart. Comp. Biochem. Physiol. A 165, 40–45.

Hickey, A.J., Renshaw, G.M., Speers-Roesch, B., Richards, J.G., Wang, Y., Farrell, A.P., Brauner, C.J., 2012. A radical approach to beating hypoxia: depressed free radical release from heart fibres of the hypoxia-tolerant epaulette shark (*Hemiscyllum ocellatum*). J. Comp. Physiol. B. 182, 91–100.

Houadjeto, M., Barman, T., Travers, F., 1991. What is the true ATPase activity of contracting myofibrils? FEBS Lett. 281, 105–107.

Houlihan, D.F., Laurent, P., 1987. Effects of exercise training on the performance, growth, and protein turnover of rainbow trout (*Salmo gairdneri*). Can. J. Fish. Aquat. Sci. 44, 1614–1621.

Houlihan, D.F., Hall, S.J., Gray, C., Noble, B.S., 1988a. Growth rates and protein turnover in Atlantic cod, *Gadus morhua*. Can. J. Fish. Aquat. Sci. 45, 951–964.

Houlihan, D.F., Agnisola, C., Lyndon, A.R., Gray, C., Hamilton, N.M., 1988b. Protein synthesis in a fish heart: responses to increased power output. J. Exp. Biol. 137, 565–587.

Houlihan, D.F., Carter, C.G., McCarthy, I.D., 1995. Protein synthesis in fish. In: Hochachka, P.W., Mommsen, T. (Eds.), Biochemistry and Molecular Biology of Fishes. vol. 4. Elsevier B.V., Amsterdam, pp. 191–220.

Hove-Madsen, L., Llach, A., Tort, L., 2000. Na^{+}/Ca^{2+}-exchange activity regulates contraction and SR Ca^{2+} content in rainbow trout atrial myocytes. Am. J. Physiol. 279, R1856–R1864.

Hubley, M.J., Locke, B.R., Moerland, T.S., 1996. The effects of temperature, pH, and magnesium on the diffusion coefficient of ATP in solutions of physiological ionic strength. Biochim Biophys Acta 1291, 115–121.

Icardo, J.M., 2017. Heart morphology and anatomy. In: Gamperl, A.K., Gillis, T.E., Farrell, A.P., Brauner, C.J. (Eds.), Fish Physiology. In: The Cardiovascular System: Morphology, Control and Function, vol. 36A. Academic Press, San Diego, pp. 1–54.

Ingwall, J.S., Balschi, J.A., 2006. Energetics of the Na^+ pump in the heart. J. Cardiovasc. Electr. 17, S127–S133.

Jayasundara, N., Gardner, L.D., Block, B.A., 2013. Effects of temperature acclimation on Pacific Bluefin tuna (*Thunnus orietalis*) cardiac transcritome. Am. J. Physiol. 305, R1010–R1021.

Jayasundara, N., Kozal, J.S., Arnold, M.C., Chan, S.S.L., Giulio, R.T., 2015a. High-throughput tissue bioenergetics analysis reveals identical metabolic allometric scaling for teleost hearts and whole organisms. PLoS One 10 (9), e0137710. http://dx.doi.org/10.1371/journal.pone.0137710.

Jayasundara, N., Tomanek, L., Dowd, W.W., Somero, G.N., 2015b. Proteomic analysis of cardiac response to thermal acclimation in the eurythermal goby fish *Gillichtys mirabilis*. J. Exp. Biol. 218, 1359–1372.

Juel, C., 1997. Lactate-proton cotransport in skeletal muscle. Physiol. Rev. 77, 321–358.

Kalinin, A., Gesser, H., 2002. Oxygen consumption and force development in turtle and trout cardiac muscle during acidosis and high extracellular potassium. J. Comp. Physiol. B 172, 145–151.

Kalinin, A.L., Costa, M.J., Rantin, F.T., Glass, M.L., 2009. Effects of temperature on cardiac function in teleost fish. In: Glass, M.L., Wood, S.C. (Eds.), Cardio-Respiratory Control in Vertebrates, Springer-Verlag, Berlin, pp. 121–160.

Kammermeier, H., 1993. Meaning of energetic parameters. Basic Res. Cardiol. 88, 380–384.

Karasiński, J., Sokalski, A., Kilarski, W., 2001. Correlation of myofibrillar ATPase activity and myosin heavy chain content in ventricular and atrial myocardium of fish heart. Fol. Histoch. Cytochem. 39, 23–28.

Kentish, J.C., 1986. The effects of inorganic phosphate and creatine phosphate on force production in skinned muscles from rat ventricle. J. Physiol. (London) 370, 585–604.

Kiceniuk, J.W., Jones, D.R., 1977. Oxygen transport system in trout (*Salmo gairdneri*) during sustained exercise. J. Exp. Biol. 69, 247–260.

Klaiman, J.M., Fenna, A.J., Shiels, H.A., Macri, J., Gillis, T.E., 2011. Cardiac remodeling in fish: strategies to maintain heart function during temperature change. PLoS One 6, e24464. http://dx.doi.org/10.1371/journal.pone.0024464.

Klaiman, J.M., Pyle, W.G., Gillis, T.E., 2014. Cold acclimation increases cardiac myofilament function and ventricular pressure generation in trout. J. Exp. Biol. 217, 4132–4140.

Kleckner, N.W., Sidell, B.D., 1985. Comparison of maximal activities of enzymes from tissues of thermally acclimated and naturally acclimatized chain pickerel (*Esox niger*). Physiol. Zool. 58, 18–28.

Korajoki, H., Vornanen, M., 2013. Temperature dependence of sarco(endo)plasmic reticulum Ca^{2+} ATPase expression in fish hearts. J. Comp. Physiol. B 183, 467–476.

Krause, S.M., Jacobus, W.E., 1992. Specific enhancement of the cardiac myofibrillar ATPase by bound creatine kinase. J. Biol. Chem. 267, 2480–2486.

Kuum, M., Kaasik, A., Joubert, F., Ventura-Clapier, R., Veksler, V., 2009. Energetic state is a strong regulator of sarcoplasmic reticulum Ca^{2+} loss in cardiac muscle: different efficiencies of different energy sources. Cardiovasc. Res. 83, 89–96.

Lague, S.L., Speers-Roesch, B., Richards, J.G., Farrell, A.P., 2012. Exceptional cardiac anoxia tolerance in tilapia (*Oreochromis* hybrid). J. Exp. Biol. 215, 1354–1365.

Lanctin, H.P., McMorran, L.E., Driedzic, W.R., 1980. Rates of glucose and lactate oxidation by the perfused isolated trout (*Salvelinus fontinalis*) heart. Can. J. Zool. 58, 1708–1711.

Land, S.C., Hochachka, P.W., 1994. Protein turnover during metabolic arrest in turtle hepatocytes: role and energy dependence. Am. J. Physiol. 266, C1028–C1036.

Legate, N.J.N., Bailey, J.R., Driedzic, W.R., 1998. Oxygen consumption in myoglobin-rich and myoglobin-poor isolated fish cardiomyocytes. J. Exp. Zool. 280, 269–276.
Lewis, J.M., Driedzic, W.R., 2007. Tissue-specific changes in protein synthesis associated with seasonal metabolic depression and recovery in the north temperate labrid, *Tautogolabrus adsperus*. Am. J. Physiol. 293, R474–R481.
Lewis, J.M., Driedzic, W.R., 2010. Protein synthesis is defended in the mitochondrial fraction of gill but not heart in cunner (*Tautogolabrus adspersus*) exposed to acute hypoxia and hypothermia. J. Comp. Physiol. B 180, 179–188.
Lewis, J.M., Costa, I., Val, A.L., Almeida-Val, V.M.F., Gamperl, A.K., Driedzic, W.R., 2007. Responses to hypoxia and recovery: repayment of oxygen debt is not associated with compensatory protein synthesis in the Amazonian cichlid, *Astronotus ocellatus*. J. Exp. Biol. 210, 1935–1943.
Lillywhite, H.B., Zippel, K.C., Farrell, A.P., 1999. Resting and maximal heart rates in ectothermic vertebrates. Comp. Biochem. Physiol. A 124, 369–382.
Llach, A., Huang, J., Sederat, F., Tort, L., Tibbits, G., Hove-Madsen, L., 2004. Effect of β-adrenergic stimulation on the relationship between membrane potential, intracellular [Ca^{2+}] and sarcoplasmic reticulum Ca^{2+} uptake in rainbow trout atrial myocytes. J. Exp. Biol. 207, 1369–1377.
MacCormack, T.J., Driedzic, W.R., 2007. The impact of hypoxia on *in vivo* glucose uptake in a hypoglycemic fish, *Myoxocephalus scorpius*. Am. J. Physiol. 292, R1033–R1042.
MacKnight, A.D.C., Leaf, A., 1977. Regulation of cellular volume. Physiol. Rev. 57, 510–573.
Macqueen, D.J., de la Serrana, D.G., Johnston, I.A., 2014. Cardiac myoglobin deficit has evolved repeatedly in teleost fishes. Biol. Lett. 10, 20140225. http://dx.doi.org/10.1098/rsbl.2014.0225.
Magnoni, L.J., Palstra, A.P., Planas, J.V., 2014. Fueling the engine: induction of AMP-activated protein kinase in trout skeletal muscle by swimming. J. Exp. Biol. 217, 1649–1652.
Marban, E., Kusuoka, H., 1987. Maximal Ca^{2+}-activated force and myofilament Ca^{2+} sensitivity in intact mammalian hearts. Differential effects of inorganic phosphate and hydrogen ions. J. Gen. Physiol. 90, 609–623.
Marcinek, D.J., Kushmerick, M.J., Conley, K.E., 2010. Lactic acidosis *in vivo*: testing the link between lactate generation and H^+ accumulation in ischemic mouse muscle. J. Appl. Physiol. 108, 1479–1486.
Meyer, R.A., 1988. A linear model of muscle respiration explains monoexponential phosphocreatine changes. Am. J. Physiol. 254, C548–C553.
Meyer, R.A., Foley, R.A., 1996. Cellular processes integrating the metabolic response to exercise. In: Rowell, L.B., Shepard, J.T. (Eds.), Handbook of Physiology. Section 12, Exercise: Regulation and Integration of Multiple Systems, Chapter 18, American Physiological Society, Bethesda, pp. 841–869.
Meyer, R.A., Sweeney, H.L., Kushmerick, M.J., 1984. A simple analysis of the phosphocreatine cycle. Am. J. Physiol. 246, C365–C377.
Milligan, C.L., Farrell, A.P., 1991. Lactate utilization by an *in situ* perfused trout heart: effects of workload and blockers of lactate transport. J. Exp. Biol. 155, 357–373.
Møller-Nielsen, T., Gesser, H., 1992. Sarcoplasmic reticulum and excitation coupling at 20 and 10°C in rainbow trout myocardium. J. Comp. Physiol. B 162, 526–534.
Mortensen, B., Gesser, H., 1999. O_2 consumption and metabolic activities in resting cardiac myocytes from rainbow trout. J. Exp. Zool. 283, 501–509.
Movafagh, S., Morad, M., 2010. L-type calcium channel as a cardiac oxygen sensor. Ann. N.Y. Acad. Sci. 1188, 153–158.
Moyes, C.D., 1996. Cardiac metabolism in high performance fish. Comp. Biochem. Physiol. A 113, 69–75.

Neely, J.R., Morgan, H.E., 1974. Relationship between carbohydrate and lipid metabolism and energy balance of heart muscle. Ann. Rev. Physiol. 36, 413–459.

Neely, J.R., Liebermeister, H., Battersby, E.J., Morgan, H.E., 1967. Effect of pressure development on oxygen consumption by isolated rat heart. Am. J. Physiol. 212, 804–814.

Nelson, J.S., Grande, T.C., Wilson, M.V., 2016. Fishes of the World. Wiley, Hoboken. 752 p.

Newsholme, E.A., Leech, A.R., 1983. Biochemistry for the Medical Sciences. John Wiley and Sons, Chichester. 952 p.

Nielsen, K.E., Gesser, H., 1984. Energy metabolism and intracellular pH in trout heart muscle under anoxia and different $[Ca^{2+}]_o$. J. Comp. Physiol. 154, 523–527.

O'Brien, K.M., Sidell, B.D., 2000. The interplay among cardiac ultrastructure, metabolism and the expression of oxygen-binding proteins in Antarctic fishes. J. Exp. Biol. 203, 1287–1297.

O'Brien, K.M., Mueller, I.A., Orczewska, J.I., Dullen, K.R., Ortego, M., 2014. Hearts of some Antarctic fishes lack mitochondrial creatine kinase. Comp. Biochem. Physiol. A 178, 30–36.

Omlin, T., Weber, J.-M., 2013. Exhausting exercise and tissue-specific expression of monocarboxylate transporters in rainbow trout. Am. J. Physiol. 304, R1036–R1043.

Opie, L.H., 1991. The Heart: Physiology and Metabolism. Raven Press, New York. 513 p.

Orchard, C.H., Kentish, J.C., 1990. Effects of changes of pH on the contractile function of cardiac muscle. Am. J. Physiol. 258, C967–C981.

Overgaard, J., Gesser, H., 2004. Force development, energy state and ATP production of cardiac muscle from turtles and trout during normoxia and severe hypoxia. J. Exp. Biol. 207, 1915–1924.

Overgaard, J., Stecyk, J.A.W., Gesser, H., Wang, T., Farrell, A.P., 2004. Effects of temperature and anoxia upon the performance of *in situ* perfused trout hearts. J. Exp. Biol. 207, 655–665.

Page, E., Earley, J., Power, B., 1974. Normal growth of ultrastructures in rat left ventricular myocardial cells. Circ. Res. 34/35 (Suppl. 2), II-12–II-16.

Patton, S., Zulak, I.M., Trams, E.G., 1975. Fatty acid metabolism via triglyceride in the salmon heart. J. Mol. Cell. Cardiol. 7, 857–865.

Pierce, V.A., Crawford, D.L., 1997. Phylogenetic analysis of thermal acclimation of the glycolytic enzymes in the genus *Fundulus*. Physiol. Zool. 70, 597–609.

Pierce, G.N., Philipson, K.D., 1985. Binding of glycolytic-enzymes to cardiac sarcolemmal and sarcoplasmic reticular membranes. J. Biol. Chem. 260, 6862–6870.

Reeds, P.J., Fuller, M.F., Nicholson, B.A., 1985. Metabolic basis of energy expenditure with particular reference to protein. In: Garrow, J.S., Halliday, D. (Eds.), Substrate and Energy Metabolism. J. Libbey, London, pp. 46–57.

Reeves, J.P., Hale, C.C., 1984. The stoichiometry of the cardiac sodium–calcium exchange system. J. Biol. Chem. 259, 7733–7739.

Robergs, R.A., Ghiasvand, F., Parker, D., 2004. Biochemistry of exercise-induced metabolic acidosis. Am. J. Physiol. 287, R502–R516.

Rodnick, K.J., Sidell, B.D., 1997. Structural and biochemical analyses of cardiac ventricular enlargement in cold-acclimated striped bass. Am. J. Physiol. 273, R252–R258.

Rodnick, K.J., Williams, S.R., 1999. Effects of body size on biochemical characteristics of trabecular cardiac muscle and plasma of rainbow trout (*Oncorhynchus mykiss*). Comp. Biochem. Physiol. A 122, 407–413.

Rodnick, K.J., Bailey, J.R., West, J.L., Rideout, A., Driedzic, W.R., 1997. Acute regulation of glucose uptake in cardiac muscle of American eel *Anguilla rostrata*. J. Exp. Biol. 200, 2871–2880.

Ruiz, M.A.M., Thorarensen, H., 2001. Genetic and environmental effects on the size of the cardiorespiratory organs in Arctic charr (*Salvelinus alpinus*). In: Abstract, Voluntary Food Intake in Fish, COST 827 Workshop. Iceland, Reykjavik.

Santer, R.M., 1985. Morphology and innervation of the fish heart. Adv. Anat. Embryol. Cell. Biol. 89, 1–99.

Sarnoff, S.J., Braunwald, G.H., Welch, G.H., Case, R.B., Stainsby, W.N., 1958. Hemodynamic determinants of oxygen consumption of the heart with special reference to the tension time index. Am. J. Physiol. 192, 148–156.

Schramm, M., Klieber, H.G., Daut, J., 1994. The energy expenditure of actomyosin-ATPase, Ca^{2+}-ATPase and Na^{+},K^{+}-ATPase in guinea pig cardiac ventricular muscle. J. Physiol. (London) 481, 647–662.

Sejersted, O.M., Sjøgaard, G., 2000. Dynamics and consequences of potassium shifts in skeletal muscle and heart during exercise. Physiol. Rev. 80, 1411–1481.

Sephton, D.H., Driedzic, W.R., 1991. Effect of acute and chronic temperature transition on enzymes of cardiac metabolism in white perch (*Morone americana*), yellow perch (*Perca flavescens*), and smallmouth bass (*Micropterus dolomieui*). Can. J. Zool. 69, 258–262.

Sephton, D.H., Driedzic, W.R., 1995. Low temperature acclimation decreases rates of protein synthesis in rainbow trout (*Oncorhynchus mykiss*) heart. Fish Physiol. Biochem. 14, 63–69.

Sephton, D., Bailey, J., Driedzic, W.R., 1990. Impact of acute temperature transition on enzyme activity levels, oxygen consumption, and exogenous fuel utilization in sea raven (*Hemitripterus americanus*) hearts. J. Comp. Physiol. B 160, 511–518.

Sepp, M., Vendelin, M., Vija, H., Birkedal, R., 2010. ADP compartmentation analysis reveals coupling between pyruvate kinase and ATPases in heart muscle. Biophys. J. 98, 2785–2793.

Sepp, M., Sokolova, N., Jugai, S., Mandel, M., Peterson, P., Vendelin, M., 2014. Tight coupling of Na^{+}/K^{+}-ATPase with glycolysis demonstrated in permeabilized rat cardiomyocytes. PLoS One 9 (6), e99413. http://dx.doi.org/10.1371/journal.pone.0099413.

Shiels, H.A., Galli, G.L.J., 2014. The sarcoplasmic reticulum and the evolution of the vertebrate heart. Physiology 29, 456–469.

Shiels, H.A., Vornanen, M., Farrell, A.P., 2002a. Effects of temperature on intracellular $[Ca^{2+}]$ in trout atrial myocytes. J. Exp. Biol. 205, 3641–3650.

Shiels, H.A., Vornanen, M., Farrell, A.P., 2002b. Temperature dependence of cardiac sarcoplasmic reticulum function in rainbow trout myocytes. J. Exp. Biol. 205, 3631–3639.

Shiels, H.A., Paajanen, V., Vornanen, M., 2006. Sarcolemmal ion currents and sarcoplasmic Ca^{2+} content in ventricular myocytes from the cold stenothermal fish, the burbot (*Lota lota*). J. Exp. Biol. 209, 3091–3100.

Shiels, H.A., Di Maio, A., Thompson, S., Block, B.A., 2011. Warm fish with cold hearts: thermal plasticity of excitation–contraction coupling in bluefin tuna. Proc. R. Soc. B 278, 18–27.

Sidell, B.D., Hazel, J.R., 1987. Temperature affects the diffusion of small molecules through cytosol of fish muscle. J. Exp. Biol. 129, 191–203.

Sidell, B.D., Stowe, D.B., Hansen, C.A., 1984. Carbohydrate is the preferred metabolic fuel of the hagfish (*Myxine glutinosa*) heart. Physiol. Zool. 57, 266–273.

Sidell, B.D., Driedzic, W.R., Stowe, D.B., Johnston, I.A., 1987. Biochemical correlations of power development and metabolic fuel preferenda in fish hearts. Physiol. Zool. 60, 221–232.

Sidell, B.D., Crockett, E.L., Driedzic, W.R., 1995. Antarctic fish tissues preferentially catabolize monoenoic fatty acids. J. Exp. Zool. 271, 73–81.

Skulachev, V.P., 1978. Membrane-linked energy buffering as the biological function of the Na/K gradient. FEBS Lett. 87, 171–179.

Smith, R.W., Houlihan, D.F., Nilsson, G.E., Brechin, J.G., 1996. Tissue-specific changes in protein synthesis rates *in vivo* during anoxia in crucian carp. Am. J. Physiol. 271, R897–R904.

Sokolova, N., Vendelin, M., Birkedal, R. (2009). Intracellular diffusion restrictions in isolated cardiomyocytes from rainbow trout. BMC Cell Biol.; 10:90. http://dx.doi.org/10.1186/1471-2121-10-90. PMID: 20017912.

Speers-Roesch, B., Treberg, J.R., 2010. The unusual energy metabolism of elasmobranch fishes. Comp. Biochem. Physiol. A 155, 417–434.

Speers-Roesch, B., Sandblom, E., Lau, G.Y., Farrell, A.P., Richards, J.G., 2010. Effects of environmental hypoxia on cardiac energy metabolism and performance in tilapia. Am. J. Physiol. 298, R104–R119.

Speers-Roesch, B., Brauner, C.J., Farrell, A.P., Hickey, A.J., Renshaw, G.M., Wang, Y.S., Richards, J.G., 2012. Hypoxia tolerance in elasmobranchs. II. Cardiovascular function and tissue metabolic responses during progressive and relative hypoxia exposures. J. Exp. Biol. 215, 103–114.

Speers-Roesch, B., Lague, S.L., Farrell, A.P., Richards, J.G., 2013. Effects of fatty acid provision during severe hypoxia on routine and maximal performance of the *in situ* tilapia heart. J. Comp. Physiol. B 183, 773–785.

Starnes, J.W., Wilson, D.F., Erecińska, M., 1985. Substrate dependence of metabolic state and coronary flow in perfused rat heart. Am. J. Physiol. 249, H799–H806.

Stecyk, J.A.W., 2017. Cardiovascular function under limiting oxygen conditions. In: Gamperl, A.K., Gillis, T.E., Farrell, A.P., Brauner, C.J. (Eds.), Fish Physiology. In: The Cardiovascular System: Development, Plasticity and Physiological Responses, vol. 36B. Academic Press, San Diego (In press).

Stecyk, J.A.W., Farrell, A.P., 2006. Regulation of the cardiorespiratory system of the common carp (*Cyprinus carpio*) during severe hypoxia at three seasonal acclimation temperatures. Physiol. Biochem. Zool. 79, 614–627.

Suga, H., 1979. Total mechanical energy of a ventricular model and cardiac oxygen consumption. Am. J. Physiol. 236, H498–H505.

Swimmer, Y., McNaughton, L., Moyes, C., Brill, R., 2004. Metabolic biochemistry of cardiac muscle in three tuna species (bigeye, *Thunnus obesus*; yellowfin, *T. albacares*; and skipjack, *Katsuwonus pelamis*) with divergent ambient temperature and oxygen tolerances. Fish Physiol. Biochem. 30, 27–35.

Terashima, K., Takeuchi, A., Sarai, N., Matsuoka, S., Shim, E.B., Leem, C.H., Noma, A., 2006. Modelling Cl^- homeostasis and volume regulation of the cardiac cell. Phil. Trans. R. Soc. A 364, 1245–1265.

Tibbits, G.F., Hove-Madsen, L., Bers, D.M., 1991. Calcium transport and the regulation of cardiac contractility in teleosts: a comparison with higher vertebrates. Can. J. Zool. 69, 2014–2019.

Tiitu, V., Vornanen, M., 2001. Cold adaptation suppresses the contractility of both atrial and ventricular muscle of the crucian carp heart. J. Fish Biol. 59, 141–156.

Tiitu, V., Vornanen, M., 2002. Regulation of cardiac contractility in a stenothermal fish, the burbot (*Lota lota*). J. Exp. Biol. 205, 1597–1606.

Tschantz, D.R., Crockett, E.L., Niewiarowski, P.H., Londraville, R.L., 2002. Cold acclimation strategy is highly variable among the sunfishes (Centrarchidae). Physiol. Biochem. Zool. 75, 544–556.

Turner, J.D., Driedzic, W.R., 1980. Mechanical and metabolic response of the perfused isolated fish heart to anoxia and acidosis. Can. J. Zool. 58, 886–889.

VanRaaij, M.T.M., VanderThillart, G.E.E.J.M., Vianen, G.J., Pit, D.S.S., Balm, P.H.M., Steffens, B., 1996. Substrate mobilization and hormonal changes in rainbow trout (*Oncorhynchus mykiss*, L) and common carp (*Cyprinus carpio*, L) during deep hypoxia and subsequent recovery. J. Comp. Physiol. B 166, 443–452.

Vornanen, M., 1994. Seasonal and temperature-induced changes in myosin heavy-chain composition of crucian carp hearts. Am. J. Physiol. 267, R1567–R1573.

Vornanen, M., 1997. Sarcolemmal influx of through L-type channels in ventricular muscle of a teleost fish. Am. J. Physiol. 272, R1432–R1440.

Vornanen, M., 1999. Na^+/Ca^{2+} exchange current in ventricular myocytes of fish heart: contribution to sarcolemmal Ca^{2+} influx. J. Exp. Biol. 202, 1763–1775.

Vornanen, M., 2017. Electrical excitability of the fish heart and its autonomic regulation. In: Gamperl, A.K., Gillis, T.E., Farrell, A.P., Brauner, C.J. (Eds.), Fish Physiology. In: The Cardiovascular System: Morphology, Control and Function, vol. 36A. Academic Press, San Diego, pp. 99–153.

Vornanen, M., Shiels, H.A., Farrell, A.P., 2002. Plasticity of excitation–contraction coupling in fish cardiac myocytes. Comp. Biochem. Physiol. A 132, 827–846.

Vornanen, M., Hassinen, M., Koskinen, H., Krasnov, A., 2005. Steady-state effects of temperature acclimation on the transcriptome of the rainbow trout heart. Am. J. Physiol. 289, R1177–R1184.

Vornanen, M., Stecyk, J.A.W., Nilsson, G.E., 2009. The anoxia-tolerant crucian carp (*Carassius carassius* L). In: Farrell, A.P., Brauner, C.J. (Eds.), In: Fish Physiology, vol. 27. Academic Press, San Diego, pp. 397–441.

van der Vusse, G.J., Glatz, J.F., Stam, H.C., Reneman, R.S., 1992. Fatty acid homeostasis in normoxic and ischemic heart. Physiol. Rev. 72, 881–940.

Wallimann, T., Wyss, M., Brdiczka, D., Nicolay, K., Eppenberger, H.M., 1992. Intracellular compartmentation, structure and function of creatine kinase isoenzymes in tissues with high and fluctuating energy demands: the ‘phosphocreatine circuit’ for cellular energy homeostasis. Biochem. J. 281, 21–40.

West, J.L., Driedzic, W.R., 1999. Mitochondrial protein synthesis in rainbow trout (*Oncorhynchus mykiss*) heart is enhanced in sexually mature males but impaired by low temperature. J. Exp. Biol. 202, 2359–2369.

West, T.G., Arthur, P.G., Suarez, R.K., Doll, C.J., Hochachka, P.W., 1993. *In vivo* utilization of glucose by heart and locomotory muscles of exercising rainbow trout (*Oncorhynchus mykiss*). J. Exp. Biol. 177, 63–79.

Wheaton, W.W., Chandel, N.S., 2011. Hypoxia 2. hypoxia regulates cellular metabolism. Am. J. Physiol. 300, C385–C393.

Wikman-Coffelt, J.R., Sievers, R.J., Coffelt, R.J., Parmley, W.W., 1983. The cardiac cycle: regulations and energy oscillations. Am. J. Physiol. 245, H354–H362.

Williams, G.S.B., Boyman, L., Lederer, W.J., 2015. Mitochondrial calcium and the regulation of metabolism in the heart. J. Mol. Cell. Cardiol. 78, 35–45.

Williamson, J.R., 1965. Glycolytic control mechanisms. I. Inhibition of glycolysis by acetate and pyruvate in the isolated perfused rat heart. J. Biol. Chem. 240, 2308–2321.

Wu, F., Beard, D.A., 2009. Roles of the creatine kinase system and myoglobin in maintaining energetic state in the working heart. BMC Syst. Biol. 3, 1–16.

Wyss, M., Kaddurah-Daouk, R., 2000. Creatine and creatine metabolism. Physiol. Rev. 80, 1107–1213.

Zammit, V.A., Newsholme, E.A., 1979. Activities of enzymes of fat and ketone body metabolism and effects of starvation on blood concentrations of glucose and fat fuels in teleost and elasmobranch fish. Biochem. J. 184, 313–322.

7

FORM, FUNCTION AND CONTROL OF THE VASCULATURE

ERIK SANDBLOM[*,1]
ALBIN GRÄNS[†]

[*]University of Gothenburg, Gothenburg, Sweden
[†]Swedish University of Agricultural Sciences, Skara, Sweden
[1]Corresponding author: erik.sandblom@bioenv.gu.se

This chapter summarizes the form, function, and control of the blood vasculature in fish. It starts with a general overview of the gross anatomy of the vascular system. We then turn to the arterial vasculature and the hemodynamic principles governing vascular resistance and arterial blood pressure, including a summary of the main local, neural, and hormonal systems controlling arterial resistance. The highly specialized vascular system of the gills (i.e., the branchial circulation) is covered in a separate section where we

The Cardiovascular System: Morphology, Control and Function, Volume 36A
FISH PHYSIOLOGY

DOI: http://dx.doi.org/10.1016/bs.fp.2017.06.001

describe their complex microvascular arrangements, as well as the control mechanisms of both the arterioarterial ("respiratory") and arteriovenous pathways. The chapter closes with a comprehensive overview of the venous vasculature. This includes the hemodynamic principles determining venous return, the importance of central venous blood pressure for cardiac performance, and the neurohumoral control mechanisms of venous capacitance and compliance. Finally, we discuss the importance of venous hemodynamic changes during integrated cardiovascular responses such as exercise and barostatic reflexes, and in response to changes in environmental variables including temperature, oxygen availability and salinity. While the chapter, by necessity, is biased toward teleosts due to the more abundant literature for that group, information on elasmobranchs, cyclostomes, and air-breathing fishes is also provided to highlight similarities and differences among the major groups of fish.

1. INTRODUCTION

The circulatory system of fish consists of a closed vascular system (i.e., blood vessels) connected to one or several hearts that generate the blood pressure which drives the blood flow through the vasculature. All living tissues are dependent on an effective supply of oxygen and nutrients, as well as the removal of carbon dioxide and metabolic waste products. This exchange function takes place in the capillary beds as blood flows through these vessels. However, as the circulatory system is shared among all tissues, an important role of the arterial vasculature is to balance tissue blood perfusion with local metabolic demands. This means that arterial resistance needs to be constantly and meticulously controlled by neural, hormonal and local factors that regulate regional blood flow to match the specific needs of each tissue, while still maintaining arterial blood pressure within a relatively narrow range. The exchange of gases with the surrounding water takes place at the gills, which represent a highly specialized part of the vascular system containing both arterioarterial ("respiratory") and arteriovenous pathways that have their own control systems. The venous vasculature, which transports blood back to the heart, is a low-pressure and high-compliance system that forms a large capacitance blood volume reservoir that can be both actively and passively mobilized. Control of venous capacitance serves a number of diverse functions including the mobilization of stressed blood volume (SBV) to maintain arterial and venous blood pressure should blood volume fall (e.g., during hemorrhage). The veins are also critical for regulating venous return and the cardiac filling pressure that adjusts cardiac output ($\dot{Q}$) during metabolically demanding situations such as exercise and increased temperature.

This chapter provides an overview of the form, function, and control of the blood vascular system of fish. Given the breadth of this topic, this chapter does not claim to provide a complete summary of all the available literature. For example, we do not cover aspects of the secondary circulation system of fishes and, thus, point you to excellent reviews on this topic (Olson and Farrell, 2011; Rummer et al., 2014). The aim is rather to offer a general summary of the main hemodynamic principles and the functional morphology of the fish vascular system, as well as to highlight the complexity of control systems affecting vascular function in this large and highly diverse group of vertebrates. As the majority of available literature concerns information on teleosts, the chapter is unarguably biased toward this group. However, numerous examples from elasmobranchs, cyclostomes and air-breathing fishes are also provided to highlight similarities, and to point out key differences among these major groups. For more detailed information on vascular control and function, the present book series has recently provided several excellent summaries for specific groups of fish including elasmobranchs (Brill and Lai, 2016), primitive fishes (Farrell, 2007a) and polar fishes (Axelsson, 2005). Detailed accounts of the vascular responses of fish to specific environmental conditions and physiological states such as environmental hypoxia (Gamperl and Driedzic, 2009) and digestion (Seth et al., 2011) are also available.

The chapter is divided into four main sections. The first gives an overview of the gross anatomy of the vascular system. In the second, we turn to the arterial system and describe the general hemodynamic principles determining arterial blood pressure and flow, as well as provide an overview of the main control systems affecting systemic arterial vascular resistance. We then provide a summary of the highly specialized microvascular arrangement of the gills, including the complex control mechanisms determining intrabranchial blood flow patterns. Finally, the chapter closes by providing a comprehensive overview of the specific hemodynamic principles and control mechanisms of the venous vasculature. This part of the circulatory system has traditionally received much less attention than the arterial circulation. However, considerable recent advances warrant a more comprehensive review of this topic. For example, our understanding has improved considerably regarding the role of changes in venous capacitance during integrated cardiovascular responses in fish (including exercise and barostatic reflexes), and during perturbations relevant from an ecophysiological and conservation perspective such as changes in temperature, salinity and water oxygen levels.

2. GROSS ANATOMY OF THE VASCULAR SYSTEM

The gross anatomy of the teleost vascular system has been comprehensively reviewed (e.g., Bushnell et al., 1992; Gamperl and Shiels, 2014; Jones and Randall, 1978; Olson, 2011c; Smith and Bell, 1975), as has the vascular anatomy of elasmobranchs (Munoz-Chapuli, 1999; Satchell, 1991) and cyclostomes (Fänge, 1972; Farrell, 2007a; Forster et al., 1991; Johansen, 1963). Moreover, Soldatov (2006) specifically reviews the microvasculature of fish. Detailed descriptions of the vascular anatomy of the salmonid gastrointestinal tract can be found in Thorarensen et al. (1991), and Seth et al. (2011) provide a comprehensive summary and discussion of the functional anatomy of the gastrointestinal vasculature. The following section provides a general description of the overall circulatory morphology of teleost fish as summarized in Fig. 1.

The arterial circulatory system in fish is fundamentally similar to that in all vertebrates in that the heart pumps blood into an aorta for distribution to the body. However, the arterial system of fish is unique in that the main gas exchange organ (i.e., the gills; see Section 4.1 for a detailed description of the gill microvasculature) is perfused in series with the systemic capillaries. This contrasts with the in-parallel arrangement of the systemic and pulmonary circuits of most other vertebrates. The arterial circulation of fish is, therefore, a one-way circular arrangement in which deoxygenated blood is pumped by the heart into a ventral aorta (VA) that branches into four pairs of afferent

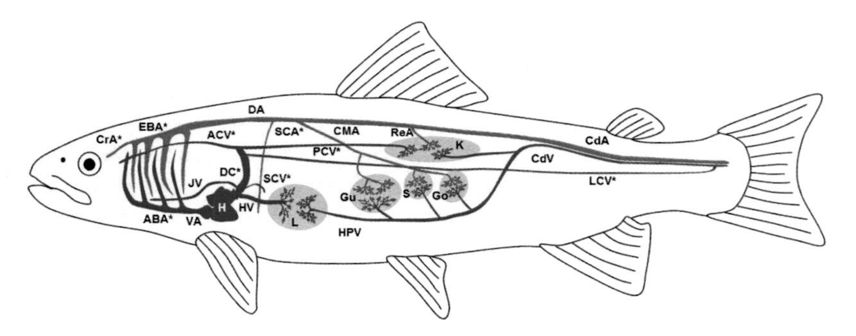

Fig. 1. Schematic representation of the main arteries and veins in a teleost fish. Abbreviations: *ABAs*, afferent branchial arteries; *ACV*, anterior cardinal vein; *CdA*, caudal artery; *CdV*, caudal vein; *CMA*, celiacomesenteric artery; *CrA*, carotid artery; *DA*, dorsal aorta; *DC*, *Ducts of Cuvier*; *EBAs*, efferent branchial arteries; *Go*, gonad; *Gu*, gut; *H*, heart; *HPV*, hepatic portal vein; *HV*, hepatic vein; *JV*, jugular vein; *K*, kidney; *L*, liver; *LCV*, lateral cutaneous vein; *PCV*, posterior cardinal vein; *ReA*, renal artery; *S*, spleen; *SCA*, subclavian artery; *SCV*, subclavian vein; *VA*, ventral aorta. Abbreviations denoted with * are bilaterally paired vessels where only one of the vessels are shown in the figure.

branchial arteries (ABAs). The blood becomes oxygenated at the lamellae of the gills and then flows into four pairs of efferent branchial arteries (EBAs). The main blood supply to the head region of the fish originates from the first EBA. This supply consists of a fine network of arteries including the important paired carotid arteries (CrAs) that supply the brain. However, some of the blood in the first EBA flows posteriorly and joins with the other three pairs of EBAs, forming the dorsal aorta (DA). The DA runs in a posterior direction in the hemal arch of the spine, with several major arteries branching off on its way to the caudal region. The first main branches are the paired subclavian arteries, which feed blood to the region around the pelvic girdle, including the pectoral locomotory muscles in labriform swimmers. The next major branch is the coeliacomesenteric artery that supplies the gut, spleen, and gonads. The last major branch is the renal artery that supplies the kidney.

The caudal end of the DA forms the caudal artery, which runs parallel to the caudal vein (CdV). The CdV brings blood to the renal portal system, which is one of two large portal venous systems found in the systemic circulation of fish. A portal system is essentially two capillary beds connected in series, where the blood passes through both of them before returning to the heart. The renal portal system, which is unique to fish, perfuses the renal parenchyma and flows around the renal tubules. The renal portal blood, together with post-glomerular arterial blood, is then collected by the posterior cardinal veins (PCVs) and carried to the *Ducts of Cuvier* (DC) which connect with the first chamber of the heart, the sinus venosus. Although the PCVs are paired, one of the vessels is often greatly reduced in size (Allen, 1905). The second portal venous system is the ubiquitous hepatic portal system. This system consists of the hepatic portal vein, which mainly collects venous outflow from the gastrointestinal tract and the visceral organs and directs it to the liver and its associated sinusoids. The outflow from the liver is through the hepatic veins (HVs) that fuse directly with the sinus venosus. The HVs are typically short and differ in number among species. The cranial portions of the body are mainly drained by the anterior cardinal veins, but also by the jugular vein and the paired subclavian veins that empty into the DC or directly into the sinus venosus. Finally, large paired lateral cutaneous veins running along the lateral line, and smaller dorsal and ventral cutaneous veins, drain the skin and the buccal and opercular cavities.

The total blood volume circulating in the vasculature of most teleosts and holostean fishes is relatively small, seldom exceeding 3%–4% of body mass. By comparison, blood volume in elasmobranchs and lampreys, as well as in many other vertebrates, is 5%–8% of body mass (Conte et al., 1963; Olson, 1992). Remarkably, the blood volume of hagfishes amounts to ~18% of body mass (Farrell, 2007a; Forster et al., 2001) and Antarctic icefishes (family *Channichthyidae*) that lack hemoglobin have blood volumes reported to be 8%–9% of body mass. These represent clear exceptions among the teleosts (see Hemmingsen, 1991; Hemmingsen and Douglas, 1970). However, comparing blood volumes among studies is somewhat problematic as the accuracy of different indicators and methods varies. For example, methods using different plasma dyes have been criticized for overestimating blood volume (Olson, 1992).

Numerous interesting deviations from the general anatomical patterns described earlier exist among the different groups of fish. Notably, air-breathing fishes possess various accessory air-breathing organs, some of which are perfused in parallel with the systemic circulation, and, thus, resemble the pattern seen in obligate air-breathing vertebrates (Olson, 1994). For example, lungfishes (subclass *Dipnoi*), which have both gills and lungs, possess a large pulmonary artery that perfuses the lungs, but the proportion of oxygen uptake from the two gas exchangers varies considerably among lungfish species, as well as with ontogeny and water oxygen availability (Burggren and Johansen, 1986). Another deviation from the basic plan outlined earlier is found in species that have arterial retia located between the gills and tissue capillaries. These are structures where one vessel splits into numerous smaller vessels, which then reunite to form a single vessel. Normally, arterial retia are interwoven with venous retia to form a countercurrent system that facilitates the exchange of gases, metabolic by-products (metabolites), or heat, and this structure is commonly referred to as a *rete mirabile* (Fänge, 1983). While these examples serve as reminders of the extraordinary diversity of adaptations found in the vascular system of fishes, the following chapter does not address these vascular specializations further. However, the fascinating anatomy and physiology of fish retial systems have recently been summarized by Stevens (2011), and the highly specialized vasculature of air-breathing fishes and its control has been reviewed on several occasions (e.g., Burggren and Johansen, 1986; Farrell, 2007a; Graham, 1997; Ishimatsu, 2012; Olson, 1994; Sandblom and Axelsson, 2011).

3. THE ARTERIAL VASCULATURE

Given the serially arranged vascular system of fish, all cardiac output ($\dot{Q}$) must pass through at least two successive high resistance vascular beds. Consequently, as the resistance of the venous system is relatively minimal (see Section 5.2.2), the sum of the branchial and the systemic (somatic and visceral) arterial resistances is largely representative of the total vascular resistance of the circulatory system of fish. Thus, a change in the resistance of one of these serially connected vascular beds will inevitably affect perfusion properties of the other. Here, we focus on the general principles governing arterial blood pressure and flow, and summarize the main control systems affecting systemic arterial resistance in fish.

3.1. Arterial Flow and Pressure

3.1.1. Pressure, Flow, and Resistance Relationships

The flow (F) of a perfect Newtonian fluid (see below) in a straight rigid tube is determined by the pressure difference (ΔP) along the tube and its resistance (R) as follows:

$$F = \frac{\Delta P}{R}$$

When applied to hemodynamics, this means that the pressure difference (ΔP, i.e., the perfusion pressure or driving pressure) along a blood vessel or across the entire circulatory system (i.e., between the ventral aorta and the central veins) drives the flow of blood or $\dot{Q}$, respectively. As the central venous blood pressure (P_{CV}) is generally close to zero (see Section 5.2.1), the upstream arterial blood pressure is the main determinant of the perfusion pressure. Blood flow is restricted by the frictional loss of energy against the vessel walls, which represents R and is determined by several different factors as outlined later.

Rearrangement of the above equation reveals that arterial blood pressure (P_A) is directly proportional to, and can be tightly regulated by, changes in $\dot{Q}$ and R according to the following equation:

$$P_A = \dot{Q} * R$$

According to Pascal's first law, all pressures act equally in all directions. Hence, any external pressure changes will act uniformly on the outside of the vessel wall. The difference between the internal and external pressures is called the *transmural pressure* and affects how vessels contract or dilate. Thus, the absolute hydrostatic water pressure on fish living at different water depths is irrelevant for the determination of flow, as the hydrostatic pressure surrounding the fish will affect all blood vessels in the same way and not affect the net driving pressure determining blood flow.

There is extensive information on *in vivo* arterial blood pressures in numerous species of fish, but somewhat less information on blood flow and resistance. The arterial blood pressure in fish is normally measured in the ventral (P_{VA}) and/or the dorsal aorta (P_{DA}). From simultaneous measurements of P_{DA} and $\dot{Q}$, it is possible to calculate the specific systemic vascular resistance (R_{sys}) using the equation above. By also measuring P_{VA}, the specific resistance of the gill vasculature (R_{gill}) can be calculated as follows:

$$R_{\text{gill}} = \frac{P_{\text{VA}} - P_{\text{DA}}}{\dot{Q}}$$

Consequently, the sum of R_{sys} and R_{gill} represents the total vascular resistance to blood flow. Table 1 summarizes resting *in vivo* values for $\dot{Q}$, P_{VA} and P_{DA}, and R_{sys} and R_{gill} of various representative fish species. From the table, it is evident that, due to the resistance of the gills, P_{DA} is typically 60%–80% of P_{VA} (i.e., the pressure generated by the heart).

3.1.2. Determinants of Vascular Resistance

To further understand how blood pressure and flow distribution are regulated, the physical factors affecting vascular resistance need to be considered. This can be achieved by applying Poiseuille's law:

$$F = \frac{\Delta P \Pi r^4}{8L\eta}$$

where F is the flow, ΔP is the pressure difference, r is the vessel radius, L is the vessel length, and η is the fluid (e.g., blood) viscosity. In principle, this law applies to nonturbulent and steady flows of Newtonian fluids through rigid tubes. In such a system the velocity profile of the fluid behaves like layers of cylinders where the outermost cylinders move slower due to shear stress, which is the force per unit area created when a tangential force (e.g., blood flow) acts on the inner surface of the tube wall (i.e., the endothelium), while the inner cylinders move gradually faster. The resulting parabolic flow profile is referred to as *laminar flow* (see Fig. 2A) and is generally the condition present in the peripheral vessels where the overall resistance of the vasculature system is primarily determined (Langille et al., 1983; Olson, 2011b).

While the total length of the blood vessels is rather constant at a given time point, the radius of the vessels and the viscosity of the blood can vary significantly and affect vascular resistance. Changes in vessel diameter are the most important and efficient means of controlling vascular resistance and even modest changes result in significant changes in blood flow. Indeed, according to Poiseuille's law, a doubling of vessel radius will increase flow 16-fold. How the diameter of blood vessels is controlled in fish is discussed in greater detail later, but first we will describe the determinants and hemodynamic effects of blood viscosity.

3.1.3. Blood Viscosity

The viscosity of blood, as with all fluids, is a measure of its resistance to deformation. Simply put, low viscosity fluids require less energy to flow through a tubular system than fluids with high viscosity. The viscosity of blood generally increases with the proportion of the blood composed of red blood

Table 1
In vivo cardiac output, arterial blood pressures, and vascular resistances of various representative fish species during "resting" conditions

	Temperature (°C)	$\dot{Q}$(mL min^{-1} kg^{-1})	P_{VA} (kPa)	P_{DA} (kPa)	R_{sys} (kPa mL^{-1} min^{-1} kg^{-1})	R_{gill} (kPa mL^{-1} min^{-1} kg^{-1})	Source
Cyclostomes							
Atlantic hagfish, *Myxine glutinosa*	11	8.7	1.0	0.8	0.11	0.05	Axelsson et al. (1990)
Broadgilled hagfish, *Eptatretus cirrhatus*	17	15.8	1.6	1.3	0.08	0.02	Forster et al. (1992)
Elasmobranchs							
Eastern shovelnose ray, *Aptychotrema rostrata*	28	39.0		3.0	0.08		Speers-Roesch et al. (2012)
Epaulet shark, *Hemiscyllium ocellatum*	28	44.0		3.0	0.07		Speers-Roesch et al. (2012)
Picked dogfish, *Squalus acanthias*	10	29.7		2.4	0.08		Sandblom et al. (2006a)
Lesser spotted dogfish, *Scyliorhinus canicula*	15	41.6	5.4	4.4	0.11	0.02	Taylor et al. (1977)
Teleosts							
European eel, *Anguilla anguilla*	15	11.5	4.9	3.0	0.26	0.17	Janvier et al. (1996)
Short-finned eel, *Anguilla australis*	17	7.0	5.1	3.1	0.27	0.18	Davie and Forster (1980)
Rainbow trout, *Oncorhynchus mykiss*	10	17.6	5.2	4.1	0.23	0.06	Kiceniuk and Jones (1977)
Chinook salmon, *Oncorhynchus tshawytscha*	13	28.5		5.2	0.18		Clark et al. (2008)

(*Continued*)

Table 1 (Continued)

	Temperature (°C)	$\dot{Q}$(mL min^{-1} kg^{-1})	P_{VA} (kPa)	P_{DA} (kPa)	R_{sys} (kPa mL^{-1} min^{-1} kg^{-1})	R_{gill} (kPa mL^{-1} min^{-1} kg^{-1})	Source
Atlantic cod, *Gadus morhua*	10	17.3	4.9	3.2	0.17	0.09	Axelsson and Nilsson (1986)
Lingcod, *Ophiodon elongatus*	10	11.2	5.1	3.8	0.34	0.12	Farrell (1982)
Sea raven, *Hemitripterus americanus*	11	18.8	3.8	3.1	0.16	0.04	Axelsson et al. (1989)
Bald notothen, *Pagothenia borchgrevinki*	0	24.4		2.3	0.12		Sandblom et al. (2012)
Winter flounder, *Pleuronectes americanus*	8	15.0		2.7	0.22		Mendonca and Gamperl (2010)

Values are obtained from unanesthetized and unrestrained animals under conditions described as "resting" or similar. Abbreviations: cardiac output ($\dot{Q}$); ventral and dorsal aortic blood pressures (P_{VA} and P_{DA}, respectively); systemic and gill vascular resistances (R_{sys} and R_{gill}, respectively). Due to the anatomical arrangement of elasmobranch fishes precluding direct measurement of total blood flow in the ventral aorta, $\dot{Q}$ for all elasmobranch species is estimated from blood flow to the two anterior pairs of afferent branchial arteries assuming that this represents 37% of total $\dot{Q}$ (Taylor et al., 1977).

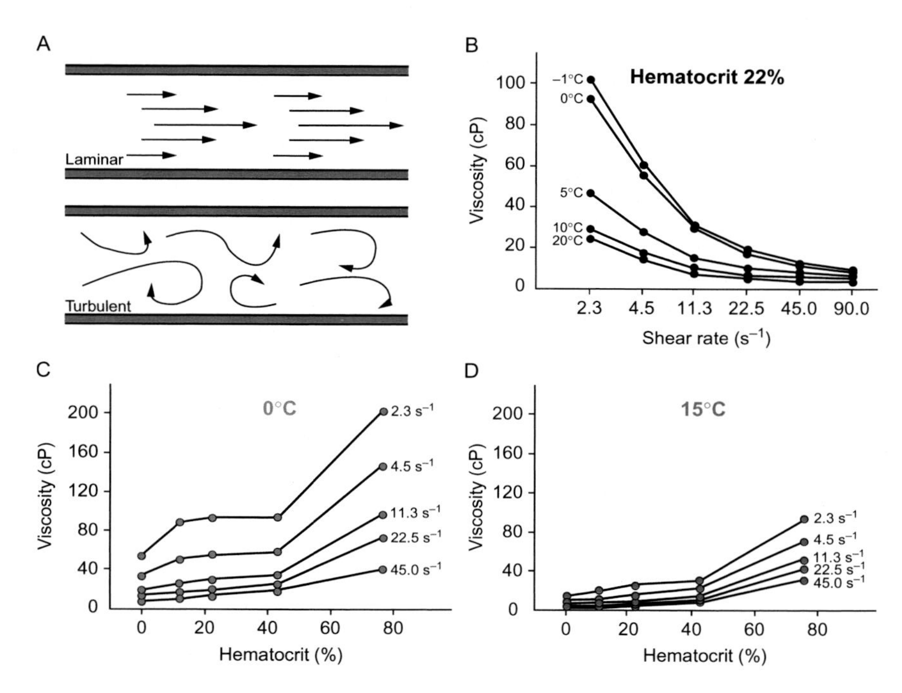

Fig. 2. Factors affecting the viscosity and flow profile of fish blood. (A) Laminar and turbulent flow profiles. Turbulent flow results when laminar flow is disrupted and may increase the viscosity. (B) The effects of shear rate on blood viscosity at different temperatures. (C and D) The effect of hematocrit on blood viscosity across different shear rates at 0°C and 15°C, respectively. Panels (B)–(D): Redrawn from Graham, M.S., Fletcher, G.L., 1983. Blood and plasma viscosity of winter flounder *Pseudopleuronectes americanus* influence of temperature red cell concentration and shear rate. Can. J. Zool. 61, 2344–2350.

cells (i.e., hematocrit) and reduced temperature (Graham and Fletcher, 1983). This relationship is exaggerated at low flow rates due to the non-Newtonian fluid properties of blood. Consequently, the viscosity of the blood is shear rate dependent and reflects the rate of change in velocity when one layer of fluid passes over an adjacent layer (Fig. 2B–D). This is because at low flow rates (i.e., low shear rates), the red blood cells tend to spread out more evenly within the vessel. This leads to an increased number of collisions between blood cells, and between blood cells and the vessel walls, and this increases the effective viscosity of the blood. In contrast, at high flow rates, the red blood cells tend to accumulate in the center of the vessel, a phenomenon known as *axial streaming*, and the viscosity is reduced as blood flow becomes more laminar with fewer collisions (Olson, 2011b). Because of the temperature-dependent viscosity and non-Newtonian properties of blood a combination of low temperature, low flow rates and high hematocrit will result in the highest viscosity (Fig. 2B–D), which in turn increases the vascular resistance and the workload on the heart. Consequently, the potential problems associated with high blood viscosity are most pronounced in fish species living in cold environments such as deep lakes, lakes that freeze, and especially in the polar regions; which is also where some rather remarkable solutions to the potential problems of high blood viscosity at low temperatures can be found. For example, the blood of icefishes (*Channichthyidae*) is completely devoid of red blood cells (erythrocytes) (Axelsson, 2005). Other red-blooded Antarctic species, such as the bald notothen (*Pagothenia borchgrevinki*), normally maintain a relatively low resting hematocrit (~15%), presumably to reduce blood viscosity. However, they can transiently increase hematocrit two- to three-fold during stress and exercise through the release of erythrocytes from the spleen (Franklin et al., 1993). This splenic capacity is far beyond any other temperate or tropical fish species studied to date (Axelsson, 2005; Franklin et al., 1993). While a low hematocrit or complete lack of red blood cells may provide a mechanism to successfully decrease blood viscosity, and thus workload on the heart, the cost of fewer erythrocytes and less hemoglobin is a lower blood oxygen carrying capacity. Thus, despite living at cold temperatures with a low metabolic rate, most Antarctic species (particularly the icefishes) exhibit relatively high routine cardiac outputs. This higher $\dot{Q}$ does reduce blood viscosity and vascular resistance by increasing shear rate (Fig. 2C–D), but can also result in a higher workload on the heart.

3.2. Control of Systemic Arterial Resistance and Blood Flow

While the arterial blood pressure generated by the heart is what drives the blood around the circulatory system, most of the regulation of blood flow distribution occurs in the tissues through local changes in peripheral arterial resistance. This is primarily achieved by controlled constriction or dilation of small arteries and arterioles, often referred to as the resistance vessels (i.e., precapillary resistance). Resistance vessels can be found both in the branchial and in the systemic arterial circulations (Olson, 2011e). Thus, by the time the blood enters the capillary beds, blood pressure has been greatly reduced, and the highly pulsatile arterial blood flow is transformed into a more even flow. The flow velocity also gradually declines at this level of the vascular system due to the greater total cross-sectional area of the arterioles and capillaries as compared with the larger conductance arteries. These hemodynamic changes prevent damage to these delicate vessels, and the greatly enhanced surface area and prolonged blood transit time of the capillary networks ensure the efficient exchange of gases and solutes.

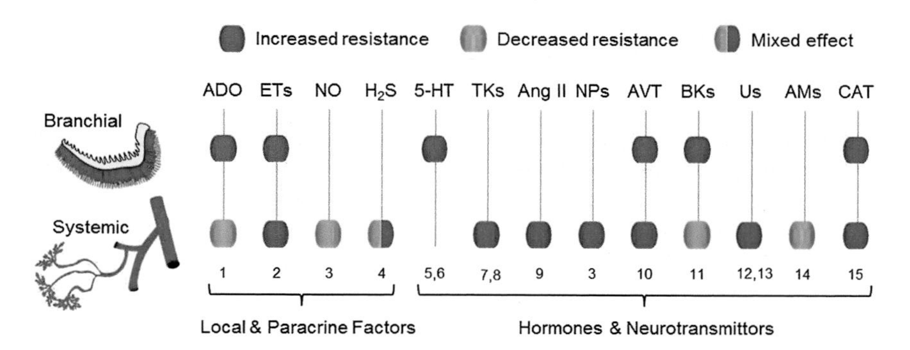

Fig. 3. The *in vivo* effects of various vasoactive substances on branchial and systemic vascular resistance in rainbow trout, *Oncorhynchus mykiss*. Abbreviations: *ADO*, adenosine; *AMs*, adrenomedullin; *Ang II*, angiotensin II; *AVT*, arginine vasotocin; *BKs*, bradykinins; *CAT*, catecholamines; *ETs*, endothelins; *NO*, nitric oxide; *NPs*, natriuretic peptides; *H_2S*, hydrogen sulfide; *5-HT*, serotonin; *TKS*, tachykinins; *Us*, urotensins. Information is obtained from the following sources: 1, Sundin and Nilsson (1996); 2, Hoagland et al. (2000); 3, Olson et al. (1997a); 4, Olson (2005); 5, Reite (1969a); 6, Sundin et al. (1995); 7, Kågström et al. (1996b); 8, Le Mével et al. (2007); 9, Bernier et al. (1999b); 10, Conklin et al. (1997); 11, Olson et al. (1997b); 12, Mimassi et al. (2000); 13, Le Mével et al., 1996; 14, Aota (1995); 15, Wood and Shelton (1980).

Blood flow through resistance vessels is controlled by numerous factors. First, tissue blood flow is regulated locally in relation to metabolic demand through the release of paracrine and autocrine factors and metabolites. Second, the arterial blood pressure is kept within relatively narrow limits to maintain constant perfusion pressure despite local changes in resistance. Finally, during large cardiovascular challenges such as exercise, blood loss or changes in environmental oxygen levels, various "remote" signaling systems that also control vascular resistance are activated. Such integrated control of vascular resistance includes vasomotor nerves of the autonomic nervous system and vasoactive hormones circulating in the plasma. Fig. 3 summarizes the *in vivo* effects of various vasoactive substances on branchial and systemic vascular resistances in rainbow trout (*Oncorhynchus mykiss*). Below is a summary of the main local, neural, and hormonal control systems affecting systemic arterial vascular resistance. A summary of the control of intrabranchial blood flow distribution and vascular resistance is provided in Section 4.2.

3.2.1. Local Regulation of Arterial Resistance

3.2.1.1. Active hyperemia. Active hyperemia refers to regionally increased blood flow that occurs when the metabolic activity of tissues or organs rises, and is caused by the accumulation of various metabolites with vasoactive properties. For example, an increase in CO_2 and H^+ within active tissues causes blood and tissue hypercarbia. In addition, tissue O_2 levels may drop (tissue hypoxia) if the metabolic demand is not sufficiently matched by increased blood flow. Indeed, both reduced pH and elevated partial pressure for CO_2 (P_{CO_2}) produce vasodilation of systemic resistance vessels (Canty and Farrell, 1985; McKendry and Perry, 2001; Smith et al., 2006), as does reduced partial pressure for O_2 (Smith et al., 2001). Other vasoactive metabolic by-products are the purines generated as intermediates in oxidative phosphorylation (ATP, ADP, and AMP) and adenosine (ADO). When infused into isolated trunk preparations of *O. mykiss*, ATP triggered vasoconstriction (Wood, 1977), while no effects were seen with ADO (Colin et al., 1979). However, when injected *in vivo*, ADO decreased systemic vascular resistance in the same species (see Fig. 3; Sundin and Nilsson, 1996).

3.2.1.2. Local Paracrine Signaling Molecules. Vasoactive paracrine factors are released by the endothelium and from vascular smooth muscle or surrounding tissues, and diffuse over short distances through the interstitial space to stimulate vascular smooth muscle. Endothelial produced molecules that cause vasodilation are known as endothelium-derived relaxing factors (EDRFs), while those that constrict the vasculature are known as endothelium-derived contracting factors. Many vasoactive paracrine factors are much more potent than most neural or hormonal messengers (Olson, 2011e).

Arachidonic acid is a polyunsaturated fatty acid present in endothelial cell membranes. It can be produced for signaling purposes and functions as a precursor for a wide range of vasoactive metabolites including prostaglandins, leukotrienes, and thromboxanes. Prostaglandins generally have vasodilatory effects on systemic vascular smooth muscle in fish (Kågström and Holmgren, 1997; Olson and Villa, 1991), while leukotrienes and thromboxane are reported to be systemic vasoconstrictors (Olson et al., 1997b; Smith et al., 2006). However, functional studies on the systemic vasoactive effects of arachidonic acid metabolites in fish are sparse.

Endothelins (ETs) are peptides released from the endothelium, and in mammals, they are among the most potent vasoconstrictors known. Indeed, ETs constrict most systemic arterial vessels in both teleosts (Hoagland et al., 2000) and elasmobranchs (Evans et al., 1996).

3.2.1.3. Gasotransmitters. Gasotransmitters including nitric oxide (NO), and the more recently described carbon monoxide (CO) and hydrogen sulfide (H_2S), are increasingly being recognized as important regulators of vascular and cardiac (see Chapter 5, Volume 36A: Imbrogno and Cerra, 2017) function in fish. Notably, research on the two latter compounds has intensified in recent years (Olson and Donald, 2009). After being formed, these gases are rapidly inactivated and, therefore, function primarily as paracrine or autocrine signaling molecules (see reviews by Donald et al., 2015; Olson, 2015; Olson and Donald, 2009; Olson et al., 2012).

In most fish species studied, NO or NO donors (e.g., sodium nitroprusside) act as vasodilators of systemic arteries and resistance vessels via activation of soluble guanylyl cyclase (Jennings et al., 2004; Olson et al., 1997a). However, unlike in mammals, most evidence indicates that an endothelial NO synthase is not present in fish and that NO does not function as an EDRF (Syeda et al., 2013: but see Imbrogno and Cerra, 2017; Chapter 5, Volume 36A). However, neural NO synthase is widespread in perivascular nerves and is believed to be the main source of vasoregulatory NO in fish (Donald et al., 2015; Olson and Donald, 2009).

Carbon monoxide is produced from the heme group of proteins such as hemoglobin, myoglobin and cytochromes by the action of the enzyme heme oxygenase that is present in both endothelial and vascular smooth muscle cells. From the few available studies on fish, CO also appears to act as a vasodilator via guanylyl cyclase and potassium channel activation. Moreover, inhibition of intrinsic CO production results in the contraction of isolated vessels, and this suggests a tonic vasodilatory action for this gasotransmitter in fish (Dombkowski et al., 2009; Olson and Donald, 2009).

Although H_2S has pronounced vasoactive effects in fish, these are highly species and vessel specific. Indeed, contraction, relaxation and multiphasic responses have been observed in isolated fish vessels (Dombkowski et al., 2004, 2005; Olson, 2005; Olson and Donald, 2009). The mechanisms underlying H_2S vasoactivity are not fully resolved, but recent studies indicate that it only acts locally and does not circulate in the plasma (Olson, 2009). Nonetheless, H_2S has also been suggested to function as a vascular O_2 sensor and, thus, a mediator of vascular responses to hypoxia (Olson, 2015; Skovgaard and Olson, 2012).

3.2.2. Neural Control of Arterial Resistance

The autonomic nervous system has a profound influence on vascular resistance in all vertebrates, including fish. It is also well established that resistance vessels are controlled by "non-adrenergic non-cholinergic (NANC)" nerves that may release a diversity of neurotransmitters such as NO, purines, and various neuropeptides. Generally, the vasoactive properties of these nerves and neurotransmitters have been much less explored in fish than the "classical" adrenergic and cholinergic control systems of the autonomic nervous system (Morris and Nilsson, 1994; Nilsson and Holmgren, 1992; Zaccone et al., 2006). Moreover, some of them (e.g., purines and NO, as discussed earlier) cannot be easily placed into distinct categories as they can act as both neurotransmitters and paracrine signaling molecules.

3.2.2.1. The Autonomic Nervous System. The functional morphology and pharmacology of the autonomic nervous system, and its general role in circulatory control in various groups of fish, have been comprehensively summarized previously (e.g., Farrell, 2007a; Nilsson, 1983; Nilsson and Holmgren,

1988; Sandblom and Axelsson, 2011). Here, we describe the specific role of the autonomic nervous system in the control of systemic arterial resistance.

The resistance vasculature of teleosts is innervated by spinal autonomic nerves that release the catecholamines adrenaline and noradrenaline, which, in turn, mediate vasoconstriction via α-adrenoreceptors and vasodilation via β-adrenoreceptors (Morris and Nilsson, 1994; Reite, 1969b; Sandblom and Axelsson, 2011). However, α-adrenoreceptors and hence vasoconstriction typically dominate in fishes. For example, *in vivo* injection of adrenaline increases systemic and gastrointestinal vascular resistance and results in arterial hypertension in most teleost species (Axelsson and Farrell, 1993; Axelsson et al., 2000; Stevens et al., 1972; Wood and Shelton, 1980). Even so, a few notable exceptions have been documented among the teleosts where catecholamine injection did not affect systemic vascular resistance, such as in the Australian eel, *Anguilla australis* (Hipkins et al., 1986), and in the bald notothen, *P. borchgrevinki* (Axelsson et al., 1994; Sandblom et al., 2012). As adrenergic nerve blockade with bretylium reduces vascular resistance and arterial blood pressure in many teleosts, the resistance vasculature must be influenced by a neural adrenergic tonus under resting conditions (Axelsson and Nilsson, 1986; Nilsson, 1994; Sandblom et al., 2010; Smith, 1978; Smith et al., 1985).

Likewise, *in vivo* administration of catecholamines typically increases arterial blood pressure and systemic vascular resistance through stimulation of α-adrenoreceptors in elasmobranchs (Capra and Satchell, 1977; Opdyke et al., 1972; Sandblom et al., 2006a), and both adrenaline and noradrenaline increase gastrointestinal vascular resistance in *Squalus acanthias* (Holmgren et al., 1992). Isolated arteries and perfused preparations from hagfish and lampreys also typically constrict in response to α-adrenergic stimulation and dilate in response to β-adrenergic stimuli (Foster et al., 2008; Reite, 1969a). Nonetheless, adrenergic innervation of the systemic arterial vasculature is poorly developed or absent in elasmobranchs and cyclostomes, and these groups of fish are generally believed to rely more on catecholamines circulating in the plasma for adrenergic vascular control (see Section 3.2.3.1). The gastrointestinal circulation of elasmobranchs may be an exception because dense networks of catecholamine-containing perivascular fibers have been observed in the gut of various species (Bjenning et al., 1989; Holmgren and Nilsson, 1983; Nilsson et al., 1975); but the functional significance of this putative control remains largely unknown. Direct innervation from cranial cholinergic neurons probably has little importance for the control of the systemic vasculature in fish, but as outlined in Section 4, it is important for controlling intrabranchial vascular resistance and blood flow distribution.

3.2.2.2. Serotonin. Serotonin (5-HT, 5-hydroxytryptamine) is a neurotransmitter whose origin can be both from enterochromaffin cells in the gut and from serotonergic nerves (Nilsson and Holmgren, 1992). Serotonin is a powerful vasoconstrictor in the branchial vasculature as discussed in Section 4.2.1, but it also affects systemic vascular resistance. Intraarterial injection of serotonin typically decreases systemic vascular resistance in various teleost species (Janvier et al., 1996; Sundin and Nilsson, 2000; Sundin et al., 1995, 1998). However, the mechanism behind the systemic serotonin-induced vasodilation remains unclear as the effects cannot be mimicked with 5-HT receptor agonists (Sundin et al., 1998) or abolished by 5-HT receptor antagonists (Janvier et al., 1996; Sundin and Nilsson, 2000). When interpreting the results of these studies, one needs to remember that serotonin is normally released locally. When injected into the circulation, it binds to all available 5-HT receptors and may cause hemodynamic effects that are very different from those observed when it is released within specific tissues. However, in a recent study on *O. mykiss*, intraarterial injection of 5-HT induced biphasic changes in P_{DA}, with an initial hypotensive response followed by hypertension (Kermorgant et al., 2014). The authors suggested that this biphasic response resulted from a combined effect of 5-HT on peripheral vascular target sites and the central nervous system when injected systemically, and that 5-HT may also exert some of its action as a circulating hormone.

3.2.2.3. Neuropeptides. Several neuropeptides are known to have vasoactive properties in fish (Morris and Nilsson, 1994; Sandblom and Axelsson, 2011; Zaccone et al., 2006). Calcitonin gene-related peptide (CGRP) is a potent vasodilator that has been confirmed in resistance vascular beds of several fish species (e.g., Jennings et al., 2007; Kågström and Holmgren, 1998; Shahbazi et al., 2009). Indeed, there is evidence that this peptide relaxes isolated arteries from *O. mykiss* and Atlantic cod (*Gadus morhua*) through activation of CGRP receptors located on the smooth muscle cells without involvement of prostaglandins or NO (Kågström and Holmgren, 1998; Shahbazi et al., 2009).

Tachykinins, such as substance P (SP) and neurokinin A (NeKA), are often colocalized in CGRP-containing neurons, and both of these tachykinins have been identified in the vasculature of various teleost and elasmobranch species (Kågström and Holmgren, 1998; Shahbazi et al., 2009; Skov and Bennett, 2004). However, the circulatory effects of tachykinins in fish are quite variable, both within and among species. For example, in *O. mykiss*, injection of tachykinins *in vivo* mainly has vasoconstrictive effects with NeKA being more potent than SP (see Fig. 3; Kågström et al., 1996b; Le Mével et al., 2007). This is a highly unusual response among vertebrates as tachykinins are normally potent vasodilators. Indeed, SP reduced both systemic and gastrointestinal vascular resistance in *G. morhua* (Jensen et al., 1991) and caused systemic vasodilation and increased gastrointestinal blood flow in spiny dogfish (*S. acanthias*) (Kågström et al., 1996a). Two native tachykinins, scyliorhinin I and II, were also tested in that study, with scyliorhinin I causing a similar vasodilation as SP, while scyliorhinin II caused a general vasoconstriction of the arterial resistance vasculature. In light of these findings, it has been suggested that two or more subtypes of tachykinin receptor subtype may be present in fish (Kågström et al., 1994, 1996a,b).

Another neuropeptide found in all vertebrates is vasoactive intestinal polypeptide (VIP). VIP decreases gastrointestinal vascular resistance in *G. morhua in vivo* and relaxes the perfused vasculature of black bullhead (*Ictalurus melas*) (Holder et al., 1983; Jensen et al., 1991). The vasodilating effect of VIP does not appear to be endothelium dependent or caused by NO production (Kågström and Holmgren, 1997). In contrast, gastrointestinal vascular resistance in *S. acanthias* increases following VIP administration *in vivo* (Holmgren et al., 1992).

3.2.3. Hormonal Control of Arterial Resistance

3.2.3.1. Circulating Catecholamines. In addition to adrenergic nerves, vascular smooth muscle may also receive adrenergic stimulation via catecholamines circulating in the plasma. As fish lack true adrenal glands, these hormones are released directly into the venous blood from chromaffin tissues corresponding to the adrenal medulla in mammals. In teleosts, the chromaffin tissue primarily lines the wall of the posterior caudal vein within the anterior part of the kidney, which is referred to as the "head kidney" (Nandi, 1961). In elasmobranchs, the main aggregations of chromaffin cells are found in structures termed "axillary bodies" that are associated with paravertebral autonomic ganglia (Nilsson, 1984). Chromaffin cells are also found in the heart and great veins of elasmobranchs (Nilsson and Holmgren, 1988; Perry and Gilmour, 1996), cyclostomes (Bernier and Perry, 1996; Epple et al., 1985), and lungfishes (Abrahamsson et al., 1979). In contrast, Antarctic fish represent a group where circulating catecholamines probably have very limited importance for vascular control. The capacity for catecholamine synthesis in the head kidney is comparatively low in a range of Antarctic species (Whiteley and Egginton, 1999).

Pre-ganglionic (cholinergic) spinal nerves innervate the chromaffin cells and provide the main control mechanism for catecholamine secretion in most fish, but hagfishes have aneural chromaffin tissue and represent a clear exception from this pattern (Greene, 1902). There are also numerous other neural, hormonal, and paracrine factors that affect the release of catecholamines in fish (see Perry and Capaldo, 2011; Reid et al., 1998). The resting plasma concentration of catecholamines is generally quite low (1–10 nM; Gamperl et al., 1994) and experiments on teleosts with the adrenergic nerve blocker bretylium indicate that the systemic adrenergic tonus at rest, as well as during moderate exercise and hypoxia, is primarily mediated by neural adrenergic mechanisms with little contribution from circulating catecholamines (Chapter 4, Volume

36A; Farrell and Smith, 2017; Axelsson and Fritsche, 1991; Axelsson and Nilsson, 1986; Fritsche and Nilsson, 1990; Sandblom et al., 2010; Smith, 1978). The branchial vasculature is generally considered to be more sensitive to changes in plasma catecholamine levels than the systemic vasculature (see Section 4.2), but during severely stressful stimuli, plasma catecholamine levels may reach levels high enough to increase systemic vascular resistance and affect arterial blood pressure (Perry and Bernier, 1999; Randall and Perry, 1992). In addition, circulating catecholamines have been suggested to be an important source for neuronal uptake, which may be a prerequisite for a sustained neural adrenergic tonus on the systemic resistance vasculature (Xu and Olson, 1993).

3.2.3.2. Renin–Angiotensin System. Angiotensin II (Ang II), which is the main bioactive product of the renin–angiotensin system (RAS), is a strong vasoconstrictor and plays an important role in arterial blood pressure control in both teleosts (Axelsson et al., 1994; Bernier et al., 1999b; Oudit and Butler, 1995a; Platzack et al., 1993) and elasmobranchs (Bernier et al., 1999a; Carroll, 1981; Opdyke and Holcombe, 1976; Opdyke et al., 1982). However, the vascular actions of Ang II appear to differ somewhat between groups. For example, in some teleosts, Ang II is tonically active and regulates arterial blood pressure under resting conditions (Butler and Oudit, 1995; Platzack et al., 1993; Tierney et al., 1995), whereas in elasmobranchs the RAS is primarily activated during hypotensive stress (Bernier et al., 1999a; Hazon et al., 1999). Moreover, while a direct stimulation of vascular smooth muscle by Ang II has been confirmed in various teleosts (Bernier et al., 1999b; Carroll, 1981; Platzack et al., 1993), Ang II primarily constricts resistance vessels in elasmobranchs indirectly by stimulating adrenergic control systems (Bernier et al., 1999a; Opdyke and Holcombe, 1976).

3.2.3.3. Arginine Vasotocin. Arginine vasotocin (AVT) is homologous to arginine vasopressin (i.e., antidiuretic hormone) in mammals and is endogenously released from the neurohypophysis primarily in response to hyperosmotic stimuli and associated cellular dehydration (Balment et al., 2006). Besides being a major endocrine regulator of water balance and osmotic homoeostasis, AVT is also a potent systemic vasoconstrictor in teleosts (Conklin et al., 1997). However, *in vivo* injection of AVT sometimes produces biphasic responses with an initial transient drop in P_{DA} followed by a slower long-lasting hypertension (Henderson and Wales, 1974; Le Mével et al., 1993). This effect probably reflects the profound branchial vasoconstrictive effects of AVT (see Section 4.2.2), where the branchial vessels initially constrict causing a rapid increase in P_{VA} and a fall in P_{DA}, whereas the slower subsequent increase in blood pressure is due to vasoconstriction of peripheral resistance vessels (see Balment et al., 2006).

3.2.3.4. Bradykinins. Bradykinins (BKs) are inflammatory mediators that have vasoactive effects when injected intravascularly in both teleosts and elasmobranchs (Conlon et al., 1995; Dasiewicz et al., 2011; Olson et al., 1997b; Shahbazi et al., 2001; Takei et al., 2001). However, their effects on systemic vascular resistance are far from unequivocal, and reports of complex multiphasic *in vivo* responses, including both vasodilation and vasoconstriction, are common. Indeed, it appears that the direct effects of BKs on vascular smooth muscle are limited or absent, and that the hemodynamic responses are mainly due to the release of vasodilatory prostaglandins and by cross-signaling with, for example, the adrenergic control systems (Dasiewicz et al., 2011; Olson et al., 1997b).

3.2.3.5. Urotensins. Urotensin I and II (UI and UII) are members of an ancient family of neuropeptides released from the caudal neurosecretory system (urophysis). *In vivo* administration of UI in *O. mykiss* causes dorsal aortic hypertension due to increased systemic vascular resistance, but this mainly appears to be an indirect α-adrenoreceptor-mediated response due to the release of catecholamines. Indeed, pre-treatment with an α-adrenoreceptor antagonist unmasked a sustained systemic vasodilation and hypotension from UI (Mimassi et al., 2000). Similar responses to UI have been reported in the elasmobranch *Scyliorhinus canicula* (Platzack et al., 1998). In contrast, UII is among the most potent vasoconstrictors known in humans (Ong et al., 2005). It is also a potent systemic vasoconstrictor in teleosts with hypertensive effects that are independent of α-adrenoreceptor activation (Le Mével et al., 1996, 2008; Nobata et al., 2011).

3.2.3.6. Cholecystokinin. Cholecystokinin (CCK) is another peptide hormone with vasoactive properties that is released during digestion in vertebrates, including fish (Jönsson et al., 2006). When injected into the blood stream of *O. mykiss, in vivo*, CCK causes increased gastrointestinal blood flow, which is probably due to relaxation of the small resistance vessels within the gastrointestinal tract (Seth et al., 2010). In addition, CCK constricts the bulbus arteriosus and prebranchial arteries, which increases the pulse pressure of the heart (Seth et al., 2010, 2014). The function of this is unknown, but it has been suggested that the increased pulse pressure may serve to open previously underperfused regions of the gill lamellae (see Farrell et al., 1980), while maintaining the overall systemic perfusion pressure that is needed to sustain the increased metabolic demand associated with digestion (Seth et al., 2014).

3.2.3.7. Adrenomedullins. The adrenomedullins (AMs) are a family of vasodilator peptide hormones that have been identified in all major groups of fish, and at least five different types exist, some of which have a head kidney origin and circulate in the plasma (Takei et al., 2013). AMs reduce overall arterial blood pressure when injected into *O. mykiss* (see Fig. 3) and Japanese eel (*Anguilla japonica*) (Aota, 1995; Nobata et al., 2008), and dilate isolated arteries from the two eel species *A. japonica* and *A. australis* (Cameron et al., 2015). The vasodilation in eel appears to be more pronounced in the systemic than in the gill vasculature, as indicated by a more pronounced dorsal aortic than ventral aortic hypotension (Nobata et al., 2008).

3.2.3.8. Natriuretic Peptides. The natriuretic peptide (NP) system is a family of peptide hormones affecting salt and volume homeostasis, and that have pronounced vasodilating properties in vertebrates, including fish (e.g., Johnson and Olson, 2008; Takei et al., 2014; Toop and Donald, 2004). Several different NPs have been identified in fish, including in ancient groups like hagfish. In teleosts, atrial natriuretic peptide (ANP), brain natriuretic peptide, ventricular natriuretic peptide, and up to four subtypes of *C* natriuretic peptide have been identified (Takei et al., 2011). While the most prominent vasodilating effects of NPs in fish may be on the branchial and venous vasculature as discussed later, they also dilate systemic resistance vessels (Johnson and Olson, 2008; Nobata et al., 2010).

4. THE BRANCHIAL VASCULATURE

The gills represent a highly specialized and complex part of the fish vascular system that is involved in a diverse range of functions including respiration, ion and acid–base regulation, and the excretion of wastes (e.g., nitrogenous compounds), as well as in the monitoring of the internal and external environments (Evans et al., 2005). The following section provides a summary of the anatomical arrangement of the branchial microvasculature, followed by an overview of the main neurohumoral control mechanisms and environmental factors affecting gill blood flow in various groups of fish. More detailed accounts on the vascular morphology of the gill, as well as the control and innervation of gill blood vessels, can be found elsewhere (Jonz and Zaccone, 2009; Morris and Nilsson, 1994; Nilsson and Sundin, 1998; Olson, 2002a,b, 2011a; Olson et al., 1981; Sundin and Nilsson, 2002), where original accounts are cited. The specific role of the gills in various physiological processes has also been reviewed on numerous occasions and some of the more recent contributions include accounts of baro- and chemoreception (Jones and Milsom, 1982; Jonz, 2011; Jonz et al., 2015; Milsom, 2012; Milsom and Burleson, 2007), hormone metabolism (Nekvasil and Olson, 1986; Olson, 2002a), respiration and ventilatory control (Gilmour, 2001; Perry and Gilmour, 2002), and the regulation of acid–base and osmotic balance (Gilmour, 2012; Goss et al., 1998).

4.1. Arrangement of the Gill Microvasculature

A schematic drawing of the gill vasculature is presented in Fig. 4. In teleosts, ABAs bring blood to each of the four pairs of gill arches, and afferent filamental arteries (AFAs) originate from these branchial arteries. These vessels carry blood up the gill filaments, from where afferent lamellar arterioles depart and distribute deoxygenated blood to one or several secondary lamellae where gas exchange takes place (see Fig. 4). Within the lamellae are "pillar cells" that provide support, and separate the epithelial layers of the lamellae to form the vascular sheet (Farrell et al., 1980), through which the blood flows

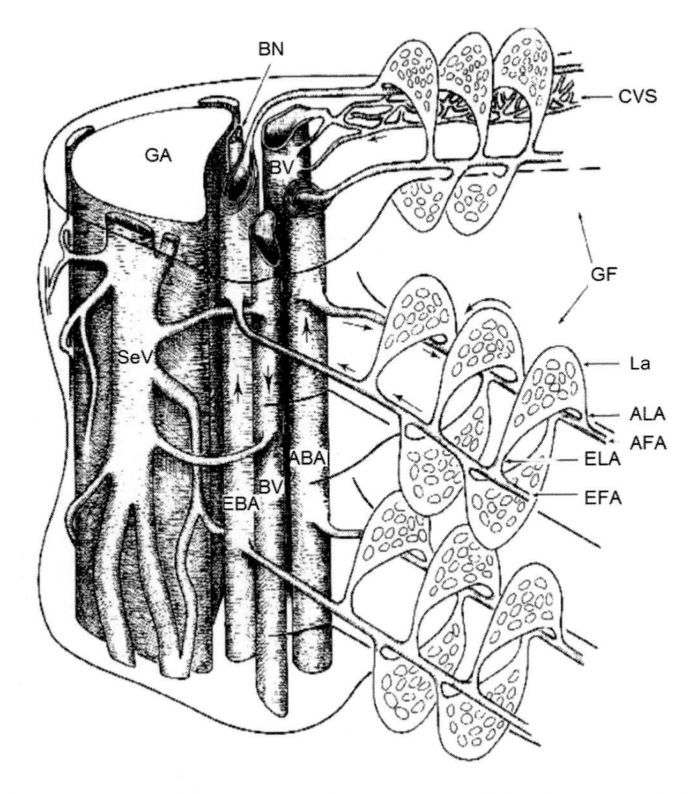

Fig. 4. Schematic diagram of the branchial microvasculature of the Atlantic cod, *Gadus morhua*. Abbreviations: *ABA*, afferent branchial artery; *AFA*, afferent filamental artery; *ALA*, afferent lamellar arteriole; *BN*, branchial nerve; *BV*, branchial veins; *CVS*, central venous system; *EBA*, efferent branchial artery; *EFA*, efferent filamental artery; *ELA*, efferent lamellar arteriole; *GA*, gill arch; *GF*, gill filaments; *La*, lamella; *SeV*, subepithelial vein. *Arrows* indicate the direction of blood flow. Filamental cartilage rods and nutritive vessels other than CVS are not shown. Modified from Sundin, L., Nilsson, S., 2002. Branchial innervation. J. Exp. Zool. 293, 232–248.

(Olson, 2002b). The blood leaves the lamellae via efferent lamellar arterioles that converge into efferent filamental arteries (EFAs). From this point, the majority of the blood flow proceeds into the so-called arterioarterial pathway; i.e., oxygenated blood enters the main systemic arterial circulation via the EBAs. This pathway is often referred to as the *respiratory pathway*. However, a smaller proportion of blood flow enters the "arteriovenous pathway" (Vogel et al., 1976; Vogel et al., 1974), which amounts to 7%–8% of $\dot{Q}$ in *O. mykiss* and *G. morhua* (Ishimatsu et al., 1988; Sundin and Nilsson, 1992). The arteriovenous pathway comprises: (i) short arteriovenous anastomoses off the EFAs; and (ii) arterioles originating from the EFA near the base of the filament that re-enter the filament to form the "nutritive" vasculature, before returning to the central venous circulation via the branchial veins (Olson, 2011a). While the arteriovenous pathway could potentially serve as a shunt for oxygenated blood directly to the lumen of the heart, the role of this pathway for cardiac oxygenation is, to the best of our knowledge, unknown. In some species, there are also prelamellar arteriovenous anastomoses whereby blood can bypass the respiratory epithelia (Olson, 2002b).

Several reviews have summarized the vascular anatomy of the gills in elasmobranchs (e.g., Butler and Metcalfe, 1988) and lampreys (e.g., Farrell, 2007a). While the overall branchial anatomy in these two groups is largely similar to the general teleost pattern outlined earlier, some important differences exist. First, the number of gill arches differ, which results in different numbers of paired branchial arteries. Elasmobranchs generally have five to seven gill arches, while lampreys have six. Another notable morphological difference is that the AFA of both elasmobranchs and lampreys is often expanded forming large *corpora cavernosa*, which essentially form a network of interconnected branchial vascular sinuses (Evans et al., 2005; Nilsson and Sundin, 1998). This structure has been ascribed several functions including: (i) to provide hydrostatic support for the gill filaments; (ii) to serve as a "windkessel," that smooths the blood pressure pulse and creates a more even blood flow through the lamellae; and (iii) a site of erythrocyte phagocytosis (Butler and Metcalfe, 1988; Evans et al., 2005). The vascular anatomy of the gills in hagfishes differs considerably from all other groups of fish (Evans et al., 2005; Farrell, 2007a; Forster, 1998). For example, hagfishes have 5–13 pairs of ABAs that perfuse specialized structures termed gill pouches where gas exchange takes place (Farrell, 2007a).

4.2. Control of Gill Blood Flow

Vasomotor control of branchial resistance and intrabranchial blood flow distribution is complex, takes place at both pre- and post-lamellar sites in the arterioarterial and arteriovenous pathways, and involves numerous vasoactive mediators including autonomic nerves, circulating hormones, and locally released factors (for reviews, see Jonz and Zaccone, 2009; Nilsson and Sundin, 1998; Olson and Donald, 2009; Sundin and Nilsson, 2002). A specific site with major importance for blood flow control in the gill is a sphincter region located at the base of the EFAs that receives dense innervation (Bailly and Dunel-Erb, 1986; Dunel-Erb and Bailly, 1986). There is also convincing evidence for an active regulation of the pillar cells within the gill lamellae, which alters intralamellar resistance and blood flow patterns (Jonz and Zaccone, 2009; Olson, 2002b; Stenslokken et al., 2006).

4.2.1. Neural Control and Circulating Catecholamines

In many teleosts, there is a well-developed spinal autonomic innervation of the gills (Donald, 1984, 1987). Indeed, catecholaminergic neurons project to the arteriovenous pathway and the sphincter region of the EFA where α-adrenoreceptors dominate, as well as to the AFA where β-adrenoreceptors dominate (Nilsson and Sundin, 1998). There is also recent evidence that adrenergic innervation allows for vasomotor control at the level of the lamellae (Jonz and Zaccone, 2009). In addition, circulating catecholamines have been suggested to be more important for controlling the gill circulation than the systemic circulation in teleosts (Nilsson and Sundin, 1998). Adrenergic stimulation (from either humorally or neurally released catecholamines) of the branchial vasculature typically favors blood flow in the respiratory arterioarterial pathway. This is mainly due to an increased vascular resistance in the arteriovenous circuit, whereas the overall resistance of the arterioarterial pathway may be reduced or unchanged (Nilsson, 1994; Sundin, 1995; Sundin and Nilsson, 1992). Consequently, this dilation of pre-lamellar vessels and constriction of post-lamellar vessels result in elevated perfusion pressure and a recruitment of perfused lamellae along the filaments (Jonz and Zaccone, 2009; Sundin and Nilsson, 2002). The teleost gill is also innervated by cholinergic cranial autonomic (vagal) vasomotor fibers, which primarily constrict the EFA sphincter by binding to muscarinic receptors (Bailly and Dunel-Erb, 1986; Jonz and Zaccone, 2009; Nilsson and Sundin, 1998). Hypoxia increases the branchial resistance *in vivo* (Holeton and Randall, 1967; Sundin and Nilsson, 1997) and *in vitro* (Pettersson and Johansen, 1982; Ristori and Laurent, 1977; Smith et al., 2001). This results, at least partly, from an increased cholinergic tone on the EFA sphincter, which can be greatly attenuated by atropine (Sundin and Nilsson, 1997).

Branchial vasomotor control from serotonergic (5-HT) nerves is also well documented in several teleost species (Sundin and Nilsson, 2002). These neurons have a post-ganglionic origin as serotonergic cell bodies are often located within the gills, which is consistent with a serotonergic cranial autonomic innervation. The 5-HT neurons primarily project to the proximal regions of the arterioarterial vasculature, including the EFA sphincter, and possibly the lamellae (Jonz and Nurse, 2003; Jonz and Zaccone, 2009; Sundin and Nilsson, 2002). Indeed, *in vivo* injection of 5-HT increases overall branchial resistance and ventral aortic blood pressure in various teleosts (Janvier et al., 1996; Sundin and Nilsson, 2000; Sundin et al., 1995, 1998). In *G. morhua*, this is primarily due to constriction of the distal efferent filamental vasculature in the respiratory pathway (Sundin et al., 1995). A serotonergic dilation of the arteriovenous pathway has also been documented in this species, which collectively leads to an increased distribution of branchial blood flow to the arteriovenous pathway (Sundin, 1995; Sundin and Nilsson, 1992).

Numerous studies suggest that an autonomic vasomotor control of the branchial vasculature is absent in elasmobranchs and cyclostomes, and that circulating hormones and local paracrine factors are more important for regulating the branchial vasculature in these groups (Brill and Lai, 2016; Taylor et al., 2009). Even so, neuropeptide-containing nerves appear to innervate the branchial vasculature of both elasmobranchs and teleosts, and so a role for neuropeptides in branchial vasomotor control cannot be excluded (Holmgren, 1995; Jonz and Zaccone, 2009). For example, nerve fibers immunoreactive for VIP and pituitary adenylate cyclase-activating polypeptide have been demonstrated in the gill vasculature of the Stinging catfish (*Heteropneustes fossilis*) (Zaccone et al., 2006), and VIP reduces the resistance of perfused brown trout (*Salmo trutta*) gills (Bolis et al., 1984). In the lesser spotted dogfish (*S. canicula*), the gill vasculature showed a dense neuropeptide Y (NPY)-like immunoreactivity, and isolated ABAs constrict in response to dogfish NPY (Bjenning et al., 1993).

4.2.2. Other Hormones and Local Factors

Blood flow in the fish gill is affected by several other hormones. For example, AVT increases overall gill resistance *in vivo* in *O. mykiss* and the American eel (*Anguilla rostrata*) (see Fig. 3; Conklin et al., 1997; Oudit and Butler, 1995b). However, at least in *O. mykiss*, it appears that the intrabranchial blood flow response to this hormone at physiologically relevant doses is to promote arterioarterial flow and reduce arteriovenous flow (Conklin et al., 1997). Similar to the effects on the systemic circulation, NPs dilate the branchial vasculature of teleosts, elasmobranchs and cyclostomes (Johnson et al., 2011; Olson et al., 1997a; Tait et al., 2009; Toop and Donald, 2004).

Various locally released and paracrine factors are also known to affect branchial blood flow. Adenosine constricts the branchial vasculature in both teleosts and elasmobranchs *in vivo* and has been suggested to favor arteriovenous perfusion during hypoxia (Colin et al., 1979; Stenslokken et al., 2004; Sundin and Nilsson, 1996; Sundin et al., 1999). Moreover, ET is a particularly potent paracrine compound and increases overall branchial resistance (Hoagland et al., 2000; Olson et al., 1991). This response is primarily due to constriction of lamellae pillar cells, which leads to a redistribution of blood flow to the marginal channels of the lamellae (Stenslokken et al., 1999, 2006; Sundin and Nilsson, 1998).

There is also recent evidence that intrabranchial blood flow patterns are affected by locally acting gasotransmitters (Olson and Donald, 2009). H_2S increases the resistance of perfused rainbow trout gills and is likely involved in hypoxic branchial vasoconstriction (Skovgaard and Olson, 2012). NO has consistent vasodilatory effects on the systemic circulation, but its effect on the branchial vasculature is variable, with dilation, constriction, and no response all being reported (see Donald et al., 2015). Indeed, NO donors had no effect on *in vivo* branchial resistance in *O. mykiss* (see Fig. 3; Olson et al., 1997a), while a dose-dependent constriction was observed in perfused gill preparations of the teleost *Anguilla anguilla* and the elasmobranch *S. acanthias* (Pellegrino et al., 2002, 2005).

5. THE VENOUS VASCULATURE

The most basal function of the veins is to serve as conduits for the blood back to the heart after it has passed through the capillary networks. However, the venous vasculature also serves as a high capacitance blood volume reservoir that can be actively and passively mobilized when needed. While arterial hemodynamics and neuroendocrine control have been studied in all major groups of fish, the venous side of the circulatory system has received much less attention. Even so, in recent years, several important advances have been made. It is now well accepted that active control of the venous circulation by both neural and hormonal systems is an integrated and important factor in the overall hemodynamics of fishes and constitutes a prerequisite for the heart's ability to generate blood flow. Most available literature on venous function in fishes still concerns *O. mykiss*, with only a few scattered studies in other species of fish. The comparative aspects of the control and function of the venous side of the fish circulatory system, therefore, represent a largely open field of research. Some interesting areas for future inquiries are discussed later.

Early summaries of venous function in fish can be found in Olson (1992) and Satchell (1991, 1992), while more recent accounts covering neuroendocrine control of venous capacitance and the role of the venous system in integrated cardiovascular challenges are provided by Olson (2011d) and Sandblom and Axelsson (2007a).

5.1. Functional Characteristics and Gravity Effects

As the vertebrate circulatory system forms a closed loop, the blood returning from the veins to the heart (i.e., venous return) must be equal to $\dot{Q}$ when averaged over time. In mammals, all of the blood leaving the heart via the aorta passes through the systemic capillary beds before flowing back to the central venous compartment via the peripheral veins (Guyton, 1963). In fish, however, nearly one-tenth of $\dot{Q}$ flows directly from the efferent side of the gills to the central venous circulation via arteriovenous pathways (see Section 4.1; Ishimatsu et al., 1988; Jonz and Zaccone, 2009; Nilsson and Sundin, 1998). This means that if $\dot{Q}$, as typically measured in the ventral aorta, is used to estimate venous return this will provide an overestimate of the systemic venous return in fish. The venous vasculature of mammals contains approximately 70%–80% of the total circulating blood volume, with most of the blood being contained in the small veins and venules (Funk et al., 2013). In fish, the volume relationship between the arterial and the venous compartments is unknown, yet the venous vasculature is generally considered to represent the main blood volume reservoir in fish (Gamperl and Shiels, 2014; Olson, 1992; Olson and Farrell, 2006; Sandblom and Axelsson, 2007a). Indeed, some ancient groups of fishes have unusually capacious venous vasculatures. Examples of this are the large central venous sinuses of elasmobranchs (Satchell, 1991) and the enigmatic subcutaneous venous sinuses of hagfishes (Forster, 1997).

Gravity has a significant impact on the blood in the circulatory system of terrestrial animals, with limbs below the heart being subject to orthostatic blood pooling. This is because, in air, the venous transmural pressure increases as a function of the distance of the blood column below the heart (Fig. 5). Numerous mechanisms to counter this gravitational impact, which may lead to the formation of edema, are therefore important in terrestrial animals. These mechanisms include: (i) the "skeletal muscle pump," which causes rhythmic compression of veins during muscular work; (ii) valves that prevent retrograde flow of venous blood and decrease the effective height of the blood column; and (iii) active and passive changes in venous vascular capacitance (Pang, 2001). In contrast, the gravitational impact on the cardiovascular system of fish is negligible (Sandblom and Axelsson, 2007a; Satchell, 1991, 1992), because blood and water have a very similar density. This means that the hydrostatic water pressure cancels out the gravitational forces acting on the blood (Fig. 5). A possible exception, however, may be in fishes with an amphibious lifestyle (e.g., amphibious mudskippers and aestivating lungfish). Studies on venous function in such taxa represent an interesting area for future research. Similar to terrestrial mammals, valves are present in fish veins, but the main difference appears to be that they are only located at the junction of

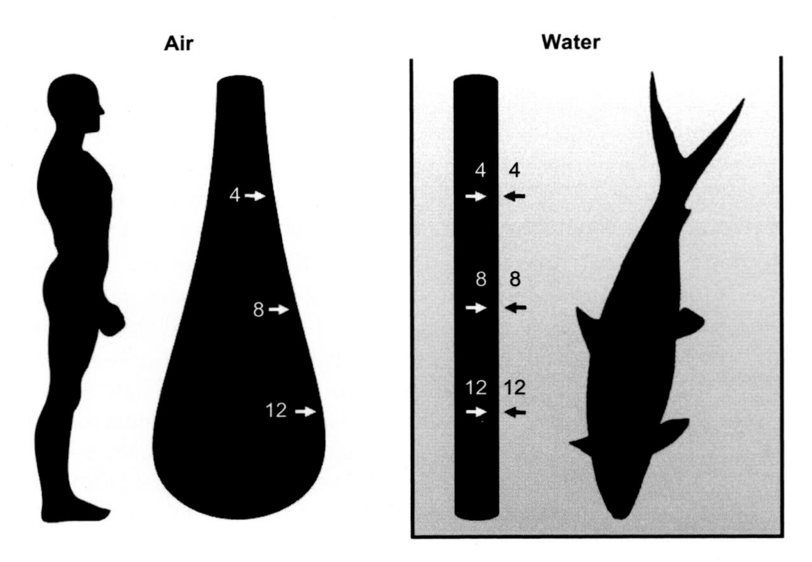

Fig. 5. Theoretical model of gravitational effects on blood vessels in the cardiovascular system. In air, pressure increases with the height of the blood column and the blood tends to pool and distend the lower parts of the vasculature (e.g., in dependent limbs). In water, the increased hydrostatic pressure with increasing water depth cancels out gravity effects on the blood column. Modified from Satchell, G.H., 1991. Physiology and Form of Fish Circulation. Cambridge University Press.

tributary vessels (i.e., they are ostial valves), while in mammals there are also valves along the length of the vessels (i.e., parietal valves). This likely represents a circulatory adaptation to tolerate the added effects of gravity in terrestrial habitats (Satchell, 1991).

5.2. Venous Hemodynamic Principles

While the mean arterial blood pressure is determined by $\dot{Q}$ and vascular resistance (as described in Section 3.1.1), the factors determining venous blood pressure and venous return are less intuitive. Here, we summarize the main hemodynamic principles of the venous circulation.

5.2.1. Central Venous Blood Pressure and Importance for Cardiac Performance

Venous blood pressure in fish is similar to, or lower than, that in other vertebrates (Table 2). In fact, in some elasmobranchs and cyclostomes, P_{CV} can be subambient (see Table 2). Within the physiological pressure range, P_{CV} is the main determinant of cardiac filling and, therefore, has important

Table 2
In vivo venous blood pressures (kpa) in various groups of fishes and other representative vertebrates

	SV/DC PCS/VC	PCV	CdV	HPV/SIV PV	Notes	Source
Cyclostomes						
Pacific lamprey, *Entosphenus tridentatus*	−0.40 to −0.10					Johansen et al. (1973)
Broadgilled hagfish, *Eptatretus cirrhatus*		0.06				Foster and Forster (2007a)
—		0.03		0.07		Foster and Forster (2007b)
—				0.04		Johnsson et al. (1996)
Atlantic hagfish, *Myxine glutinosa*		0.10				Johnsson et al. (1996)
Elasmobranchs						
Lesser spotted dogfish, *Scyliorhinus canicula*	−0.45					Short et al. (1977)
Picked dogfish, *Squalus acanthias*	−0.10 to 0.00					Johansen and Hanson (1967)
—	−0.08 to −0.07					Sandblom et al. (2006a)
—	−0.08 to −0.06					Sandblom et al. (2009b)
—	~−0.2		0.1 to 0.4		Fish fixated	Capra and Satchell (1977)
Leopard shark, *Triakis semifasciata*	0.20–0.26				Pericardium cannulated	Lai et al. (1990)
Air-breathing fishes						
Marbled lungfish, *Protopterus aethiopicus*	~0.0–0.4					Johansen et al. (1968)
Marbled swamp eel, *Synbranchus marmoratus*	~0.25–0.47					Skals et al. (2006)
Water-breathing teleosts						
Japanese eel, *Anguilla japonica*	0.13	0.12				Chan and Chow (1976)
European sea bass, *Dicentrarchus labrax*	0.11					Sandblom et al. (2005)
Rainbow trout, *Oncorhynchus mykiss*	0.2–0.47				Pericardium opened	Conklin et al. (1997); Hoagland et al. (2000); Olson et al. (1997a); Olson and Hoagland (2008); Perry et al. (1999); Zhang et al. (1998)
—	−0.06 to 0.07					Altimiras and Axelsson (2004); Sandblom and Axelsson (2005a); Sandblom and Axelsson (2006); Sandblom and Axelsson (2007b); Sandblom et al. (2006b); Seth et al. (2008)
—		0.19				Kiceniuk and Jones (1977)
—			0.55		Pericardium opened	Wood and Shelton (1980)
—				1.06–1.20		Stevens and Randall (1967)
Chinook salmon, *Oncorhynchus tshawytscha*	0.01					Clark et al. (2008)

(*Continued*)

Table 2 (Continued)

	SV/DC PCS/VC	PCV	CdV	HPV/SIV PV	Notes	Source
Bald notothen, *Pagothenia borchgrevinki*	0.08–0.09					Sandblom et al. (2008)
—	0.07–0.11					Sandblom et al. (2009a)
Starry flounder, *Platichthys stellatus*			0.28			Wood et al. (1979)
Winter flounder, *Pseudopleuronectes americanus*			0.40–0.53			Cech et al. (1976)
—			0.40			Cech et al. (1977)
Other vertebrates						
Cane toad, *Rhinella marina*	0.52					Withers et al. (1988)
South American rattlesnake, *Crotalus durissus*	~0.4					Skals et al. (2005)
Savannah monitor, *Varanus exanthematicus*	0.57					Munns et al. (2004)
Brown rat, *Rattus norvegicus*	0.28			0.85		Tabrizchi et al. (1993)
Man, *Homo sapiens*	0.20			1.1		Pang (2001)

Values are obtained from unanesthetized and unrestrained animals under conditions described as "resting" or similar. The pericardium is assumed to be intact unless otherwise stated. Abbreviations: *CdV*, caudal vein; *DC*, Ducts of Cuvier; *HPV*, hepatic portal vein; *PCS*, posterior cardinal sinus; *PCV*, posterior cardinal vein; *PV* (in *Protopterus*), pulmonary vein; *SIV*, supraintestinal vein; *SV*, sinus venosus; *VC*, vena cava.

implications for stroke volume (V_S) and $\dot{Q}$. At a given heart rate (f_H), an increase in cardiac filling pressure increases V_S by increasing the end-diastolic volume and the contraction force of the myocardium through the Frank–Starling mechanism. This mechanism relies on the intrinsic property of myocardial tissue where increased myocardial stretch (i.e., cardiac preload) leads to increased myocardial contraction force (Farrell and Jones, 1992; Olson and Farrell, 2006). Indeed, studies using *in situ* perfused heart preparations where the (venous) filling pressure can be manipulated by the experimenter and the resultant change in cardiac performance recorded have verified the stimulatory effects of increases in filling pressure on cardiac performance in all major groups of fish (see Sandblom and Axelsson, 2007a). Several studies have reported that the ejection fraction of both elasmobranch and teleost hearts is comparatively high, typically around 80%–100% (Coucelo et al., 2000; Franklin and Davie, 1992; Lai et al., 1990). This implies that, unlike in mammals, the capacity to increase V_S through reductions in end-systolic volume may be relatively limited for most fishes, and that cardiac filling pressure may be of added importance for determining V_S in fish (Coucelo et al., 2000; Sandblom and Axelsson, 2007a). However, there are undoubtedly exceptions to this pattern. For example, the bald notothen *P. borchgrevinki* exhibits a three-fold increase in $\dot{Q}$ and a doubling of V_S after being chased (Franklin et al., 2007), but the same exercise/stress protocol only results in a very small (0.02 kPa) increase in P_{CV}. This indicates that the pronounced increase in V_S may instead be due to reduced end-systolic volume (Sandblom et al., 2008). While both atrial and ventricular muscles in fish exhibit a pronounced Frank–Starling mechanism (Farrell and Jones, 1992), atrial work is likely more directly affected by the P_{CV} *in vivo* because the only way of filling this chamber is through direct inflow from the central veins, whereas ventricular filling is due to a combination of direct venous inflow and atrial contraction (Lai et al., 1998; Olson and Farrell, 2006; see Chapter 4, Volume 36A: Farrell and Smith, 2017). Nonetheless, changes in P_{CV} that affect atrial filling and contraction performance will translate into altered filling and performance of the ventricle (Farrell, 1984; Lai et al., 1990, 1998).

Although $\dot{Q}$ is mechanistically linked to P_{CV} through the Frank–Starling mechanism, it is important to consider that the activity of the heart itself affects P_{CV} and cardiac filling *in vivo*. This is primarily due to changes in f_H that affect cardiac filling time, and a phenomenon known as *vis-á-fronte* (cardiac suction) filling (see below). For example, decreases in f_H prolong filling time and lead to pooling of venous blood in the central veins, which in turn increases P_{CV} and V_S. Altimiras and Axelsson (2004) elegantly demonstrated this mechanism by showing that $\dot{Q}$ in *O. mykiss* was maintained relatively constant through compensatory changes in P_{CV} and V_S when the cardiac contraction frequency was pharmacologically manipulated *in vivo* using the bradycardic agent zatebradine (Fig. 6). This also means that increases in f_H reduce diastolic filling time, which in turn may reduce P_{CV} and V_S unless there are vascular (venous) compensatory adjustments. For example, during situations when $\dot{Q}$ increases through tachycardia (e.g., exercise and warming), there are often profound reductions in venous capacitance (e.g., increased venous tone). This mobilizes blood toward the central venous compartment to maintain or increase P_{CV} and V_S. We will return to the implications of this mechanism in Section 5.4.1.

The flow of blood into the cardiac chambers is mainly determined by the pressure difference between the central veins and the cardiac chambers. As the heart is positioned in a more or less rigid pericardial cavity, ventricular contraction leads to a reduction in the intrapericardial pressure, and this distends the atrium and enhances the flow of blood into the heart. This filling mechanism is referred to as *vis-á-fronte* ("force from in front"), while the remaining kinetic and potential energy that creates the upstream pressure in the central veins is termed *vis-á-tergo* ("force from behind") filling (Satchell, 1991). The hydraulic coupling between ventricular contraction and cardiac filling through the *vis-á-fronte* mechanism is particularly pronounced in some

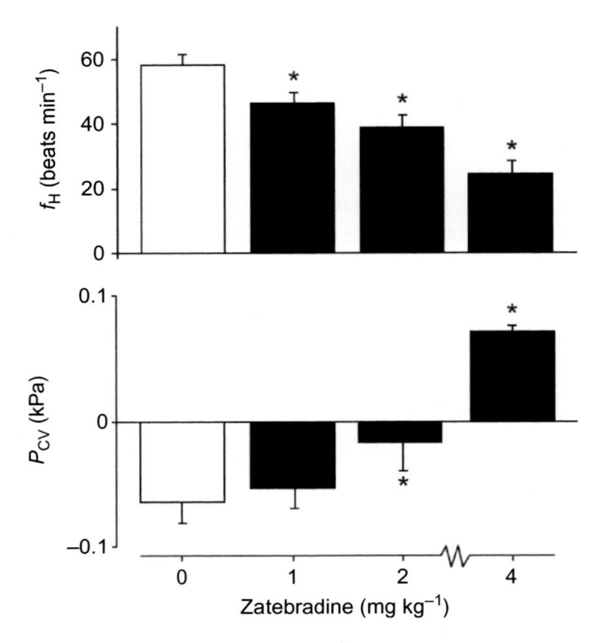

Fig. 6. Effect of changes in heart rate (f_H) on central venous blood pressure (P_{CV}) in rainbow trout (*Oncorhynchus mykiss*). *Open bars* represent untreated (resting) conditions and *filled bars* represent increasing dosages of the bradycardic agent zatebradine. *Asterisks* denote a significant ($P \leq 0.05$) change from the untreated condition. Modified from Altimiras, J., Axelsson, M., 2004. Intrinsic autoregulation of cardiac output in rainbow trout (*Oncorhynchus mykiss*) at different heart rates. J. Exp. Biol. 207, 195–201.

elasmobranchs that have a very rigid pericardial cavity and normally exhibit relatively strong subambient routine venous pressures (see Table 2; Fig. 7). Even so, this mechanism is undoubtedly also important for cardiac filling in species that may not routinely exhibit subambient venous pressures. Indeed, the Starling curve in *O. mykiss* is right-shifted when the pericardium is opened (Farrell et al., 1988), and *in vivo* P_{CV} typically attains positive values in elasmobranchs (Fig. 7) or becomes more positive in species such as *O. mykiss* that normally exhibit slightly positive P_{CV}, if the integrity of the pericardium is surgically destroyed (i.e., a pericardioectomy is performed) (Sandblom and Axelsson, 2006, 2007a). The mean circulatory filling pressure (MCFP, see section 5.2.2) and the SBV typically increase after pericardioectomy, which suggests that such increases in P_{CV} are, at least in part, due to active reductions in venous capacitance in both elasmobranchs and teleosts (Sandblom and Axelsson, 2007a; Sandblom et al., 2006a, 2009b).

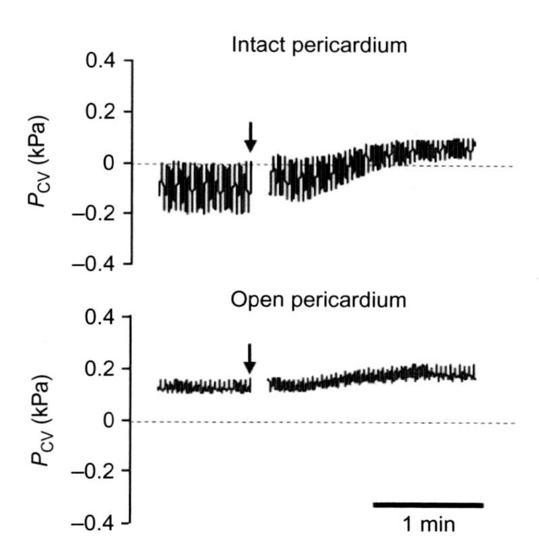

Fig. 7. Effects of pericardioectomy and intraarterial adrenaline injection (5 μg kg^{-1} body mass, at *vertical arrow*) on central venous blood pressure (P_{CV}) in the elasmobranch *Squalus acanthias.* P_{CV} is strongly pulsatile and subambient in fish with an intact pericardium (*upper panel*), whereas pressure is less pulsatile and positive in fish with opened pericardium (*lower panel*). The hypertensive response to adrenaline is less pronounced in fish with an open pericardium. Modified from Sandblom, E., Axelsson, M., Farrell, A.P., 2006. Central venous pressure and mean circulatory filling pressure in the dogfish *Squalus acanthias*: adrenergic control and role of the pericardium. Am. J. Physiol. 291, R1465–R1473.

5.2.2. Venous Capacitance, Mean Circulatory Filling Pressure, and Venous Resistance

Vascular capacitance reflects the pressure–volume relationship in the circulatory system. In the mammalian circulatory system, *in vivo* vascular capacitance is primarily determined by the capacitance of the venous system. This is because approximately 70%–80% of the total blood volume resides in the veins, with the majority of this volume being contained in the highly compliant peripheral small veins and venules (Pang, 2001; Rothe, 1993). The veins in the gastrointestinal circulation (i.e., liver, spleen, and small and large intestine) may be particularly important in this regard as they contain around one-fourth of the total blood volume, have a very high compliance (*C*), and are highly reactive to vasoactive substances and baroreflex stimulation (Haase and Shoukas, 1991, 1992; Shoukas and Bohlen, 1990). In fish, the blood volume distribution between arteries and veins and the potential role of the gastrointestinal circulation as a high capacitance blood volume reservoir are not known. However, it is generally accepted that the venous circulation also has

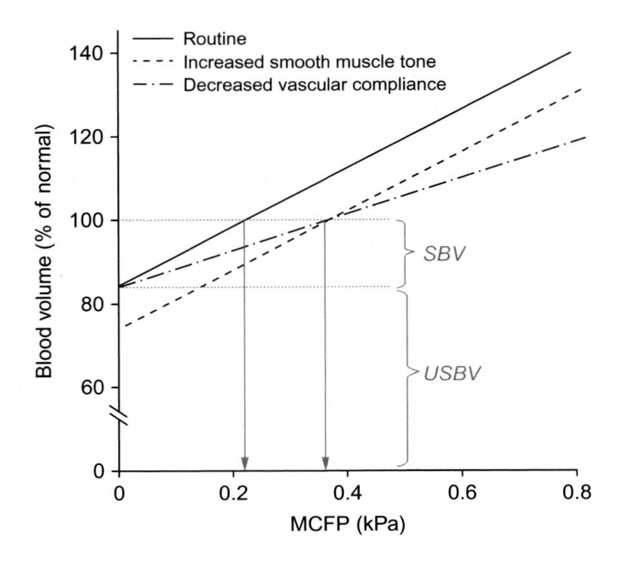

Fig. 8. Schematic diagram of vascular capacitance curves. The slope of the curve represents the vascular compliance and the y-axis intercept is the unstressed blood volume (USBV). The remainder of the blood volume that creates intravascular pressure is the stressed blood volume (SBV). As demonstrated by the two *vertical arrows*, the mean circulatory filling pressure (MCFP) at normal blood volume (i.e., at 100%) can increase either through increased smooth muscle (vascular) tone, which reduces the USBV and causes a rightward parallel displacement of the capacitance curve (*hatched line*), or through a decrease in vascular compliance which rotates the curve clockwise (*dotted line*). Modified from Sandblom, E., Axelsson, M., 2007. The venous circulation: a piscine perspective. Comp. Biochem. Physiol. A. 148, 785–801.

an important capacitance function in the piscine circulatory system (Olson and Farrell, 2006; Sandblom and Axelsson, 2007a).

Venous capacitance is determined by venous compliance and smooth muscle (vascular) tone and can be described graphically by plotting a vascular capacitance curve (see Fig. 8). The slope of the capacitance curve represents C, which is the ratio of the change in volume (ΔV) that results from a change in blood pressure (ΔP) according to the formula:

$$C = \frac{\Delta V}{\Delta P}$$

Thus, blood vessels (or vascular beds) where a large increase in blood volume only results in a small pressure increase have a high compliance. The compliance of large conducting veins and arteries from *O. mykiss* has been determined, with the veins having a compliance that is approximately

20- to 30-fold greater than that of the arteries (Conklin and Olson, 1994b). This arteriovenous compliance ratio is comparable to that in mammals (Funk et al., 2013) and is consistent with the view that the venous vasculature has an important capacitance function in the fish cardiovascular system. The intercept of the capacitance curve with the *y*-axis (i.e., at zero blood pressure) is the unstressed blood volume (USBV), while the remaining portion of total blood volume (i.e., the volume above the USBV along the *y*-axis) is the SBV. The USBV is the proportion of the total blood volume that fills the vascular space without stretching the vessel walls and creating pressure, whereas the SBV is the hemodynamically active part of total blood volume that dilates the vessel walls and creates pressure. Increased venous smooth muscle tone typically leads to a parallel downward displacement of the capacitance curve (see Fig. 8). This means that blood in the unstressed compartment is mobilized into the stressed compartment, which results in greater pressure at a given blood volume. Nonetheless, venous smooth muscle tone and compliance can be independently altered (Sandblom and Axelsson, 2007a).

Changes in venous capacitance can be determined by measuring the MCFP, which is the average pressure in the circulation when the blood flow is stopped. In practice, MCFP can be measured by transiently stopping $\dot{Q}$ by inflating a surgically implanted vascular occluder and then recording the central venous plateau pressure (see Fig. 9). The measurement of MCFP *in vivo* should ideally be accomplished relatively fast (typically within ~10 s) to avoid barostatic responses that will result in a reflex constriction of the vasculature and lead to an overestimate of MCFP (for further discussions on this methodology, see Sandblom and Axelsson, 2007a). Although MCFP provides an estimate of venous capacitance at constant blood volume, it is generally not possible to determine if a change in capacitance is due to altered USBV, *C*, or some combination of these factors. This information can only be obtained by measuring MCFP at different blood volumes and by constructing *in vivo* vascular capacitance curves where MCFP is plotted against total blood volume, as discussed earlier. Such experiments are technically challenging for a number of reasons. First, the blood volume is typically estimated rather than actually measured, and any change in blood volume from the experimental or pharmacological treatment protocol will affect MCFP and the shape of the capacitance curve. Second, it is very challenging to rapidly change the blood volume of a live animal. Typically, large-bore catheters are required to quickly withdraw or infuse blood (i.e., from a donor fish) to minimize the effects of compensatory reflex responses affecting venous tone. So far, *in vivo* vascular capacitance metrics have only been determined for *O. mykiss*, with reported routine USBV ranging between 13.3 and 26.0 mL kg M_b^{-1} and vascular compliances ranging from 12.8 to 25.5 mL kPa^{-1} kg M_b^{-1} (Conklin et al., 1997;

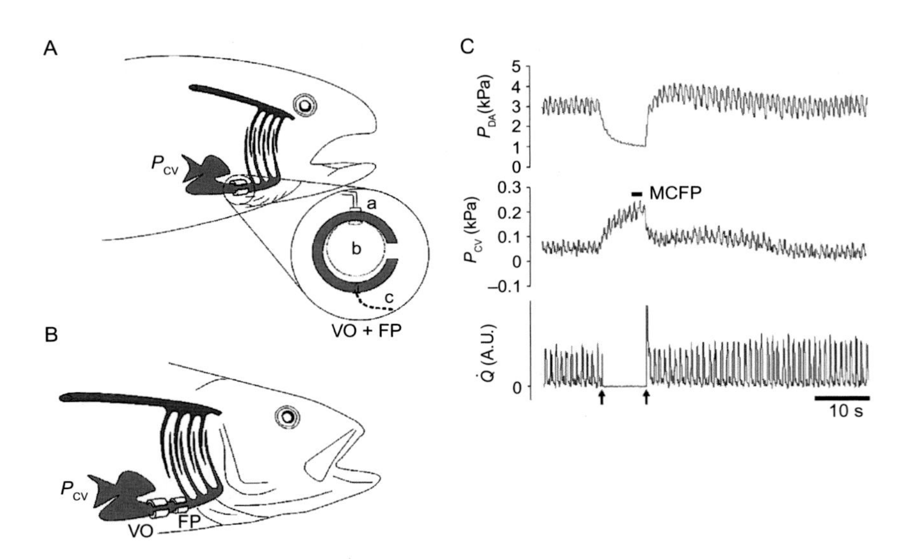

Fig. 9. Schematic illustration of (A) rainbow trout (*Oncorhynchus mykiss*) and (B) European sea bass (*Dicentrarchus labrax*) showing the placement of a perivascular blood flow probe (FP) and a vascular occluder (VO) for measurements of mean circulatory filling pressure (MCFP). Due to the limited space in *O. mykiss*, a combined flow probe and vascular occluder is used. A cross section of the probe used in trout is magnified to illustrate the PE-50 catheter (a) which is connected to an inflatable latex balloon (b) and Doppler flow crystal with lead (c). Due to the different morphology of the vasculature and opercular cavity in the sea bass, a separate occluder and flow probe can be placed in series on the same vessel. The central venous pressure (P_{CV}) is measured with a catheter in the *Ducts of Cuvier* in both species. (C) Representative raw traces of dorsal aortic pressure (P_{DA}), P_{CV}, and cardiac output ($\dot{Q}$) during ventral aorta occlusion (between *vertical arrows*) during an MCFP measurement in *O. mykiss.* Modified from Sandblom, E., Axelsson, M., 2005. Baroreflex mediated control of heart rate and vascular capacitance in trout. J. Exp. Biol. 208, 821–829; Sandblom, E., Farrell, A.P., Altimiras, J., Axelsson, M., Claireaux, G., 2005. Cardiac preload and venous return in swimming sea bass (*Dicentrarchus labrax* L.). J. Exp. Biol. 208, 1927–1935.

Hoagland et al., 2000; Olson et al., 1997a; Sandblom and Axelsson, 2006; Zhang et al., 1995, 1998).

Since the activity of the beating heart increases the pressure in the arterial circulation, while it tends to reduce P_{CV}, MCFP is lower than arterial and capillary pressure, but higher than P_{CV}. In mammals, MCFP is considered to be an accurate estimate of the upstream venous pressure in the peripheral small veins and venules (Guyton et al., 1955; Pang, 2001; Rothe, 1993). The pressure gradient that drives venous return (ΔP_V) can then be calculated as:

$$\Delta P_V = \text{MCFP} - P_{CV}$$

Assuming that $\dot{Q}$ is equivalent to systemic venous return (which may be a slight overestimate for fish as outlined earlier), or if venous return has been directly measured, the resistance to venous return (RVR) can be calculated as follows:

$$\mathrm{RVR} = \frac{(\mathrm{MCFP} - P_{\mathrm{CV}})}{\text{Venous return}}$$

Indeed, using $\dot{Q}$ as a proxy for venous return, Zhang et al. (1995) estimated that RVR accounts for only ~2% of the total systemic vascular resistance in *O. mykiss.*

It is important to consider that changes in venous capacitance have different hemodynamic functions and consequences depending on the associated changes in cardiac activity and peripheral vascular resistance. For example, if everything else is constant, a reduced venous capacitance (e.g., reduced USBV) will elevate MCFP and lead to a mobilization of blood from the venous periphery to the central venous compartment, which in turn will increase P_{CV}, V_{S}, and $\dot{Q}$ via the Frank–Starling mechanism. However, if the f_{H} increases simultaneously with a reduction in the venous capacitance, as often occurs during physiological states requiring elevated blood convection (e.g., exercise), the increased activity of the heart may reduce P_{CV} despite a reduced venous capacitance (Sandblom and Axelsson, 2007a; Sheriff et al., 1993). In this situation, the role of the reduced venous capacitance may instead be to increase the venous pressure gradient to match venous return to $\dot{Q}$. In fact, if venous capacitance did not decrease in concert with elevated f_{H}, the ability of the heart to increase $\dot{Q}$ would be severely compromised as cardiac filling and V_{S} would decline.

5.2.3. Passive Venous Changes Affecting Tissue Blood Volume

Another important consideration with regard to changes in venous blood volume is that the blood volume within compliant veins is passively affected by changes in flow rate (Sandblom and Axelsson, 2007a; Satchell, 1992). This effect can be understood by applying Poiseuille's law (see Section 3.1.2), which essentially states that the pressure difference along a vascular segment will increase when flow increases. For example, when $\dot{Q}$ and the overall blood flow in the vascular system increases, a decreased venous capacitance may be important to prevent passive distension and blood pooling in the venous periphery and to maintain cardiac filling pressure. Similarly, if the upstream arterial resistance of a vascular bed increases, the pressure in the downstream venous vasculature will decrease and cause the compliant peripheral venous vessels to passively recoil. The contained blood volume will then decrease and be transferred to other parts of the circulation, including the central veins

(Hainsworth, 1986; Rothe, 2006). This mechanism has been proposed to be important for mobilizing blood from the peripheral microcirculation to the central circulation during hemorrhage (Olson et al., 2003). It may also be important for transferring blood from the visceral circulation to increase or maintain P_{CV} during exercise and hypoxia, both of which are conditions that include considerable reductions in the arterial flow to the gastrointestinal tract (Altimiras et al., 2008; Axelsson and Fritsche, 1991; Behrens et al., 2012; Eliason and Farrell, 2014; Gräns et al., 2009).

5.3. Neuroendocrine Control of Venous Capacitance

Venous capacitance in fishes is actively controlled by a range of neuroendocrine factors, and in this section, we begin by summarizing the main control systems affecting venous compliance and tone. We then provide examples of the diversity of venous capacitance responses and their control during various integrated cardiovascular responses and reflexes.

5.3.1. Neural Control Systems and Circulating Catecholamines

Catecholaminergic control systems (primarily adrenaline and noradrenaline) are by far the most extensively studied with regard to venous function in fish, yet available information is restricted to a handful species, and even then it can be conflicting. Nonetheless, there is evidence for venous vasomotor control involving both α-adrenergic constriction and β-adrenergic dilation in all major groups of fish (Sandblom and Axelsson, 2007a).

Some early studies on teleosts reported changes in P_{CV} following the injection of catecholamines or other adrenergic drugs. For example, pressure in the cardinal vein and sinus venosus of *A. japonica* increased in a dose-dependent manner in response to both adrenaline and noradrenaline and decreased in response to the β-adrenoreceptor agonist isoprenaline (Chan and Chow, 1976). Similarly, caudal venous pressure showed a dose-dependent increase in *O. mykiss* following adrenaline and noradrenaline injection, whereas both the β-adrenoreceptor agonist isoproterenol and the α-adrenoreceptor agonist phenylephrine had few or variable effects (Wood and Shelton, 1980). A limitation of these earlier studies was that it was difficult to determine if the changes in venous pressures were due to direct venospecific changes (i.e., changes in venous capacitance) or were secondary to changes in cardiac function or passive effects of changes in arteriolar resistance and regional blood flows.

More recent studies have characterized the vessel-specific effects of catecholamines on isolated large conducting veins, and several studies have assessed the venous capacitance responses to catecholamines in various teleost species. For example, adrenaline and noradrenaline increase tension in isolated segments of large veins from *O. mykiss*, while C is unaffected (Conklin and Olson, 1994b). These effects were blocked by the α-adrenoreceptor antagonist phentolamine, but unaffected by propranolol, indicating a dominant role for α-adrenoreceptors. Similarly, isolated vein segments from Gar, *Lepisosteous* spp., constrict in response to adrenaline (Conklin et al., 1996). *In vivo* injection of adrenaline and phenylephrine increase P_{CV} and MCFP in the swamp eel (*Synbranchus marmoratus*). This latter result indicates that the venous vasculature constricts through an α-adrenergic mechanism, and that this mobilizes blood to the central venous compartment and raises P_{CV}. Isoproterenol injection *in vivo* significantly reduced P_{CV}, but only a weak trend for decreased MCFP was observed (Skals et al., 2006). The Antarctic bald notothen (*P. borchgrevinki*) exhibits similar, but smaller, increases in both P_{CV} and MCFP in response to injection with adrenaline and phenylephrine (Sandblom et al., 2009a). We are only aware of one study on fish that has examined the effects of exogenous catecholamines on USBV and vascular compliance *in vivo* (Zhang et al., 1998). That study reported that adrenaline infusion increased P_{CV} and decreased venous capacitance through reductions of both USBV and C in *O. mykiss*. Interestingly, infusion of noradrenaline at doses up to 10.4 nmol min^{-1} kg M_b^{-1} did not affect P_{CV} or any indices of venous capacitance. Moreover, Johnsson et al. (2001) showed that a range of neuropeptide-containing perivascular nerves were associated with large veins in *G. morhua* and *O. mykiss*. The functional significance of this putative neural control of the venous vasculature remains to be explored.

In elasmobranchs, venous capacitance is also controlled by adrenergic mechanisms. *In vivo* injections of adrenaline and phenylephrine increase P_{CV} and MCFP in *S. acanthias*, while isoproterenol decreases the same variables (Sandblom et al., 2006a). However, the functional significance of this *in vivo* vasomotor control in elasmobranchs is less clear. For example, exposure of *S. acanthias* to 30 min of severe hypoxia (water oxygen tension = 2.5 kPa) caused significant increases in plasma adrenaline (from 9 to 57 nM) and noradrenaline (from 2 to 41 nM), but these elevated endogenous catecholamine levels were ineffective in producing a significant increase in venous tone as indicated by an unchanged MCFP (Sandblom et al., 2009b).

Despite the unusually large blood volume (~18% of body mass) and the large venous sinuses of hagfishes (Farrell, 2007a; Forster, 1997; Forster et al., 2001), there is very limited information on venous function in cyclostomes, and we are not aware of any study examining the control of *in vivo* venous capacitance in this group. Nonetheless, Foster et al. (2008) performed detailed examinations of isolated vessel responses to adrenergic agonists in the New Zealand hagfish (*Eptatretus cirrhatus*) and reported that conducting veins constricted in response to adrenaline at low concentrations, whereas partial relaxation was observed at higher concentrations. This was suggested to result from the differential stimulation of α- and β-adrenoreceptors because phenylephrine caused a clear dose-dependent constriction and the application of isoprenaline resulted in a dose-dependent relaxation of all veins studied. Thus, it may not be surprising that the *in vivo* injection of noradrenaline causes a pressor response in the PCV of *E. cirrhatus* (Foster and Forster, 2007a). Importantly, the results from elasmobranchs and hagfish indicate that adrenergic vasomotor control of the venous vasculature probably evolved early in the vertebrate lineage.

There is still uncertainty whether an adrenergic tonus acts on the venous vasculature under routine conditions. Experiments using the α-adrenergic antagonist prazosin in *O. mykiss*, the European sea bass (*Dicentrarchus labrax*), and the Antarctic fish *P. borchgrevinki* have not affected resting MCFP, whereas P_{CV} increases (Sandblom and Axelsson, 2005a, 2006; Sandblom et al., 2005, 2006b, 2008, 2009a). Further, the blockade was certainly effective in these studies, as the venous responses to exercise and transient hypoxia were abolished or strongly attenuated after prazosin treatment. Similarly, while a neural adrenergic tonus on the arterial resistance vasculature has been demonstrated in several teleost species (see Section 3.2.2.1), adrenergic nerve blockade with bretylium has no effect on routine P_{CV}, MCFP, SBV, or C in *O. mykiss* (Sandblom and Axelsson, 2006). These results contrast with the pioneering study on catecholaminergic control of venous capacitance by Zhang and co-workers, where P_{CV} and SBV decreased in *O. mykiss* following prazosin treatment, indicating the release of a venous α-adrenergic vasomotor tone (Zhang et al., 1998). The reasons for these discrepancies are not clear, but it has been suggested that time-dependent increases in circulating blood volume caused by reduced capillary filtration pressure after α-adrenergic blockade may explain these conflicting results (Sandblom and Axelsson, 2007a). Importantly, the study by Zhang and co-workers assessed venous capacitance within 40 min after administration of the blocker, whereas more than 1.5 h was allowed before recordings began in the other studies. Thus, in the longer protocols, it is possible that the fish's blood volume had increased, and masked any venodilation, leaving MCFP unaffected. It also appears likely that the opening of the pericardium in earlier studies may have led to an elevated routine adrenergic tone on the venous vasculature, and therefore, a greater vascular dilation following α-adrenergic blockade. A limitation in all of these studies, however, is that blood volume was not directly determined. Blood volume measurements after adrenergic receptor blockade would provide valuable information and help to resolve these questions.

Another unresolved question regarding catecholaminergic control of the venous capacitance vasculature relates to the relative importance of humorally vs neurally released catecholamines. While we are not aware of any study that has examined the pattern of innervation of the venous microcirculation in fishes, early studies using Falck–Hillarp fluorescence histochemistry provided evidence for catecholamine-containing perivascular nerves surrounding large veins in various fish species (Holmgren, 1977; Nilsson, 1983). During moderate hypoxia (water oxygen tension: $\sim 9\,\text{kPa}$) exposure in *O. mykiss*, both circulating and neurally released catecholamines appear to contribute to the active reduction of venous capacitance because adrenergic nerve blockade with bretylium partially abolished venous capacitance changes, whereas the general α-adrenoreceptor blockade with prazosin nearly completely eliminated all responses (Sandblom and Axelsson, 2006). Indeed, the rapidity (within 10–20 s) of the compensatory increase in MCFP following baroreceptor unloading (i.e., through ventral aortic occlusion) in some teleost species argues for a functional adrenergic innervation of the venous capacitance vasculature (Sandblom and Axelsson, 2005a; Skals et al., 2006; Zhang et al., 1995).

5.3.2. Other Vasoactive Hormones

While catecholamines are the most studied hormonal regulator of the venous vasculature, there is some information on how other vasoactive hormones affect venous function in fish. Endocrine systems that have received some attention with regards to their effects on venous vasomotor control in fish include: hormones released in response to hypotension/hypovolemic stress such as those of the RAS and AVT (Nishimura et al., 1979; Olson, 1992); as well as the NPs that are released in response to hypertension/hypervolemia (Johnson et al., 2011; Olson, 1992; Toop and Donald, 2004).

5.3.2.1. Renin–Angiotensin System. The RAS is an important modulator of venous tone in mammals (Pang and Tabrizchi, 1986; Rothe and Maass-Moreno, 2000; Tabrizchi et al., 1992), but there is mixed evidence for its venoregulatory role in fish. Blockade of angiotensin-converting enzyme does not affect venous capacitance curves in rainbow trout *in vivo*, but it should be noted that blood volume changes were not directly measured in these studies (Olson et al., 1997a; Zhang et al., 1995). Moreover, large isolated veins from *O. mykiss* are either refractory or exhibit an endothelium-dependent relaxation to Ang II *in vitro* (i.e., due to the local release of prostaglandins), and the C of isolated veins is not affected by this hormone (Conklin and Olson, 1994a,b; Zhang et al., 1995). However, P_{CV} and MCFP slowly increase in swimming *O. mykiss* after general α-adrenergic blockade, a response that can be abolished by pre-treatment with the angiotensin-converting enzyme inhibitor enalapril. This latter result suggests that there is an endogenous activation of RAS and a venospecific role of Ang II during extreme hypotensive stress in exercising fish (Sandblom and Axelsson, 2007a; Sandblom et al., 2006b). Indeed, *in vivo* injection of Ang II into the Antarctic teleost *P. borchgrevinki* increases P_{CV} and MCFP, and this pressor response can only be partially blocked by prazosin. This suggests that there is a direct effect of Ang II on the venous capacitance vasculature (Sandblom et al., 2009a). Thus, it is possible that the response of the venous microcirculation to Ang II, which likely has the primary capacitive function, may be fundamentally different to that in large conductance veins.

5.3.2.2. Arginine Vasotocin. The neurohypophysial hormone AVT has clear venopressor effects in *O. mykiss* and increases P_{CV}, MCFP, and SBV *in vivo* (Conklin et al., 1997). It also constricts isolated systemic veins *in vitro* in this species (Conklin and Olson, 1994b; Conklin et al., 1999). However, hepatic, intestinal, and ovarian veins from the gar (*Lepisosteus* spp.) relax in response to AVT (Conklin et al., 1996). An interesting possibility that remains to be tested is that this relaxation of large conducting veins in the gastrointestinal tract (particularly the HVs) may be instrumental in reducing outflow resistance from the gastrointestinal circulation and in mobilizing blood from the splanchnic to the central venous compartment to raise $\dot{Q}$ and counter systemic hypotension (see Johansen and Hanson, 1967; Satchell, 1992).

5.3.2.3. Natriuretic Peptides. NPs induce pronounced venodilation in fish. For example, *in vivo* infusion of both homologous and heterologous NPs in *O. mykiss* reduces P_{CV} and MCFP (Johnson et al., 2011; Olson et al., 1997a). It appears that these venous effects are mainly due to increased *C*, while the SBV is unchanged (Farrell and Olson, 2000; Olson et al., 1997a). ANP is released from the heart in response to increased cardiac stretch (i.e., cardiac preload) (Cousins and Farrell, 1996; Cousins et al., 1997), and NPs have been ascribed a "cardioprotective" role due to their overall vasodilating properties. This vasodilation reduces cardiac afterload by lowering branchial, and possibly, systemic vascular resistances and decreases cardiac preload by increasing venous capacitance and reducing P_{CV} (Farrell and Olson, 2000; Johnson and Olson, 2008).

5.3.3. Gasotransmitters and Locally Released Vasoactive Agents

Venous responses to gasotransmitters in fish have, to our knowledge, only been characterized for NO and H_2S, with the former being the more extensively studied. The vascular actions of NO in fish appear to be quite diverse and variable among taxa (see Donald and Broughton, 2005; Donald et al., 2015; Olson and Donald, 2009; Olson et al., 2012), and the venous effects of NO are no exception. For example, while NO and NO donors such as sodium nitroprusside (SNP) generally relax isolated conducting veins from various teleosts (Jennings et al., 2004; Olson and Villa, 1991; Olson et al., 1997a), *in vivo* infusion of SNP in *O. mykiss* has relatively minor effects on P_{CV} and venous capacitance. This suggests that NO may primarily affect resistance arterioles (Olson et al., 1997a). In contrast, a NO-mediated constriction of elasmobranch conducting veins has been reported (Evans, 2001), whereas precontracted vein segments from *Lepisosteus* spp. were refractory to SNP (Conklin et al., 1996). The gasotransmitter H_2S produces a triphasic relaxation–constriction–relaxation response in precontracted anterior cardinal vein segments from steelhead trout (Dombkowski et al., 2004).

Other factors released locally from the vascular endothelium are known to affect the venous vasculature through paracrine or autocrine mechanisms. Indeed, both heterologous and homologous trout ET-1 increase venous pressure in *O. mykiss*. This effect is thought to be primarily due to a reduction in C, with only a minor contribution from increased venous tone as the SBV is unchanged (Hoagland et al., 2000).

5.4. Venous Function in Integrated Cardiovascular Responses

5.4.1. Elevated Metabolic Demand (Exercise and Elevated Temperature)

Exercise and elevated temperature typically result in increased rates of oxygen consumption and $\dot{Q}$. With warming, the increase in $\dot{Q}$ is almost entirely through tachycardia (Ekström et al., 2014; Farrell et al., 2009; Gamperl, 2011; Gräns et al., 2013; Keen and Gamperl, 2012), while there is substantial variability among species and environmental conditions regarding the relative contributions of V_S and f_H in modulating $\dot{Q}$ during exercise (Farrell, 2011; Farrell et al., 2009; Gamperl and Shiels, 2014; see Chapter 4, Volume 36A: Farrell and Smith, 2017). Here, we summarize recent findings on the integrated venous hemodynamic responses and the underlying control mechanisms associated with cardiovascular changes during exercise and increases in temperature.

In species that exhibit a large increase in V_S during exercise (e.g., salmonids), an increase in P_{CV} is required to raise end-diastolic volume. However, in species where $\dot{Q}$ is increased primarily through tachycardia, an increase in P_{CV} may serve to maintain V_S by increasing the pressure gradient between the central venous compartment and the heart as cardiac filling time diminishes (Farrell et al., 2009; Sandblom and Axelsson, 2007a). Indeed, a comparison of the hemodynamic responses of *O. mykiss* and European sea bass (*D. labrax*) to sustained swimming reveals that increased P_{CV} is common irrespective of the mechanisms utilized to increase $\dot{Q}$. In *D. labrax*, the increase in $\dot{Q}$ was exclusively the result of an increase in f_H, whereas in *O. mykiss* $\dot{Q}$ primarily increased through an enhancement of V_S (Sandblom et al., 2005, 2006b). Similarly, the Antarctic fish, *P. borchgrevinki*, increases P_{CV} following forced exercise (i.e., a chasing stress), although this increase is small compared to that seen in temperate teleosts (Sandblom et al., 2008). We are only aware of one study that has recorded P_{CV} in an exercising elasmobranch. That study reported that the cardinal sinus pressure in the leopard shark (*Triakis semifasciata*) increased from resting values of 0.2–0.3 kPa (min–max values) to 0.3–0.5 kPa at moderate swimming speeds (Lai et al., 1990). While this response is consistent with the general pattern observed in swimming teleosts, it would be interesting to examine the venous blood pressure response to swimming in an elasmobranch species such as *S. acanthias* which has subambient P_{CV} at rest (Sandblom et al., 2006a). There is convincing evidence that the increase in P_{CV} during exercise is mediated by active reductions in venous capacitance (Sandblom and Axelsson, 2007a). For example, MCFP in *D. labrax* increases from a resting value of 0.27 kPa to 0.31 and then 0.40 kPa when swimming at 1 and 2 body lengths s^{-1}, respectively (Sandblom et al., 2005). Adrenergic control systems are undoubtedly important for these responses because the increase in MCFP during sustained swimming in *D. labrax* (Sandblom et al., 2005) and enforced exercise in *P. borchgrevinki* (Sandblom et al., 2008) is abolished by pretreatment with prazosin.

The venous hemodynamic effects of changes in water temperature have been examined in various teleost species and in one elasmobranch. P_{CV} is unchanged in trout during acute warming from 10°C to 16°C (Sandblom and Axelsson, 2007b), and in the winter flounder (*Pseudopleuronectes americanus*) when acutely warmed by 5°C from acclimation temperatures of 5°C, 10°C, and 15°C (Cech et al., 1976). However, at least in *O. mykiss*, an active reduction in the venous capacitance occurs, as indicated by elevated MCFP. This serves to mobilize blood to the central venous compartment and, thus, maintain cardiac filling pressure and compensate for reduced cardiac filling time as f_H increases with temperature (Fig. 10). Similarly, MCFP increased slightly in the elasmobranch, *S. acanthias*, during warming from 10°C to 16°C, such that P_{CV} was unchanged despite a pronounced tachycardia (Sandblom et al., 2009b). Few studies have examined venous hemodynamic responses in fish at extreme temperatures. In Chinook salmon (*Oncorhynchus tshawytscha*), P_{CV} remained unchanged at approximately 0.01 kPa at intermediate temperatures (13–21°C), but increased significantly to 0.19 kPa at 25°C; this temperature is close to the upper lethal temperature limit for this species (Clark et al., 2008). Because f_H and $\dot{Q}$ were highest at 25°C, it appears that this increase in P_{CV} reflects an adaptive response, where actively reduced venous capacitance serves to improve cardiac function through increased filling pressure, rather than a passive effect of venous blood pooling due to cardiac failure at extreme temperatures. Antarctic cold-water species may represent an exception to the general pattern of a decreased venous capacitance with warming,

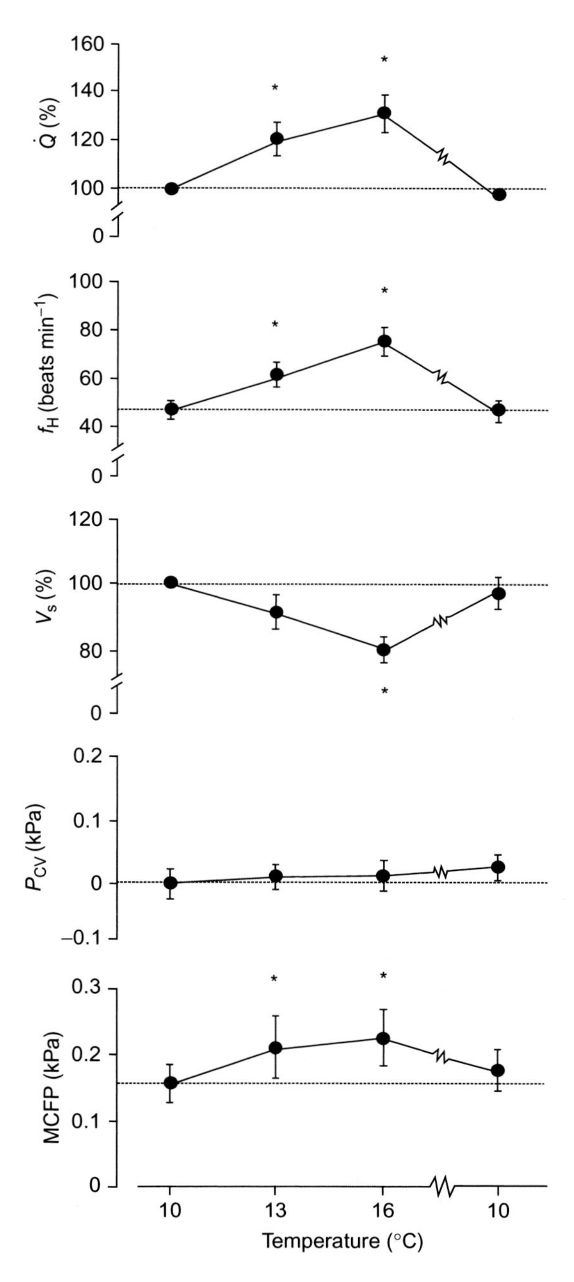

Fig. 10. Cardiovascular responses to acute warming in rainbow trout (*Oncorhynchus mykiss*). The variables are cardiac output ($\dot{Q}$), heart rate (f_H), stroke volume (V_S), central venous blood pressure (P_{CV}) and mean circulatory filling pressure (MCFP). *Asterisks* denote a significant ($P \leq 0.05$) increase from the initial 10°C value. Modified from Sandblom, E., Axelsson, M., 2007. Venous hemodynamic responses to acute temperature increase in the rainbow trout (*Oncorhynchus mykiss*). Am. J. Physiol. Regul. Integr. Comp. Physiol. 292, R2292–R2298.

because MCFP was unchanged when *P. borchgrevinki* were acutely warmed from 0°C to 2.5°C and then to 5°C, which must be considered an extreme thermal challenge for this stenothermal cold-water species. Consequently, P_{CV} decreased as f_H increased, which would negatively impact the heart's ability to maintain V_S at high temperatures in this species (Sandblom et al., 2008).

To summarize, the benefits of a reduction in the venous capacitance during exercise and warming in fish are at least twofold. First, as blood flow increases, it may be important to increase the tone in the venous vasculature to prevent passive flow-mediated pooling of blood in the compliant peripheral venous vessels. Moreover, the decrease in venous capacitance shifts blood into the central venous compartment, which maintains or increases V_S depending on the associated f_H response. Nonetheless, particularly during warming, the contributions of active neurohumoral and myogenic mechanisms, and the relative importance of passive and active changes in contained blood volume in different vascular beds, are largely unresolved. Finally, it is not known if the observed changes in MCFP during exercise and warming are due to changes in *C* or SBV. These questions can best be resolved by constructing vascular capacitance curves during sustained swimming, and at different temperatures. These are challenging, but feasible, experiments for the ambitious fish physiologist.

5.4.2. Barostatic Reflexes

There is evidence that venous capacitance is actively controlled through various homeostatic cardiovascular reflexes in fish. The baroreflex (i.e., barostatic reflex) originates from afferent sensory nerve endings (i.e., baroreceptors) located in the branchial vasculature (Ristori, 1970; Ristori and Dessaux, 1970; Sundin and Nilsson, 2002) and functions to buffer changes in arterial blood pressure (Bagshaw, 1985; Jones and Milsom, 1982; Van Vliet and West, 1994). The efferent pathways of the reflex involve both cardiac and vascular responses, with the former leading to reduced f_H and $\dot{Q}$ through increased vagal inhibition of the heart in response to branchial hypertension (Sandblom and Axelsson, 2011). Conversely, baroreceptor unloading (i.e., branchial hypotension) results in the opposite cardiac response (i.e., tachycardia and increased $\dot{Q}$), as well as increased arterial resistance and a reflex constriction of the venous vasculature. The latter is indicated by elevated MCFP following a brief (30 s) mechanical occlusion of the ventral aorta to unload branchial baroreceptors (Sandblom and Axelsson, 2005a). The venous vasomotor response appears to be exclusively mediated through the stimulation of α-adrenoreceptors as it can be fully blocked with prazosin. Interestingly, mechanical occlusion of $\dot{Q}$ in the dogfish (*S. acanthias*) does not induce a reflex venous constriction (Sandblom et al., 2006a, 2009b), and elasmobranchs typically tolerate a gravitational challenge in air (with maintained gill ventilation) very poorly (Ogilvy and DuBois, 1982). This suggests that barostatic reflex control of the venous vasculature is limited or absent in elasmobranchs, which is consistent with the general view that adrenergic neural vasomotor control is poorly developed in this taxa (Brill and Lai, 2016; Sandblom and Axelsson, 2011).

5.4.3. Reflex Responses to Aquatic Hypoxia and Air Breathing

Hypoxia is a frequently occurring or chronic condition in many aquatic environments (Diaz and Breitburg, 2009), and various cardiorespiratory reflex responses have, therefore, evolved in fish to deal with this environmental challenge (see Gamperl and Driedzic, 2009). While most teleosts and elasmobranchs exhibit a pronounced bradycardia when exposed to sudden hypoxia (i.e., hypoxic bradycardia), the adaptive role of this response is not yet fully understood (Farrell, 2007b; Perry and Desforges, 2006). Even so, $\dot{Q}$ typically only drops slightly or remains unchanged due to a compensatory increase in V_S in both teleosts and elasmobranchs subjected to hypoxia (Butler and Taylor, 1971, 1975; Perry and Desforges, 2006; Sandblom and Axelsson, 2005b, 2006; Short et al., 1977; Taylor et al., 1977). Indeed, there is recent evidence for active reflex constriction of the venous capacitance vasculature to mobilize blood to the central venous compartment and raise the cardiac filling pressure in teleosts during hypoxia (Perry et al., 1999; Sandblom and Axelsson, 2005b, 2006). For example, *in vivo* vascular capacitance decreases in *O. mykiss* during brief (3 min) moderate hypoxia (water O_2 tension: ~9 kPa) due to a reduced USBV. This venous vasomotor response is likely mediated by α-adrenoreceptor stimulation from both neural and circulating catecholamines, because the nerve-blocking agent bretylium only partially abolishes the response, while general α-adrenergic blockade with prazosin abolishes the response nearly completely (Sandblom and Axelsson, 2006). Interestingly, P_{CV} in the elasmobranch *S. acanthias* also increases slightly with 30 min of exposure to severe hypoxia (water O_2 tension: 2.5 kPa), but MCFP is unchanged despite significant increases in circulating catecholamines (Sandblom et al., 2009b). Again, this finding suggests limited reflex control of the veins in elasmobranchs.

In the bimodally respiring Marbled swamp eel (*S. marmoratus*), aquatic hypoxia triggers air breathing, which involves reflex tachycardia and increased arterial blood flow (Skals et al., 2006). Interestingly, this tachycardia is associated with a pronounced rise in cardiac filling pressure associated with an active reduction in the venous capacitance (i.e., elevated MCFP). This response compensates for the reduced filling time associated with the elevation in f_H and leads to elevated V_S and $\dot{Q}$ during air breathing (see Fig. 11). This study is important as it is the first to reveal a significant role of reflex venous vasomotor control during air breathing in a bimodally respiring fish.

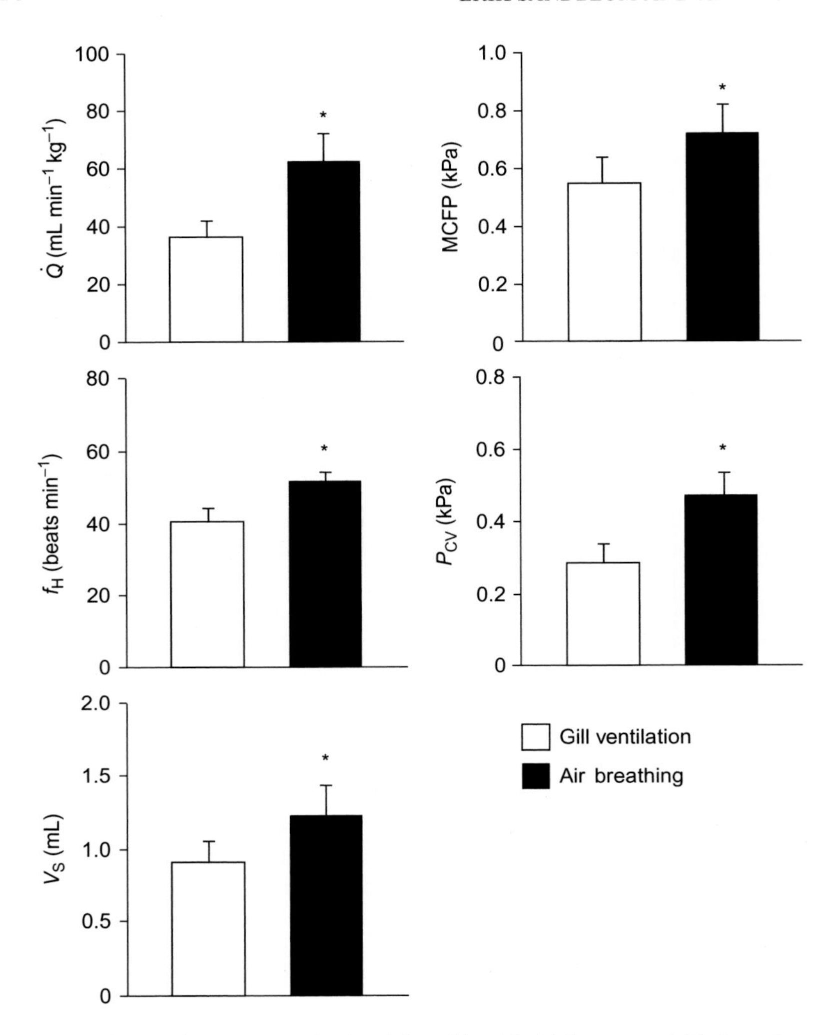

Fig. 11. Cardiovascular parameters in the air-breathing Marbled swamp eel (*Synbranchus marmoratus*) during gill ventilation in normoxia (*open bars*) and during air-breathing in hypoxia (*filled bars*). The variables are cardiac output ($\dot{Q}$), heart rate (f_H), mean circulatory filling pressure (MCFP), central venous blood pressure (P_{CV}), and stroke volume (V_S). *Asterisks* denote significant ($P \leq 0.05$) changes during air breathing. Modified from Skals, M., Skovgaard, N., Taylor, E.W., Leite, C.A., Abe, A.S., Wang, T., 2006. Cardiovascular changes under normoxic and hypoxic conditions in the air-breathing teleost *Synbranchus marmoratus*: importance of the venous system. J. Exp. Biol. 209, 4167–4173.

5.4.4. Changes in Water Salinity

Increases in salinity are generally expected to have a volume-depleting effect, which would be predicted to result in reductions in the total circulating blood volume and P_{CV} in fish, unless they are compensated for through changes in venous capacitance (Olson, 1992). Indeed, euryhaline fishes can tolerate large changes in salinity and exhibit a range of physiological acclimatory responses, including cardiovascular changes, when facing this osmoregulatory challenge. For example, rainbow trout exhibits a two-fold increase in gastrointestinal blood flow when acclimated to seawater, which is matched by an equivalent increase in $\dot{Q}$ through elevated V_S (Brijs et al., 2015, 2016). The pronounced increase in V_S in seawater is somewhat surprising given the volume-depleting effects of increased salinity, and previous reports of reduced blood volume in seawater-acclimated trout that would be expected to reduce venous pressure (Olson and Hoagland, 2008). However, a recent study showed that acclimation of trout to seawater results in a marked increase in P_{CV}, which may explain the elevated V_S (Fig. 12; Brijs et al., 2017). While the neurohumoral mechanisms behind this response are presently unknown, the above results suggest that active changes in venous capacitance and vasomotor tone are fundamentally important when euryhaline fishes acclimate to different water salinities.

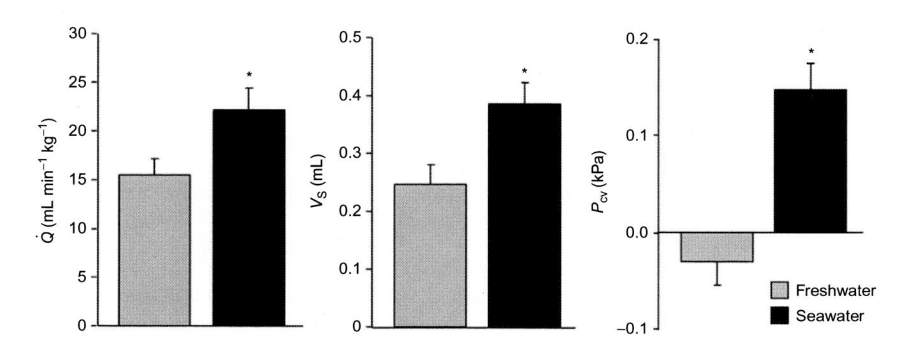

Fig. 12. Routine cardiovascular parameters in freshwater (*gray bars*) and seawater (*black bars*) acclimated rainbow trout (*Oncorhynchus mykiss*). The variables are cardiac output ($\dot{Q}$), stroke volume (V_S), and central venous blood pressure (P_{CV}). The increase in $\dot{Q}$ with seawater acclimation was exclusively mediated by increased S_V, while the heart rate remained unchanged. *Asterisks* denote significant ($P \leq 0.05$) differences in cardiovascular parameters between water salinities. Modified from Brijs, J., Sandblom, E., Dekens, E., Näslund, J., Ekström, A., Axelsson, M., 2017. Cardiac remodeling and increased central venous pressure underlie elevated stroke volume and cardiac output of seawater-acclimated rainbow trout. Am. J. Physiol. Reg. Integr. Physiol. 312, R31–R39..

ACKNOWLEDGMENTS

The authors work is supported by the Swedish Research Council (E.S.) and the Swedish Research Council for Environment, Agricultural Sciences and Spatial Planning (FORMAS) (E.S. and A.G.). The authors would like to thank an anonymous referee and the editors of this volume for constructive comments on an earlier version of this chapter.

REFERENCES

Abrahamsson, T., Holmgren, S., Nilsson, S., Pettersson, K., 1979. On the chromaffin system of the African lungfish, *Protopterus aethiopicus*. Acta. Physiol. Scand. 107, 135–139.

Allen, W.F., 1905. The Blood-Vascular System of the Loricati, the Mail-Cheeked Fishes. vol. VII. Washington Academic of Sciences, Washington, pp. 27–157.

Altimiras, J., Axelsson, M., 2004. Intrinsic autoregulation of cardiac output in rainbow trout (*Oncorhynchus mykiss*) at different heart rates. J. Exp. Biol. 207, 195–201.

Altimiras, J., Claireaux, G., Sandblom, E., Farrell, A.P., McKenzie, D.J., Axelsson, M., 2008. Gastrointestinal blood flow and postprandial metabolism in swimming sea bass *Dicentrarchus labrax*. Physiol. Biochem. Zool. 81, 663–672.

Aota, S., 1995. Cardiovascular effects of adrenomedullin in teleost fishes. Braz. J. Med. Biol. Res. 28, 1223–1226.

Axelsson, M., 2005. The circulatory system and its control. In: Farrell, A.P., Steffensen, J.F. (Eds.), Physiology of Polar Fishes, vol. 22. Academic Press, London, pp. 239–280.

Axelsson, M., Farrell, A.P., 1993. Coronary blood flow in vivo in the coho salmon (*Oncorhynchus kisutch*). Am. J. Physiol. 264, R963–R971.

Axelsson, M., Fritsche, R., 1991. Effects of exercise, hypoxia and feeding on the gastrointestinal blood flow in the Atlantic cod *Gadus morhua*. J. Exp. Biol. 158, 181–198.

Axelsson, M., Nilsson, S., 1986. Blood pressure control during exercise in the Atlantic cod, *Gadus morhua*. J. Exp. Biol. 126, 225–236.

Axelsson, M., Driedzic, W.R., Farrell, A.P., Nilsson, S., 1989. Regulation of cardiac output and gut blood flow in the sea raven, *Hemitripterus americanus*. Fish Physiol. Biochem. 6, 315–326.

Axelsson, M., Farrell, A.P., Nilsson, S., 1990. Effects of hypoxia and drugs on the cardiovascular dynamics of the Atlantic hagfish *Myxine glutinosa*. J. Exp. Biol. 151, 297–316.

Axelsson, M., Davison, B., Forster, M., Nilsson, S., 1994. Blood pressure control in the Antarctic fish *Pagothenia borchgrevinki*. J. Exp. Biol. 190, 265–279.

Axelsson, M., Thorarensen, H., Nilsson, S., Farrell, A.P., 2000. Gastrointestinal blood flow in the red Irish lord, *Hemilepidotus hemilepidotus*: long-term effects of feeding and adrenergic control. J. Comp. Physiol. 170, 145–152.

Bagshaw, R.J., 1985. Evolution of cardiovascular baroreceptor control. Biol. Rev. 60, 121–162.

Bailly, Y., Dunel-Erb, S., 1986. The sphincter of the efferent filament artery in teleost Gills. 1. Structure and parasympathetic innervation. J. Morphol. 187, 219–237.

Balment, R.J., Lu, W., Weybourne, E., Warne, J.M., 2006. Arginine vasotocin a key hormone in fish physiology and behaviour: a review with insights from mammalian models. Gen. Comp. Endocrinol. 147, 9–16.

Behrens, J.W., Axelson, M., Neuenfeldt, S., Seth, H., 2012. Effects of hypoxic exposure during feeding on SDA and postprandial cardiovascular physiology in the Atlantic cod, *Gadus morhua*. PLoS One 7, 1–8.

Bernier, N., Perry, S., 1996. Control of catecholamine and serotonin release from the chromaffin tissue of the Atlantic hagfish. J. Exp. Biol. 199, 2485–2497.

Bernier, N.J., Gilmour, K.M., Takei, Y., Perry, S.F., 1999a. Cardiovascular control via angiotensin II and circulating catecholamines in the spiny dogfish, *Squalus acanthias*. J. Comp. Physiol. B. 169, 237–248.

Bernier, N.J., McKendry, J.E., Perry, S.F., 1999b. Blood pressure regulation during hypotension in two teleost species: differential involvement of the renin-angiotensin and adrenergic systems. J. Exp. Biol. 202, 1677–1690.

Bjenning, C., Driedzig, W., Holmgren, S., 1989. Neuropeptide Y-like immunoreactivity in the cardiovascular nerve plexus of the elasmobranchs *Raja erinacea* and *Raja radiata*. Cell Tissue Res. 255, 481–486.

Bjenning, C., Hazon, N., Balasubramaniam, A., Holmgren, S., Conlon, J.M., 1993. Distribution and activity of dogfish Npy and peptide YY in the cardiovascular system of the common dogfish. Am. J. Physiol. Regul. Integr. Comp. Physiol. 264, R1119–R1124.

Bolis, L., Mandolfino, M., Marino, D., Rankin, J.C., 1984. Vascular actions of vasoactive intestinal polypeptide and beta-endorphin in isolated perfused gills of the brown trout, *Salmo trutta* L. Mol. Physiol. 5, 221–226.

Brijs, J., Axelsson, M., Gräns, A., Pichaud, N., Olsson, C., Sandblom, E., 2015. Increased gastrointestinal blood flow: an essential circulatory modification for euryhaline rainbow trout (*Oncorhynchus mykiss*) migrating to sea. Sci. Rep. 5, 10430.

Brijs, J., Gräns, A., Ekstrom, A., Olsson, C., Axelsson, M., Sandblom, E., 2016. Cardiorespiratory upregulation during seawater acclimation in rainbow trout: effects on gastrointestinal perfusion and postprandial responses. Am. J. Physiol. Regul. Integr. Comp. Physiol. 310, R858–R865.

Brijs, J., Sandblom, E., Dekens, E., Näslund, J., Ekström, A., Axelsson, M., 2017. Cardiac remodeling and increased central venous pressure underlie elevated stroke volume and cardiac output of seawater-acclimated rainbow trout. Am. J. Physiol. Reg. Integr. Physiol. 312, R31–R39.

Brill, R.W., Lai, C.H., 2016. Elasmobranch cardiovascular system. In: Shadwick, R.E., Farrell, A.P., Brauner, C.J. (Eds.), Fish Physiology—Physiology of Elasmobranch Fishes, Internal Processes, vol. 34B. Elsevier, Cambridge, USA, pp. 1–82.

Burggren, W.W., Johansen, K., 1986. Circulation and respiration in lungfishes (Dipnoi). J. Morphol. 190 (S1), 217–236.

Bushnell, P.G., Farrell, A.P., Jones, D.R., 1992. The arterial system. In: Hoar, W.S., Randal, D.J., Farrell, A.P. (Eds.), Fish Physiology: The Cardiovascular System, vol. XII. Academic Press Inc., London, pp. 89–139.

Butler, P.J., Metcalfe, J.D., 1988. Cardiovascular and respiratory systems. In: Shuttleworth, T.J. (Ed.), Physiology of Elasmobranch Fishes. Springer-Verlag, Berlin, Heidelberg, New York, London, Paris, Tokyo, pp. 1–47.

Butler, D.G., Oudit, G.Y., 1995. Angiotensin-I- and -III-mediated cardiovascular responses in the freshwater North American eel, *Anguilla rostrata*: effect of Phe^8 deletion. Gen. Comp. Endocrinol. 97, 259–269.

Butler, P.J., Taylor, E.W., 1971. Response of the dogfish (*Scyliorhinus canicula* L) to slowly induced and rapidly induced hypoxia. Comp. Biochem. Physiol. 39A, 307–323.

Butler, P.J., Taylor, E.W., 1975. The effect of progressive hypoxia on respiration in the dogfish (*Scyliorhinus canicula*) at different seasonal temperatures. J. Exp. Biol. 63, 117–130.

Cameron, M.S., Nobata, S., Takei, Y., Donald, J.A., 2015. Vasodilatory effects of homologous adrenomedullin 2 and adrenomedullin 5 on isolated blood vessels of two species of eel. Comp. Biochem. Phys. A. 179, 157–163.

Canty, A.A., Farrell, A.P., 1985. Intrinsic regulation of flow in an isolated tail preparation of the ocean pout (*Macrozoarces americanus*). Can. J. Zool. 63, 2013–2020.

Capra, M.F., Satchell, G.H., 1977. The differential haemodynamic responses of the elasmobranch, *Squalus acanthias*, to the naturally occurring catecholamines, adrenaline and noradrenaline. Comp. Biochem. Physiol. C 58, 41–47.

Carroll, R.G., 1981. Vascular response of the dogfish and sculpin to angiotensin II. Am. J. Physiol. 240, R139–R143.

Cech Jr., J.J., Bridges, D.W., Rowell, D.M., Balzer, P.J., 1976. Cardiovascular responses of winter flounder, *Pseudopleuronectes americanus* (Walbaum), to acute temperature increase. Can. J. Zool. 54, 1383–1388.

Cech Jr., J.J., Rowell, D.M., Glasgow, J.S., 1977. Cardiovascular responses of the winter flounder *Pseudopleuronectes americanus* to hypoxia. Comp. Biochem. Physiol. A **123**, 123–123, 125.

Chan, D.K., Chow, P.H., 1976. The effects of acetylcholine, biogenic amines and other vasoactive agents on the cardiovascular functions of the Eel, *Anguilla japonica*. J. Exp. Zool. 196, 13–26.

Clark, T.D., Sandblom, E., Cox, G.K., Hinch, S.G., Farrell, A.P., 2008. Circulatory limits to oxygen supply during an acute temperature increase in the Chinook salmon (*Oncorhynchus tshawytscha*). Am. J. Physiol. 295, R1631–R1639.

Colin, D.A., Kirsch, R., Leray, C., 1979. Haemodynamic effects of adenosine on gills of the trout (*Salmo gairdneri*). J. Comp. Physiol. B. 130, 325–330.

Conklin, D.J., Olson, K.R., 1994a. Angiotensin II relaxation of rainbow trout vessels in vitro. Am. J. Physiol. 266, R1856–R1860.

Conklin, D.J., Olson, K.R., 1994b. Compliance and smooth muscle reactivity of rainbow trout (*Oncorhynchus mykiss*) vessels in vitro. J. Comp. Physiol. 163, 657–663.

Conklin, D.J., Mick, N.W., Olson, K.R., 1996. Arginine vasotocin relaxation of gar (*Lepisosteous* spp.) hepatic vein in vitro. Gen. Comp. Endocrinol. 104, 52–60.

Conklin, D., Chavas, A., Duff, D., Weaver, L., Zhang, Y., Olson, K.R., 1997. Cardiovascular effects of arginine vasotocin in the rainbow trout *Oncorhynchus mykiss*. J. Exp. Biol. 200, 2821–2832.

Conklin, D.J., Smith, M.P., Olson, K.R., 1999. Pharmacological characterization of arginine vasotocin vascular smooth muscle receptors in the trout (*Oncorhynchus mykiss*) in vitro. Gen. Comp. Endocrinol. 114, 36–46.

Conlon, J.M., Platzack, B., Marra, L.E., Youson, J.H., Olson, K.R., 1995. Isolation and biological activity of [Trp 5] bradykinin from the plasma of the phylogenetically ancient fish, the bowfin and the longnosed gar. Peptides 16, 485–489.

Conte, F.P., Wagner, H.H., Harris, T.O., 1963. Measurement of blood volume in fish (*Salmo gairdneri gairdneri*). Am. J. Physiol. Regul. Integr. Comp. Physiol. 205, 533.

Coucelo, J., Joaquim, N., Coucelo, J., 2000. Calculation of volumes and systolic indices of heart ventricle from *Halobatrachus didactylus*: echocardiographic noninvasive method. J. Exp. Zool. 286, 585–595.

Cousins, K.L., Farrell, A.P., 1996. Stretch-induced release of atrial natriuretic factor from the heart of rainbow trout (*Oncorhynchus mykiss*). Can. J. Zool. 74, 380–387.

Cousins, K.L., Farrell, A.P., Sweeting, R.M., Vesely, D.L., Keen, J.E., 1997. Release of atrial natriuretic factor prohormone peptides 1-30, 31-67 and 99-126 from freshwater- and seawater-acclimated perfused trout (*Oncorhynchus mykiss*) hearts. J. Exp. Biol. 200, 1351–1362.

Dasiewicz, P.J., Conlon, J.M., Anderson, W.G., 2011. Cardiovascular and vasoconstrictive actions of skate bradykinin in the little skate, *Leucoraja erinacea* (Elasmobranchii). Gen. Comp. Endocrinol. 174, 89–96.

Davie, P.S., Forster, M.E., 1980. Cardiovascular responses to swimming in eels. Comp. Biochem. Physiol. A 67, 367–373.

Diaz, R.J., Breitburg, D.L., 2009. The hypoxic environment. In: Richards, J.G., Farrell, A.P., Brauner, C.J. (Eds.), Hypoxia, vol. 27. Elsevier, Cambridge, USA, pp. 1–23.

Dombkowski, R.A., Russell, M.J., Olson, K.R., 2004. Hydrogen sulfide as an endogenous regulator of vascular smooth muscle tone in trout. Regul. Integr. Comp. Physiol. 286, R678–R685.

Dombkowski, R.A., Russell, M.J., Schulman, A.A., Doellman, M.M., Olson, K.R., 2005. Vertebrate phylogeny of hydrogen sulfide vasoactivity. Am. J. Physiol. Regul. Integr. Comp. Physiol. 288, R243–R252.

Dombkowski, R.A., Whitfield, N.L., Motterlini, R., Gao, Y., Olson, K.R., 2009. Effects of carbon monoxide on trout and lamprey vessels. Am. J. Physiol. Regul. Integr. Comp. Physiol. 296, R141–R149.

Donald, J., 1984. Adrenergic innervation of the gills of brown and rainbow trout, *Salmo trutta* and *Salmo gairdneri*. J. Morphol. 182, 307–316.

Donald, J.A., 1987. Comparative study of the adrenergic innervation of the teleost gill. J. Morphol. 193, 63–73.

Donald, J.A., Broughton, B.R., 2005. Nitric oxide control of lower vertebrate blood vessels by vasomotor nerves. Comp. Biochem. Physiol. A 142, 188–197.

Donald, J.A., Forgan, L.G., Cameron, M.S., 2015. The evolution of nitric oxide signalling in vertebrate blood vessels. J. Comp. Physiol. B 185, 153–171.

Dunel-Erb, S., Bailly, Y., 1986. The sphincter of the efferent filament artery in teleost gills. 2. sympathetic innervation. J. Morphol. 187, 239–246.

Ekström, A., Jutfelt, F., Sandblom, E., 2014. Effects of autonomic blockade on acute thermal tolerance and cardioventilatory performance in rainbow trout, *Oncorhynchus mykiss*. J. Therm. Biol. 44, 47–54.

Eliason, E.J., Farrell, A.P., 2014. Effect of hypoxia on specific dynamic action and postprandial cardiovascular physiology in rainbow trout (*Oncorhynchus mykiss*). Comp. Biochem. Physiol. A 171, 44–50.

Epple, A., Hilliard, R.W., Potter, I.C., 1985. The cardiovascular chromaffin cell system of the southern hemisphere lamprey *Geotria australis*. J. Morphol. 183, 225–231.

Evans, D.H., 2001. Vasoactive receptors in abdominal blood vessels of the dogfish shark, *Squalus acanthias*. Physiol. Biochem. Zool. 74, 120–126.

Evans, D.H., Gunderson, M., Cegelis, C., 1996. ET_B-type receptors mediate endothelin stimulated contraction in the aortic vascular smooth muscle of the spiny dogfish shark, *Squalus acanthias*. J. Comp. Physiol. B 165, 659–664.

Evans, D.H., Piermarini, P.M., Choe, K.P., 2005. The multifunctional fish gill: dominant site of gas exchange, osmoregulation, acid-base regulation, and excretion of nitrogenous waste. Physiol. Rev. 85, 97–177.

Fänge, R., 1972. The circulatory system. In: Hardisty, M.W., Potter, I.C. (Eds.), The Biology of Lampreys, vol. 2. Academic Press, London, pp. 241–259.

Fänge, R., 1983. Gas exchange in fish swim bladder. Rev. Physiol. Biochem. Pharmacol. 97, 111–158.

Farrell, A.P., 1982. Cardio vascular changes in the unanesthetized lingcod *Ophiodon-Elongatus* during short-term progressive hypoxia and spontaneous activity. Can. J. Zool. 60, 933–941.

Farrell, A.P., 1984. A review of cardiac performance in the teleost heart: intrinsic and humoral regulation. Can. J. Zool. 62, 523–536.

Farrell, A.P., 2007a. Cardiovascular systems in primitive fishes. In: McKenzie, D.J., Farrell, A.P., Brauner, C.J. (Eds.), Fish Physiology, vol. 26. Elsevier, Cambridge, USA, pp. 53–120.

Farrell, A.P., 2007b. Tribute to P. L. Lutz: a message from the heart—why hypoxic bradycardia in fishes? J. Exp. Biol. 210, 1715–1725.

Farrell, A.P., 2011. Integrated cardiovascular responses of fish to swimming. In: Farrell, A.P. (Ed.), Encyclopedia of Fish Physiology. Elsevier Inc., Amsterdam.

Farrell, A.P., Jones, D.R., 1992. The heart. In: Hoar, W.S., Randal, D.J., Farrell, A.P. (Eds.), Fish Physiology. The Cardiovascular System, vol. XII. Elsevier, Cambridge, USA, pp. 1–88.

Farrell, A.P., Olson, K.R., 2000. Cardiac natriuretic peptides: a physiological lineage of cardioprotective hormones? Physiol. Biochem. Zool. 73 (1), 1–11.
Farrell, A.P., Smith, F., 2017. Cardiac form, function and physiology. In: Gamperl, A.K., Gillis, T.E., Farrell, A.P., Brauner, C.J. (Eds.), Fish Physiology. In: The Cardiovascular System: Morphology, Control and Function, vol. 36A. Academic Press, San Diego, pp. 155–264.
Farrell, A.P., Sobin, S.S., Randall, D.J., Crosby, S., 1980. Intralamellar blood-flow patterns in fish gills. Am. J. Physiol. Regul. Integr. Comp. Physiol. 239, R428–R436.
Farrell, A.P., Johansen, J.A., Graham, M.S., 1988. The role of the pericardium in cardiac performance of the trout (Salmo gairdneri). Physiol. Zool. 61 (3), 213–221.
Farrell, A.P., Eliason, E.J., Sandblom, E., Clark, T.D., 2009. Fish cardiorespiratory physiology in an era of climate change. Can. J. Zool. 87, 835–851.
Forster, M., 1997. The blood sinus system of hagfish: its significance in a low-pressure circulation. Comp. Biochem. Physiol. A 116A, 239–244.
Forster, M.E., 1998. Cardiovascular function in hagfishes. Biol. Hagfishes, 237–258.
Forster, M.E., Axelsson, M., Farrell, A.P., Nilsson, S., 1991. Cardiac function and circulation in hagfishes. Can. J. Zool. 69, 1985–1992.
Forster, M.E., Davison, W., Axelsson, M., Farrell, A.P., 1992. Cardiovascular responses to hypoxia in the hagfish, *Eptatretus cirrhatus*. Respir. Physiol. 88, 373–386.
Forster, M.E., Russell, M.J., Hambleton, D.C., Olson, K.R., 2001. Blood and extracellular fluid volume in whole body and tissues of the Pacific hagfish, *Eptatretus stouti*. Physiol. Biochem. Zool. 74, 750–756.
Foster, J.M., Forster, M.E., 2007a. Changes in plasma catecholamine concentration during salinity manipulation and anaesthesia in the hagfish *Eptatretus cirrhatus*. J. Comp. Physiol. B. 177, 41–47.
Foster, J.M., Forster, M.E., 2007b. Effects of salinity manipulations on blood pressures in an osmoconforming chordate, the hagfish, *Eptatretus cirrhatus*. J. Comp. Physiol. B. 177, 31–39.
Foster, J.M., Forster, M.E., Olson, K.R., 2008. Different sensitivities of arteries and veins to vasoactive drugs in a hagfish, *Eptatretus cirrhatus*. Comp. Biochem. Physiol. C 148, 107–111.
Franklin, C.E., Davie, P.S., 1992. Dimensional analysis of the ventricle of an in situ perfused trout heart using echocardiography. J. Exp. Biol. 166, 47–60.
Franklin, C.E., Davison, W., Mckenzie, J.C., 1993. The role of the spleen during exercise in the Antarctic teleost, *Pagothenia borchgrevinki*. J. Exp. Biol. 174, 381–386.
Franklin, C.E., Davison, W., Seebacher, F., 2007. Antarctic fish can compensate for rising temperatures: thermal acclimation of cardiac performance in *Pagothenia borchgrevinki*. J. Exp. Biol. 210, 3068–3074.
Fritsche, R., Nilsson, S., 1990. Autonomic nervous control of blood pressure and heart rate during hypoxia in the cod *Gadus morhua*. J. Comp. Physiol. 160, 287–292.
Funk, D.J., Jacobsohn, E., Kumar, A., 2013. The role of venous return in critical illness and shock—Part I: physiology. Crit. Care Med. 41, 255–262.
Gamperl, A.K., 2011. Integrated responses of the circulatory system to temperature. In: Farrell, A.P. (Ed.), Encyclopedia of Fish Physiology. Elsevier Inc., Amsterdam, pp. 1197–1205.
Gamperl, A.K., Driedzic, W.R., 2009. Cardiovascular function and cardiac metabolism. In: Richards, J.G., Farrell, A.P., Brauner, C.J. (Eds.), Hypoxia, vol. 27. Elsevier, Cambridge, USA, pp. 301–360.
Gamperl, A.K., Shiels, H., 2014. The cardiovascular system. In: Evans, D.H., Claiborne, J.B., Currie, S. (Eds.), The Physiology of Fishes. CRC Press, Boca Raton, FL.
Gamperl, A.K., Vijayan, M.M., Boutilier, R.G., 1994. Epinephrine, norepinephrine, and cortisol concentrations in cannulated seawater-acclimated rainbow trout (*Oncorhynchus mykiss*) following black-box confinement and epinephrine injection. Fish. Biol. 45, 313–324.

Gilmour, K.M., 2001. The CO_2/pH ventilatory drive in fish. Comp. Biochem. Physiol. A 130, 219–240.
Gilmour, K.M., 2012. New insights into the many functions of carbonic anhydrase in fish gills. Respir. Physiol. Neurobiol. 184, 223–230.
Goss, G.G., Perry, S.F., Fryer, J.N., Laurent, P., 1998. Gill morphology and acid-base regulation in freshwater fishes. Comp. Biochem. Physiol. A 119, 107–115.
Graham, J.B., 1997. Air-Breathing Fishes: Evolution, Diversity, and Adaptations. Academic Press, New York.
Graham, M.S., Fletcher, G.L., 1983. Blood and plasma viscosity of winter flounder Pseudopleuronectes americanus influence of temperature red cell concentration and shear rate. Can. J. Zool. **61**, 2344–2350.
Gräns, A., Axelsson, M., Olsson, C., Höjesjö, J., Pitsillides, K., Kaufman, R., Cech Jr., J.J., 2009. A fully implantable multi-channel biotelemetry system for measurement of blood flow and temperature: a first evaluation in the green sturgeon. Hydrobiologia 619, 11–25.
Gräns, A., Seth, H., Axelsson, M., Sandblom, E., Albertsson, F., Wiklander, K., Olsson, C., 2013. Effects of acute temperature changes on gut physiology in two species of sculpin from the west coast of Greenland. Polar Bio. 36, 775–785.
Greene, C.W., 1902. Contributions to the physiology of the California hagfish, *Polistotrema stouti*. II. The absence of regulative nerves for the systemic heart. Am. J. Physiol. 6, 318–324.
Guyton, A.C., 1963. Circulatory Physiology: Cardiac Output and Its Regulation. W.B. Saunders Company, Philadelphia and London.
Guyton, A.C., Lindsey, A.W., Kaufmann, B.N., 1955. Effect of mean circulatory filling pressure and other peripheral circulatory factors on cardiac output. Am. J. Physiol. 180, 463–468.
Haase, E.B., Shoukas, A.A., 1991. Carotid sinus baroreceptor reflex control of venular pressure-diameter relations in rat intestine. Am. J. Physiol. 260, H752–H758.
Haase, E.B., Shoukas, A.A., 1992. Blood volume changes in microcirculation of rat intestine caused by carotid sinus baroreceptor reflex. Am. J. Physiol. 263, H1939–H1945.
Hainsworth, R., 1986. Vascular capacitance: its control and importance. Rev. Physiol. Biochem. Pharmacol. 105, 101–173.
Hazon, N., Tierney, M.L., Takei, Y., 1999. Renin-angiotensin system in elasmobranch fish: a review. J. Exp. Zool. 284, 526–534.
Hemmingsen, E.A., 1991. Respiratory and cardiovascular adaptations in hemoglobin-free fish: resolved and unresolved problems. In: di Prisco, G., Maresca, B., Tota, B. (Eds.), Biology of Antarctic Fish. Springer-Verlag, Berlin, Heidelberg, New York, London, Paris, Tokyo, Hong Kong, Barcelona, Budapest, pp. 191–203.
Hemmingsen, E.A., Douglas, E.L., 1970. Respiratory characteristics of hemoglobin-free fish *Chaenocephalus aceratus*. Comp. Biochem. Physiol. 33, 733.
Henderson, I.W., Wales, N.A., 1974. Renal diuresis and antidiuresis after injections of arginine vasotocin in the freshwater eel (*Anguilla anguilla* L.). J. Endocrin. 61, 487–500.
Hipkins, S.F., Smith, D.G., Evans, B.K., 1986. Lack of adrenergic control of dorsal aortic blood-pressure in the resting eel, Anguilla australis. J. Exp. Zool. A 238, 155–166.
Hoagland, T.M., Weaver Jr., L., Conlon, J.M., Wang, Y., Olson, K.R., 2000. Effects of endothelin-1 and homologous trout endothelin on cardiovascular function in rainbow trout. Am. J. Physiol. 278, R460–R468.
Holder, F.C., Vincent, B., Ristori, M.T., Laurent, P., 1983. Vascular perfusion of an intestinal segment in the catfish (*Ictalurus melas*, R): demonstration of the vasoactive effects of mammalian VIP and of gastrointestinal extracts from teleost fish. C. R. Séances Acad. Sci. III Sci. de la vie 296, 783.
Holeton, G.F., Randall, D.J., 1967. Changes in blood pressure in the rainbow trout during hypoxia. J. Exp. Biol. 46, 297–305.

Holmgren, S., 1977. Regulation of the heart of a teleost, *Gadus morhua*, by autonomic nerves and circulating catecholamines. Acta Physiol. Scand. 99, 62–74.
Holmgren, S., 1995. Neuropeptide control of the cardiovascular system in fish and reptiles. Braz. J. Med. Biol. Res. 28, 1207–1216.
Holmgren, S., Nilsson, S., 1983. Bombesin-, gastrin/CCK-, 5-hydroxytryptamine-, neurotensin-, somatostatin-, and VIP-like immunoreactivity and catecholamine fluorescence in the gut of the elasmobranch, *Squalus acanthias*. Cell Tissue Res. 234, 595–618.
Holmgren, S., Axelsson, M., Farrell, A.P., 1992. The effect of catecholamines, substance P and vasoactive intestinal polypeptide on blood flow to the gut in the dogfish *Squalus acanthias*. J. Exp. Biol. 168, 161–175.
Imbrogno, S., Cerra, M.C., 2017. Hormonal and autacoid control of cardiac function. In: Gamperl, A.K., Gillis, T.E., Farrell, A.P., Brauner, C.J. (Eds.), Fish Physiology. In: The Cardiovascular System: Morphology, Control and Function, vol. 36A. Academic Press, San Diego, pp. 265–315.
Ishimatsu, A., 2012. Evolution of the cardiorespiratory system in air-breathing fishes. Aqua-BioSci. Monogr. 5, 1–28.
Ishimatsu, A., Iwama, G.K., Heisler, N., 1988. In vivo analysis of partitioning of cardiac output between systemic and central venous sinus circuits in rainbow trout: a new approach using chronic cannulation of the branchial vein. J. Exp. Biol. 137, 75–88.
Janvier, J.J., Peyraud-Waïtzenegger, M., Soulier, P., 1996. Effects of serotonin on the cardio-circulatory system of the European eel (*Anguilla anguilla*) in vivo. J. Comp. Physiol. B. 166, 131–137.
Jennings, B.L., Broughton, B.R., Donald, J.A., 2004. Nitric oxide control of the dorsal aorta and the intestinal vein of the Australian short-finned eel *Anguilla australis*. J. Exp. Biol. 207, 1295–1303.
Jennings, B.L., Bell, J.D., Hyodo, S., Toop, T., Donald, J.A., 2007. Mechanisms of vasodilation in the dorsal aorta of the elephant fish, *Callorhinchus milii* (Chimaeriformes: Holocephali). J. Comp. Physiol. B. 177, 557–567.
Jensen, J., Axelsson, M., Holmgren, S., 1991. Effect of substance P and vasoactive intestinal polypeptide on gastrointestinal blood flow in the Atlantic cod *Gadus morhua*. J. Exp. Biol. 156, 361–374.
Johansen, K., 1963. The cardiovascular system of *Myxine glutinosa* L. In: Brodal, A., Fänge, R. (Eds.), The Biology of Myxine. Gröndahl & Sön, Oslo.
Johansen, K., Hanson, D., 1967. Hepatic vein sphincters in elasmobranchs and their significance in controlling hepatic blood flow. J. Exp. Biol. 46, 195–203.
Johansen, K., Lenfant, C., Hanson, D., 1968. Cardiovascular dynamics in the lungfishes. Z. Vgl. Physiol. 59, 157–186.
Johansen, K., Lenfant, C., Hanson, D., 1973. Gas exchange in the lamprey, *Entosphenus tridentatus*. Comp. Biochem. Physiol. A 44, 107–119.
Johnson, K.R., Olson, K.R., 2008. Comparative physiology of the piscine natriuretic peptide system. Gen. Comp. Endocrinol. 157, 21–26.
Johnson, K.R., Hoagland, T.M., Olson, K.R., 2011. Endogenous vascular synthesis of B-type and C-type natriuretic peptides in the rainbow trout. J. Exp. Biol. 214, 2709–2717.
Johnsson, M., Axelsson, M., Davison, W., Forster, M., Nilsson, S., 1996. Effects of preload and afterload on the performance of the in situ perfused portal heart of the New Zealand hagfish *Eptatretus cirrhatus*. J. Exp. Biol. 199, 401–405.
Johnsson, M., Axelsson, M., Holmgren, S., 2001. Large veins in the Atlantic cod (*Gadus morhua*) and the rainbow trout (*Oncorhynchus mykiss*) are innervated by neuropeptide-containing nerves. Anat. Embryol. 204, 109–115.

Jones, D.R., Milsom, W.K., 1982. Peripheral receptors affecting breathing and cardiovascular function in non-mammalian vertebrates. J. Exp. Biol. 100, 59–91.

Jones, D.R., Randall, D.J., 1978. The respiratory and circulatory systems during exercise. In: Hoar, W., Randall, D. (Eds.), Fish Physiology. Academic Press, New York, London, pp. 425–492.

Jönsson, E., Forsman, A., Einarsdottir, I.E., Egner, B., Ruohonen, K., Björnsson, B.T., 2006. Circulating levels of cholecystokinin and gastrin-releasing peptide in rainbow trout fed different diets. Gen. Comp. Endocrinol. 148, 187–194.

Jonz, M.G., 2011. Oxygen sensing in fish. In: Farrell, A.P. (Ed.), Encyclopedia of Fish Physiology. Elsevier Inc., Amsterdam, pp. 871–878.

Jonz, M.G., Nurse, C.A., 2003. Neuroepithelial cells and associated innervation of the zebrafish gill: a confocal immunofluorescence study. J. Comp. Neurol. 461, 1–17.

Jonz, M.G., Zaccone, G., 2009. Nervous control of the gills. Acta Histochem. 111, 207–216.

Jonz, M.G., Zachar, P.C., Da Fonte, D.F., Mierzwa, A.S., 2015. Peripheral chemoreceptors in fish: a brief history and a look ahead. Comp. Biochem. Physiol. A 186, 27–38.

Kågström, J., Holmgren, S., 1997. Vip-induced relaxation of small arteries of the rainbow trout, *Oncorhynchus mykiss*, involves prostaglandin synthesis but not nitric oxide. J. Auton. Nerv. Syst. 63, 68–76.

Kågström, J., Holmgren, S., 1998. Calcitonin gene-related peptide (CGRP), but not tachykinins, causes relaxation of small arteries from the rainbow trout gut. Peptides 19, 577–584.

Kågström, J., Axelsson, M., Holmgren, S., 1994. Cardiovascular responses to scyliorhinin I and II in the rainbow trout, *Oncorhynchus mykiss*, in vivo and in vitro. J. Exp. Biol. 191, 155–166.

Kågström, J., Axelsson, M., Jensen, J., Farrell, A.P., Holmgren, S., 1996a. Vasoactivity and immunoreactivity of fish tachykinins in the vascular system of the spiny dogfish. Am. J. Physiol. 270, R585–R593.

Kågström, J., Holmgren, S., Olson, K.R., Conlon, J.M., Jensen, J., 1996b. Vasoconstrictive effects of native tachykinins in the rainbow trout, *Oncorhynchus mykiss*. Peptides 17, 39–45.

Keen, A.N., Gamperl, A.K., 2012. Blood oxygenation and cardiorespiratory function in steelhead trout (*Oncorhynchus mykiss*) challenged with an acute temperature increase and zatebradine-induced bradycardia. J. Therm. Biol. 37, 201–210.

Kermorgant, M., Lancien, F., Mimassi, N., Le Mevel, J.C., 2014. Central ventilatory and cardiovascular actions of serotonin in trout. Respir. Physiol. Neurobiol. 192, 55–65.

Kiceniuk, J.W., Jones, D.R., 1977. The oxygen transport system in trout (*Salmo gairdneri*) during sustained exercise. J. Exp. Biol. 69, 247–260.

Lai, N.C., Shabetai, R., Graham, J.B., Hoit, B.D., Sunnerhagen, K.S., Bhargava, V., 1990. Cardiac function of the Leopard shark *Triakis semifasciata*. J. Comp. Physiol. B 160, 259–268.

Lai, N.C., Graham, J.B., Dalton, N., Shabetai, R., Bhargava, V., 1998. Echocardiographic and hemodynamic determinations of the ventricular filling pattern in some teleost fishes. Physiol. Zool. 71, 157–167.

Langille, B.L., Stevens, E.D., Anantaraman, A., 1983. Cardiovascular and respiratory flow dynamics. In: Webb, P.W., Weihs, D. (Eds.), Fish Biomechanics. Praeger, New York, pp. 92–139.

Le Mével, J.-C., Pamantung, T.-F., Mabin, D., Vaudry, H., 1993. Effects of central and peripheral administration of arginine vasotocin and related neuropeptides on blood pressure and heart rate in the conscious trout. Brain Res. 610, 82–89.

Le Mével, J.C., Olson, K.R., Conklin, D., Waugh, D., Smith, D.D., Vaudry, H., Conlon, J.M., 1996. Cardiovascular actions of trout urotensin II in the conscious trout, *Oncorhynchus mykiss*. Am. J. Physiol. 271, R1335–R1343.

Le Mével, J.C., Lancien, F., Mimassi, N., Conlon, J.M., 2007. Ventilatory and cardiovascular actions of centrally administered trout tachykinins in the unanesthetized trout. J. Exp. Biol. 210, 3301–3310.

Le Mével, J.-C., Lancien, F., Mimassi, N., Leprince, J., Conlon, J.M., Vaudry, H., 2008. Central and peripheral cardiovascular, ventilatory, and motor effects of trout urotensin-II in the trout. Peptides 29, 830–837.

McKendry, J.E., Perry, S.F., 2001. Cardiovascular effects of hypercarbia in rainbow trout (*Oncorhynchus mykiss*): a role for externally oriented chemoreceptors. J. Exp. Biol. 204, 115–125.

Mendonca, P.C., Gamperl, A.K., 2010. The effects of acute changes in temperature and oxygen availability on cardiac performance in winter flounder (*Pseudopleuronectes americanus*). Comp. Biochem. Physiol. A 155, 245–252.

Milsom, W.K., 2012. New insights into gill chemoreception: receptor distribution and roles in water and air breathing fish. Respir. Physiol. Neurobiol. 184, 326–339.

Milsom, W.K., Burleson, M.L., 2007. Peripheral arterial chemoreceptors and the evolution of the carotid body. Respir. Physiol. Neurobiol. 157, 4–11.

Mimassi, N., Shahbazi, F., Jensen, J., Mabin, D., Conlon, J.M., Le Mével, J.-C., 2000. Cardiovascular actions of centrally and peripherally administered trout urotensin-I in the trout. Am. J. Physiol. Regul. Integr. Comp. Physiol. 279, R484–R491.

Morris, J.L., Nilsson, S., 1994. The circulatory system. In: Nilsson, S., Holmgren, S. (Eds.), Comparative Physiology and Evolution of the Autonomic Nervous System, vol. 4. Harwood Academic Publisher, Chur, pp. 193–246.

Munns, S.L., Hartzler, L.K., Bennett, A.F., Hicks, J.W., 2004. Elevated intra-abdominal pressure limits venous return during exercise in *Varanus exanthematicus*. J. Exp. Biol. 207, 4111–4120.

Munoz-Chapuli, R., 1999. Circulatory system: anatomy of the peripheral circulatory system. In: Hamlett, W.C. (Ed.), Sharks, Skates and Rays: The Biology of Elasmobranch Fishes. The Johns Hopkins University Press, Baltimore, MD, pp. 198–217.

Nandi, J., 1961. New arrangement of interrenal and chromaffin tissues of teleost fishes. Science 134, 389–390.

Nekvasil, N.P., Olson, K.R., 1986. Extraction and metabolism of circulating catecholamines by the trout gill. Am. J. Physiol. 250, R526–R531.

Nilsson, S., 1983. Autonomic Nerve Function in the Vertebrates. Springer-Verlag, Berlin, Heidelberg, New York.

Nilsson, S., 1984. Adrenergic control systems in fish. Mar. Biol. Lett. 5, 127–146.

Nilsson, S., 1994. Evidence for adrenergic nervous control of blood pressure in teleost fish. Physiol. Zool. 67, 1347–1359.

Nilsson, S., Holmgren, S., 1988. The autonomic nervous system. In: Shuttleworth, T.J. (Ed.), Physiology of Elasmobranch Fishes. Springer-Verlag, Berlin, Heidelberg, New York, London, Paris, Tokyo, pp. 143–169.

Nilsson, S., Holmgren, S., 1992. Cardiovascular control by purines, 5-hydroxytryptamine, and neuropeptides. Fish Physiol. 12, 301–341.

Nilsson, S., Sundin, L., 1998. Gill blood flow control. Comp. Biochem. Physiol. 119, 137–147.

Nilsson, S., Holmgren, S., Grove, D.J., 1975. Effects of drugs and nerve stimulation on the spleen and arteries of two species of dogfish, *Scyliorhinus canicula* and *Squalus acanthias*. Acta Physiol. Scand. 95, 219–230.

Nishimura, H., Lunde, L.G., Zucker, A., 1979. Renin response to hemorrhage and hypotension in the aglomerular toadfish *Opsanus tau*. Am. J. Physiol. 237, H105–H111.

Nobata, S., Ogoshi, M., Takei, Y., 2008. Potent cardiovascular actions of homologous adrenomedullins in eels. Am. J. Physiol. Regul. Integr. Comp. Physiol. 294, R1544–R1553.

Nobata, S., Ventura, A., Kaiya, H., Takei, Y., 2010. Diversified cardiovascular actions of six homologous natriuretic peptides (ANP, BNP, VNP, CNP1, CNP3, and CNP4) in conscious eels. Am. J. Physiol. Regul. Integr. Comp. Physiol. 298, R1549–R1559.

Nobata, S., Donald, J.A., Balment, R.J., Takei, Y., 2011. Potent cardiovascular effects of homologous urotensin II (UII)-related peptide and UII in unanesthetized eels after peripheral and central injections. Am. J. Physiol. Regul. Integr. Comp. Physiol. 300, R437–R446.
Ogilvy, C.S., DuBois, A.B., 1982. Effects of tilting on blood pressure and fluid pressures of bluefish and smooth dogfish. Am. J. Physiol. 242, R70–R76.
Olson, K.R., 1992. Blood and extracellular fluid volume regulation: role of the renin angiotensin system, kallikrein-kinin system, and atrial natriuretic peptides. In: Hoar, D.J.R.W.S., Farrell, A.P. (Eds.), In: Fish Physiology, vol. XIIB. Academic Press, San Diego, New York, London, pp. 135–254.
Olson, K.R., 1994. Circulatory anatomy in bimodally breathing fish. Am. Zool. 34, 280–288.
Olson, K.R., 2002a. Gill circulation: regulation of perfusion distribution and metabolism of regulatory molecules. J. Exp. Zool. 293, 320–335.
Olson, K.R., 2002b. Vascular anatomy of the fish gill. J. Exp. Zool. 293, 214–231.
Olson, K.R., 2005. Vascular actions of hydrogen sulfide in nonmammalian vertebrates. Antioxid. Redox. Signal. 7, 804–812.
Olson, K.R., 2009. Is hydrogen sulfide a circulating "gasotransmitter" in vertebrate blood? Biochim. Biophys. Acta 1787, 856–863.
Olson, K.R., 2011a. Branchial anatomy. In: Farrell, A.P. (Ed.), Encyclopedia of Fish Physiology: From Genome to Environment. Elsevier, Amsterdam, pp. 1095–1103.
Olson, K.R., 2011b. Circulatory System Design: Roles and Principles. Academic Press, San Diego.
Olson, K.R., 2011c. Design and Physiology of Arteries and Veins. Academic Press, San Diego.
Olson, K.R., 2011d. Physiology of capacitance vessels. In: Farrell, A.P. (Ed.), Encyclopedia of Fish Physiology. Elsevier Inc., Amsterdam, pp. 1111–1118.
Olson, K.R., 2011e. Physiology of resistance vessels. In: Farrell, A.P. (Ed.), Encyclopedia of Fish Physiology. Elsevier Inc., Amsterdam, pp. 1111–1118.
Olson, K.R., 2015. Hydrogen sulfide as an oxygen sensor. Antioxid. Redox. Signal. 22, 377–397.
Olson, K.R., Donald, J.A., 2009. Nervous control of circulation—the role of gasotransmitters, NO, CO, and H_2S. Acta Histochem. 111, 244–256.
Olson, K.R., Farrell, A.P., 2006. The cardiovascular system. In: Evans, D.H., Claiborne, J.B. (Eds.), The Physiology of Fishes. CRC Press, Boca Raton.
Olson, K.R., Farrell, A.P., 2011. Design and physiology of capillaries and secondary circulation | Secondary Circulation and Lymphatic Anatomy. In: Farrell, A.P. (Ed.), Encyclopedia of Fish Physiology. Elsevier Inc., Amsterdam, pp. 1161–1168.
Olson, K.R., Hoagland, T.M., 2008. Effects of freshwater and saltwater adaptation and dietary salt on fluid compartments, blood pressure, and venous capacitance in trout. Am. J. Physiol. 294, R1061–R1067.
Olson, K.R., Villa, J., 1991. Evidence against nonprostanoid endothelium-derived relaxing factor(s) in trout vessels. Am. J. Physiol. 260, R925–R933.
Olson, K.R., Flint, K.B., Budde Jr., R.B., 1981. Vascular corrosion replicas of chemobaroreceptors in fish: the carotid labyrinth in Ictaluridae and Clariidae. Cell Tissue Res. 219, 535–541.
Olson, K.R., Duff, D.W., Farrell, A.P., Keen, J., Kellogg, M.D., Kullman, D., Villa, J., 1991. Cardiovascular effects of endothelin in trout. Am. J. Physiol. 260, H1214–H1223.
Olson, K.R., Conklin, D.J., Farrell, A.P., Keen, J.E., Takei, Y., Weaver Jr., L., Smith, M.P., Zhang, Y., 1997a. Effects of natriuretic peptides and nitroprusside on venous function in trout. Am. J. Physiol. 273, R527–R539.
Olson, K.R., Conklin, D.J., Weaver Jr., L., Duff, D.W., Herman, C.A., Wang, X., Conlon, J.M., 1997b. Cardiovascular effects of homologous bradykinin in rainbow trout. Am. J. Physiol. 272, R1112–R1120.

Olson, K.R., Kinney, D.W., Dombkowski, R.A., Duff, D.W., 2003. Transvascular and intravascular fluid transport in the rainbow trout: revisiting Starling's forces, the secondary circulation and interstitial compliance. J. Exp. Biol. 206, 457–467.

Olson, K.R., Donald, J.A., Dombkowski, R.A., Perry, S.F., 2012. Evolutionary and comparative aspects of nitric oxide, carbon monoxide and hydrogen sulfide. Respir. Physiol. Neurobiol. 184, 117–129.

Ong, K.L., Lam, K.S.L., Cheung, B.M.Y., 2005. Urotensin II: its function in health and its role in disease. Cardiovasc. Drugs Ther. 19, 65–75.

Opdyke, D.F., Holcombe, R., 1976. Response to angiotensins I and II and to AI-converting-enzyme inhibitor in a shark. Am. J. Physiol. 231, 1750–1753.

Opdyke, D.F., McGreehan, J.R., Messing, S., Opdyke, N.E., 1972. Cardiovascular responses to spinal cord stimulation and autonomically active drugs in *Squalus acanthias*. Comp. Biochem. Physiol. 42, 611–620.

Opdyke, D.F., Wilde, D.W., Holcombe, R.F., 1982. Effect of angiotensin ii on vascular resistance in whole body perfused dogfish. Comp. Biochem. Physiol. C 73, 45–50.

Oudit, G.Y., Butler, D.G., 1995a. Angiotensin II and cardiovascular regulation in a freshwater teleost, *Anguilla rostrata* LeSueur. Am. J. Physiol. 269, R726–R735.

Oudit, G.Y., Butler, D.G., 1995b. Cardiovascular effects of arginine vasotocin, atrial natriuretic peptide, and epinephrine in freshwater eels. Am. J. Physiol. 268, R1273–R1280.

Pang, C.C., 2001. Autonomic control of the venous system in health and disease: effects of drugs. Pharmacol. Ther. 90, 179–230.

Pang, C.C., Tabrizchi, R., 1986. The effects of noradrenaline, B-HT 920, methoxamine, angiotensin II and vasopressin on mean circulatory filling pressure in conscious rats. Br. J. Pharmacol. 89, 389–394.

Pellegrino, D., Sprovieri, E., Mazza, R., Randall, D.J., Tota, B., 2002. Nitric oxide-cGMP-mediated vasoconstriction and effects of acetylcholine in the branchial circulation of the eel. Comp. Biochem. Physiol. A Mol. Integr. Physiol. 132, 447–457.

Pellegrino, D., Tota, B., Randall, D.J., 2005. Adenosine/nitric oxide crosstalk in the branchial circulation of *Squalus acanthias* and *Anguilla anguilla*. Comp. Biochem. Physiol. A 142, 198–204.

Perry, S.F., Bernier, N.J., 1999. The acute humoral adrenergic stress response in fish: facts and fiction. Aquaculture 177, 285–295.

Perry, S.F., Capaldo, A., 2011. The autonomic nervous system and chromaffin tissue: neuroendocrine regulation of catecholamine secretion in non-mammalian vertebrates. Auton. Neurosci. 165, 54–66.

Perry, S.F., Desforges, P.R., 2006. Does bradycardia or hypertension enhance gas transfer in rainbow trout (*Oncorhynchus mykiss*)? Comp. Biochem. Physiol. A Mol. Integr. Physiol. 144, 163–172.

Perry, S.F., Gilmour, K.M., 1996. Consequences of catecholamine release on ventilation and blood oxygen transport during hypoxia and hypercapnia in an elasmobranch (*Squalus acanthias*) and a teleost (*Oncorhynchus mykiss*). J. Exp. Biol. 199, 2105–2118.

Perry, S.F., Gilmour, K.M., 2002. Sensing and transfer of respiratory gases at the fish gill. J. Exp. Zool. 293, 249–263.

Perry, S.F., Fritsche, R., Hoagland, T.M., Duff, D.W., Olson, K.R., 1999. The control of blood pressure during external hypercapnia in the rainbow trout (*Oncorhynchus mykiss*). J. Exp. Biol. 202, 2177–2190.

Pettersson, K., Johansen, K., 1982. Hypoxic vasoconstriction and the effects of adrenaline on gas exchange efficiency in fish gills. J. Exp. Biol. 97, 263–272.

Platzack, B., Axelsson, M., Nilsson, S., 1993. The renin-angiotensin system in blood pressure control during exercise in the cod *Gadus morhua*. J. Exp. Biol. 180, 253–262.

Platzack, B., Schaffert, C., Hazon, N., Conlon, J.M., 1998. Cardiovascular actions of dogfish urotensin I in the dogfish, *Scyliorhinus canicula*. Gen. Comp. Endocrinol. 109, 269–275.

Randall, D.J., Perry, S.F., 1992. Catecholamines. In: Hoar, W.S., Randall, D.J., Farrell, A.P. (Eds.), In: Fish Physiology, vol. XIIB. Academic Press, San Diego, New York, London, pp. 255–300.

Reid, S.G., Bernier, N.J., Perry, S.F., 1998. The adrenergic stress response in fish: control of catecholamine storage and release. Comp. Biochem. Physiol. C Pharmacol. Toxicol. Endocrinol. 120, 1–27.

Reite, O.B., 1969a. The evolution of vascular smooth muscle responses to histamine and 5-hydroxytryptamine. I. Occurrence of stimulatory actions in fish. Acta Physiol. Scand. 75, 221–239.

Reite, O.B., 1969b. The evolution of vascular smooth muscle responses to histamine and 5-hydroxytryptamine. II. Appearance of inhibitory actions of 5-hydroxytryptamine in amphibians. Acta Physiol. Scand. 77, 36–51.

Ristori, M.T., 1970. Réflexe de barosensibilité chez un poisson téléostéen (*Cyprinus carpio* L.). C. R. Seances Soc. Biol. Fil. 164, 1512–1516.

Ristori, M.T., Dessaux, G., 1970. Sur l'existence d'un gradient de sensibilité dans les récepteurs branchiaux de *Cyprinus carpio* L. C. R. Seances Soc. Biol. 164, 1517–1519.

Ristori, M.T., Laurent, P., 1977. Action de l'hypoxie sur le système vasculaire branchial de la tête perfusée de truite. C. R. Seances Soc. Biol. 171, 809–813.

Rothe, C.F., 1993. Mean circulatory filling pressure: its meaning and measurement. J. Appl. Physiol. 74, 499–509.

Rothe, C.F., 2006. Point: active venoconstriction is/is not important in maintaining or raising end-diastolic volume and stroke volume during exercise and orthostasis. J. Appl. Physiol. 101, 1262–1266. discussion 1265–1256, 1270.

Rothe, C.F., Maass-Moreno, R., 2000. Active and passive liver microvascular responses from angiotensin, endothelin, norepinephrine, and vasopressin. Am. J. Physiol. 279, H1147–H1156.

Rummer, J.L., Wang, S., Steffensen, J.F., Randall, D.J., 2014. Function and control of the fish secondary vascular system, a contrast to mammalian lymphatic systems. J. Exp. Biol. 217, 751–757.

Sandblom, E., Axelsson, M., 2005a. Baroreflex mediated control of heart rate and vascular capacitance in trout. J. Exp. Biol. 208, 821–829.

Sandblom, E., Axelsson, M., 2005b. Effects of hypoxia on the venous circulation in rainbow trout (*Oncorhynchus mykiss*). Comp. Biochem. Physiol. A 140, 233–239.

Sandblom, E., Axelsson, M., 2006. Adrenergic control of venous capacitance during moderate hypoxia in the rainbow trout (*Oncorhynchus mykiss*): role of neural and circulating catecholamines. Am. J. Physiol. Regul. Integr. Comp. Physiol. 291, R711–R718.

Sandblom, E., Axelsson, M., 2007a. The venous circulation: a piscine perspective. Comp. Biochem. Physiol. A 148, 785–801.

Sandblom, E., Axelsson, M., 2007b. Venous hemodynamic responses to acute temperature increase in the rainbow trout (*Oncorhynchus mykiss*). Am. J. Physiol. Regul. Integr. Comp. Physiol. 292, R2292–R2298.

Sandblom, E., Axelsson, M., 2011. Autonomic control of circulation in fish: a comparative view. Auton. Neurosci. 165, 127–139.

Sandblom, E., Farrell, A.P., Altimiras, J., Axelsson, M., Claireaux, G., 2005. Cardiac preload and venous return in swimming sea bass (*Dicentrarchus labrax* L.). J. Exp. Biol. 208, 1927–1935.

Sandblom, E., Axelsson, M., Farrell, A.P., 2006a. Central venous pressure and mean circulatory filling pressure in the dogfish *Squalus acanthias*: adrenergic control and role of the pericardium. Am. J. Physiol. 291, R1465–R1473.

Sandblom, E., Axelsson, M., McKenzie, D.J., 2006b. Venous responses during exercise in rainbow trout, *Oncorhynchus mykiss*: α-adrenergic control and the antihypotensive function of the renin-angiotensin system. Comp. Biochem. Physiol. A 144, 401–409.

Sandblom, E., Axelsson, M., Davison, W., 2008. Enforced exercise, but not acute temperature elevation, decreases venous capacitance in the stenothermal Antarctic fish *Pagothenia borchgrevinki*. J. Comp. Physiol. B 178, 845–851.

Sandblom, E., Axelsson, M., Davison, W., 2009a. Circulatory function at sub-zero temperature: venous responses to catecholamines and angiotensin II in the Antarctic fish *Pagothenia borchgrevinki*. J. Comp. Physiol. B 179, 165–173.

Sandblom, E., Cox, G.K., Perry, S.F., Farrell, A.P., 2009b. The role of venous capacitance, circulating catecholamines, and heart rate in the hemodynamic response to increased temperature and hypoxia in the dogfish. Am. J. Physiol. Regul. Integr. Comp. Physiol. 296, R1547–R1556.

Sandblom, E., Olsson, C., Davison, W., Axelsson, M., 2010. Nervous and humoral catecholaminergic control of blood pressure and cardiac performance in the Antarctic fish *Pagothenia borchgrevinki*. Comp. Biochem. Physiol. A 156, 232–236.

Sandblom, E., Davison, W., Axelsson, M., 2012. Cold physiology: postprandial blood flow dynamics and metabolism in the Antarctic fish *Pagothenia borchgrevinki*. PLoS One 7, 1–8.

Satchell, G.H., 1991. Physiology and Form of Fish Circulation. Cambridge University Press, Cambridge.

Satchell, G.H., 1992. The venous system. In: Hoar, W.S. (Ed.), The Cardiovascular System, Part A. In: vol. XII. Academic Press, London, pp. 141–183.

Seth, H., Sandblom, E., Holmgren, S., Axelsson, M., 2008. Effects of gastric distension on the cardiovascular system in rainbow trout (*Oncorhynchus mykiss*). Am. J. Physiol. Regul. Integr. Comp. Physiol. 294, R1648–R1656.

Seth, H., Gräns, A., Axelsson, M., 2010. Cholecystokinin as a regulator of cardiac function and postprandial gastrointestinal blood flow in rainbow trout (*Oncorhynchus mykiss*). Am. J. Physiol. Regul. Integr. Comp. Physiol. 298, R1240–R1248.

Seth, H., Axelsson, M., Farrell, A.P., 2011. The circulation and metabolism of the gastrointestinal tract. In: Grosell, M., Farrell, A.P., Brauner, C.J. (Eds.), Fish Physiology: The Multifunctional Gut of Fish. vol. 30. Academic Press, London.

Seth, H., Axelsson, M., Gräns, A., 2014. The peptide hormone cholecystokinin modulates the tonus and compliance of the bulbus arteriosus and pre-branchial vessels of the rainbow trout (*Oncorhynchus mykiss*). Comp. Biochem. Physiol. A Mol. Integr. Physiol. 178, 18–23.

Shahbazi, F., Conlon, J.M., Holmgren, S., Jensen, J., 2001. Effects of cod bradykinin and its analogs on vascular and intestinal smooth muscle of the Atlantic cod, *Gadus morhua*. Peptides 22, 1023–1029.

Shahbazi, F., Holmgren, S., Jensen, J., 2009. Cod CGRP and tachykinins in coeliac artery innervation of the Atlantic cod, *Gadus morhua*: presence and vasoactivity. Fish Physiol. Biochem. 35, 369–376.

Sheriff, D.D., Zhou, X.P., Scher, A.M., Rowell, L.B., 1993. Dependence of cardiac filling pressure on cardiac output during rest and dynamic exercise in dogs. Am. J. Physiol. 265, H316–H322.

Short, S., Butler, P.J., Taylor, E.W., 1977. The relative importance of nervous, humoral and intrinsic mechanisms in the regulation of heart rate and stroke volume in the dogfish (*Scyliorhinus canicula*). J. Exp. Biol. 70, 77–92.

Shoukas, A.A., Bohlen, H.G., 1990. Rat venular pressure-diameter relationships are regulated by sympathetic activity. Am. J. Physiol. 259, H674–H680.

Skals, M., Skovgaard, N., Abe, A.S., Wang, T., 2005. Venous tone and cardiac function in the South American rattlesnake *Crotalus durissus*: mean circulatory filling pressure during adrenergic stimulation in anaesthetised and fully recovered animals. J. Exp. Biol. 208, 3747–3759.

Skals, M., Skovgaard, N., Taylor, E.W., Leite, C.A., Abe, A.S., Wang, T., 2006. Cardiovascular changes under normoxic and hypoxic conditions in the air-breathing teleost *Synbranchus marmoratus*: importance of the venous system. J. Exp. Biol. 209, 4167–4173.
Skov, P.V., Bennett, M.B., 2004. Structural basis for control of secondary vessels in the long-finned eel *Anguilla reinhardtii*. J. Exp. Biol. 207, 3339–3348.
Skovgaard, N., Olson, K.R., 2012. Hydrogen sulfide mediates hypoxic vasoconstriction through a production of mitochondrial ROS in trout gills. Am. J. Physiol. Regul. Integr. Comp. Physiol. 303, R487–R494.
Smith, D.G., 1978. Neural regulation of blood pressure in rainbow trout (*Salmo gairdneri*). Can. J. Zool. 56, 1678–1683.
Smith, L.S., Bell, G.R., 1975. A practical guide to the anatomy and physiology of pacific salmon. In: Canadian Fisheries and Marine Service Miscellaneous Publication, vol. 27. Department of the Environment, Fisheries and Marine Service, Ottawa, ON, Canada, pp. 1–21.
Smith, D.G., Nilsson, S., Wahlqvist, I., Eriksson, B.M., 1985. Nervous control of the blood pressure in the Atlantic cod, *Gadus morhua*. J. Exp. Biol. 117, 335–347.
Smith, M.P., Russell, M.J., Wincko, J.T., Olson, K.R., 2001. Effects of hypoxia on isolated vessels and perfused gills of rainbow trout. Comp. Biochem. Physiol. 130, 171–181.
Smith, M.P., Dombkowski, R.A., Wincko, J.T., Olson, K.R., 2006. Effect of pH on trout blood vessels and gill vascular resistance. J. Exp. Biol. 209, 2586–2594.
Soldatov, A.A., 2006. Organ blood flow and vessels of microcirculatory bed in fish. J. Evol. Biochem. Physiol. 42, 243–252.
Speers-Roesch, B., Brauner, C.J., Farrell, A.P., Hickey, A.J., Renshaw, G.M., Wang, Y.S., Richards, J.G., 2012. Hypoxia tolerance in elasmobranchs. II. Cardiovascular function and tissue metabolic responses during progressive and relative hypoxia exposures. J. Exp. Biol. 215, 103–114.
Stenslokken, K.O., Sundin, L., Nilsson, G.E., 1999. Cardiovascular and gill microcirculatory effects of endothelin-1 in Atlantic cod: evidence for pillar cell contraction. J. Exp. Biol. 202, 1151–1157.
Stenslokken, K.O., Sundin, L., Renshaw, G.M., Nilsson, G.E., 2004. Adenosinergic and cholinergic control mechanisms during hypoxia in the epaulette shark (*Hemiscyllium ocellatum*), with emphasis on branchial circulation. J. Exp. Biol. 207, 4451–4461.
Stenslokken, K.O., Sundin, L., Nilsson, G.E., 2006. Endothelin receptors in teleost fishes: cardiovascular effects and branchial distribution. Am. J. Physiol. Regul. Integr. Comp. Physiol. 290, R852–R860.
Stevens, E.D., 2011. The retia. In: Farrell, A.P. (Ed.), Encyclopedia of fish physiology: from genome to environment. Academic Press, San Diego, pp. 1119–1131.
Stevens, E.D., Randall, D.J., 1967. Changes in blood pressure, heart rate and breathing rate during moderate swimming activity in rainbow trout. J. Exp. Biol. 46, 307–315.
Stevens, E.D., Bennion, G.R., Randall, D.J., Shelton, G., 1972. Factors affecting arterial pressure and blood flow from the heart in intact, unrestrained lingcod, *Ophiodon elongatus*. Comp. Biochem. Physiol. 43A, 681–695.
Sundin, L.I., 1995. Responses of the branchial circulation to hypoxia in the Atlantic cod, *Gadus morhua*. Am. J. Physiol. 268, R771–R778.
Sundin, L., Nilsson, S., 1992. Arterio-venous branchial blood flow in the Atlantic cod *Gadus morhua*. J. Exp. Biol. 165, 73–84.
Sundin, L., Nilsson, G.E., 1996. Branchial and systemic roles of adenosine receptors in rainbow trout: an in vivo microscopy study. Am. J. Physiol. Regul. Integr. Comp. Physiol. 271, R661–R669.
Sundin, L., Nilsson, G.E., 1997. Neurochemical mechanisms behind gill microcirculatory responses to hypoxia in trout: in vivo microscopy study. Am. J. Physiol. 272, R576–R585.

Sundin, L., Nilsson, G.E., 1998. Endothelin redistributes blood flow through the lamellae of rainbow trout gills. J. Comp. Physiol. B. 168, 619–623.

Sundin, L., Nilsson, G.E., 2000. Branchial and circulatory responses to serotonin and rapid ambient water acidification in rainbow trout. J. Exp. Zool. 287, 113–119.

Sundin, L., Nilsson, S., 2002. Branchial innervation. J. Exp. Zool. 293, 232–248.

Sundin, L., Nilsson, G.E., Block, M., Lofman, C.O., 1995. Control of gill filament blood flow by serotonin in the rainbow trout, *Oncorhynchus mykiss*. Am. J. Physiol. 268, R1224–R1229.

Sundin, L., Davison, W., Forster, M., Axelsson, M., 1998. A role of 5-HT2 receptors in the gill vasculature of the Antarctic fish *Pagothenia borchgrevinki*. J. Exp. Biol. 201, 2129–2138.

Sundin, L., Axelsson, M., Davison, W., Forster, M.E., 1999. Cardiovascular responses to adenosine in the Antarctic fish *Pagothenia borchgrevinki*. J. Exp. Biol. 202 (Pt 17), 2259–2267.

Syeda, F., Hauton, D., Young, S., Egginton, S., 2013. How ubiquitous is endothelial NOS? Comp. Biochem. Physiol. A 166, 207–214.

Tabrizchi, R., King, K.A., Pang, C.C., 1992. Direct and indirect effects of angiotensin II on venous tone in conscious rats. Eur. J. Pharmacol. 219, 141–145.

Tabrizchi, R., Lim, S.L., Pang, C.C., 1993. Possible equilibration of portal venous and central venous pressures during circulatory arrest. Am. J. Physiol. 264, H259–H261.

Tait, L.W., Simpson, C.W., Takei, Y., Forster, M.E., 2009. Hagfish natriuretic peptide changes urine flow rates and vascular tensions in a hagfish. Comp. Biochem. Physiol. C Toxicol. Pharmacol. 150, 45–49.

Takei, Y., Tsuchida, T., Li, Z.H., Conlon, J.M., 2001. Antidipsogenic effects of eel bradykinins in the eel *Anguilla japonica*. Am. J. Physiol. Regul. Integr. Comp. Physiol. 281, R1090–R1096.

Takei, Y., Inoue, K., Trajanovska, S., Donald, J.A., 2011. B-type natriuretic peptide (BNP), not ANP, is the principal cardiac natriuretic peptide in vertebrates as revealed by comparative studies. Gen. Comp. Endocrinol. 171, 258–266.

Takei, Y., Ogoshi, M., Nobata, S., 2013. Exploring new CGRP family peptides and their receptors in vertebrates. Curr. Protein Pept. Sci. 14, 282–293.

Takei, Y., Hiroi, J., Takahashi, H., Sakamoto, T., 2014. Diverse mechanisms for body fluid regulation in teleost fishes. Am. J. Physiol. Regul. Integr. Comp. Physiol. 307, R778–R792.

Taylor, E.W., Short, S., Butler, P.J., 1977. The role of the cardiac vagus in the response of the dogfish *Scyliorhinus canicula* to hypoxia. J. Exp. Biol. 70, 57–75.

Taylor, E.W., Leite, C.A., Levings, J.J., 2009. Central control of cardiorespiratory interactions in fish. Acta Histochem. 111, 257–267.

Thorarensen, H., McLean, E., Donaldson, E.M., Farrell, A.P., 1991. The blood vasculature of the gastrointestinal tract in Chinook *Oncorhynchus tshawytscha* Walbaum and Coho *Oncorhynchus kisutch* Walbaum Salmon. J. Fish Biol. 38, 525–532.

Tierney, M.L., Luke, G., Cramb, G., Hazon, N., 1995. The role of the renin-angiotensin system in the control of blood pressure and drinking in the European eel, *Anguilla anguilla*. Gen. Comp. Endocrinol. 100, 39–48.

Toop, T., Donald, J.A., 2004. Comparative aspects of natriuretic peptide physiology in nonmammalian vertebrates: a review. J. Comp. Physiol. B. 174, 189–204.

Van Vliet, B.N., West, N.H., 1994. Phylogenetic trends in the baroreceptor control of arterial blood pressure. Physiol. Zool. 67, 1284–1304.

Vogel, W., Vogel, V., Schlote, W., 1974. Ultrastructural study of arteriovenous anastomoses in gill filaments of *Tilapia mossambica*. Cell Tissue Res. 155, 491–512.

Vogel, W., Vogel, V., Pfautsch, M., 1976. Arteriovenous anastomoses in rainbow-trout gill filaments—scanning and transmission electron-microscopic study. Cell Tissue Res. 167, 373–385.

Whiteley, N.M., Egginton, S., 1999. Antarctic fishes have a limited capacity for catecholamine synthesis. J. Exp. Biol. 202, 3623–3629.

Withers, P.C., Hillman, S.S., Simmons, L.A., Zygmunt, A.C., 1988. Cardiovascular adjustments to enforced activity in the anuran amphibian, *Bufo marinus*. Comp. Biochem. Physiol. A Physiol. 89, 45–49.

Wood, C.M., 1977. Cholinergic mechanisms and the response to ATP in the systemic vasculature of the rainbow trout. J. Comp. Physiol. B 122, 325–345.

Wood, C.h.M., Shelton, G., 1980. Cardiovascular dynamics and adrenergic responses of the rainbow trout in vivo. J. Exp. Biol. 87, 247–270.

Wood, C.M., McMahon, B.R., McDonald, D.G., 1979. Respiratory, ventilatory, and cardiovascular responses to experimental anaemia in the Starry flounder, *Platichthys stellatus*. J. Exp. Biol. 82, 139–162.

Xu, H.Y., Olson, K.R., 1993. Significance of circulating catecholamines in regulation of trout splanchnic vascular resistance. J. Exp. Zool. A Ecol. Genet. Physiol. 267, 92–96.

Zaccone, G., Mauceri, A., Fasulo, S., 2006. Neuropeptides and nitric oxide synthase in the gill and the air-breathing organs of fishes. J. Exp. Zool. A Ecol. Genet. Physiol. 305, 428–439.

Zhang, Y., Jenkinson, E., Olson, K.R., 1995. Vascular compliance and mean circulatory filling pressure in trout: effects of ACE inhibition. Am. J. Physiol. 268, H1814–H1820.

Zhang, Y., Weaver Jr., L., Ibeawuchi, A., Olson, K.R., 1998. Catecholaminergic regulation of venous function in the rainbow trout. Am. J. Physiol. 274, R1195–R1202.

INDEX

Note: Page numbers followed by "*f*"and "*t*" refer to figures and tables, respectively.

B

C

D

E

F

G

H

I

T

U

V

W

OTHER VOLUMES IN THE FISH PHYSIOLOGY SERIES

Printed in the United States
By Bookmasters